St. Olaf College

JUN 10 1994

Science Library

PCR TECHNOLOGY
Current Innovations

Edited by

Hugh G. Griffin
Annette M. Griffin

CRC Press
Boca Raton   Ann Arbor   London   Tokyo

Library of Congress Cataloging-in-Publication Data

PCR technology : current innovations / edited by Hugh G. Griffin, Annette M. Griffin.
      p. cm.
   Includes bibliographical references and index.
   ISBN 0-8493-8674-8
   1. Polymerase chain reaction—Methodology. I. Griffin, Hugh G.
II. Griffin, Annette M.
QP606.D46P362 1994
574.87'3282—dc20   93-37561
                                                                                                                    CIP

   This book contains information obtained from authentic and highly regarded sources. Reprinted material is quoted with permission, and sources are indicated. A wide variety of references are listed. Reasonable efforts have been made to publish reliable data and information, but the author and the publisher cannot assume responsibility for the validity of all materials or for the consequences of their use.
   Neither this book nor any part may be reproduced or transmitted in any form or by any means, electronic or mechanical, including photocopying, microfilming, and recording, or by any information storage or retrieval system, without prior permission in writing from the publisher.
   All rights reserved. Authorization to photocopy items for internal or personal use, or the personal or internal use of specific clients, may be granted by CRC Press, Inc., provided that $.50 per page photocopied is paid directly to Copyright Clearance Center, 27 Congress Street, Salem, MA 01970 USA. The fee code for users of the Transactional Reporting Service is ISBN 0-8493-8674-8/94/$0.00+$.50. The fee is subject to change without notice. For organizations that have been granted a photocopy license by the CCC, a separate system of payment has been arranged.
   CRC Press, Inc.'s consent does not extend to copying for general distribution, for promotion, for creating new works, or for resale. Specific permission must be obtained in writing from CRC Press for such copying.
   Direct all inquiries to CRC Press, Inc., 2000 Corporate Blvd., N.W., Boca Raton, Florida 33431.

© 1994 by CRC Press, Inc.

No claim to original U.S. Government works
International Standard Book Number 0-8493-8674-8
Library of Congress Card Number 93-37561
Printed in the United States of America 1 2 3 4 5 6 7 8 9 0
Printed on acid-free paper

PREFACE

The technique of polymerase chain reaction (PCR)* was first described in 1985. This ingenious tool has had an enormous impact on biological research that can probably be compared to the development of recombinant DNA technology in the 1970s. One of the most important features of the PCR technique is its simplicity. Thus, it can be used to generate meaningful results by scientists who have little or no familiarity with molecular biology techniques; for instance, scientists working in the fields of zoology, botany, environmental science, and forensic science.

Following the appreciation of the true potential of PCR, an explosion of applications of this technique soon occurred. This book brings together a selection of the most widely used applications including the generation and detection of genetic mutations, diagnosis of clinical disease, detection of food-borne pathogens, and the determination of genetic relatedness of plant and animal species. Other chapters deal more closely with the needs of the scientist involved in basic research tasks and cover topics such as the purification and cloning of PCR products, sequencing PCR products, primer design, generation of labeled probes, and screening of lambda and cosmid libraries. To aid the beginner, chapters on primer design, choice of polymerase, and precautions necessary to avoid false positives in PCR are included.

This essential reference book should serve both as the foundation of basic instruction for the scientist new to PCR and a source of updated applications for those already familiar with the basic method.

H. G. Griffin
A. M. Griffin

---

* The polymerase chain reaction (PCR) is covered by patents owned by Hoffmann-La Roche, Inc. A license is required to use the PCR process.

THE EDITORS

Dr. Hugh G. Griffin and **Dr. Annette M. Griffin** are both senior scientists at the B.B.S.R.C. Institute of Food Research, Norwich Research Park, Colney, Norwich England.

Dr. Hugh Griffin received his training at Trinity College, University of Dublin obtaining a BA (mod) degree in Microbiology in 1983 and a Ph.D. in 1986. He has worked at Washington University, St. Louis, MO, the Babraham Institute, Cambridge, England and the Institute for Animal Health, Huntingdon, Cambridgeshire. His current major research interests relate to the molecular biology of Lactic Acid Bacteria.

Dr. Annette Griffin graduated in 1982 from University College Cork, Cork, Ireland with a B.Sc. (hons) degree in Biochemistry. She obtained her Ph.D. degree in 1986 from Trinity College, Dublin, Ireland for her work on Semliki Forest virus. She has studied the molecular biology of infectious laryngotracheitis virus at the Institute for Animal Health, Huntingdon, Cambridgeshire. Her current research interests are in the molecular biology of bacterial exopolysaccharide biosynthesis.

Both editors have published extensively on various aspects of molecular biology and have acted as editors on a number of books related to DNA sequencing and computer analysis of sequence data. They are married and have three children.

CONTRIBUTORS

Robert P. Adams
Plant Biotechnology Center
Baylor University
Waco, TX

S. F. An
University of Oxford
Nuffield Department of Pathology and
  Bacteriology
John Radcliffe Hospital
Oxford, U.K.

Asim K. Bej
Department of Biology
University of Alabama at Birmingham
Birmingham, AL

Bruce Budowle
FBI Laboratory
Washington, D.C.

Jean-Paul Charlieu
Institut de Biologie
Montpellier, France

Ernesto d'Aloja
Immunohematology Laboratory
Department of Forensic Medicine
Catholic University of Sacred Heart
Rome, Italy

Tigst Demeke
Agriculture Canada
Lethbridge Research Station
Lethbridge, Alberta, Canada

Marina Dobosz
Immunohematology Laboratory
Department of Forensic Medicine
Catholic University of Sacred Heart
Rome, Italy

K. A. Fleming
University of Oxford
Nuffield Department of Pathology and
  Bacteriology
John Radcliffe Hospital
Oxford, U.K.

Hugh Griffin
Genetics and Microbiology Department
BBSRC Institute of Food Research
Norwich Research Park
Colney, Norwich, U.K.

Barbara Grubinska
Department of Anatomy
School of Medicine
West Virginia University
Morgantown, WV

Daniel D. Jones
Department of Biology
University of Alabama at Birmingham
Birmingham, AL

Gregory W. Konat
Department of Anatomy
School of Medicine
West Virginia University
Morgantown, WV

Iwona Laszkiewicz
Department of Anatomy
School of Medicine
West Virginia University
Morgantown, WV

K. C. Patrick Lee
Department of Chemical Engineering
Purdue University
West Lafayette, IN

Andrew M. Lew
The Walter and Eliza Hall Institute of
  Medical Research
Melbourne, Victoria, Australia

Y-M. D. Lo
University of Oxford
Nuffield Department of Pathology and
  Bacteriology
John Radcliffe Hospital
Oxford, U.K.

Meena H. Mahbubani
Department of Biology
University of Alabama at Birmingham
Birmingham, AL

Vikki M. Marshall
The Walter and Eliza Hall Institute of
 Medical Research
Melbourne, Victoria, Australia

Michael McClelland
California Institute for Biological Research
La Jolla, CA

J. O'D. McGee
University of Oxford
Nuffield Department of Pathology and
 Bacteriology
John Radcliffe Hospital
Oxford, U.K.

Louis M. Mezei
Promega Corporation
Madison, WI

Michael Panaccio
Victoria Institute of Animal Sciences
Attwood, Victoria, Australia

Vincenzo L. Pascali
Immunohematology Laboratory
Department of Forensic Medicine
Catholic University of Sacred Heart
Rome, Italy

K. Peter Pauls
Crop Science Department
University of Guelph
Guelph, Ontario, Canada

Leena Peltonen
Department of Human Molecular Genetics
National Public Health Institute
Helsinki, Finland

Lawrence A. Presley
FBI Laboratory
Washington, D.C.

David Ralph
California Institute for Biological Research
La Jolla, CA

Antti Sajantila
Department of Human Molecular Genetics
National Public Health Institute
Helsinki, Finland

Andrew D. Sharrocks
Department of Biochemistry and Genetics
The Medical School
University of Newcastle upon Tyne
Newcastle, U.K.

Anand K. Srivastava
Department of Molecular Microbiology
Washington University School of Medicine
St. Louis, MO

Jörg Stappert
Max-Planck Institute for Immunobiology
Freiburg, Germany

Douglas R. Storts
Promega Corporation
Madison, WI

Bernard Y. Tao
Biochemical and Food Process Engineering
Department of Agricultural Engineering
Purdue University
West Lafayette, IN

Anthony B. Troutt
Immunix Research and Development
 Corporation
Seattle, WA

Andrey B. Vartapetian
Belozersky Institute of Physico-Chemical
 Biology
Moscow State University
Moscow, Russia

Richard C. Wiggins
Department of Anatomy
School of Medicine
West Virginia University
Morgantown, WV

E. P. H. Yap
University of Oxford
Nuffield Department of Pathology and
  Bacteriology
John Radcliffe Hospital
Oxford, U.K.

Kangfu Yu
Department of Molecular Biology and
  Genetics
University of Guelph
Guelph, Ontario, Canada

TABLE OF CONTENTS

Chapter 1
PCR as a Technique Used Daily in Molecular Biology 1
Jean-Paul Charlieu

Chapter 2
The Design of Primers for PCR 5
Andrew D. Sharrocks

Chapter 3
Purification of PCR Products 13
Louis M. Mezei and Douglas R. Storts

Chapter 4
Cloning PCR Products 21
Louis M. Mezei and Douglas R. Storts

Chapter 5
Amplification of Unknown Flanking DNA by Single-Specific-Primer PCR 29
Jörg Stappert

Chapter 6
Generation of Labeled DNA Probes by PCR 37
*Gregory W. Konat, Iwona Laszkiewicz, Barbara Grubinska,
and Richard C. Wiggins*

Chapter 7
Nonisotopic Probe Generation by PCR 43
Y-M. D. Lo, E. P. H. Yap, S. F. An, J. O.'D. McGee, and K. A. Fleming

Chapter 8
Direct PCR Screening of Lambda and Cosmid Libraries 53
Hugh G. Griffin

Chapter 9
Methods for Generating Multiple Site-Directed Mutations *In Vitro* 59
Jörg Stappert

Chapter 10
Mutagenesis by PCR 69
Bernard Y. Tao and K. C. Patrick Lee

Chapter 11
Direct Automated DNA Sequencing of ds and ss PCR Products 85
Vikki M. Marshall and Andrew M. Lew

Chapter 12
Distinction between Almost-Identical DNA Sequences by
Polymerase Chain Reaction 101
Jean-Paul Charlieu

Chapter 13
Detection of Mutations by PCR 107
E. P. H. Yap and J. O'D. McGee

Chapter 14
Mapped Restriction Site Polymorphisms (MRSPs) in PCR Products for Rapid
Identification and Classification of Genetically Distinct Organisms 121
David Ralph and Michael McClelland

Chapter 15
Detection of Polymorphic DNA Sequences at the 3' End of Alu Repeats by PCR 133
Jean-Paul Charlieu

Chapter 16
Ligation-Anchored PCR 141
Anthony B. Troutt

Chapter 17
PCR-Limiting Dilution Analysis 147
Anthony B. Troutt

Chapter 18
Direct PCR from Whole Blood Using Formamide and Low Temperatures 151
Michael Panaccio and Andrew M. Lew

Chapter 19
Immuno-PCR: A Generic Method for Purifying Target for PCR 159
Andrew M. Lew and Michael Panaccio

Chapter 20
Non-isotopic Single-Strand Conformation Polymorphism (SSCP)
Analysis of PCR Products 165
E. P. H. Yap and J. O'D. McGee

Chapter 21
The Use of PCR-RAPD Analysis in Plant Taxonomy and Evolution 179
Tigst Demeke and Robert P. Adams

Chapter 22
Optimization of DNA-Extraction and PCR Procedures for Random Amplified
Polymorphic DNA (RAPD) Analysis in Plants 193
Kangfu Yu and K. Peter Pauls

Chapter 23
The Use of RAPD Analysis to Tag Genes and Determine Relatedness in
Heterogeneous Plant Populations using Tetraploid Alfalfa as an Example 201
Kangfu Yu and K. Peter Pauls

Chapter 24
DNA Recombination in the Course of PCR 215
Andrey B. Vartapetian

Chapter 25
Thermostable DNA Polymerases for *In Vitro* DNA Amplifications 219
Asim K. Bej and Meena H. Mahbubani

Chapter 26
PCR in Sequence-Tagged Site (STS) Content Genome Mapping 239
Anand K. Srivastava

Chapter 27
False-Positives and Contamination in PCR 249
E. P. H. Yap, Y.-M. D. Lo, K. A. Fleming, and J. O'D. McGee

Chapter 28
The Application of PCR-Based Technologies to Forensic Analysis 259
Lawrence A. Presley and Bruce Budowle

Chapter 29
Application of PCR-Amplified DNA Markers in Identification of Individuals 277
Antti Sajantila and Leena Peltonen

Chapter 30
PCR in Forensic Science 289
Vincenzo L. Pascali, Marina Dobosz, and Ernesto d'Aloja

Chapter 31
Applications of Polymerase Chain Reaction Methodology in Clinical Diagnostics 307
Meena H. Mahbubani and Asim K. Bej

Chapter 32
Applications of the Polymerase Chain Reaction (PCR) *In Vitro* DNA-Amplification
Method in Environmental Microbiology 327
Asim K. Bej and Meena H. Mahbubani

Chapter 33
Detection of Foodborne Microbial Pathogens Using Polymerase
Chain Reaction Methods 341
Daniel D. Jones and Asim K. Bej

Index 363

Chapter 1

PCR AS A TECHNIQUE USED DAILY IN MOLECULAR BIOLOGY

Jean-Paul Charlieu

The polymerase chain reaction (PCR) is a very powerful technique in molecular biology and is widely used today for an increasing number of applications. Several are presented in this volume, however many of these were developed for a specific purpose. In this chapter, PCR approaches for more general uses will be presented. Through a few examples, a PCR approach to a problem will be compared to more "classical" approaches in order to show that the former is often easier and faster to perform than the latter. Since PCR generally reduces the number of steps of an analysis and therefore the number of products and enzymes required, the economic character is also taken into account.

The characterization of cloned DNA fragments can sometimes be very time consuming, despite the availability of an increasing number of kits allowing a more rapid utilization of some techniques of molecular biology (e.g., labeling of DNA, molecular hybridization, purification of plasmids).

The determination of the size of DNA fragments inserted into plasmids, for example, requires the growth of bacterial clones followed by the preparation, purification, and enzymatic hydrolysis of plasmid DNAs. In addition, problems may occur at each step of such an analysis: (1) bacteria can grow insufficiently to obtain enough material; (2) the prepared DNA may not be pure enough; or (3) contamination with chemicals (e.g., phenol, chloroform) may inhibit the activity of restriction enzymes resulting in partial hydrolysis of the DNA. When double digestion is necessary to release the cloned fragment, it may be necessary to change the incubation buffer of enzymes that are not compatible. PCR provides a simple way to determine the size of DNA fragments inserted into plasmid vectors. PCR primers can be designed from the vector sequence on both side of the cloning site (Figure 1A). For plasmids of the pUC series or derived from it, the M13 "universal" and "reverse" primers can be used (M13 primer = 5' GTAAAACGACGGCCAGT 3'; reverse primer = 5' AACAGCTATGACCATG 3'). A one-step PCR study can be achieved according to the scheme of Figure 1B. The analysis of PCR products in an agarose gel (Figure 1C) allows the direct determination of the insert sizes. These PCR products can also be purified from the gel and sequenced, without the need of producing single-stranded templates from the bacterial strains.[1]

The screening of a library for a given sequence can be performed either by hybridization or by PCR. Here again, the PCR approach is cheaper, easier, and faster. For screening by molecular hybridization, clones have to be first plated and grown (usually overnight), then transferred onto a solid support (nitrocellulose or nylon). In case of bacterial or yeast clones, colonies are then lysed and the DNA bound to the membrane is denatured. A probe must be labeled either radioactively or using biotinylated or digoxigenin-linked nucleotides by nick translation or oligolabeling. After hybridization and washing, the positive clones are detected by autoradiography or enzymatic immunodetection. All these steps must then be repeated at least once in order to obtain pure clones. The PCR strategy consists of amplifying a DNA fragment known as "sequence tagged site" (STS),[2] which is characterized and localized in the genome using pools of clones as templates.[3] The secondary PCR screening is then performed on individual clones of the positive pools. In this PCR approach again, it is not necessary to prepare the DNA templates. Note that the two PCR applications described above can be easily automated.

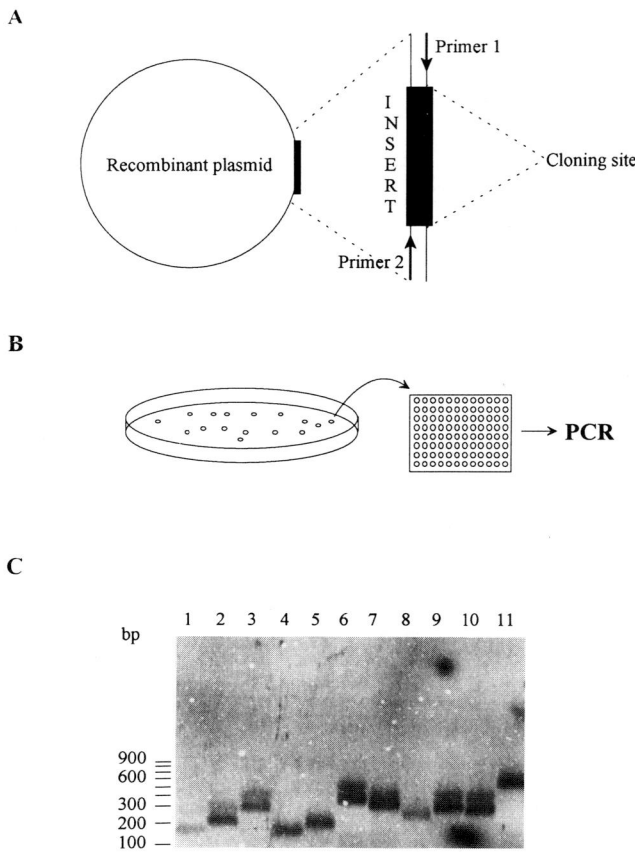

Figure 1. Determination of the insert size of recombinant plasmids by PCR. (A) Position of PCR primers designed from the vector sequence, on both sides of the cloning site. (B) A single colony of each clone is picked up (with a sterile toothpick) from the plate, inoculated in 10 µl of PCR mix containing the Taq DNA polymerase buffer, 20 pmol of each primer, 0.8 mM of dNTP (total), and 1 U of Taq DNA polymerase, and 30 cycles of PCR consisting in a denaturation step at 92°C for 10 s, an annealing step at 60°C (for M13 and reverse primers) for 30 s, and an elongation step at 72°C for 1 min are performed. (C) 2 to 5 µl of PCR products are electrophoresed in a 1.5% agarose gel and detected by ethydium bromide staining. PCR amplification of the plasmid vector without insert (lane 1) provides a control of the basic size.

STSs can also serve to characterize overlapping clones, without the establishment and comparison of restriction maps of each clone. This PCR approach is described in Figure 2.

Nelson et al.[4] have developed a method to PCR amplify DNA fragments that are comprised between two Alu repeats in the human genome ("Alu PCR", Figure 3). Alu PCR can be performed with yeast artificial chromosomes (YACs) or cosmids to obtain in a simple and rapid way "Alu fingerprints", which are very useful for the construction of contigs.[5]

Alu PCR can also be performed for other purposes. One of them is the characterization of somatic hybrid cell lines. Cytogenetic identification of the human chromosome(s) present on a rodent genomic background is the classical approach to this problem but is quite long and difficult. In addition, small fragments of chromosomes can escape cytogenetic detection. The Alu PCR pattern specific for each human chromosome allows a molecular characterization of the human DNA in rodent/human somatic hybrid cells.[6]

Probes for *in situ* and Southern blot hybridization are generally obtained by labeling a cloned DNA fragment by nick translation or oligolabeling. PCR is an alternative method used

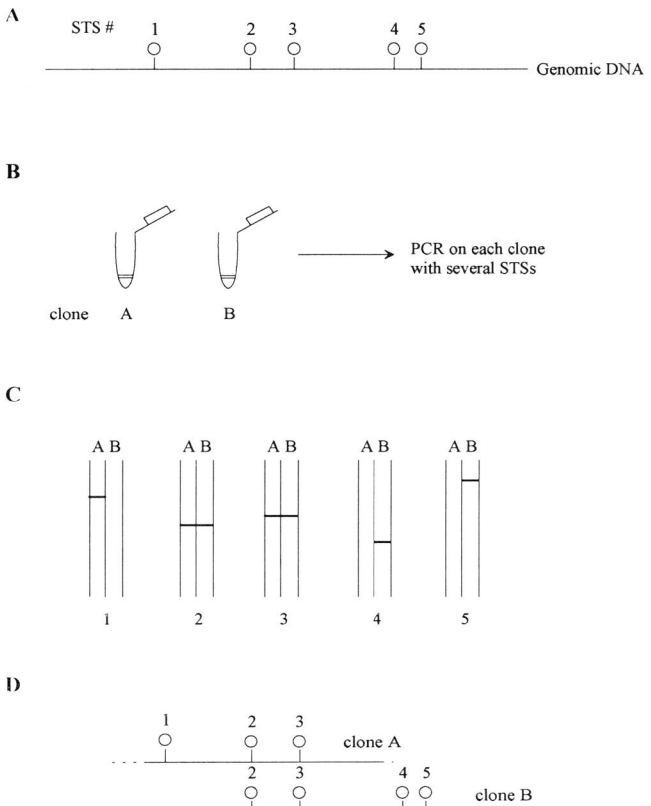

Figure 2. Construction of a contig with STSs. (A) Contiguous STSs are chosen in a given region of the genome. (B) Using PCR, YACs or cosmids are tested for the presence of each STS. If the PCR products are of different sizes, multiplex PCR can be performed. (C) The PCR products are analyzed (presence or absence of the amplified band in each clone) by the appropriate method. (D) In this example, STSs #2 and #3 are common to clones A and B. These clones therefore overlap in the region containing these STSs.

to produce a fragment to be used as a probe from genomic DNA. An improvement to this method is to incorporate a labeled precursor in the *in vitro* synthesized DNA fragment. In this case, the nucleotide mix for PCR should contain a 1:10 molar ratio between the unlabeled nucleotide (0.02 mM) and the labeled precursor (0.2 mM). In place of a radioactive precursor, it is possible to use cold labeled nucleotides such as biotin-16-dUTP or DIG-11-dUTP. These are particularly useful as *in situ* probes since they are stable for up to one year and can be stored for several experiments.

Alu PCR is also another way to produce probes for *in situ*[7,8] or Southern blot[9] hybridization from a YAC clone or from somatic hybrid cells (painting probes for *in situ* hybridization).[10-12]

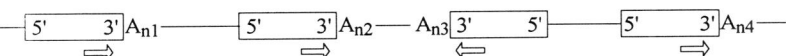

Figure 3. Alu PCR. Alu sequences are found interspersed throughout the human genome.[13] These repeats are polarized (the 3′ end contains a [dA]-rich extension)[14] and can be oriented in any sense.[15] A primer (represented by an arrow) directed toward the 3′ end of Alu sequences will allow the *in vitro* amplification of DNA fragments contained between two "tail-to-tail" Alu repeats.

This can be performed directly from a yeast colony or cultured cells without the DNA preparation step.

The PCR approach cannot substitute for all others currently used in molecular biology, but it provides a good alternative in many cases, and can be used for the study of most problems. Note also that PCR does not need heavy equipment and that the thermocycler can be used for purposes other than PCR, such as incubation of samples at the appropriate temperature for enzymatic reactions and sequencing using the chain-termination method.

REFERENCES

1. **Casanova, J.-L., Pannetier, C., Jaulin, C., and Kourilsky, P.,** Optimal conditions for directly sequencing double-stranded PCR products with sequenase, *Nucleic Acids Res.,* 18, 4028, 1990.
2. **Olson, M., Hood, L., Cantor, C. R., and Botstein, D.,** A common language for physical mapping of the human genome, *Science,* 245, 1434, 1989.
3. **Kwiatkowski T. J., Jr., Zoghbi, H., Ledbetter, S. A., Ellison, K. A., and Chinault, A. C.,** Rapid identification of yeast artificial chromosome clones by matrix pooling and crude lysate PCR, *Nucleic Acids Res.,* 18, 7191, 1990.
4. **Nelson, D. L., Ledbetter, S. A., Corbo, L., Victoria, M. F., Ramirezsolis, R., Webster, T. D., Ledbetter, D. H., and Caskey, C. T.,** Alu polymerase chain reaction — a method for rapid isolation of human specific sequences from complex DNA sources, *Proc. Natl. Acad. Sci. U.S.A.,* 86, 6686, 1989.
5. **Chumakov, I., Rigault, P., Guillou, S., Ougen, P., Billaut, A., Guasconi, G., Gervy, P., LeGall, I., Soularue, P., Grinas, L., Bougueleret, L., Bellanné-Chantelot, C., Lacroix, B., Barillot, E., Gesnoin, P., Pook, S., Vaysseix, G., Frelat, G., Schmitz, A., Sambucy, J.-L., Bosch, A., Estivill, X., Weissenbach, J., Vignal, A., Riethman, H., Cox, D., Patterson, D., Gardiner, K., Hattori, M., Sakaki, Y., Ichikawa, H., Ohki, M., LePaslier, D., Heilig, R., Antonorakis, S., and Cohen, D.,** Continuum of overlapping clones spanning the entire human chromosome 21q, *Nature,* 359, 380, 1992.
6. **Ledbetter, S. A., Garcia-Heras, J., and Ledbetter, D. H.,** "PCR caryotype" of human chromosomes in somatic cell hybrids, *Genomics,* 8, 614, 1990.
7. **Lengauer, C., Green, E., and Cremer, T.,** Fluorescence *in situ* hybridization of YAC clones after Alu-PCR amplification, *Genomics,* 13, 826, 1992.
8. **Baldini, A., Ross, M., Nizetic, D., Vatcheva, R., Lindsay, E., Lehrach, H., and Siniscalco, M.,** Chromosomal assignment of human YAC clones by fluorescent *in situ* hybridization: use of single-yeast-colony PCR and multiple labeling, *Genomics,* 14, 181, 1992.
9. **Charlieu, J.-P., Laurent, A.-M., Orti, R., Viegas-Péquignot, E., Bellis, M., and Roizès, G.,** A 37-kilobases DNA fragment common to the pericentromeric region of chromosomes 13 and 21 and to the inactive centromere on chromosome 2, *Genomics,* 15, 576, 1992.
10. **Desmaze, C., Zucman, J., Delattre, O., Thomas, G., and Aurias, A.,** *In situ* hybridization of PCR amplified inter-Alu sequences from a hybrid cell line, *Hum. Genet.,* 88, 541, 1992.
11. **Lichter, P., Ledbetter, S. A., Ledbetter, D. H., and Ward, D. C.,** Fluorescent *in situ* hybridization with Alu and L1 polymerase chain reaction probes for rapid characterization of human chromosomes,
12. **Cotter, F. E., Hampton, G. M., Nasipuri, S., Bodmer, W. F., and Young, B. D.,** Rapid isolation of human chromosome-specific DNA probes from a somatic cell hybrid, *Genomics,* 7, 257, 1990.
13. **Deininger, P. L.,** SINEs Short interspersed repeated DNA elements in higher eucaryotes, in *Mobile DNA,* Howe, M. and Berg, D., Eds., ASM Press, Washington, DC, 619–636.
14. **Weiner, A. M., Deininger, P. L., and Efstratiadis, A.,** Nonviral transposon: genes, pseudogenes and transposable elements generated by the reverse flow of genetic information. *Annu. Rev. Biochem.,* 55, 631, 1986.
15. **Slagel, V., Flemmington, E., Traina-Dorge, V., Bradshaw, H., and Deininger, P. L.,** Clustering and subfamily relationships of the Alu family in the human genome, *Mol. Biol. Evol.,* 4, 19, 1987.

Chapter 2

THE DESIGN OF PRIMERS FOR PCR

Andrew D. Sharrocks

TABLE OF CONTENTS

I. Introduction 5

II. General Rules for Primer Design 6

III. Modifications to PCR Primer Design 7
    A. Gene Amplification 7
    B. Gene Manipulation 7

IV. Computer-Aided PCR Primer Design 9

V. Conclusions 10

References 10

I. INTRODUCTION

The development of the polymerase chain reaction (PCR) has revolutionized the field of molecular biology. PCR has been used for a plethora of applications, many of which are covered in this book. These applications involve both novel procedures (e.g., gene amplification from nanograms of genomic DNA[1,2]) and modifications of existing methods (e.g., site-directed mutagenesis[3,4]). Although many variables need to be optimized in the design of PCR-based procedures for each of these various applications, the most critical parameter in all cases is the correct designing of PCR primers. Indeed, the correct choice of PCR primers often dictates the success or failure of the PCR amplification. Careful design of primers can therefore save valuable research time, and in addition, can lead to significant savings in costs as the primers usually represent the most expensive component in a PCR.

One of the principal considerations in the design of a PCR protocol is to obtain unique, specific products as dictated by the selected primers. The first step in PCR primer design is to ensure this specificity. However, after specificity has been assured, further manipulation of the PCR primer design is possible. This allows the introduction of novel genetic information into the product. Such alterations range from single point mutations to the tagging of products with new coding sequences or regulatory elements. By careful and thoughtful primer design, specific products can be produced from a PCR with a multitude of possible engineered features.

Most of the rules for primer design are empirical with no guarantee of success. However, careful adherence to these rules will significantly increase the probability of a successful PCR. Computer programs are especially useful in the assessment of the basic parameters governing primer design. Even so, some primer pairs that fulfill all known criteria still fail to work for obscure reasons. This chapter provides a guide to reducing the possibility of an unsuccessful PCR but simultaneously demonstrates the flexibility that can be incorporated into primer design for product manipulation.

II. GENERAL RULES FOR PRIMER DESIGN

Successful primer design and hence successful PCRs rely on the unique annealing of the two primers to the template with both high specificity and high efficiency. This ensures that only the desired product is synthesized. The problem of non-specific amplifications is intensified during the early rounds, when amplifications are performed on very small quantities of target DNA, which is often immersed in an excess of non-specific sequences. Correct annealing at this stage is imperative or errors will be compounded throughout the ensuing PCR.

Several parameters must be carefully considered in order to ensure correct annealing (Table 1). The first of these is to choose primers that have a sequence unique within the region to be amplified. The most important region to check is at the 3' end of the primer as this is where synthesis of the PCR product begins. Such a procedure is tedious when executed manually but is handled easily by computer programs (see below). The second parameter to consider is the inclusion of a G/C residue at the 3' end of the primer. This "G.C clamp" helps to ensure correct annealing at the 3' end due to the strong hydrogen bonding utilized by G/C base pairs.

Primers should also be designed with no self-homology. Such self-homology can lead to partially double-stranded "snap back" structures that render the primer incapable of hybridizing to the template. A general rule of thumb is that no self-homology involving four contiguous base pairs should be present in the primer. A related parameter is that primers should show no homology to their antisense counterparts. Formation of partial hybrids between primer pairs can lead to the formation of "primer-dimers" in the ensuing PCR. Elimination of this artifact is essential as this by-product can easily swamp a PCR. Particular care should again be taken in removing complementarity between the 3' ends of the two primers.

The primer base composition should also be closely monitored. In general a G/C content between 45 and 55% should be selected to direct specific binding yet allow efficient melting during the PCR. Efforts should also be made to keep the base composition close to that exhibited by the amplified region. In addition, the base distribution of the primers should be random, with polypurine and polypyrimidine tracts avoided. Nucleotide sequence repeats should also be avoided in primer design. The target sequence-specific part of primers should ideally be between 18 and 25 bases long. It is also important to have primer pairs with similar melting temperatures (T_m). This can be accurately calculated using the nearest-neighbor method with the formula: $T_m^{primer} = \Delta H/[\Delta S + R \ln(c/4)] - 273.15°C + 16.6 \log_{10}[K^+]$ where ΔH and ΔS are the enthalpy and entropy for helix formation, respectively, R is the molar gas constant, and c is the concentration of probe.[5] However, approximate T_ms can be calculated manually using the simpler formula $T_m = 2AT + 4GC$.[6] An equity in primer T_ms ensures simultaneous annealing of the primers. The calculated T_m can then be incorporated into the PCR protocol to optimize specific binding. This can be exploited when two sets of primer pairs with different matched T_ms are used in a single PCR to amplify different specific fragments.

An additional parameter that can be incorporated into primer design is to ensure that the T_m of the amplified region between the primers is low enough to ensure 100% melting at 92°C. This can be calculated using the formula $T_m = 81.5 + 16.6 (\log_{10}[K^+]) + 0.41 (\%G + C) - 675/\text{length}$.[7] This reduces to $T_m = 59.9 + 0.41 (\%G + C) - 675/\text{length}$ at standard PCR conditions containing 50 mM KCl. Inefficient melting will ultimately lead to a reduced yield from the PCR. Finally it is useful to design primers whose annealing sites are spaced between 100 and 600 bp. This distance allows efficient synthesis of product during the PCR.[8]

Adherence to the above parameters in designing primers for a PCR helps ensure specificity of the product. Such specificity is essential in applications where the non-target DNA is in great excess over the target DNA. Many of these PCR applications, covered in this volume, include procedures involving amplifications from genomic DNA, gene libraries, and whole cells.

TABLE 1.
General Rules for PCR Primer Design

Parameter	Optimum values
1. Unique oligonucleotide sequence	
2. G.C clamp at the 3' end	1–2 G/C nucleotides
3. No self-complementarity	≤3 contiguous bases
4. No complementarity to antisense counterpart	≤3 contiguous bases
5. Random base distribution and composition	45–55% G/C content
6. Primer length	18–25 bases
7. Match primer T_ms	
8. Distance and composition of intraprimer sequence	100–600 bases apart

III. MODIFICATIONS TO PCR PRIMER DESIGN

A. GENE AMPLIFICATION

The use of PCR in gene amplification has applications in gene cloning and diagnostic techniques. Both these applications involve the amplification of specific sequences from a vast excess of non-specific DNA. In general the rules described above must be followed to give specific product amplification. However certain extra criteria must be considered (Table 2).

It is advisable to choose primer pairs that are unique not only within the target gene but also in the genome of the organism under investigation. As the entire nucleotide sequence of any given genome is not yet known, searches must obviously be limited to known gene sequences deposited in databases. Computer assistance is essential for this procedure.

In PCRs performed on cDNA templates, it is beneficial to design primer pairs consisting of primers that lie in different exons. In this way, contamination from genomic DNA can be easily diagnosed. This is of extreme importance if low transcript levels are to be detected. The siting of PCR primers in different exons has the additional benefit of being able to also lead to information about differential transcript splicing, a task that is extremely laborious using conventional cDNA cloning techniques.

PCR is often used for cloning genes that have largely unknown sequences. Clues to a sequence are gained from sequencing peptides derived from the purified gene product or alternatively by deriving consensus peptide sequences either from other members of gene families or from related species. Pairs of degenerate PCR primers can then be designed based on these peptide sequences and used in gene amplifications.[9,10] One important consideration when designing such primers is to site the 3' end at the most complementary region of the deduced sequence. This will help ensure accurate primer annealing and the correct initiation of product synthesis. The use of multiple combinations of primer pairs is often necessary to successfully amplify the desired gene. Slight alterations in the position of the primers may bring success, but rules governing this type of analysis are largely empirical. It should be emphasized that the PCR profile must be altered to ensure annealing of degenerate primers, which will have low T_ms.

B. GENE MANIPULATION

PCR not only allows specific amplification of products but also allows the simultaneous manipulation of the cloned DNA sequence. One application that takes advantage of the latter is the introduction of restriction sites into the product for cloning.[2] Site-directed mutagenesis of the PCR product and the insertion of larger DNA fragments encoding either regulatory elements or additional coding sequences are also frequently used modifications.

Restriction enzyme recognition sequences can be engineered into PCR primers either individually or alternatively in groups on a 5' extension known as the adaptor.[2] Commonly

TABLE 2.
Modifications of Primer Design for Specific Applications

Application	Parameters to be optimized
Gene amplification	Test uniqueness in genome
cDNA amplifications	Site primers in different exons
Degenerate primers	Site 3' end in least ambiguous region
Gene manipulation	
Restriction site engineering	Locate recognition site >4 bases from 5' end
Product tagging	Retain 15 — 18 specific bases at 3' end
	Test adaptors for non-specific hybridization
"Two-step" site-directed mutagenesis	Site 5' end of primer next to a T residue or in a codon wobble position
	Locate mismatches in the center of the primer
	Add 1.5 extra specific bases per mismatch

engineered restriction sites include those for Nde1 and Nco1, which are used to allow the direct cloning of genes into expression vectors using either the natural or artificial ATG start codons. In addition PCR can be used to introduce restriction sites with concomitant silent mutations in the coding sequence. These sites can subsequently be used in domain swapping experiments. One important parameter to consider in designing these primers is that restriction enzyme cleavage is inefficient when its site is located toward the end of a DNA fragment. It is therefore advisable to position restriction sites at least 4 bp from the 5' end of the primer. This will ensure efficient product cleavage by most restriction enzymes.[11]

The 5' adaptors on primers can also be used to introduce larger stretches of additional sequence onto the products. Such sequences include promoter sequences (allowing direct *in vitro* transcription of PCR products[12,13]) and epitope tags (allowing antibody screening of proteins encoded by PCR products). In designing such primers, it is advisable to retain at least 15 to 18 contiguous nucleotides on the gene-specific 3' end of the primer to help ensure correct amplification. The primer design parameters outlined in Table 1 must be strictly adhered to within this gene-specific region to maintain primer specificity. However, it is also essential to verify that the desired adaptor sequence does not have significant similarity to sequences elsewhere in the target DNA.

Primer modifications can also be introduced to permit site-directed mutagenesis at either single or multiple residues. One of the most commonly used methods utilizes two vector-specific primers in conjunction with a single mutagenic primer.[4] In this method, two PCR reactions are carried out. The first PCR contains one vector-specific primer and the mutagenic primer. The second PCR contains the second vector-specific primer and the whole of the first reaction as the second primer.

Several important considerations must be taken into account in designing mutagenic primers for this procedure. The mismatched nucleotide(s) must be located toward the center of the primer. Ideally the mutations would be located toward the 5' end of the primer so that the 3' end is highly specific and therefore anneals correctly during the first PCR. However as the product of the first PCR becomes the primer for the second PCR, the 3' end of the antisense strand must also be allowed to anneal correctly. Hence in practice, mismatches should not be incorporated too close to the 5' end of the original mutagenic primer. When multiple mismatches are to be introduced into a single primer, the primer T_m and specificity will be reduced. To compensate for this reduction, extra specific residues must be added. In general, it is advisable to add 1.5 bases to a primer per mismatch. Successful mutagenic primers range from between 20 and 24 nucleotides long, containing single mismatches, up to 36 nucleotides long, with eight mismatches.

Another parameter affecting the choice of primer for this mutagenic approach is that Taq polymerase often catalyzes the non-template directed addition of a residue (often but not always adenine) to the 3' end of the strand.[13] As the synthesized strands from the first PCR are used directly as a primer in the second PCR, incorrect bases are frequently inserted into the product. To circumvent this, special consideration must be given to the location of the 5' end of the primer. By siting the 5' end immediately adjacent to a T residue, the deleterious effect of the addition of an adenine to the opposite strand can be nullified.[15] However, as residues other than adenine are often added and sequences around the location of the desired mutation may often lack appropriately positioned T residues, a more commonly applicable rule is to site the 5' end of the primer adjacent to a nucleotide in the wobble position of a codon.[16] This will often result in no change to the protein product. By combining these two strategies, mutagenic primers can be designed that direct only the desired changes to protein coding sequences. Adherence to the above rules is also necessary for other PCR methods involving multiple primers and several PCR reactions to resynthesize DNA fragments.[17,18]

IV. COMPUTER-AIDED PCR PRIMER DESIGN

PCR primer design relies on the accurate assessment of a plethora of standard parameters (Table 1). A series of largely empirical considerations can then be considered to design primers for specific applications or product modifications (Table 2). Many of the latter category of parameters can be determined manually. However, it is extremely laborious and time consuming to optimize all the basic parameters for selecting primer pairs. Several computer programs are available to facilitate this analysis.[19,20] These programs search for primers based on optimum sizes and accurate T_ms calculated using the nearest-neighbor method.[5] Additional searches are done to eliminate primers that have non-standard base composition and are capable of forming hairpins. Special attention is given to ensuring the "uniqueness" of the primer sequence within the target DNA, with care given to ensure stable and specific annealing of the 3' end of the primer to the DNA. Once suitable primers have been selected, "matching pairs" are selected that have compatible T_ms, are spaced correctly, and are not likely to form primer-dimers.

The program *OLIGO* (National BioSciences) is a purpose-designed package for either PC-based or Macintosh computers and can be used for general oligonucleotide design as well as for the selection of PCR primer pairs. Both versions are user friendly and perform all the basic tasks in primer design (Table 1). In addition, this program calculates the annealing temperature of PCR reactions, taking into account the stability of the PCR product[8,20] as well as the primer T_m values. Optimization of this parameter can increase both the specificity and yield of product. The PC version of *OLIGO* also has the additional feature of being able to back-translate protein sequences. This can facilitate the design of degenerate oligonucleotides used in cloning (Table 2).

Programs for primer design are also available from Intelligenetics but are incorporated into more extensive DNA analysis packages. These programs are also available in two versions, *PCGene* for PC-based and *GeneWorks* for Macintosh computers. Again, both versions are user friendly and able to adequately perform the basic steps in primer design (Table 1). In common with the *OLIGO* program, these programs are able to calculate the optimum annealing temperature of the PCR reaction and use it in the selection of primer pairs. Both *PCGene* and *GeneWorks* allow the determination of the uniqueness of the primer pairs in the whole of the *GenBank* database. This task is obviously beyond manual methods, and for many applications it is an essential parameter to consider in primer design. In addition, *PCGene* has the advantage of allowing a primer to be inserted manually and subsequently analyzed, and then a compatible primer can be selected. This function is of considerable importance when primers that contain modifications are to be used in the PCR (Table 2).

V. CONCLUSIONS

Careful oligonucleotide primer design is an essential step in obtaining a successful PCR amplification. A plethora of parameters have been identified that must be optimized to improve the chances of success in a PCR reaction. Particular care must be taken to ensure the correct annealing of the 3' end of the primer. This is a task that can often be accomplished manually. However, it is obviously impractical to consider all the necessary parameters using manual methods alone. Nevertheless many manually designed primer pairs function well whereas some primer pairs that appear to fulfill all known criteria fail to produce a PCR product. This emphasizes that we have yet to learn all there is to know about rules governing PCR primer design. As more rules are developed, computer programs will become an absolute necessity to cope with this analysis. Indeed, the use of a computer program in designing primers is already strongly advised to simplify this process. One should not, however, lose sight of the fact that the more elaborate primer manipulations are still best done with manual methods due to their empirical nature. I therefore strongly recommend the use of computer assistance, allied to some manual adjustments, many of which can be done within the programs themselves. This combined approach should pay dividends in ensuring consistently successful PCR amplifications.

REFERENCES

1. Saiki, K., Gelfand, D. H., Stoffel, S., Scharf, S. J., Higuchi, R., Horn, G. T., Mullis, K. B., and Erlich, H. A., Primer-directed enzymatic amplification of DNA with a thermostabile DNA polymerase, *Science*, 219, 487, 1988.
2. Scharf, S. J., Horn, G. T., and Erlich, H., A., Direct cloning and sequence analysis of enzymatically amplified genomic sequences, *Science*, 233, 1076, 1986.
3. Ito, W., Ishiguro, H., and Kurosawa, Y., A general method for introducing a series of mutations into cloned DNA using the polymerase chain reaction, *Gene*, 102, 67, 1991.
4. Landt, O., Grunert, H.-P., and Hahn, U., A general method for rapid site-directed mutagenesis using the polymerase chain reaction, *Gene*, 96, 125, 1990.
5. Breslauer, K. J., Frank, R., Blocker, H., and Markey, L. A., Predicting DNA duplex stability from the base sequence, *Proc. Natl. Acad. Sci. U.S.A.*, 83, 3746, 1986.
6. Suggs, S. V., Hirose, T., Miyake, E. H., Kawashima, M. J., Johnson, K. I., and Wallace, R. B., Using purified genes, in *ICN-UCLA Symp. Developmental Biology*, Vol. 23, Brown, D. D., Ed., Academic Press, New York, 1981, 683.
7. Baldino, F., Jr., Chesselet, M.-F., and Lewis, M. E., High-resolution *in situ* hybridization histochemistry, *Methods Enzymol.*, 168, 761, 1989.
8. Rychlik, W., Spencer, W. J., and Rhoads, R. E., Optimisation of the annealing temperature for DNA amplification *in vitro*, *Nucleic Acids Res.*, 18, 6409, 1990.
9. Lathe, R., Synthetic oligonucleotide probes deduced from amino acid sequence data: theoretical and practical considerations, *J. Mol. Biol.*, 183, 1, 1985.
10. Sambrook, J., Fritsch, E. F., and Maniatis, T., *Molecular Cloning: A Laboratory Manual*, 2nd ed., Cold Spring Harbor Laboratory Press, Cold Spring Harbor, NY, 1989.
11. Kaufman, D. L. and Evans, G. A., Restriction endonuclease cleavage at the termini of PCR products, *Biotechniques*, 9, 304, 1990.
12. Stoflet, E. S., Koeberl, D. D., Sarkar, G., and Sommer, S. S., Genomic amplification with transcript sequencing, *Science*, 239, 491, 1988.
13. Kain, K. C., Orlandi, P. A., and Lanar, D. E., Universal promoter for gene expression without cloning: expression PCR, *Biotechniques*, 10, 366, 1991.
14. Mole, S. E., Iggo, R. D., and Lane, D. P., Using the polymerase reaction to modify expression plasmids for epitope mapping, *Nucleic Acids Res.*, 17, 3319, 1989.
15. Kuipers, O. P., Boot, H. J., and de Vos, W. M., Improved site-directed mutagenesis method using PCR, *Nucleic Acids Res.*, 19, 4558, 1991.
16. Sharrocks, A. D. and Shaw, P. E., Improved primer design for PCR-based, site-directed mutagenesis, *Nucleic Acids Res.*, 20, 1147, 1992.
17. Higuchi, R., Krummel, B., and Saiki, R. K., A general method for *in vitro* preparation and specific mutagenesis of DNA fragments: study of protein and DNA interactions, *Nucleic Acids Res.*, 16, 7351, 1988.

18. **Ho, S. N., Hunt, H. D., Horton, R. M., Pullen, J. K., and Pease, L. R.,** Site-directed mutagenesis by overlap extension using the polymerase chain reaction, *Gene,* 77, 51, 1989.
19. **Lowe, T., Sharefkin, J., Yang, S. Q., and Dieffenbach, C. W.,** A computer program for selection of oligonucleotide primers for polymerase chain reactions, *Nucleic Acids Res.,* 18, 1757, 1990.
20. **Rychlik, W. and Rhoads, R. D.,** A computer program for choosing optimal oligonucleotides for filter hybridisation, sequencing and *in vitro* amplification of DNA, *Nucleic Acids Res.,* 17, 8543, 1989.

Chapter 3

PURIFICATION OF PCR* PRODUCTS

Louis M. Mezei and Douglas R. Storts

TABLE OF CONTENTS

I. Introduction 13
   A. DNA Size vs. Recovery 14

II. Materials and Methods 15
   A. Preparation of PCR Product 15
   B. Direct Purification Protocol 15
   C. Gel Purification Protocol 15

III. Results
   A. Primer Removal 16
   B. Primer-Dimer Removal 16
   C. Loading Capacity 16
   D. Sample Volume 17
   E. Comparison to Ultrafiltration Methods 17
   F. Elimination of Non-specific Amplification Products 18

IV. Discussion 19

References 19

I. INTRODUCTION

Amplification of DNA by the polymerase chain reaction (PCR) has become a core technology in molecular biology. However, purification of amplified DNA has been a stumbling block. Contaminating primers and oligonucleotides are difficult to remove and can adversely affect subsequent applications of the amplified product.

A number of factors should be considered when choosing a purification system for PCR product. Most importantly, the system must be able to do the following:

- Remove single-stranded oligonucleotide primers
- Remove short (less than about 50 bp) double-stranded oligonucleotides ("primer-dimers")
- Retain larger (greater than about 200 bp) double-stranded DNA (PCR products)
- Remove Taq DNA polymerase, buffer salts, dNTPs, and other reaction components

Other traits are also important. The method should be able to do the following:

- Provide an acceptable yield of high-purity DNA
- Accept a wide range of DNA concentrations

---

* The polymerase chain reaction (PCR) process for amplifying nucleic acid is covered by U.S. Patents 4,683,195 and 4,683,202, assigned to Hoffman-La Roche. Patents pending in other countries.

- Accept a wide range of volumes
- Concentrate the DNA
- Provide quick purifications and be easy to perform
- Adapt to protocols for elimination of non-specific amplification products

Several methods are commonly used for purification of PCR-amplified DNA. One method employs phenol-chloroform extraction followed by an ethanol precipitation.[1] This approach efficiently removes mineral oil, Taq DNA polymerase, buffer salts, and dNTPs, but may not eliminate contaminating primers and primer-dimers.

The phenol-chloroform method is very tedious and time consuming and requires the use of hazardous organic chemicals. Moreover, the method often results in substantial loss of product. This loss is especially apparent when purifying short PCR-amplified product (e.g., 200 to 500 bp).

A second technique that has become popular for PCR product purification is exemplified by Amicon® Centricon* microconcentrators. This technique is based on the ultrafiltration of a sample through a molecular weight cutoff filter membrane.[2] With this technique, PCR product is placed in the sample reservoir of a microconcentrator. The device is then centrifuged to drive solvents and low-molecular-weight solutes through the membrane and into a filtrate cup. DNA and protein above the membrane's molecular weight cutoff are retained in a sample reservoir.

As shown below, however, ultrafiltration does not efficiently eliminate unwanted low-molecular-weight PCR by-products. As with the phenol-chloroform extraction technique, the ultrafiltration method is tedious and time consuming.

The above methods are widely used for the purification of PCR products. However, they retain all products greater than the exclusion limit, which can be a significant drawback when the procedures are being used to clean up PCR product from amplifications in which the desired PCR product contains a high background of non-specific product (as exhibited by a smear on a gel).

DNA purification resins (such as Promega's Wizard™ PCR Preps system), anion exchange macroporous silica gel resins (such as Diagen's QIAGEN PCR purification spin kits), and glassmilk (such as BIO 101's Geneclean® II) provide alternatives to these approaches. These purification methods are simple to perform, routinely take less than 15 to 20 min, and result in high yields of purified product.

Furthermore, these approaches can be readily adapted to gel purification protocols for the removal of non-specific amplification products. For example, the Wizard™ PCR Preps DNA purification system facilitates the rapid purification of PCR products from low-gelling/melting-temperature agarose gels with recoveries routinely greater than 60%. Because the desired fragment is isolated from an agarose gel, non-specific amplification products, truncated products, primers, and primer-dimers do not interfere with subsequent manipulations.

The studies described below illustrate some of the performance characteristics of these approaches to PCR product purification. The studies were performed using Wizard™ PCR Preps as the example system. Many of the alternative matrices may offer similar performance characteristics.

A. DNA SIZE VS. RECOVERY

The most important trait of a purification system for PCR product is to be able to remove short double-stranded oligonucleotides (less than about 75 bp) while retaining a substantial portion of larger PCR-amplified product (greater than about 200 bp). Primer-dimer formation is a well-recognized problem when using PCR to amplify target DNA templates. Primer-

---

* Amicon® and Centricon are trademarks of Amicon Div., W.R. Grace & Co.

Purification of PCR Products

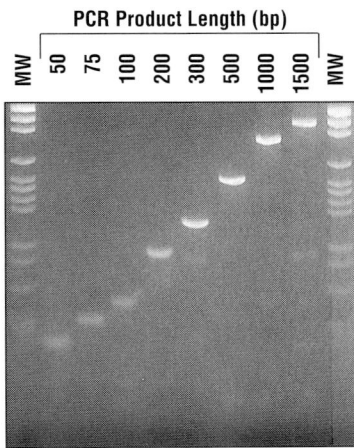

Figure 1. Amplification products used for evaluations of size vs. recovery. ^{32}P end-labeled PCR products were separated by electrophoresis on a 3% NuSieve®/1% SeaKem® gel. The lengths of the DNA products are given in base pairs. Lanes labeled MW contain molecular weight markers.

dimers are generated as a result of the extension of a template formed by overlap of the primers. The number of bases in a primer-dimer is approximately the total number of bases in the two amplification primers being used. For example, if two 25-mer oligonucleotides are being used, a primer-dimer approximately 45 to 50 bases long could be generated. Primer-dimers often adversely affect subsequent processing of the PCR product. Therefore, they must be removed.

II. MATERIALS AND METHODS

A. PREPARATION OF PCR PRODUCT

A family of ^{32}P end-labeled amplification products (50, 75, 100, 200, 300, 500, 1000, and 1500 bp) were used to evaluate the size-dependent recovery of the purification system. A sample of each reaction was separated on a 3% NuSieve®/1% SeaKem®* gel to verify the size of the products (Figure 1).

B. DIRECT PURIFICATION PROTOCOL

To purify amplified DNA, 100 to 650 µl of PCR reaction products, 100 µl of direct purification buffer (50 mM KCl, 10 mM Tris-HCl [pH 8.8 at 25°C], 1.5 mM MgCl$_2$, 0.1% Triton X-100), and 1 ml DNA purification resin were mixed in 12- by 75-mm polypropylene tubes. The purification resin consists of a chaotropic, silica-based medium containing components to enhance rehydration. The resin-bound DNA preparations were collected in mini-columns, washed once with 2 ml of column wash buffer (80% isopropanol), air dried for 5 min on a vacuum manifold, and eluted with 50 µl of water.

C. GEL PURIFICATION PROTOCOL

Two PCR reactions were performed to amplify 300- and 500-bp target sequences. Magnesium concentrations were adjusted to increase the prevalence of non-specific amplification products. Aliquots of the amplifications were separated by electrophoresis on a 1.2% SeaPlaque® agarose gel using standard protocols.[1] The 300- and 500-bp DNA bands were excised from the gel and transferred to 1.5-ml microcentrifuge tubes. The tubes were incubated at 70°C until

---

* NuSieve®, SeaKem®, and SeaPlaque® are trademarks of FMC BioProducts.

TABLE 1.
DNA Size vs. Recovery: Comparison of the Direct Purification and Ultrafiltration Methods

DNA size dsDNA	Direct purification method % recovery (Average of 6 samples)	Ultrafiltration % recovery (Average of 2 samples)
1500 bp	95.9%	90.5%
1000	107.9	75.4
500	98.4	70.2
300	99.1	66.7
200	68.6	63.8
100	8.1	39.3
75	3.2	25.3
50	1.9	18.7
ssDNA		
73 bases	1.1%	28.4%
45	1.5	3.2
29	1.0	7.6

the agarose was completely melted, then mixed with 1 ml DNA purification resin. Bound DNA was collected, washed, and eluted as described in the Section II.B.

III. RESULTS

A. PRIMER REMOVAL

The presence of residual oligonucleotide primers can adversely affect subsequent processing of PCR product. Three oligonucleotide primers, containing 29, 45, and 73 bases, were synthesized and end-labeled with ^{32}P using T4 polynucleotide kinase. Recoveries of these oligonucleotide primers were evaluated using the protocol described above for evaluating DNA size vs. recovery. As shown in Table 1, more than 97% of the primers were removed by the DNA purification system.

B. PRIMER-DIMER REMOVAL

To evaluate the efficacy of the DNA purification system at removing primer-dimers from PCR product while retaining high yields of the PCR product of interest, the PCR products described above were processed using the direct purification protocol. The resulting eluates were then separated on 3% NuSieve®/1% SeaKem® gels, the ^{32}P-labeled bands were excised, and the percent recovery determined by liquid scintillation counting. As shown in Figure 2, nearly 100% of the PCR products greater than 300 bp were recovered, while more than 97% of the 50- and 75-bp products (primer-dimers) were removed.

C. LOADING CAPACITY

Ideally, a PCR purification system should accommodate the widest possible range of DNA concentrations. That is, the system must be able to capture as little as 50 ng of product, while still being able to retain large quantities (greater than 10 µg).

To evaluate the loading capacity of the DNA purification resin, the 500-bp PCR product described above was diluted to concentrations of 0.05, 1, 2, 4, and 8 µg/200 µl with PCR mix containing no additional product. Samples (200 µl) of each dilution were processed using the standard DNA purification protocol, the eluates were separated by gel electrophoresis, the

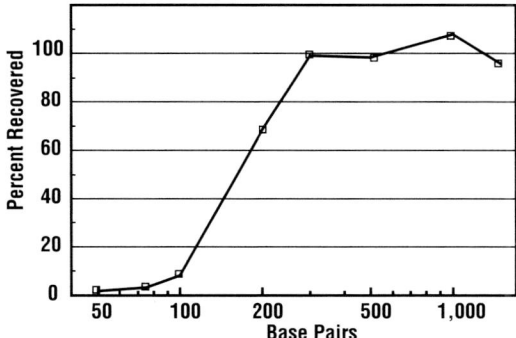

Figure 2. DNA size vs. recovery. PCR products of varying lengths (50 to 1500 bp) were purified using the direct purification protocol. Recoveries were determined by liquid scintillation counting of excised gel bands.

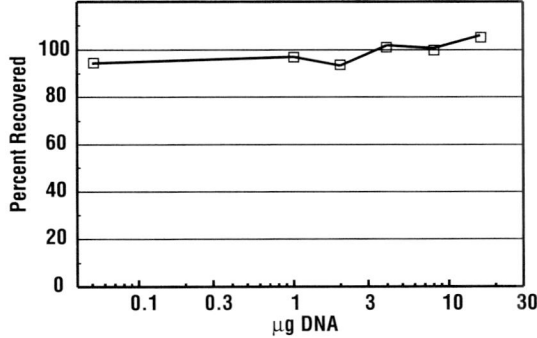

Figure 3. DNA concentration vs. recovery. Varying concentrations (50 ng to 16 µg) of 500bp PCR product were purified using the direct purification protocol. Recoveries were determined by liquid scintillation counting of excised gel bands.

bands were excised, and the percent remaining was determined by liquid scintillation counting. As shown in Figure 3, the DNA purification system is effective over a wide range of DNA concentrations.

D. SAMPLE VOLUME

The ability of a purification method to tolerate a substantial variation in sample volume is also important. PCR typically is performed using 100 µl reaction volumes. If more than one amplification is performed with a single sample, it would be convenient to be able to pool the reaction products and purify them collectively.

To evaluate the effect of sample volume on recovery, 3 µg aliquots of the 500-bp PCR product were added to each of a series of tubes. Then, appropriate volumes of PCR mix were added, to bring the final volumes in the tubes to 100, 150, 250, 350, 450, and 650 µl. As shown in Figure 4, high yields were obtained from all volumes studied. Because sample volumes of up to 650 µl can be loaded and elution can be performed with as little as 50 µl, the DNA purification system can be used to both purify and concentrate PCR product.

E. COMPARISON TO ULTRAFILTRATION METHODS

As discussed above, the use of ultrafiltration microconcentrators has become popular for purifying PCR product. To evaluate the performance of ultrafiltration microconcentrators, the

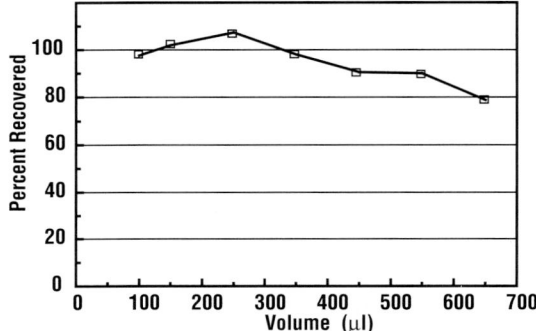

Figure 4. Volume added vs. recovery. Varying volumes of PCR reaction mix were added to tubes containing 3 µg of 500-mer PCR product. The amount indicated on the *x* axis represents the total volume processed. Recoveries were determined by liquid scintillation counting of excised gel bands.

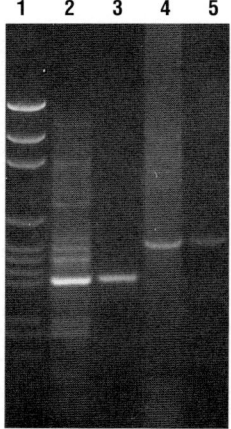

Figure 5. Purification of PCR products from low-melting agarose gels. Aliquots of gel purified PCR products were electrophoresed adjacent to equivalent amounts of unpurified PCR reactions on a 1.2% SeaPlaque® agarose gel. Lane 1, molecular weight markers; lane 2, 300-bp PCR product containing non-specific amplification products (unpurified); lane 3, gel purified 300-bp PCR product; lane 4, 500-bp PCR product containing non-specific amplification products (unpurified); lane 5, gel purified 500-bp PCR product.

recovery of PCR products and primer oligonucleotides was tested as described above for the DNA purification system. As shown in Table 1, ultrafiltration retained a substantial portion of the DNA in the primer-dimer range of molecular weights and a substantial portion of large-molecular-weight primers.

Furthermore, the ultrafiltration microconcentrator methods are limited by centrifuge rotor size and require up to 2 h to purify ten samples. In contrast, with the DNA purification system large numbers of samples can be conveniently run, and the procedure takes about 15 min to perform.

F. ELIMINATION OF NON-SPECIFIC AMPLIFICATION PRODUCTS

To assess the effectiveness of the DNA purification system at recovering high-purity PCR product from agarose gels, aliquots of two PCR reactions containing non-specific bands were separated in a 1.2% SeaPlaque agarose gel, and the desired bands were excised and purified as described in the Section II.C. Equivalent DNA concentrations of the crude reaction mixes and the gel purified products were electrophoresed on a 1.2% SeaPlaque® agarose gel. As shown in Figure 5, more than 60% of the desired 300- and 500-bp PCR products were recovered, while non-specific amplification products were effectively removed.

IV. DISCUSSION

The speed, ease of use, and flexibility of resin-based purification systems greatly simplify purification of PCR-amplified DNA. PCR product is effectively purified away from contaminants, including primer-dimers and amplification primers. The purified DNA can be used directly for cloning, DNA sequencing, transcription, labeling, and other applications, and it greatly improves the effectiveness of these downstream processes.

Resin-based purification systems are readily adapted to gel purification protocols. PCR amplification frequently yields non-specific amplification products (as evidenced by smearing in agarose gels). These non-specific products may interfere with subsequent manipulations. Optimization of the amplification conditions may reduce the level of non-specific amplification, but is time consuming. An alternate approach consists of low-melting agarose gel isolation of the desired PCR product, followed by purification with a resin-based system.

REFERENCES

1. **Sambrook, J., Fritsch, E. F., and Maniatis, T.,** *Molecular Cloning: A Laboratory Manual,* 2nd ed., Cold Spring Harbor Laboratory Press, Cold Spring Harbor, NY, 1989.
2. *Centricon Microconcentrator Technical Manual,* Amicon Division, W.R. Grace & Co., Beverly, MA, 1991.

Chapter 4

CLONING PCR* PRODUCTS

Louis M. Mezei and Douglas R. Storts

TABLE OF CONTENTS

I. Introduction 21

II. Materials and Methods 23
    A. Preparation of a T Vector 23
    B. Generation of PCR Products 24
    C. PCR Product Cleanup 24
    D. Ligation of T-Vector and PCR Product 24
    E. Transformation of Ligated PCR:T Vector 25

IV. Discussion 26

References 27

I. INTRODUCTION

Primed amplification by the polymerase chain reaction (PCR) can be used to rapidly produce large amounts of specific DNA from complex sources containing cDNA, genomic, YAC, phage, cosmid, and/or plasmid DNA. However, difficulties are frequently encountered when cloning the PCR products.

Restriction endonuclease sites are often incorporated into the primers used for amplification so that cleavage of the product will create overhangs that can be ligated to an equivalently cut vector.[1] Unfortunately, there are some drawbacks to this approach:

- When recognition sequences are located within a few base pairs of the end of a DNA fragment, many restriction endonucleases have difficulty or fail to cleave at the site.[2] Proposed explanations for this deficiency are as follows:[3]

  - Taq DNA polymerase binds to the terminal site, thereby preventing access of the restriction endonuclease to the recognition site.
  - Having a limited number of bases at a terminal restriction recognition site is insufficient to allow stable association with, and cutting by, certain restriction endonucleases.
  - Taq DNA polymerase is inefficient at fully extending certain terminal sequences and generates frayed ends that cannot be cleaved by the restriction endonuclease.

- If secondary cleavage sites are located within the PCR product, the sites negate this approach to cloning.
- The addition of at least 12 bases to the 5′ end of the primer is usually required, which substantially increases the cost of synthesis and the complexity of amplification problems.

---

* The polymerase chain reaction (PCR) process for amplifying nucleic acid is covered by U.S. Patents 4,683,195 and 4,683,202, assigned to Hoffmann-LaRoche. Patents pending in other countries.

Attempts to clone PCR products as blunt-ended fragments have also proven inefficient, due to the template-independent terminal transferase activity of Taq DNA polymerase (among others). This activity results in the addition of a single unpaired nucleotide at the 3' end of the fragment, leaving "ragged" ends on the amplified DNA.[4,5] This nucleotide is almost exclusively an adenosine, due to the strong preference of the polymerase for dATP.[4] Thus, to efficiently clone PCR products as blunt-ended fragments, further enzymatic processing is required.

One approach uses an enzyme with 3' → 5' exonuclease activity to remove the 3' overhang.[6] Another approach is to use the Klenow fragment of *Escherichia coli* DNA polymerase I to fill in the overhang. Both approaches can overcome this problem to some extent, but cloning efficiencies are often low.

The template-independent activity of Taq DNA polymerase can be exploited to create a cloning scheme that has the efficiency of sticky-end cloning, while requiring no additional enzymatic modification of the PCR product. The method employs a plasmid containing a single T overhang. Essentially three different methods have been used to create this overhang on the vector:

- Digestion with either *Xcm* I or *Hph* I restriction endonuclease to yield 3' unpaired deoxythymidine residues at both ends[7,8]
- Addition of a single 3' overhang thymidine residue using dideoxythymidine triphosphate (ddTTP) and terminal transferase[9]
- Blunt-end cutting of the plasmid and addition of a single 3' overhang thymidine residue using deoxythymidine triphosphate (dTTP) and the template-independent terminal transferase activity of Taq DNA polymerase[10]

The example protocols described below are similar to the ones used to prepare the Promega pGEM® T vector from pGEM-5Zf(+). Yields are typically 30 to 50%, based on the starting mass of plasmid DNA. The method describes the preparation of a T vector by digesting plasmid DNA with a restriction enzyme that produces blunt ends and adding a 3' terminal thymidine to both ends. These single 3' T overhangs at the insertion site greatly improve the efficiency of ligation of PCR product into the plasmid. More specifically, ligation of the overhangs takes advantage of the template-independent addition of a single deoxyadenosine to the 3' end of PCR product noted above.

The preparation, ligation, and transformation methods were developed as a result of extensive optimization studies. If other vectors are used, the protocol should be considered as a starting point and further optimized for the vector being used. Other plasmid vectors containing blunt-end cut sites within the multiple cloning site (such as Stratagene's Bluescipt®, pUC18, pUC19, other Promega pGEM® vectors, etc.) should work similarly.

The pGEM-5Zf(+) vector exemplifies some of the desirable attributes to consider when choosing a starting material for a T vector. For example, this vector contains the origin of replication of the filamentous phage f1. For induction of single-stranded DNA, bacterial cells containing recombinants are infected with an appropriate helper phage. The plasmid then enters the f1 replication mode, and the resulting single-strand DNA is exported from the cell as an encapsulated virus particle. The plasmid also contains both T7 and SP6 RNA polymerase promoters flanking a multiple cloning region within the α-peptide coding region of the enzyme β-galactosidase. Insertional inactivation of the α-peptide allows recombinant clones to be directly identified by color screening on indicator plates. The multiple cloning region is unique and includes several restriction sites for inserting foreign DNA fragments. The central linker contains cut sites for restriction enzymes that produce 5' overhangs or blunt ends (sensitive to exonuclease III). These enzyme cut sites are flanked on both sides by blocks of restriction sites that generate 3' overhangs (resistant to exonuclease III).

II. MATERIALS AND METHODS

A. PREPARATION OF A T VECTOR

I. Digestion with blunt-end restriction enzyme

   NOTE: CHOOSE A RESTRICTION ENZYME WITH NO STAR OR CONTAMINATING NUCLEASE OR PHOSPHATASE ACTIVITY.

   Sample digestion (20 µg):
   1. Prepare the following in an autoclaved 1.5-ml microcentrifuge tube. (**NOTE: NON-AUTOCLAVED TUBES OFTEN CONTAIN INHIBITORS THAT ARE REMOVED BY THE AUTOCLAVING PROCESS.**)

10 × restriction enzyme buffer	20 µl
Acetylated-BSA (1 mg/ml)	20 µl
1 µg/µl plasmid DNA	20 µl
Deionized H$_2$O	to final volume 200 µl
Restriction endonuclease	200 U

   2. Incubate 4 h at 37°C. Store at 4°C until phenol-chloroform extraction. For extended storage (more than 8 h), freeze at –70°C.

II. Phenol-chloroform extraction

   1. Add NaCl to a final concentration of 0.5 M (1/10 volume of 5 M NaCl).
   2. Add an equal volume of fresh phenol chloroform:isoamyl alcohol (50:49:1) saturated with 0.5 M NaCl.
   3. Mix well by inversion.
   4. Spin for 5 min at 12,000 G.
   5. Remove the aqueous (upper) layer and place in a fresh microcentrifuge tube.
   6. Add an equal volume of chloroform: isoamyl alcohol (24:1).
   7. Mix well by inversion.
   8. Spin for 5 min at 12,000 G.
   9. Repeat steps 5 through 8.
   10. Remove the aqueous (upper) layer and place in a fresh microcentrifuge tube.
   11. Add 0.5 vol of 7.5 M ammonium acetate. Mix.
   12. Add 3 vol 100% ethanol, mix, and place on ice for 30 min.
   13. Spin 30 min at 12,000 G. Remove and discard supernatant.
   14. Add 1 × the original volume of prechilled 70% ethanol. Mix.
   15. Spin 30 min at 12,000 G. Remove and discard supernatant.
   16. Dry the pellets at room temperature. Store at –70°C until reconstitution.

III. Reconstitution and quality control

   1. Resuspend pellets to a concentration of about 1 µg/µl in **STERILE** 1 × TE (10 mM Tris, 1 mM EDTA, pH 7.3 to 7.5), assuming a 100% recovery.
   2. Check the concentration and quality by analyzing samples diluted 1/100 in TE on a spectrophotometer. Record absorbance at 250, 260, and 280 nm. Calculate ratios and concentrations as follows:

      a. Compute 260:250 ratio. The value should be 1.10 ± 0.05.
      b. Compute 260:280 ratio. The value should be >1.80.
      c. Concentration (in micrograms per milliliter) = dilution factor × A260 × 50.

3. Dilute to a final concentration of about 0.3 to 0.4 µg/µl in **STERILE** 1 × TE.
4. Run a 2% agarose gel with 200 ng of uncut plasmid DNA as a control. If the digestions appear complete (i.e., there is no trace of uncut DNA), then proceed to next step. If not, add more enzyme and repeat steps 1 and 2 above.

IV. Adding the "T"

IMPORTANT: USE A GRADE OF Taq DNA POLYMERASE THAT LACKS 5'→3' EXONUCLEASE ACTIVITY. (Sequencing grades of Taq DNA polymerase from some vendors lack this activity.)

1. Prepare the following in a microcentrifuge tubes:

10 × Taq buffer (500 mM KCl, 100 mM Tris-HCl, pH 9.0, 1.0% Triton X-100)	2 µl
25 mM MgCl$_2$	2 µl
1 µg/µl digested plasmid DNA	1 µl
Deionized H$_2$O	to final volume 20 µl
dTTP (100 mM)	0.15 µl
Taq DNA polymerase (low 5'→3' exonuclease)	1 U

2. Place tube(s) in an incubation block (e.g., a thermal cycler), and place weight on caps to prevent tubes from popping open.
3. Incubate for 3 h at 72°C.
4. Cool to 4°C. Freeze at –70°C for extended storage (i.e., > 8 h). (At this point, the terminal T is labile to exonuclease activity.)
5. Extract with phenol-chloroform and ethanol precipitate as in step II above.
6. Dry using a vacuum desiccator for about 1.5 h.
7. Store at –70°C until reconstitution.

V. Reconstitution and final quality control

1. Resuspend pellets to a concentration of about 100 ng/µl in **STERILE** 1 × TE (10 mM Tris, 1 mM EDTA, pH 7.3 to 7.5), assuming a 30% recovery.
2. Determine the DNA concentration by fluorescent assay[11] on duplicate 10-µl samples or estimate by agarose gel electrophoresis.
3. Adjust to 50 ng/µl with **STERILE** 1 × TE.

B. GENERATION OF PCR PRODUCTS

Use standard amplification protocols to generate PCR product. Purify and adjust the concentration of the product such that 2 µl can be used in a 10-µl reaction to achieve a 1:1 molar ratio. (See below.)

C. PCR PRODUCT CLEANUP

PCR product to be ligated can be gel purified, purified directly using a resin (such as Promega's Wizard™ PCR Preps resin), or unpurified (see Chapter 3). If smearing is observed on a gel, the bands to be cloned should be excised from low-melt agarose gel and purified with resin. If distinct bands of the PCR product occur, primer-dimer should be removed using resin directly from the PCR reaction mix. The use of crude PCR product is acceptable. However the number of white colonies containing the relevant insert may be reduced due to preferential incorporation of lower-molecular-weight PCR product or incorporation of extraneous product.

D. LIGATION OF T VECTOR AND PCR PRODUCT

To obtain the optimal ratio of T vector to PCR product, a 1:1 molar ratio should be used. (Ratios of 8:1 to 1:8 have been successfully used.) To calculate the amount of PCR product, use the size of the T vector (e.g., 3003 for pGEM® T vector). The concentration of PCR product may be estimated from agarose gels or by fluorescent assay.[11] The T vector is at a concentration of 50 ng/μl. The following is an example calculation:

$$\frac{\text{ng of T vector} \times \text{kb size of insert}}{\text{kb size of T vector}} \times \text{molar ratio of } \frac{\text{insert}}{\text{T vector}} = \text{ng insert}$$

For example, for a 0.5-Kb insert, 50 ng of a 3 Kb T-Vector, and a 1:1 molar ratio:

$$\frac{50 \text{ ng T vector} \times 0.5 \text{ kb insert}}{3 \text{ kb T vector}} \times \frac{1}{1} = 8.3 \text{ ng insert}$$

When choosing a T4 DNA ligase preparation, it is essential that it has low nuclease and phosphatase activity. The most common cause for poor ligation/transformation efficiency has been traced to poor lots of ligase.

1. Set up a ligation reaction as follows. Use autoclaved 0.5-ml tubes known to have low DNA-binding capacity.

T4 DNA ligase 10 × buffer (300 mM tris-HCl, pH 7.8, 100 mM MgCl$_2$ 100 mM DTT, 10 mM ATP)	1 μl
T vector (50 ng)	1 μl
PCR product*	X μl
T4 DNA ligase (1 Weiss unit/μl)	1 μl
Deionized water	to final volume 10 μl

2. Incubate for 3 h at 15°C.
3. Heat for 10 min at 70 to 72°C. Refrigerate.
4. Transform into high-efficiency competent cells.

E. TRANSFORMATION OF LIGATED PCR:T VECTOR

Transformation is performed using high-efficiency competent cells. Host strains should be compatible with the blue/white color screening and appropriate antibiotic selection. The efficiency of the competent cells used should be at least 1×10^8 cfu/ml. Transformed strains should be grown on LB (Luria-Bertaini) plates supplemented with 100 μg/ml ampicillin, or other appropriate antibiotic. Luria-Bertaini medium is as follows:

   10 g/l Bacto-tryptone
   5 g/l Bacto-yeast extract
   5 g/l NaCl, pH 7.0

Add 15 g agar to 1 liter of LB medium. Autoclave and allow the medium to cool to 50°C before adding antibiotic, IPTG, and X-GAL. IPTG and X-GAL can be added in either of two ways:

- They can be added directly to the medium before plates are poured.
- 100 μl 100 mM IPTG and 20 μl 50 mg/ml X-GAL can be spread over the surface of the plate and allowed to absorb for 30 min at 37°C just prior to use.

If the first method is being used, add the IPTG and X-GAL to final concentrations of 0.5 mM and 80 μg/ml, respectively. Pour 30 to 35 ml of medium into 85-mm petri dishes. Allow the agar to harden. Store at 4°C for up to one month or at room temperature for up to one week.

---

* To evaluate the quality of the T vector, it is often desirable to ligate and transform a sample containing no PCR product. This sample serves as a negative control to ensure that the number of false-positive white colonies is less than 1% and that 3′ terminal thymidine has been efficiently added to both ends.

NOTE: FOR BEST RESULTS, DO NOT USE PLATES MORE THAN 30 DAYS OLD.

1. Prepare LB/IPTG/X-GAL plates containing an appropriate antibiotic or bring to room temperature.
2. Add 2 µl of ligated PCR:T vector to sterile 1.5-ml microcentrifuge tube(s) on ice. (Note: Tubes from some manufacturers bind DNA and therefore decrease the white colony count. We have successfully used autoclaved Sarstedt tubes.)
3. Remove tube(s) of frozen, high-efficiency competent cells from –70°C storage and place in an ice bath until just thawed (about 5 min). Mix by *gently* flicking tube.
4. *CAREFULLY* aliquot 50 µl of cells per tube prepared in step 2. *THE CELLS ARE VERY FRAGILE.*
5. *GENTLY* flick tubes to mix and place back on ice for 20 min.
6. Heat shock cells for 45 to 50 s at exactly 42°C.
7. Return tubes to ice for 2 min.
8. Add LB broth to tubes. Since fewer transformants are recovered when cloning large inserts than with small inserts, use the following guidelines to produce reasonable white colony counts:

   For about 1500-bp PCR products: 450 µl LB
   For about 500-bp PCR products: 1.4 ml LB
   For negative control (ligation without PCR product): 450 µl LB

9. Gently invert tubes to mix. Incubate 1 h at 37°C.
10. Mix tubes and spread 50 µl of cells onto LB/IPTG/X-GAL plates containing an appropriate antibiotic.
11. Incubate overnight (24 h) at 37°C. White (recombinant) colonies contain inserts. Longer incubations or storage at 4°C may be used to facilitate blue/white screening. In general, colonies containing β-galactosidase activity grow poorly relative to cells lacking this activity. After overnight growth, blue colonies are pinpoint in size, while white colonies are approximately 1 mm in diameter.

IV. DISCUSSION

The single 3′-thymidine overhangs at the insertion site greatly improve the efficiency of ligation of a PCR product into the plasmid. The procedure described here typically produces about 200 colonies. Approximately 30% of the colonies are white, and 80% to 90% of the white colonies contain the relevant insert.

Decreased performance can occur if poor-quality ligase, restriction endonuclease, or Taq DNA polymerase containing 5′ → 3′ exonuclease activity is used. If any of the enzymes are contaminated with exonuclease, phosphatase, or other restriction enzymes, plasmids can be damaged by these contaminants and reduce ligation efficiency, yield fewer white colonies, and/or produce white or light-blue colonies, which are indistinguishable from clones containing inserts.[12]

The number of blue colonies can be substantially reduced or eliminated by dephosphorylating the vector with alkaline phosphatase after the 3′-thymidine has been added. This step eliminates self-ligation and reduces the number of blue colonies. However, if a dephosphorylated T vector is used, the primers used for PCR amplification must be phosphorylated prior to use.

REFERENCES

1. **Scharf, S. J., Horn, G. T., and Erlich, H. A.**, Direct cloning and sequence analysis of enzymatically amplified genomic sequences, *Science,* 233, 1076, 1986.
2. **Kaufman, D. L. and Evans, G. A.**, Restriction endonuclease cleavage at the termini of PCR products, *BioTechniques,* 9, 304, 1990.
3. **Jung, V., Pestka, S. B., and Pestka, S.**, Efficient cloning of PCR generated DNA containing terminal restriction endonuclease recognition sites, *Nucleic Acids Res.,* 18, 6156, 1990.
4. **Clark, J. M.**, Novel non-template nucleotide addition reactions catalyzed by procaryotic and eucaryotic DNA polymerases, *Nucleic Acids Res.,* 16, 9677, 1988.
5. **Mole, S. E., Iggo, R. D., and Lane, D. P.**, Using the polymerase chain reaction to modify expression plasmids for epitope mapping, *Nucleic Acids Res.,* 17, 3319, 1989.
6. **Hemsley, A., Arnheim, N., Toney, M. D., Cortopassi, G., and Galas, D. J.**, A simple method for site-directed mutagenesis using the polymerase chain reaction, *Nucleic Acids Res.,* 17, 6545, 1989.
7. **Mead, D. A., Pey, N. K., Herrnstadt, C., Marcil, R. A., and Smith, L. M.**, A universal method for the direct cloning of PCR amplified nucleic acid, *Bio/technology,* 9, 657, 1991.
8. **Kovalic, D., Kwak, J., and Weisblum, B.**, General method for direct cloning of DNA fragments generated by the polymerase chain reaction, *Nucleic Acids Res.,* 19, 4560, 1991.
9. **Holton, T. A. and Graham, M. W.**, A simple and efficient method for direct cloning of PCR products using ddT-tailed vectors, *Nucleic Acids Res.,* 19, 1156, 1991.
10. **Marchuk, D., Drumm, M., Saulino, A., and Collins, F.**, Construction of T-vectors, a rapid and general system for direct cloning of unmodified PCR Products, *Nucleic Acids Res.,* 19, 1154, 1991.
11. **Haff, L. and Mezei, L. M.**, Measurement of PCR amplification by fluorescence, *Amplifications,* 1, 8, 1989.
12. **Murray, E., Singer, K., Cash, K., and Williams, R.**, Cloning-qualified blunt end restriction enzymes: causes and cures for light blue colonies, *Promega Notes,* 41, 1, 1993.

Chapter 5

AMPLIFICATION OF UNKNOWN FLANKING DNA BY SINGLE-SPECIFIC-PRIMER PCR

Jörg Stappert

TABLE OF CONTENTS

I. Introduction 29
   A. Methods of Amplifying Flanking Genomic DNA Sequences 30
   B. Methods of Amplifying Flanking cDNA Sequences 32

II. Materials and Methods 33
   A. Oligonucleotides 33
   B. 3' End Blocking of the Anchor Oligonucleotide 34
   C. First-Strand Synthesis 34
   D. Oligonucleotide Ligation to Single-Stranded cDNA 34
   E. Amplify 5' End of cDNA with the PCR 34

III. Discussion 34

References 35

I. INTRODUCTION

In order for amplification to occur, PCR requires primer annealing sites on each end of the target sequence; until recently, knowledge of the sequences initially flanking the target sequence was also needed. When the sequence is known in only one region, the requirement of a primer complementary to both ends of the segment to be amplified poses a problem.

Various methods to circumvent this problem have been developed to amplify segments of DNA with known DNA sequence on only one side. As these techniques have been developed in two different directions, they can be subdivided into two different groups.

The first are methods that allow amplification of genomic DNA, such as inverse PCR, vectorette PCR, or panhandle PCR.[1-4] These methods have been successfully used for the determination of viral integration sites, determination of transposon integration sites, amplification of cis-regulatory elements, mapping of introns in genomic DNA from cDNA clones, and isolation of terminal sequences from yeast artificial chromosome (YAC) clones.[5-7]

In contrast, techniques like RACE-PCR (rapid amplification of cDNA ends) and LA-PCR (ligation-anchored PCR) have been developed in order to determine the 5' unknown region of a given cDNA.[8,9] These techniques have yielded a broad spectrum of applications, for example, constructing PCR-directed cDNA libraries from very small amounts of starting material, analysis of T-cell receptor and immunoglobulin V regions, cloning of developmentally regulated mRNAs detected with gene-trap vectors, and characterization of alternative splice products and alternative promotor usage.[10-13]

Some of the currently published methods mentioned above will be summarized in the following sections, and a detailed description of LA-PCR will be given in Section II.

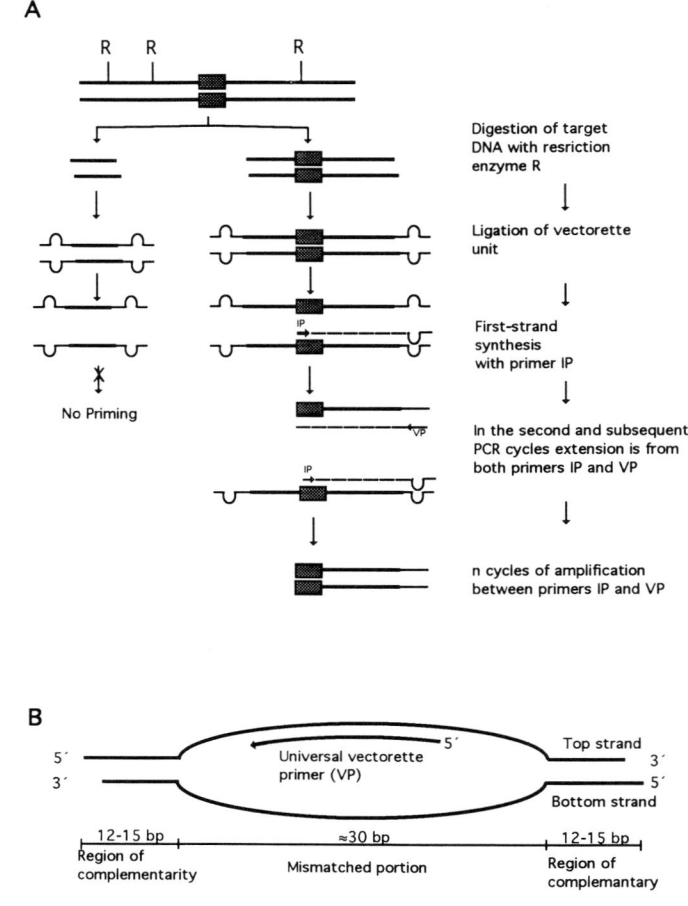

Figure 1. (A) Schematic diagram of vectorette PCR; known flanking DNA sequence is marked by hatched boxes; and vectorettes are designed as loops; (B) schematic of a vectorette unit.

A. METHODS OF AMPLIFYING FLANKING GENOMIC DNA SEQUENCES

Besides inverse PCR (also designated inside-out PCR), which was originally employed by three different laboratories and which can now be considered the classical strategy to amplify genomic DNA on one or both sides of a known nucleotide sequence, two other methods have been described recently.

The first technique, termed *vectorette PCR*, has been used to sequence the termini of YAC clone inserts[5] as well as to do genomic walking.[6] The principle of this method is shown in Figure 1A. It consists of three basic steps: (1) digestion of target DNA with a suitable restriction enzyme, (2) ligation of suitable synthetic oligonucleotides onto the digested DNA, and (3) PCR using a specific primer and a primer directed toward the synthetic oligonucleotide. In the first step, target DNA is digested with a range of restriction enzymes to maximize the chance of obtaining DNA fragments suitable for amplification. Preferably restriction enzymes generating 5' overhangs should be used in order to facilitate ligation. In the second step, synthetic DNAs called *vectorettes* are ligated to all digested fragments. These adaptors are only partially double-stranded and contain a central mismatched region to avoid first-strand synthesis by the vectorette primer (VP) (Figure 1B). The vectorette PCR primer has the same sequence as the bottom strand of this mismatched region and therefore has no comple-

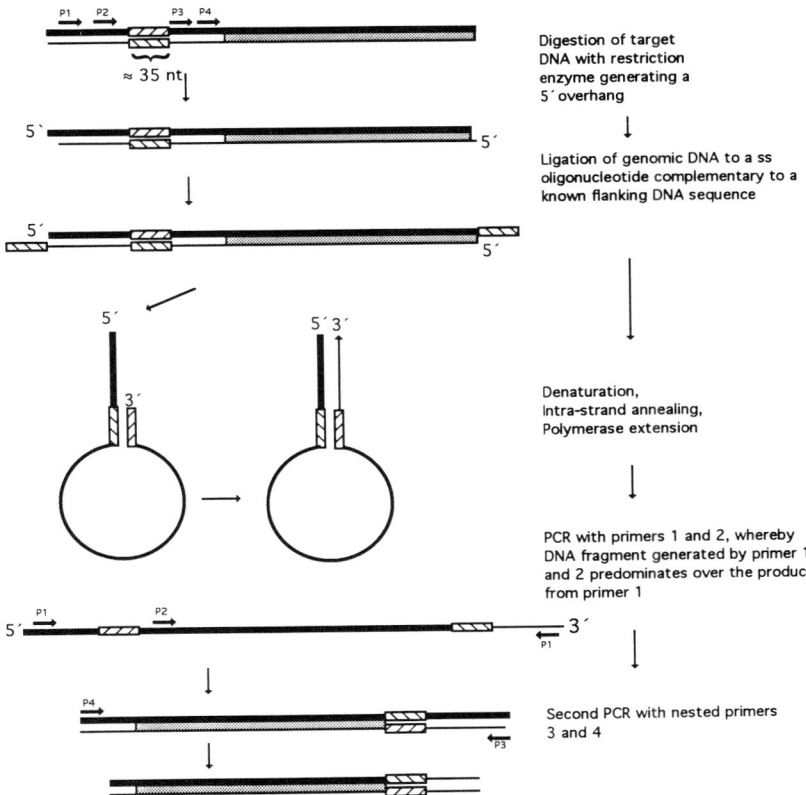

Figure 2. Strategy of panhandle PCR. The two complementary DNA strands are marked by thin or thick lines. Striped boxes represent the internal region of DNA complementary, and stippled areas indicate unknown genomic DNA.

mentary sequence to anneal to in the first cycle of PCR of the third step. Consequently, only the known, initiating primer (IP), which is directed toward the sequence of interest, will prime DNA synthesis.

The second approach to amplify unknown genomic flanking DNA, by Jones and Winistorfer,[4] is termed *panhandle PCR*, with reference to the intermediate form of the template, which resembles a pan with a handle. The basic strategy is outlined in Figure 2.

After the digestion of genomic DNA with a restriction enzyme that leaves a 5′ overhang, a single-stranded oligonucleotide is ligated to each 3′ end of digested fragments. This oligonucleotide is designed to be complementary to the known region of DNA upstream of the unknown region of interest. By using denaturation and annealing conditions favoring intrastrand annealing, genomic DNA, which contains the complement of the ligated oligonucleotide, will form a stem-loop structure. The resulting recessed 3′ end can prime DNA polymerization using the known sequence as the template. The stem structure built up by the known complementary DNA sequences will be the starting point for the following PCR. One of the two primers used to amplify the unknown DNA region, which is located in the pan, is homologous to the region upstream from the annealing site for the oligonucleotide (primer P1), while the second primer (P2) is homologous to the region located between the ligated oligonucleotide annealing site and the unknown flanking DNA. The product generated by

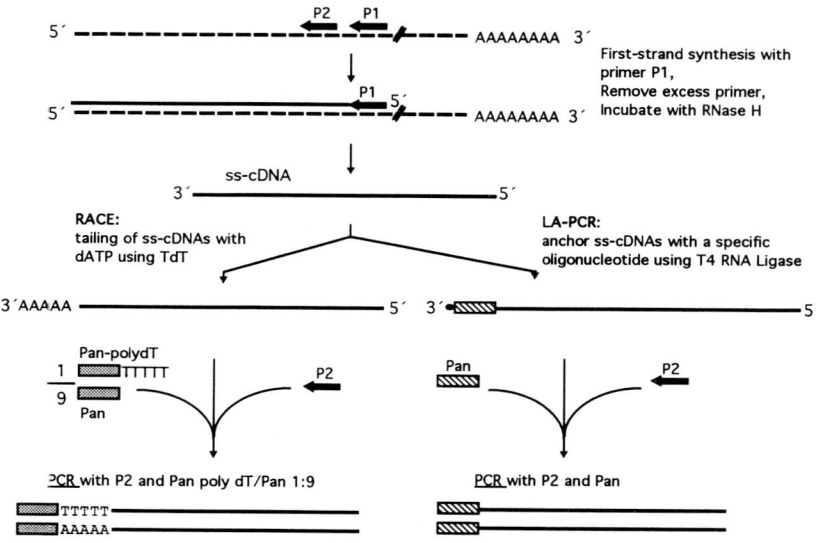

Figure 3. Schematic diagram of RACE and LA-PCR. Primers Pan-polydT and Pan are marked by stippled and striped boxes. The anchor oligonucleotide ligated to the first-strand cDNA is blocked at its 3'-end, symbolized by a solid circle.

primers P1 and P2 predominates over the product resulting from the annealing of primer P1 to both DNA ends due to their complementarity. In a second round of amplification, two nested primers (P3, P4) are used, thereby permitting consistent amplification of DNA flanking one side of a known DNA sequence. This technique enabled the authors to amplify DNA fragments of up to 2.3 kb directly from complex human genomic DNA.

B. METHODS OF AMPLIFYING FLANKING cDNA SEQUENCES

The second group comprises methods that allow specific amplification of cDNA, where the 5' sequence of the molecule of interest is unknown. This can be done either by homopolymer tailing of the cDNA or by linking a short single-stranded oligonucleotide to first-strand cDNA.

The strategy of methods like anchored or single-sided PCR or RACE is based on the addition of a homopolymeric $dG^{10,12}$ or $dA^{1,14}$ tail to the 3' end of the cDNA, using terminal deoxyribonucleotide transferase (TdT). Reverse transcription of the mRNA is primed with a specific oligonucleotide (P1), or with an oligonucleotide tailored with a poly (dT) stretch, thus taking advantage of the poly (A) sequence at the 3' end of most mRNAs. In the following steps, subsequent amplification of the unknown flanking region is achieved by using the specific primer (P2) and a second primer (Pan-polydT) containing a defined anchor sequence attached to a homopolymer sequence complementary to the tail, and finally, reamplification using primer P2 specific for the cDNA of interest and primer Pan specific for the anchor (see Figure 3).

However, this strategy has important limitations relating primarily to the use of homopolymeric tails. The TdT reaction is difficult to control and additionally has a low efficiency. The number of nucleotides added to the 3' end of the first-strand cDNA greatly influences the effectiveness of the tail as a PCR primer template. Furthermore, primers containing a homopolymeric tail of dT or dC at the 3' end bind non-specifically to the template DNA, thus generating non-specific amplification, a phenomenon that can prevent the isolation of low-abundance mRNA species. Even the use of nested primers does not prevent the amplification of non-specific fragments, making it necessary to sequence individual clones to obtain the desired cDNA fragment.

In fact, these problems have supported the renaissance of T4 RNA ligase, an enzyme capable of joining single-stranded DNA molecules.[15] Thanks to improvement of the reaction conditions for this enzyme, different groups have described methods using T4 RNA ligase to facilitate amplification of flanking 5' cDNA sequences. These techniques, termed SLIC (single-strand ligation to single-stranded cDNA)[16] and LA-PCR,[9] are based on covalent linkage of an anchor oligonucleotide to first-strand cDNAs, primed with the specific primer P1. These anchored cDNAs are then amplified by using one primer specific for the anchor (Pan) and another specific primer for a sequence within the molecule of interest (P2) (Figure 3).

This method has been successfully used in the amplification of different cDNAs using RNA equivalent to as few as 100 cells and has also been tested for generating cDNA libraries from low amounts of RNA. It overcomes the drawbacks imposed by homopolymeric tailing.

With this approach, we succeeded in cloning the missing 5' part of the mouse plakoglobin cDNA, a fragment 400 bp in length. Additionally, we adapted this method to amplify 5' unknown cDNA of developmentally regulated mouse genes identified by the gene-trap approach.

To use this method successfully, the following precautions should be taken to target the ligation of the anchor primer specifically to the 3' end of the single-stranded cDNA:

1. To prevent self-ligation of the anchor oligonucleotide, it should be blocked at its 3' end by adding ddATP.
2. To minimize ligation of the oligonucleotide used to prime the single-stranded cDNA synthesis either to the anchor primer or to the ss-cDNA, it should not have a 5' phosphate group, and it must be removed before the ligation reaction is done.
3. To maximize the chance of amplifying the desired DNA sequence, a second PCR should be performed using a pair of nested primers.

In the following protocol, T4 RNA ligase-supported amplification of unknown cDNA by single-specific primer PCR is described. Practical approaches are also given exclusively for the method itself. Commonly used techniques like the preparation of total RNA or polyA+ RNA, elution of small fragments after polyacryamide gel electrophoresis, etc., should be done using standard protocols.[17]

II. MATERIALS AND METHODS

A. OLIGONUCLEOTIDES

All oligonucleotides were synthesized on an Applied Biosystems DNA synthesizer and then purified using a C-18 reversed-phase high-pressure liquid-chromatography (HPLC) column. The nucleotide sequence of the anchor primer we used is as follows: 5' GTAGGAATTCGGGTTGTAGGGAGGTCGACATTGCC 3'. Underlined sequences represent restriction sites for *Eco*RI and *Sal*I. Initiation of the reverse transcriptase reaction was done using a specific primer for mouse plakoglobin 17 nucleotides in length, having a GC content of 50%. This primer annealed 400 bp downstream of the assumed start codon. It was also used in the first PCR in combination with the primer AnOut, which is complementary to the last 23 nucleotides of the anchor oligonucleotide. The second PCR was done using two nested primers, one complementary to the first 23 nucleotides of the anchor primer and a second specific plakoglobin primer with a 5' added-on sequence containing a restriction site for *Sal*I.

B. 3' END BLOCKING OF THE ANCHOR OLIGONUCLEOTIDE

The anchor oligonucleotide was synthesized with a 5' phosphate extremity and was blocked at its 3' end, as described by Dumas et al.:[16]

60 pmol oligonucleotide were tailed with 2',3'-dideoxy-adenosine-triphosphate (ddATP) in a reaction volume of 30 µl containing 100 mM potassium cacodylate, pH 7.2; 2.0 mM $CoCl_2$; 0.2 mM DTT; 0.05 µg/µl BSA together with 100 µM ddATP, [α-P^{32}] ddATP (2.5 µCi of the Amersham 3000 Ci/mM solution), and 25 U of TdT (Boehringer Mannheim). The reaction was incubated for 60 min at 37°C and was then terminated by heating for 10 min at 75°C. Samples were then electrophoresed on a polyacrylamide gel. After autoradiography, the band referred to as was excised and eluted, The blocked oligonucleotide was resuspended in TE to a final concentration of 2 pmol/µl.

C. FIRST STRAND SYNTHESIS

Precipitate RNA (0.2 µg of polyA^+ or 5.0 µg of total RNA), wash with 70% ethanol, and air dry. Add 17 µl dd H_2O, 2 µl Hybridization Buffer (10 × Hybridization Buffer is: 0.1 M Tris-HCl, pH 7.5; 300 mM NaCl; and 10 mM EDTA) and 1 µl of specific primer (6 pmol total). Denature at 90°C for 3 min and anneal at 60°C for 20 min. To initiate reverse transcription, add the following mix and incubate reaction at 42°C for 2 h: 5 µl RT Buffer (10 × RT Buffer is: 500 mM Tris-Cl, pH 8.3; 100 mM DTT; 60 mM $MgCl_2$; 400 mM KCl; and 10 mM dNTPs), 0.5 µl (20 U) RNase Inhibitor (Promega Biotec, Madison), 23 µl ddH_2O, and 1 µl AMV Reverse Transcritase (40 U). Terminate the reaction with 1 µl of 0.5 M EDTA, pH 8.0. To remove RNA, incubate the reaction mixture with 4 U RNase H for 30 min at 37°C. Remove the proteins by extraction with phenol:chloroform (1:1) twice and with chloroform once.

To remove excess specific primer, dilute the mixture to 300 µl and apply it to a Millipore Ultrafree-MC spin filter (30K NMWL polysulfone filter units), which has been coated with 0.01%, gelatin, following the manufacturer's instructions. If necessary, concentrate the retained volume.

D. OLIGONUCLEOTIDE LIGATION TO SINGLE-STRANDED cDNA

To ligate the anchor oligonucleotide to the single-stranded cDNA, set up the following reaction in a total volume of 20 µl: concentrated retentate, 2 µl Ligase Buffer (10 × Ligase Buffer is: 500 mM Tris-HCl, pH 8.0; 100 mM $MgCl_2$; 100 µg/ml BSA; 10 mM hexamine cobalt chloride; and 1 mM ATP), 25% PEG 8000 (w/v), 1 µl anchor oligonucleotide (2 pmol), and 1 µl (10 U) T4 RNA Ligase (Biolab) and perform ligation at 22°C overnight.

E. AMPLIFY 5' END OF cDNA WITH THE PCR

For the first PCR, use between 1/4 and 1/20 of the ligated first-strand DNA from step D. After 10 to 20 c with the outer primers, remove a small aliquot and do a second PCR using the inner primers for 30 to 40 c. Test 1/10 of the final PCR product by separating on a TBE-Agarose gel. Additionally, products may be screened by hybridizing to an internal oligo after electrophoresis through agarose.

III. DISCUSSION

One big advantage of the method described is the possibility of adding an oligonucleotide with a defined length and base composition to the unknown 3' end of the first-strand cDNA. This greatly diminishes mispriming of the oligonucleotides complementary to the anchor sequence, a problem frequently arising in techniques based on the addition of homopolymeric tails to the unknown 3' end. A critical point still remaining in methods like SLIC or LA-PCR, is the efficient removal of excess primer prior to the ligation of the anchor-oligonucleotide to the cDNA. If the concentration of the specific primer is too high, it can compete with the single-stranded cDNA for the ligation of the anchor primer. Besides the approach described in the protocol, it should also be possible to remove non-extended primers by using 5' biotinylated oligonucleotides.[18,19] Before reverse transcription, these primers are immobilized on Dynal streptavidin magnetic beads. After completing first-strand cDNA synthesis, unprimed oligonucleotide could then be removed by the 3'-5' exonuclease activity of T4 DNA polymerase. The coupling of the specific primer to magnetic beads would also facilitate the subsequent reaction steps.

A second important step is to completely hydrolyze the RNA after reverse transcription, since T4 RNA ligase can use both single-stranded DNA and single-stranded RNA as substrates. Thus any RNA present in the ligation reactions will act as a competitive inhibitor of anchor-cDNA joining.

Interestingly Tessier et al.,[20] reinvestigating the ligation of oligodeoxyribonucleotides by T4 RNA ligase, pointed out that the yield of ligation decreased from 67 to 49% when the length of the acceptor oligonucleotide increased from 25 to 40 nucleotides. Due to this fact, one should avoid using primer longer than 30 nucleotides. In contrast, Hofmann and Brian[21] described the construction of concatemers resulting from head-to-tail ligation of the first-stranded cDNAs, which were up to 200 nucleotides in length. If it should become possible to ligate bigger ss-cDNA fragments, thus combining multimers, this method would be greatly improved.

REFERENCES

1. **Silver, J. and Keerikatte, V.,** Novel use of polymerase chain reaction to amplify cellular DNA adjacent to an integrated provirus, *J. Virol.,* 63, 1924, 1989.
2. **Ochman, H., Gerber, A. S., and Hartl, D. L.,** Genetic application of an inverse polymerase chain reaction, *Genetics,* 120, 621, 1988.
3. **Triglia, T., Peterson, M. G., and Kemp, D. J.,** A procedure for *in vitro* amplification of DNA segments that lie outside the boundaries of known sequences, *Nucleic Acids Res.,* 16, 8186, 1988.
4. **Jones, D. H. and Winistorfer, S. C.,** Sequence specific generation of a DNA panhandle permits PCR amplification of unknown flanking DNA, *Nucleic Acids Res.,* 20, 595, 1992.
5. **Riley, J., Butler, R., Ogilvie, D., Finnear, R., Jenner, D., Powell, S., Anand, R., Smith, J. C., and Markham, A. F.,** A novel, rapid method for the isolation ot terminal sequences from yeast artificial chromosome (YAC) clones, *Nucleic Acids Res.,* 18, 2887, 1990.
6. **Arnold, C. and Hodgson, I. J.,** Vectorette PCR: A novel approach to genomic walking, *PCR Meth. Appl.,* 1, 39, 1991.
7. **Ochman, H., Medhora, M. M., Garza, D., and Hartl, D. L.,** in *PCR Protocols: A Guide to Methods and Applications,* Innis, M. A., Gelfand, D. H., Sninsky, J. J., and White, T. J., Eds., Academic Press, San Diego, CA, 1990, 219.
8. **Frohman, M. A., Dush, M. K., and Martin, G. R.,** Rapid production of full-length cDNAs from rare transcripts: amplification using a single gene-specific oligonucleotide primer, *Proc. Natl. Acad. Sci. U.S.A.,* 85, 8998, 1988.
9. **Troutt, A. B., McHeyzer-Williams, M. G., Pulendran, B., and Nossal, G. J. V.,** Ligation-anchored PCR: a simple amplification technique with single-sided specificity, *Proc. Natl. Acad. Sci. U.S.A.,* 89, 9823, 1992.
10. **Loh, E. Y., Elliott, J. F., Cwirla, S., Lanier, L. L., and Davis, M. M.,** Polymerase chain reaction with single-sided specificity: analysis of T cell receptor δ chain, *Science,* 243, 217, 1989.
11. **Pascoe, W. S., Kemler, R., and Wood, S. A.,** Genes and functions: trapping and targeting in embryonic stem cells, *Biochim. Biophys. Acta,* 1114, 209, 1992.
12. **Belyavsky, A., Vinogradova, T., and Rajewski, K.,** PCR-based cDNA library construction: general cDNA libraries at the level of a few cells, *Nucleic Acids Res.,* 17, 2919, 1989.
13. **Schaefer, B. C., Woisetschlaeger, M., Strominger, J. L., and Speck, S. H.,** Exclusive expression of Epstein-Barr virus nuclear antigen 1 in Burkitt lymphoma arises from a third promotor, distinct from the promotors used in latently infected lymphocytes, *Proc. Natl. Acad. Sci. U.S.A.,* 88, 6550, 1991.

14. **Ohara, O., Dorit, R. L., and Gilbert, W.,** One sided polymerase chain reaction: the amplification of cDNA, *Proc. Natl. Acad. Sci. U.S.A.,* 86, 5673, 1989.
15. **Snopek, T. J., Sugino, A., Agarwal, K. L., and Cozarelli, N. R.,** Catalysis of DNA joining by bacteriophage T4 RNA ligase, *Biochem. Biophys. Res. Commun.,* 68, 417, 1976.
16. **Dumas, J., Edwards, M., Delort, J., and Mallet, J.,** Oligodeoxyribonucleotide ligation to single-stranded cDNAs: a new tool for cloning ends of mRNAs and for constructing cDNA libraries by *in vitro* amplification, *Nucleic Acids Res.,* 19, 5227, 1991.
17. **Maniatis, T.,** *Molecular Cloning: A Laboratory Manual,* 2nd ed., Cold Spring Harbor Laboratory Press, Cold Spring Harbor, New York, 1989.
18. **Lee, Y. H. and Vacquir, V.,** Reusable cDNA libraries coupled to magnetic beads, *Anal. Biochem.,* 206, 206, 1992.
19. **Lambert, K. N. and Williamson, V. M.,** cDNA library construction from small amounts of RNA using paramagnetic beads and PCR, *Nucleic Acids Res.,* 21, 775, 1993.
20. **Tessier, D. C., Brousseau, R., and Vernet, T.,** Ligation of single-stranded oligodeoxyribonucleotides by T4 RNA ligase, *Anal. Biochem.,* 158, 171, 1986.
21. **Hofmann, M. A. and Brian, D. A.,** A PCR-enhanced method for determining the 5' end sequence of mRNAs, *PCR Methods Appl.,* 1, 43, 1991.

Chapter 6

GENERATION OF LABELED DNA PROBES BY PCR

Gregory W. Konat, Iwona Laszkiewicz, Barbara Grubinska,
and Richard C. Wiggins

TABLE OF CONTENTS

I. Introduction 37

II. Materials and Methods 37
   A. Reagents and Equipment 37
   B. DNA Template 38
   C. Radioactive ssDNA Hybridization Probes (Protocol 1) 38
   D. Digoxigenin-Labeled ssDNA Hybridization Probes (Protocol 2) 39
   E. Singly End-Labeled dsDNA Probes (Protocol 3) 39

III. Discussion 40
   A. ssDNA Hybridization Probes 40
   B. Singly End-Labeled Probes 41

References 42

I. INTRODUCTION

The PCR technology can be employed to generate two types of commonly used DNA probes, namely, (1) probes used for nucleic acid hybridization, and (2) singly end-labeled probes used for studies of protein-DNA interactions. The labeling can be achieved by either incorporation of labeled dNTPs into the extending strand (internal labeling), or by extension of labeled primer (end labeling). There are several PCR-based protocols available for the generation of such probes labeled with isotopes,[1-6] biotin,[7-9] digoxigenin,[4,10-13] bromide,[14] or sulfonylated poly-cytosine.[15] As compared to conventional procedures, the PCR technology offers several substantial advantages, such as versatility of sequence selection, improved efficiency of labeling, and simplicity of procedures. In addition, the procedures require small amounts of DNA template, and even short probes can be efficiently labeled.

Three PCR-based techniques have been developed in our laboratory for the generation of high-efficiency ssDNA hybridization probes of both isotope-labeled and digoxigenin-labeled varieties, as well as for the generation of singly end-labeled dsDNA probes. The respective protocols are provided here, and their applicability and modifications are discussed.

II. MATERIALS AND METHODS

A. REAGENTS AND EQUIPMENT

We routinely use GeneAmpT™* DNA Amplification Reagent Kit (Perkin Elmer Cetus, Norwalk, CT). Taq polymerase has a concentration of 5 U/µl, and we recommend diluting this

---

* Registered trademark of Perkin Elmer Cetus Corp., Norwalk, CT.

stock solution fourfold with 1X PCR buffer before use to provide for better accuracy of the enzyme aliquoting. The ^{32}P-α-dCTP (>3000 Ci/mmol) and ^{32}P-γ-ATP (>4000 Ci/mmol) are from ICN Biochemicals (Irvine, CA). Digoxigenin-11-dUTP (DIG-dUTP), Quick Spin™* G-25 and Quick Spin™ G-50 columns, and polynucleotide kinase are from Boehringer Mannheim (Indianapolis, IN). Oligonucleotide primers (20- or 21-mers) are custom synthesized and HPLC purified by commercial suppliers. All other reagents are of the highest quality obtained from different suppliers. The PCR amplification is performed in a DNA Thermal Cycler (Perkin Elmer Cetus, Norwalk, CT) in standard 0.5-ml Eppendorf tubes.

B. DNA TEMPLATE

A fragment of approximately 500 bp is amplified by the conventional two-primer PCR from an appropriate DNA, e.g., genomic digest, cloned inserts, reversely transcribed RNA, etc. The DNA is subsequently purified by 1% agarose gel electrophoresis followed by Elutip-d®** (Schleicher & Schuell, Keen, NH) elution using the manufacturer's protocol. Although this procedure yields highly purified templates, alternative methods efficiently removing unincorporated dNTPs and primers can also be employed. The dsDNA template is quantitated spectrophotometrically at 260 nm. When preparing the template from complex nucleic acid mixtures, i.e., genomic digest or total RNA, it is a good practice to confirm the identity of the amplified fragment by sequence or restriction analysis. We occasionally observe that in such reactions multiple amplification products comigrate as an apparent well-defined, tight band on agarose gels.

C. RADIOACTIVE ssDNA HYBRIDIZATION PROBES (PROTOCOL 1)

1. Prepare 25 µl reaction mixture (final concentrations are indicated in parentheses):

   2.5 µl of 10X PCR buffer (10 mM Tris-HCl, 50 mM KCl, 1.5 mM MgCl$_2$, 0.01% gelatin, pH 8.3)
   0.5 µl of 1 mM mixture of each dATP, dTTP, and dGTP (20 µM)
   1.0 µl of 25 ng/µl dsDNA template (1 ng/µl)
   2.5 µl of 10 µM primer (antisense; 1 µM)
   15.0 µl of ^{32}P-dCTP (150 µCi, 3000 Ci/mmol; 2 µM)
   0.5 µl of 1.25 units/µl Taq polymerase (25 U/ml)
   3.0 µl of double-distilled sterile water

For convenience, a premix of buffer, template, primer and unlabeled dNTPs can be prepared, aliquoted and stored at −20°C.

2. Overlay the mix with mineral oil (approximately 25 µl), and process the samples through 30 cycles at the following profile:

   Denaturation for 2 min at 94°C
   Annealing for 2 min at appropriate temperature***
   Extension for 5 min at 72°C

3. Remove oil and dilute the reaction mix with 25 µl of water. Purify the probe from unincorporated nucleotides by passing the mix through a Quick Spin™ G-50 column, or by other alternative methods.

---

*   Registered trademark of Boehringer Mannheim Corp., Indianapolis, IN.
**   Registered trademark of Schleicher & Schuell, Inc., Keene, NH.
***   The temperature has to be determined for a particular template/primer set.

D. DIGOXIGENIN-LABELED ssDNA HYBRIDIZATION PROBES (PROTOCOL 2)

1. Prepare 100 µl reaction mixture (final concentrations are indicated in parentheses):

    10.0 µl of 10X PCR buffer (10 mM Tris-HCl, 50 mM KCl, 1.5 mM MgCl$_2$, 0.01% gelatin, pH 8.3)
    8.0 µl of 1.25 mM mixture of each dATP, dCTP, dGTP, and dTTP/DIG-dUTP 9:1* (100 µM)
    16.0 µl of 25 ng/µl dsDNA template (4 ng/µl)
    30.0 µl of 10 µM primer (antisense; 3 µM)
    1.5 µl of 5 units/µl Taq polymerase (75 units/ml)
    34.5 µl of double-distilled sterile water

For convenience, a premix of buffer, template, primer, and dNTPs can be prepared, aliquoted, and stored at –20°C.

2. Overlay the mix with mineral oil (approximately 25 µl), and process the samples through 45 cycles at the following profile:

    Denaturation for 1.5 min at 94°C
    Annealing for 1.5 min at appropriate temperature**
    Extension for 2 min at 72°C

3. Remove oil, add 0.1 vol of 4 M LiCl and 3 vol of ethanol, and place the tube at –20°C for 30 min. Spin down the DNA for 20 min at 12,000 g, and wash the pellet with 100 µl 70% ethanol. Air dry the pellet, dissolve it in TE buffer, and store at –20°C.

E. SINGLY END-LABELED dsDNA PROBES (PROTOCOL 3)

1. Prepare 10 µl reaction mixture (final concentrations are indicated in parentheses):

    1.0 µl of 10X kinase buffer (50 mM Tris-HCl, 10 mM MgCl$_2$, 5 mM DTT, 0.1 mM spermine, pH 7.5)
    1.0 µl of 10 µM primer #1 (1 µM)
    4.5 µl of γ-P^{32} ATP (50 µCi, 4000 Ci/mmol; 1.25 µM)
    0.5 µl of 10 U/µl polynucleotide kinase (0.5 U/µl)
    3.0 µl of double-distilled sterile water

2. Incubate for 30 min at 37°C, and then stop the reaction by heating the tube for 2 min at 90°C.
3. Add 5 µl double-distilled sterile water, and purify the labeled primer on water-equilibrated Quick Spin™ (G-25 column). Measure the volume of the eluate. This step is necessary to remove kinase buffer which is not compatible with the PCR reaction.
4. Prepare 25 µl reaction mixture (final concentrations are indicated in parentheses):

    2.5 µl of 10X PCR buffer (10 mM Tris-HCl, 50 mM KCl, 1.5 mM MgCl$_2$, 0.01% gelatin, pH 8.3)
    4.0 µl of 1.25 mM mixture of each dATP, dCTP, dGTP, and dTTP (200 µM)

---

* The ratio of dTTP to DIG-dUTP can be increased up to 2:1 (see discussion in Section III).
** The temperature has to be determined for a particular template/primer set.

1.0 µl of 25 ng/µl dsDNA template (1 ng/µl)
Total eluate from step 3 (labeled primer #1; approximately 0.4 µM)
1.0 µl of 10 µM unlabeled primer #2 (0.4 µM)
0.5 µl of 1.25 units/µl Taq polymerase (25 U/ml)
Double-distilled sterile water to 25 µl

For convenience, a premix of buffer, template, unlabeled primer, and dNTPs can be prepared, aliquoted, and stored at −20°C.

5. Overlay the mix with mineral oil (approximately 25 µl), and process the samples through 25 cycles at the following profile:

   Denaturation for 1.5 min at 94°C
   Annealing for 1.5 min at appropriate temperature*
   Extension at 72°C for 2 min

6. Remove oil and purify the probe by nondenaturing polyacrylamide gel electrophoresis. Depending on the length of the probe, standard 5 to 15% gels can be used.

III. DISCUSSION

These protocols represent three different formats of the PCR reaction. For example, Protocols 1 and 2 employ an arithmetic (linear) amplification paradigm in which the production of ssDNA is proportional to the number of thermal cycles. Labeled dNTP is incorporated into the growing strand during the extension phase of the PCR reaction (internal labeling). In contrast, Protocol 3 employs exponential amplification to generate dsDNA, and labeled primer is used. Furthermore, Protocol 1 differs from Protocols 2 and 3 as it employs highly reduced concentrations of dNTPs. Because of the idiokinetics of these reactions, all the parameters were individually optimized to provide for maximal amplification/labeling.[4]

However, as is often the case for PCR, the conditions should be tried and eventually optimized for individual template/primer sets. This pertains obviously to the annealing temperature and/or time. Hot start and/or incremental elongation of the extension time may also be beneficial for some applications. Furthermore, some DNAs may require DMSO or other agents of choice to attain efficient amplification/labeling. The concentration of magnesium is another parameter to be individually optimized as it profoundly affects both the quality (average length) and the quantity (yield) of the probes.[4] In general, the optimal conditions found for template amplification should also be optimal for generation of the probes.

Although the probes can be amplified from vectors or mixtures containing the sequence of interest, we recommend preparing the discrete template first. This additional step may profoundly improve the quality of probes. Once prepared, the template may be used to generate shorter probes by using internal primers.

A. ssDNA HYBRIDIZATION PROBES

The detection sensitivity of dsDNA probes generated by PCR is comparable to that of riboprobes.[2] The ssDNA probes offer an advantage over the dsDNA probes by further augmenting the detection sensitivity, as these probes do not reanneal with themselves in solution, and thence, their availability for interaction with the complementary sequences of the target is increased. For example, ssDNA probes provide an approximately eightfold increase in the signal intensity, as compared to dsDNA PCR probes in Northern analysis.[3] All in all,

* The temperature has to be determined for a particular template/primer set.

the use of ssDNA probes results in 27-fold increase in the signal intensity, as compared with random-primed probes generated with the same amount of ^{32}P-dCTP.[4]

Although extension of 1 min/kb of template DNA is recommended for standard PCR,[16] we found that in the case of radiolabeling (Protocol 1), a 5-min extension time ensures high proportion of full-length probe at conditions that are suboptimal in regard to dNTP, and especially, dCTP concentration. Under the optimized conditions, approximately 50% yield of ^{32}P-dCTP incorporation is reached after 30 cycles of amplification, which is comparable to the efficiency of incorporation observed for dsDNA probes.[1] Only slight increase in the incorporation yield can be achieved by further cycling. For some applications (e.g., *in situ* hybridization), probes labeled with ^{35}S or ^{3}H in lieu of ^{32}P can be prepared by the same protocol.

In the nonradioactive probe paradigm (Protocol 2) the template concentration is increased fourfold as compared to Protocol 1 to maximize the production of the ssDNA probe at a high dNTP concentration. The optimal concentration of both *Taq* polymerase and the primer is three times higher, whereas the concentration of dNTPs is 50% lower than the standard concentrations.[16] Approximately 8 µg of uniformly full-length digoxigenin-labeled ssDNA probe (60% incorporation) can be generated under these conditions in a 100-µl reaction using 45 cycles. The specific labeling of these probes can be further augmented by increasing the ratio of DIG-dUTP to dTTP up to 1:2.[13] The DIG-labeled probes preserve their activity for at least one year when stored at $-20°C$.[11]

Since ssDNA is generated by arithmetic amplification, its yield is proportional to the amount of template. Thus, relatively high concentrations of the template, as compared to dsDNA PCR probes, have to be used to optimize the isotope incorporation. Excessive amounts of template, however, are undesirable as they result in a higher proportion of unlabeled DNA and consequently may reduce the intensity of hybridization signal through annealing with the labeled probe.

Optimized reaction conditions (as exemplified in Protocol 1) provide for, not only a higher incorporation rate into the ssDNA probes, but also the production of full-length probes, which may extend their application to techniques other than solid-phase hybridization, e.g., S1 protection assay. S1 mapping using ssDNA probes has been reported,[5,17] proving this stratagem. Furthermore, the ssDNA probes will undoubtedly be valuable whenever a discrete and/or strand-specific hybridization is required, for example, in genomic footprinting analysis using indirect end-labeling paradigm.[18]

For some applications, e.g., *in situ* hybridization, shorter probes may be desired. Such ssDNA probes can be generated by prior cutting of the template with appropriate restriction enzymes. Either the shortened template may be purified or the whole digest may be used directly for the amplification reaction, provided that the restriction buffer is compatible with the PCR buffer. Shorter probes can also be generated on the intact template by employing internal primers. Alternatively, the probes themselves can be fragmented by alkaline treatment. This strategy applies also to digoxigenin-labeled probes, as the digoxigenin-nucleotide linkage is resistant to alkali.

B. SINGLY END-LABELED PROBES

The unique feature of the PCR methodology to amplify discrete regions of DNA is particularly useful for generating probes for studies of protein-DNA interactions, e.g., mobility shift assay, methylation interference assay, or *in vitro* footprinting. DNA fragments of any desired length, limited only by the constraints of the PCR amplification, and 5' labeled on either strand, can be produced based on selected sequences, thus eliminating the dependence on the availability of restriction endonuclease cleavage sites. As the templates can be prepared from genomic DNA or reverse transcribed mRNA, the procedure also obviates the need for cloning. Furthermore, probes from PCR-mutated templates can be rapidly prepared by this simple protocol.

The specific radioactivity of the PCR-generated probes (Protocol 3) is generally about twofold higher than that of probes labeled by conventional techniques (e.g., kination or fill-in) when the same amount of isotope is used. If desired, the labeling may be further augmented by increasing the concentration of primers (both labeled and unlabeled) in the PCR reaction. This, however, requires scaling up the primer kination reaction (see Protocol 3) and adding an extra step of concentrating the labeled primer after elution.

A similar protocol, in which both the kination and amplification reactions are carried out in a single tube, was recently published.[6] This simplification, however, may be offset by a lower yield of the probe as compared to our protocol. For example, we found that with a variety of templates our protocol generates approximately twice as much probe as Hooft van Huijsduijnen's protocol,[6] while it requires only half the isotope. Nevertheless, the single-tube protocol[6] may be a convenient alternative technique, when the yield of the probe is not a crucial factor.

REFERENCES

1. **Jansen, R. and Ledley, F. D.**, Production of discrete high specific activity DNA probes using the polymerase chain reaction, *Gene Anal. Techn.*, 6, 79, 1989.
2. **Schowalter, D. B. and Sommer, S. S.**, The generation of radiolabeled DNA and RNA probes with polymerase chain reaction, *Anal. Biochem.*, 177, 90, 1989.
3. **Bednarczuk, T. A., Wiggins, R. C., and Konat, G.**, Generation of high efficiency, single-stranded DNA hybridization probes by PCR, *BioTechniques*, 10, 478, 1991.
4. **Konat, G., Laszkiewicz, I., Bednarczuk, T. A., Kanoh, M., and Wiggins, R. C.**, Generation of radioactive and nonradioactive ssDNA hybridization probes by polymerase chain reaction, *Technique*, 3, 64, 1991.
5. **Blakeley, M. S. and Carman, M. D.**, Generation of an S1 probe using arithmetic polymerase chain reaction, *BioTechniques*, 10, 52, 1991.
6. **Hooft van Huijsduijnen, R. A. M.**, PCR-generated probes for the study of DNA-protein interactions, *BioTechniques*, 12, 830, 1992.
7. **Day, P. J. R., Bevan, I. S., Gurney, S. J., Young, L. S. and Walker, M. R.**, Synthesis *in vitro* and application of biotinylated DNA probes for human papilloma virus type 16 by utilizing the polymerase chain reaction, *Biochem. J.*, 267, 119, 1990.
8. **Saiki, R. K., Walsh, P. S., Levenson, C. H., and Erlich, H. A.**, Genetic analysis of amplified DNA with immobilized sequence-specific oligonucleotide probes, *Proc. Natl. Acad. Sci. U.S.A.*, 86, 6230, 1989.
9. **Lo, Y.-M. D., Mehal, W. Z., and Fleming, K. A.**, Rapid production of vector-free biotinylated probes using the polymerase chain reaction, *Nucleic Acids Res.*, 16, 8719, 1988.
10. **Liesack, W., Menke, M. A. O. H., and Stackebrandt, E.**, Rapid generation of vector-free digoxigenin-dUTP labeled probes for nonradioactive hybridization using the polymerase chain reaction (PCR) method, *Syst. Appl. Microbiol.*, 13, 255, 1990.
11. **Lion, T. and Haas, O. A.**, Nonradioactive labeling of probe with digoxigenin by polymerase chain reaction, *Anal. Biochem.*, 188, 335, 1990.
12. **Lanzillo, J. J.**, Preparation of digoxigenin-labeled probes by the polymerase chain reaction, *BioTechniques*, 8, 621, 1990.
13. **Emanuel, J. R.**, Simple and efficient system for synthesis of non-radioactive nucleic acid hybridization probes using PCR, *Nucleic Acids Res.*, 19, 2790, 1991.
14. **Tabibzadeh, S., Bhat, U. G., and Sun, X.**, Generation of nonradioactive bromodeoxyuridine labeled DNA probes by polymerase chain reaction, *Nucleic Acids Res.*, 19, 2783, 1991.
15. **Uchimura, Y., Ishida, H., Asada, K., Mukai, H., and Kato, I.**, Nonradioactive labeling with chemically modified cytosine tails by polymerase chain reaction, *Gene*, 108, 103, 1991.
16. **Saiki, R. K.**, The design and optimization of the PCR, in *PCR Technology, Principles and Applications for DNA Amplification*, Erlich, H. A., Ed., Stockton Press, New York, 1989, chap. 1.
17. **Sharrocks, A. D. and Hornby, D. P.**, S1 nuclease transcript mapping using Sequenase-derived single-stranded probes, *BioTechniques*, 10, 426, 1991.
18. **Reik, A., Schutz, G., and Stewart, A. F.**, Glucocorticoid are required for establishment and maintenance of an alteration in chromatin structure: induction leads to a reversible disruption of nucleosomes over an enhancer, *EMBO J.*, 10, 2569, 1991.

Chapter 7

NONISOTOPIC PROBE GENERATION BY PCR

Y-M. D. Lo, E. P. H. Yap, S. F. An, J. O'D. McGee, and K. A. Fleming

TABLE OF CONTENTS

I. Introduction 43

II. Materials and Methods 44
    A. Reagents for PCR 44
    B. Incorporation of Biotin-11-dUTP 44
    C. Filter Hybridization Using Biotinylated Probe 45
    D. Incorporation of Digoxigenin-11-dUTP 45
    E. Production of Single-Stranded Digoxigenin-Labeled Probe 46
    F. Filter Hybridization Using Digoxigenin-Labeled Probe 46
    G. *In situ* Hybridization (ISH) 46

III. Results 47
    A. Model System 47
    B. Biotinylated Probes 47
    C. Digoxigenin-Labeled Probes 48

IV. Discussion 50

Acknowledgments 52

References 52

I. INTRODUCTION

Probes for nucleic acids are important for many molecular biological techniques. With the continual refinement of nonisotopic probe production and detection technology, many investigators prefer nonradioactively to radioactively labeled probes. Nonisotopically labeled probes have the advantages of being relatively nonhazardous, stable for long periods (e.g., one year), and easily produced in large quantities.

Probe labeling is conventionally performed by nick translation[1] or random priming.[2] In many applications, cloned DNA sequences have to be separated from the vector sequences to avoid potential cross-hybridization between vector sequences and the target nucleic acid. The generation of these vector-free probes normally requires plasmid preparation, restriction digestion of the plasmid DNA, and recovery of the insert by techniques such as preparative gel electrophoresis followed by electroelution. These procedures are time consuming and relatively inefficient.

The discovery that the *Taq* polymerase can utilize deoxyribonucleotide triphosphate analogues such as biotin-11-dUTP as substrate in the polymerase chain reaction (PCR)[3] has enabled the rapid production of large quantities of nonisotopically labeled vector-free probes. In this chapter, we outline the production of biotin- and digoxigenin-labeled DNA probes using PCR.

TABLE 1.
Sequence of PCR Primers

Sequence name	Sequence
PCR1	5'GATTGAGATCTTCTGCGACGC3'
PCR2	5'GAGTGTGGATTCGCACTCCTC3'
BB1660	5'CTTGTTGACAAGAATCCTCAC3'
BB1661	5'GATGGGATGGGAATACA3'
BB1666	5'CAGAGTCTAGACTCGTGG3'
BB1667	5'ACAAACGGGCAACATACCTTG3'
BB1671	5'GACATACTTTCCAATCAATAG3'
T3 promoter	5'ATTAACCCTCACTAAAGGGA3'
SP6 promoter	5'GATTTAGGTGACACTATAG3'
T7 promoter	5'TAATACGACTCACTATAGGG3'

II. MATERIALS AND METHODS

A. REAGENTS FOR PCR

Reaction buffer for the PCR and *Taq* polymerase are available from Perkin Elmer Cetus (Emeryville, CA). Biotin-11-dUTP in powder form is available from Sigma (Dorset, U.K.) and should be dissolved in 100 mM Tris, 0.1 mM EDTA, pH 7.5 to make a 0.3-mM solution before use. Digoxigenin-11-dUTP (dig-11-dUTP) is obtained from Boehringer Mannheim (East Sussex, U.K.). PCR primers are available from British Biotechnology, Ltd. (Abingdon, U.K.). Sequences of primers are listed in Table 1.

The PCR is set up essentially as described[4] in 100-µl reaction volumes containing 30 to 100 pmol of each primer; 200 µM each of dATP, dCTP, and dGTP; and 2.5 to 5 U Ampli*Taq* DNA polymerase. Variations of TTP and either biotin-11-dUTP or dig-11-dUTP are discussed below, and thermal profiles are described in the relevant sections; 100 µl of paraffin oil (BDH, Merck Ltd., Dorset, U.K.) is used to prevent excessive evaporation. Ampliwax® (Perkin Elmer Cetus, Emeryville, CA) is an alternative vapor barrier that has the advantages of automation and reducing contamination with the hot-start technique.

B. INCORPORATION OF BIOTIN-11-DUTP

1. Set up the PCR reagent mix. Use TTP at 150 µM and biotin-11-dUTP at 50 µM (i.e., a ratio of 3:1). Primers PCR1 and PCR2 are used as a model system to amplify a 185-bp fragment from the hepatitis B virus (HBV) genome.
2. Add target DNA. For the production of vector-free probe from plasmid DNA, the amount of plasmid should be kept low at 0.2 fmol (approximately 1 ng of an 8-kb plasmid).
3. Start thermal cycling with a 10-min incubation at 94°C. In our example of a plasmid containing a full length HBV insert (pHBV130[5]), this initial thermal denaturation is followed by 25 c, consisting of 94°C for 2 min, 55°C for 2 min, and 72°C for 3 min.
4. Remove as much mineral oil as is possible.
5. Add 1 µl of glycogen (20 mg/ml) (Boehringer Mannheim, East Sussex, U.K.), 10 µl of 2.5-M sodium acetate, pH 5.2, and 220 µl of ethanol. Leave the mixture at –20°C overnight or –70°C for 1 h.
6. Collect the precipitated labeled PCR product by spinning at maximum speed (11,600 g) on a Micro Centaur (MSE, Sussex, U.K.) for 15 min. Remove the supernatant and vacuum desiccate.
7. Dissolve the precipitate in 100 µl of 10-mM Tris; 1-mM EDTA, pH 8.0.

8. Run 5 µl on an agarose gel. The biotinylated PCR product exhibits slightly reduced electrophoretic mobility compared with a nonbiotinylated probe.
9. The efficiency of labeling may be checked by running 1 µl of the PCR product on an alkaline denaturing agarose gel, Southern blotting to a nitrocellulose filter (BA 85, Schleicher & Schuell, Keene, NH), and visualizing the biotin label by using a streptavidin/alkaline phosphatase detection system[6] (see below).
10. Store probe at either –20 or –70°C in aliquots.

C. FILTER HYBRIDIZATION USING BIOTINYLATED PROBE

As a guideline we use 50 ng of an 185-bp biotinylated PCR product in 2 ml of hybridization mix for hybridization to every 50 cm² of filter.

1. Following standard hybridization procedures,[6] wash the filters three times for 5 min each with 2X TBS (1X TBS = 0.15 M NaCl; 0.015 M Tris, pH7.2), 0.1% (w/v) SDS at 22°C.
2. Wash three times for 5 min each with 0.5X TBS, 0.1% SDS at 22°C.
3. Wash three times for 15 min each with 0.5X TBS, 0.1% SDS at 60°C. Prior heating of the washing solution to 60°C is essential.
4. Block the filters with blocking buffer (0.1 M Tris; 0.1 M NaCl; 3 mM MgCl$_2$; 0.5% Tween 20, pH 7.5) for 60 to 90 min.
5. Dilute streptavidin to 2 µg/ml in incubation buffer AP7.5 (0.1 M Tris; 0.1 M NaCl; 3 mM MgCl$_2$, pH 7.5), 0.05% Tween 20.
6. Incubate the filters with diluted streptavidin for 10 min with gentle rocking. As a guideline, use 3 to 4 ml of diluted streptavidin per 100 cm² filter area.
7. Wash the filters three times with blocking buffer for 5 min each.
8. Make up biotinylated alkaline phosphatase to 1 µg/ml in incubation buffer.
9. Incubate the filters with biotinylated alkaline phosphatase for 10 min with gentle rocking.
10. Repeat the three washes with blocking for 3 min each.
11. Wash filters twice with substrate buffer AP 9.0 (0.1 M Tris, 0.1 M NaCl, 0.1 M MgCl$_2$, pH 9.0) for 3 min each.
12. Make up nitro blue tetrazolium (NBT) (Sigma, Dorset, U.K.) and 5-bromo-4-chloro-3-indolyl phosphate (BCIP) (Sigma, Dorset, U.K.) stock solutions by dissolving 1 mg NBT in 40 µl 70% dimethylformamide and 2 mg BCIP in 40 µl 100% dimethylformamide.
13. Make up the substrate solution by adding 36 and 40 µl stock solutions of NBT and BCIP, respectively, to 12 ml of substrate buffer at 22°C.
14. Incubate the filters with substrate solution for up to 20 h in plastic petri dishes or sealed plastic hybridization bags in the dark. Monitor the color development periodically.
15. Following color development, wash the filters extensively with water.

D. INCORPORATION OF DIGOXIGENIN-11-DUTP

1. Set up the PCR reagent mix with 200 µM each of dATP, dGTP, and dCTP. Use TTP at 60 µM and digoxigenin-11-dUTP at 20 µM (i.e., a ratio of 3:1).
2. Add the target DNA. For the production of vector-free probe from plasmid DNA the amount of plasmid should be low, at 0.2 fmol (approximately 1 ng of an 8-kb plasmid).
3. Start thermal cycling with a 10-min incubation at 94°C. In our example of the plasmid pHBV130[5] and primer combinations mentioned in Section III on results, this initial thermal denaturation is followed by 30 c consisting of 94°C for 1.5 min, 60°C for 1.5 min, and 72°C for 3 min.
4. Remove as much mineral oil as is possible.

5. Run 10 μl on an agarose gel. The digoxigenin-labeled PCR product should exhibit decreased electrophoretic mobility compared with a non-digoxigenin-labeled PCR product.
6. The efficiency of labeling may be checked by running 1 μl of the PCR product on an alkaline denaturing agarose gel and Southern blotting to a nitrocellulose filter (BA 85, Schleicher & Schuell, Keene, NH). Detection of the digoxigenin label is detailed below.
7. Store probe at either −20 or −70°C in aliquots.

E. PRODUCTION OF SINGLE-STRANDED DIGOXIGENIN-LABELED PROBE

For the production of single-stranded probes, two rounds of PCR are performed:

1. Set up the first round PCR using 200 μM TTP without dig-11-dUTP.
2. Following the first round, centrifuge the product through a Centricon-30 membrane (Anachem, Luton, U.K.) to remove the free primers.
3. Set up a second round of linear amplification using only one primer (either the same or an internal one), 20 μM dig-11-dUTP, and 60 μM TTP. Use 60 ng of the first-round product the template for the second round. Perform 30 c.

F. FILTER HYBRIDIZATION USING DIGOXIGENIN-LABELED PROBE

As a guideline, digoxigenin-labeled PCR probes are used at a concentration of 50 ng/ml of hybridization mix.

1. Following standard hybridization procedures,[7] wash the filters twice for 5 min each at 22°C with 2X SSC (1X SSC = 0.3 M NaCl and 0.03 M sodium citrate), 0.1% (w/v) SDS.
2. Wash twice for 15 min each at 65°C with 0.1X SSC, 0.1% SDS.
3. Block the filters for 1 h at 22°C with 1X AP7.5 (0.1 M Tris; 0.1 M NaCl and 3 mM MgCl$_2$, pH 7.5), 0.5% Tween 20.
4. Dilute anti-digoxigenin alkaline phosphatase, Fab fragment (Boehringer Mannheim, East Sussex, U.K.) to 150 mU/ml (1:5000) with 1X AP7.5, 0.05% Tween 20, 2% bovine albumin (Sigma, Dorset, U.K.).
5. Incubate the filters with anti-digoxigenin for 30 min at 22°C.
6. Wash twice for 5 min each with 1X AP7.5, 0.05% Tween 20.
7. Wash twice for 5 min each with 1X AP7.5.
8. Wash twice for 5 min each with 1X AP9.0 (0.1 M Tris; 0.1 M NaCl; 0.1 M MgCl$_2$, pH 9.0).
9. Make up nitro blue tetrazolium (NBT) (Sigma, Dorset, U.K.) and 5-bromo-4-chloro-3-indolyl phosphate (BCIP) (Sigma, Dorset, U.K.) stock solutions by dissolving 1 mg NBT in 40 μl 70% dimethylformamide and 2 mg BCIP in 40 μl 100% dimethylformamide.
10. Develop the filters with substrate solution: 36 μl NBT solution and 40 μl BCIP solution diluted in 12 ml of AP9.0 (0.1 M Tris; 0.1 M NaCl; 0.1 M MgCl$_2$, pH 9.0).

G. *IN SITU* HYBRIDIZATION (ISH)

1. Put 5-μm sections onto 1% silane-treated slides using autoclaved water, and dry overnight at room temperature.
2. Heat the section for 10 min at 80°C. Then dewax in Citroclear (HD Supplies, Aylesbury, U.K.) three times for 5 min each, followed by industrial methylated spirits three times for 5 min each. Hydrate in autoclaved water.
3. Protease digest to access nucleic acid target. This step is crucial; too much digestion results in loss of morphology and nucleic acid (especially mRNA), and too little results in reduced sensitivity and higher background. Incubate the slide in Protease 8 (Sigma,

Dorset, U.K.) in PBS (0.012 M Na$_2$HPO$_4$; 0.04 M KH$_2$PO$_4$; 0.15 M NaCl, pH 7.2) at 37°C with gentle shaking. Concentration and time depend on tissue type and fixation. Typically we use 50 mg/ml for 20 min at 22°C for a DNA target and 5 mg/ml for 20 min at 22°C for RNA.

4. Rinse the slide in TBS (0.15 M NaCl; 0.015 M Tris, pH 8.0).
5. For a DNA target, denature the slide/section in autoclaved water at 95°C for 20 min. For RNA, 70% formamide in 2 × SET (1 × SET = 0.15 M NaCl, 0.02 M Tris, 1 mM EDTA, pH 7.8), 37°C, 20 min.
6. Rinse in ice cold water for 3 min and then air dry.
7. Prepare mix:
    100 µl dH$_2$O
    100 µl 0.1% SDS
    100 µl 10 mg/ml polyvinyl pyrrolidone
    400 µl 50% dextran sulfate in 10 × SET
    1000 µl deionized formamide
   Mix and filter through 0.2-µm filter. Divide into 170-µl aliquots and store –20°C. Before use, add 10µl of 10-mg/ml salmon sperm DNA and 20 µl of 10-µg/ml probe.
8. Denature the probe/mix at 95°C for 20 min. As a guideline, we use PCR-generated digoxigenin-labeled probes at a concentration of 10 ng per section for *in situ* hybridization.
9. Apply 7 to 16 µl of the denatured probe/mix to a coverslip, invert the section onto the probe/mix, and seal with silicone grease.
10. For DNA, denature the probe/section again by placing it on a 95°C hot plate for 10 min.
11. Hybridize at 37°C for 2 to 24 h in a moist petri dish. The time required depends on the amount of target and the concentration of probe; higher probe concentration requires less time.
12. Remove coverslips in TBS/0.5% Triton X100, and rinse three times for 5 min each at 37°C in the same solution. Then rinse three times for 5 min each in 0.5 × TBS at 65°C.
13. Block with 15% dried milk in AP7.5 (0.1 M Tris; 0.1 M NaCl; 3 mM MgCl$_2$, pH 7.5) containing 0.5% Triton X100, for 20 min at 37°C.
14. Rinse and apply alkaline phosphatase conjugated anti-digoxigenin (Boehringer Mannheim, East Sussex, U.K.) diluted 1:750 in AP7.5 containing 2% BSA and 0.5% Triton X100, for 30 min at 37°C.
15. Rinse three times in AP7.5 containing 0.5% Triton X100 for 5 min each, then three times in AP9.0 (0.1 M Tris; 0.1 M NaCl; 0.1 M MgCl$_2$, pH 9.0) for 5 min each.
16. Develop in NBT/BCIP as for filters (typically 30 min to overnight).

III. RESULTS

A. MODEL SYSTEM

For our study on the incorporation of nonisotopic labels by PCR, we used a model system based on the amplification of specific genomic fragments of the hepatitis B virus (HBV) from a plasmid containing a full-length HBV insert (pHBV130[5]). Thus, PCR1 and PCR2 amplify a 185-bp fragment from the core-polymerase region of the HBV genome. BB1666/BB1667 and BB1660/BB1661 amplify, respectively, a 233- and a 401-bp fragment from the surface antigen gene. BB1666/BB1671 and BB1660/BB1671 amplify, respectively, a 749- and a 777-bp product from the surface-polymerase region.

B. BIOTINYLATED PROBES

Figure 1, track 3, shows a 185-bp biotinylated HBV probe labeled using PCR. Note the reduction in electrophoretic mobility compared with a nonbiotinylated PCR product. Track 4

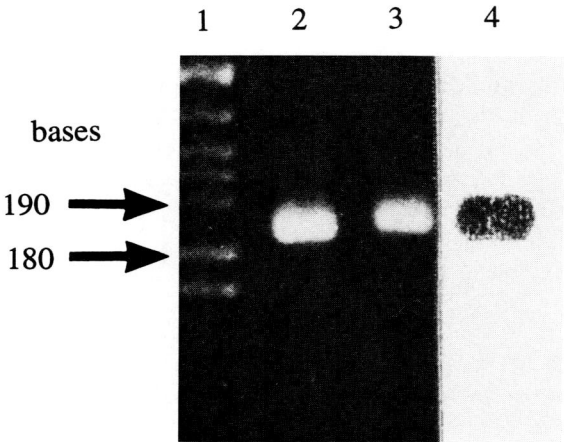

Figure 1. Production of a 185-bp biotinylated HBV probe using PCR. Track 1, pBR322 DNA digested with *Msp*I (marker); track 2, nonbiotinylated PCR product; track 3, biotinylated PCR product; track 4, Southern blot of biotinylated PCR product (alkaline denaturing gel).

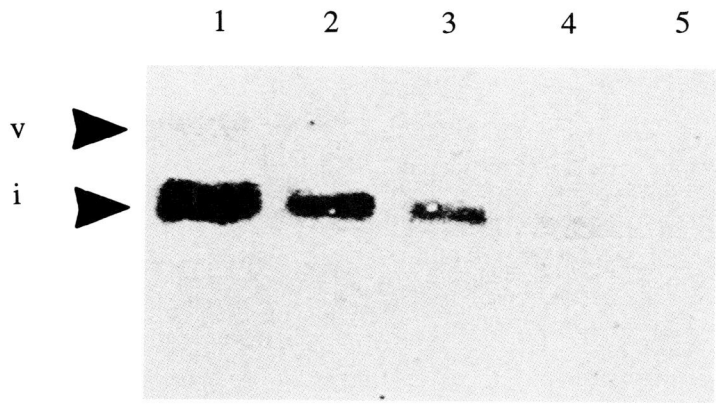

Figure 2. Sensitivity and insert specificity of the 185-bp PCR-generated biotinylated probe using serial dilutions of pHBV130 restricted with *Xho*I to release insert. (v) position of vector; (i) position of insert. The amounts of insert sequence in the tracks are as follows: track 1, 1 ng; track 2, 100 pg; track 3, 10 pg; track 4, 1 pg; track 5, 0.1 pg.

shows the visualization of the biotin label following Southern blotting and detection using a streptavidin/alkaline phosphatase system.

The sensitivity of the biotinylated probe is illustrated in Figure 2. Under optimal conditions, 1 pg of target DNA could be detected. From Figure 2 it can be seen that the probe is virtually vector free.

C. DIGOXIGENIN-LABELED PROBES

Figure 3 shows a dilution series of pHBV130 plasmid DNA. It can be seen that a 233-bp digoxigenin-labeled double-stranded DNA probe can detect between 10 and 1 pg of target sequence.

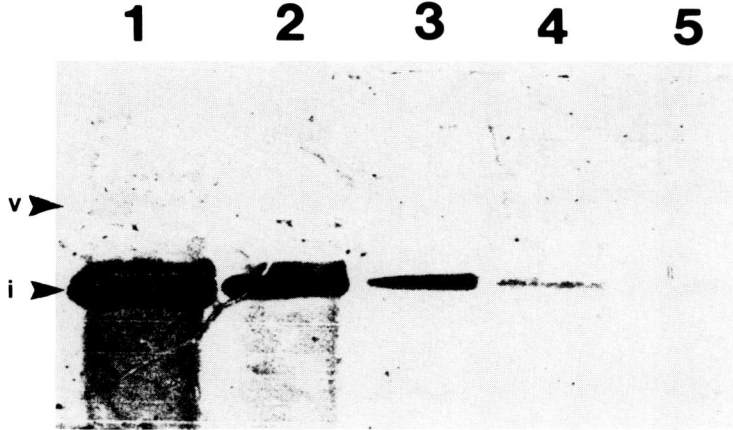

Figure 3. Sensitivity and insert specificity of the 233-bp PCR-generated digoxigenin-labeled probe using serial dilutions of pHBV130 restricted with *XhoI* to release insert. (v) position of vector; (i) position of insert. The amounts of insert sequence in the tracks are as follows: track 1, 5 ng; track 2, 1 ng; track 3, 100 pg; track 4, 10 pg; track 5, 1 pg.

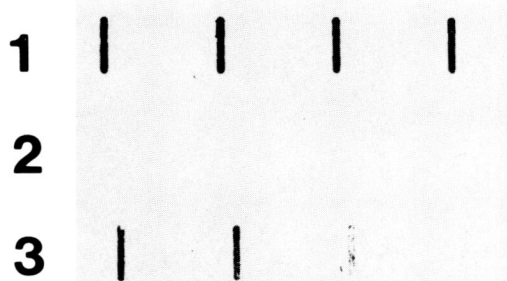

Figure 4. RNA slot-blot hybridization using digoxigenin-labeled probes. Total cellular RNA from Alexander cell line is diluted and blotted onto nitrocellulose membrane and hybridized with single-stranded probes (749 bases antisense strand and 401 bases sense strand) as well as double-stranded probe (777-bp). Filter 1, antisense strand probe; filter 2, sense strand probe; filter 3, double-stranded probe. From left to right the amount of total RNA is as follows: 30 µg, 16 µg, 8 µg, and 4 µg.

The application of single- and double-stranded probes in RNA slot-blot hybridization is shown in Figure 4. The ability of the PCR to produce strand-specific probes is seen in Figure 4, where hybridization is seen only for the antisense strand and double-stranded probes, but not for the sense strand probes.

The use of double- and single-stranded digoxigenin-labeled PCR probes in *in situ* hybridization is shown in Figure 5. It can be seen that the single-stranded probe is more sensitive than the double-stranded one.

We have investigated the relationship between probe length and hybridization sensitivity. By constructing PCR primers to various regions of the HBV genome, probes from 151 to 777 bp have been studied in dot-blot hybridization and *in situ* hybridization.[7] Longer probes are found to give a stronger signal. When two or three small probes that have overlapping sequences are used simultaneously, a summation of sensitivity is obtained (results not shown).

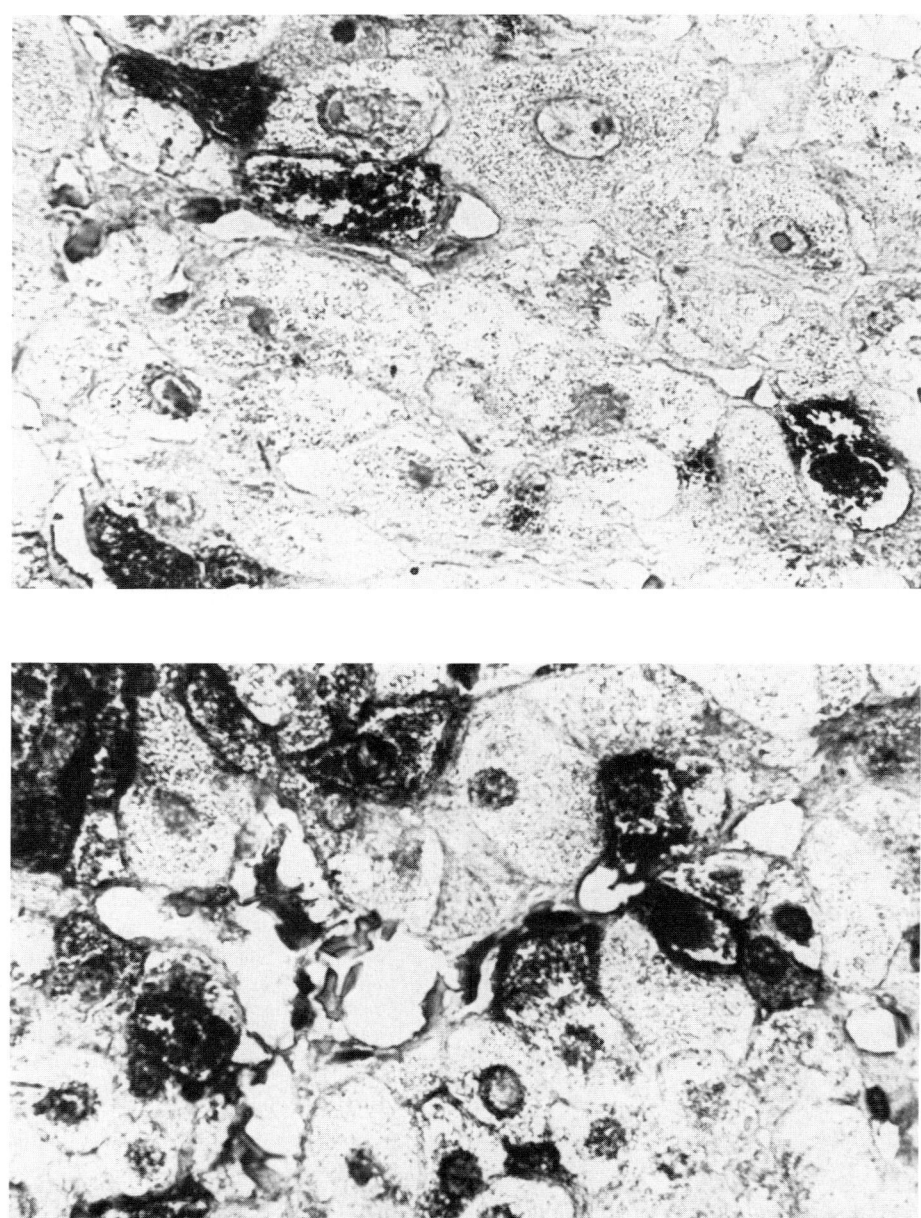

Figure 5. *In situ* hybridization of formalin-fixed, paraffin-embedded sections of chronic HBV-infected liver using digoxigenin-labeled probes. (a) double-stranded 777-bp probe, (b) single-stranded 749-base probe.

IV. DISCUSSION

We, and others, have demonstrated the feasibility of incorporating nonisotopic labels during DNA amplification by PCR.[3,8-13] The PCR-based approach for probe labeling is capable of producing vector-free probes very rapidly, and large numbers of probes can be generated in one experiment and stored for extended periods. The fractional substitution of nonisotopic labels can be much higher in PCR-labeled probes than in those labeled by nick translation.

One disadvantage of this approach is that it cannot be applied to label probes longer than a few kilobases unless multiple primer pairs are used. In addition, sequence information for construction of primers is essential for PCR.

There are PCR-based approaches for probe labeling other than the incorporation of modified deoxynucleotide triphosphate analogues. For example, it is possible to perform PCR with a primer labeled with biotin at its 5' end.[14] However, these 5'-labeled probes have reduced sensitivity at the detection step since only a single-label molecule is available for detection, compared with the multiple-label molecules incorporated for internally labeled probes. However, 5'-labeled probes have the theoretical advantage that steric hindrance during the hybridization step is minimized. Weier et al.[10] have demonstrated that internally labeled biotinylated probes have reduced T_m, which needed to be taken into consideration in hybridization experiments.

To maximize the efficiency of incorporation of biotin-11-dUTP and digoxigenin-11-dUTP during PCR, the ratio of the deoxynucleotide triphosphate analogue and TTP should be carefully controlled. For biotin-11-dUTP, we use TTP and biotin-11-dUTP at concentrations of 150 and 50 μM, respectively. For digoxigenin-11-dUTP incorporation, the concentrations of TTP and digoxigenin-11-dUTP are kept at 60 and 20 μM, respectively.

When the starting material is plasmid DNA and if vector-free probe is required, it is important to use a relatively small amount of plasmid (typically, 0.2 fmol or about 1 ng of an 8-kb plasmid). The use of larger amounts of plasmid will result in a significant amount of labeled vector sequence. During the early PCR cycles, when the number of PCR product molecules is small compared with the number of starting template molecules, the primers will bind to the original plasmid, and the *Taq* polymerase will extend the primers for a variable distance, even into the vector sequence. This will result in a *linear* accumulation of labeled vector sequence. Thus, if the starting plasmid concentration is high, the products of this linear amplification will form a significant proportion of the final amplification products. However, if the starting plasmid concentration is low, the linear amplification products will be insignificant when compared to the *exponentially* produced PCR product.

Where PCR probes have to be synthesized for sequences that have already been cloned into an expression vector, it is possible to use universal primers to the RNA polymerase promoter sites for amplification. Primer sequences to the T3, T7, and SP6 promoter sites are listed in Table 1. Use of these primers obviates the synthesis of a specific primer pair for each probe. For example, the T3 and SP6 primers can be used as a pair in PCR for insert flanked by the T3 promoter on one side and the SP6 promoter on the other. The presence of these bacteriophage-specific promoters and the polycloning site sequence (comprising multiple endonuclease recognition sites) flanking the insert should not affect probe hybridization specificity. The size limitation of about 2 kb for PCR-generated probes remains, and both single strands can be produced from the same clone by asymmetric PCR.

Apart from producing probes from plasmid DNA, PCR can also be used to produce labeled probes from genomic DNA. However, due to the sequence complexity of genomic DNA, we recommend that the PCR products of genomic amplification be subjected to gel electrophoresis and electroelution before PCR labeling. High background may result if this prior gel purification step is not performed.

We have also produced single-stranded digoxigenin-labeled probes by performing linear amplification using one primer on PCR products. PCR products used as templates in this labeling reaction should have been previously centrifuged through a Centricon-30 membrane to remove unused primers. In our experience, though probes produced in this fashion are single stranded, denaturation prior to hybridization seems to improve sensitivity; this is probably the result of secondary structures in the labeled single-stranded probes. We find that single-stranded probes are at least twofold more sensitive than double-stranded probes of the same size (probably due to the absence of re-annealing), with the added advantage of strand specificity. Furthermore, single-stranded probes give much less background staining than do double-stranded probes of the same size in *in situ* hybridization. In other words, single-stranded DNA probes labeled by PCR seem to combine the advantages of double-stranded DNA probes (e.g., ease of handling and stability) with those of single-stranded RNA probes (e.g., increased sensitivity).[15] We envisage, therefore, that PCR-based probe-labeling techniques will be more widely employed in the future.

ACKNOWLEDGMENTS

We thank the Wellcome Trust, Foulkes Foundation Fellowship, Rhodes Trust, Cancer Research Campaign, and the Foundation of the Study of Infant Deaths for financial support. We are grateful to British Biotechnology, Ltd., for supplying PCR primers.

REFERENCES

1. **Rigby, P. W. J., Dieckmann, M., Rhodes, C., and Berg, P.**, Labelling deoxyribonucleic acid to high specific activity *in vitro* by nick translation with DNA polymerase I, *J. Mol. Biol.*, 113, 237, 1977.
2. **Feinberg, A. P. and Vogelstein, B.**, A technique for radiolabelling DNA restriction endonuclease fragments to high specific activity, *Anal. Biochem.*, 132, 6, 1983.
3. **Lo, Y-M. D., Mehal, W. Z., and Fleming, K. A.**, Rapid production of vector-free biotinylated probes using the polymerase chain reaction, *Nucleic Acids Res.*, 16, 8719, 1988.
4. **Saiki, R. K., Gelfand, D. H., Stoffel, S., Scharf, S. J., Higuchi, R., Horn, G. T., Mullis, K. B., and Erlich, H. A.**, Primer-directed enzymatic amplification of DNA with a thermostable DNA polymerase, *Science*, 239, 487, 1988.
5. **Gough, N. M. and Murray, K.**, Expression of the hepatitis B virus surface, core and e antigen genes by stable rat and mouse cell lines, *J. Mol. Biol.*, 162, 43, 1982.
6. **Chan, V. T. W., Fleming, K. A., and McGee, J. O'D.**, Detection of subpicogram quantities of specific DNA sequences on blot hybridisation with biotinylated probes, *Nucleic Acids Res.*, 13, 8083, 1985.
7. **Boehringer Mannheim**, *Biochemica-Applications Manual: DNA Labeling and Detection Nonradioactive*, Boehringer Mannheim GmbH Biochemica, East Sussex, U.K., 1989, 4.
8. **An, S. F., Franklin, D., and Fleming, K. A.**, Generation of digoxigenin-labeled double-stranded and single-stranded probes using the polymerase chain reaction, *Mol. Cell. Probes*, 6, 193, 1992.
9. **Day, P. J. R., Bevan, I. S., Gurney, S. J., Young, L. S., and Walker, M. R.**, Synthesis *in vitro* and application of biotinylated DNA probes for human papilloma virus type 16 by utilizing the polymerase chain reaction, *Biochem. J.*, 267, 119, 1990.
10. **Weier, H. U. G., Segraves, R., Pinkel, D., and Gray, J. W.**, Synthesis of Y chromosome-specific labelled DNA probes by *in vitro* DNA amplification, *J. Histochem. Cytochem.*, 38, 421, 1990.
11. **Forghani, B., Hurst, J. W., and Shell, G. R.**, Detection of the human immunodeficiency virus genome with a biotinylated DNA probe generated by polymerase chain reaction, *Mol. Cell. Probes*, 5, 221, 1991.
12. **Lion, T. and Haas, O. A.**, Nonradioactive labelling of probes with digoxigenin by polymerase chain reaction, *Anal. Biochem.*, 188, 335, 1990.
13. **Taveira, N. C., Ferreira, M. O. S., and Pereira, J. M.**, Detection of HIV1 proviral DNA by PCR and hybridization with digoxigenin labelled probes, *Mol. Cell. Probes*, 6, 265, 1992.
14. **Saiki, R. K., Walsh, P. S., Levenson, C. H., and Erlich, H. A.**, Genetic analysis of amplified DNA with immobilized sequence-specific oligonucleotide probes, *Proc. Natl. Acad. Sci. U.S.A.*, 86, 6230, 1989.
15. **Sturzl, M. and Roth, W. K.**, "Run-off" synthesis and application of defined single-stranded DNA hybridization probes, *Anal. Biochem.*, 185, 164, 1990.

Chapter 8

DIRECT PCR SCREENING OF LAMBDA AND COSMID LIBRARIES

Hugh G. Griffin

TABLE OF CONTENTS

I. Introduction 53

II. Materials and Methods 54
   A. Bacterial Strains, Plasmids, and Lambda Libraries 54
   B. PCR Primers 54
   C. PCR Reaction Conditions 55
   D. Direct PCR Screening of Lambda Libraries 55
   E. Direct PCR Screening of Cosmid or Plasmid Libraries 56

III. Results 56

IV. Discussion 56

Acknowledgment 57

References 57

I. INTRODUCTION

The polymerase chain reaction (PCR) is a new and extremely powerful technology that has been developed during the last few years and has revolutionized many areas of science. PCR has changed the manner in which biological research is conducted, and valuable applications of this new technology have already been found in areas as diverse as clinical diagnosis, immunology, food safety, environmental microbiology, forensic science, and molecular evolution. The rapidity, simplicity, and convenience of the technique ensure that new adaptations and applications are being continually developed, and PCR is by now one of the most widely used technologies in modern biological science.

The identification and isolation of chromosomal genes from a lambda or cosmid library is an important procedure in the molecular biology laboratory. Traditionally, libraries have been screened by filter hybridization using DNA probes that have been labeled with a radioactive marker.[1-3] However, filter hybridization is a time-consuming process, that often results in false positives and has the potential safety risk of exposure to radioactivity. Recently, the polymerase chain reaction has been used to amplify DNA directly from plaques and to reduce the amount of filter hybridization necessary when screening cDNA or genomic libraries.[4,5] These techniques have been adapted to produce a method for the direct PCR screening of lambda libraries, obviating the need for any filter hybridization with labeled probes.[6] This strategy is also applicable to cosmid and plasmid libraries.

The development of a rapid PCR-based method for the identification and isolation of genes from libraries greatly facilitates genetic and molecular biology research and demonstrates the power and versatility of PCR. The technique described here has been used to isolate a number of genes from bacterial lambda libraries in a much shorter time than would have been possible using traditional techniques.[6]

II. MATERIALS AND METHODS

A. BACTERIAL STRAINS, PLASMIDS, AND LAMBDA LIBRARIES

We have successfully used this technique with λ EMBL-3, and λ ZAP II libraries (Stratagene, Cambridge, U.K.), but any lambda library should work. Similarly, any cosmid or plasmid library is suitable. λ EMBL-3 and λ Zap II have insert size ranges of 9 to 23 and 0 to 10 kbp, respectively. In this laboratory, *Escherichia coli* strain MC1022[7] is used as a host for plasmids and strain LE392[1] is used to propagate lambda phage, but other appropriate strains can also be used.[1] *Escherichia coli* strains are grown in L broth or on L agar.[8] Plasmids pUC3 and pUC8[9] are used as vectors in *E. coli*.

B. PCR PRIMERS

A vast amount of sequence data is now available in the international databases and the size of the databases is growing at an ever-increasing rate. This means that the sequence of genes analogous to the one of interest may well be available. Areas of conservation are often present between analogous genes from different organisms. Even genes from diverse organisms may share small regions of homology, for example, in the region encoding the active site of an enzyme. Computer-generated multiple alignments[10] of both the predicted amino acid sequence and the nucleic acid sequence of analogous genes from different organisms can help identify areas of similarity.[6] A codon usage table for the organism from which the gene is being sought can further aid the design of PCR primers. This information may already be available, or it can be generated from the known sequences of genes from that organism.

Degenerate pools of oligonucleotide primers are designed from the areas of greatest homology. The codon usage data are used to decide which codons are most likely to be employed, and in this way the amount of degeneracy in the primer is reduced. However, even highly degenerate primers (those consisting of a mixture of several hundred different oligonucleotides) will work in a PCR reaction.

In many cases the N terminal amino acid sequence of the particular protein is available. The traditional way to screen a library is by end labeling a degenerate oligonucleotide designed from the amino acid sequence.[1] Apart from the hazards associated with radioactivity, this method is technically difficult to perform. In particular, finding the correct hybridization and washing conditions is arduous, and false positives are common. However, by designing two degenerate oligonucleotide primers, one from each end of the available amino acid sequence, it is possible to perform a PCR reaction and generate a product equivalent in length to that predicted from the N terminal amino acid sequence. The design of the primers should be based on codon usage data for that organism. In most cases the amount of N terminal sequence available will be small, and the PCR product will therefore be about 60 to 100 bp long. This size of fragment is better visualized on an 8% polyacrylamide gel rather than on an agarose gel. It is important to ensure that primer-dimers do not form, as they could lead to misleading bands on the gel.

Careful choice of the 3' base of a PCR primer is essential as it is unlikely that the reaction will work if this base does not hybridize to the target DNA. Ideally the primer should be chosen so that the final two or three bases at the 3' end are homologous to the target DNA. The annealing temperature of the reaction also requires some consideration. For a short oligonucleotide each A-T base pair contributes approximately 2°C and each G-C base pair contributes about 4°C to the melting temperature (T_m). The T_m is therefore empirically determined by the formula: $T_m = 2(A+T) + 4(G+C)$. Thus a perfectly matched 20-mer with a 50% G-C content has a T_m of 60°C [(2×10)+(4×10)]. A mismatch will destabilize the helix and

TABLE 1.
A General Purpose PCR Buffer

Component	Stock concentration	Amount of stock per ml 4x buffer	Concentration in 4x buffer	Concentration in 1x buffer
Tris-HCl (pH 8.3)[a]	1 M	40 µl	40 mM	10 mM
KCl	1 M	200 µl	200 mM	50 mM
MgCl$_2$	1 M	8 µl	8 mM	2 mM
Gelatin	1%	40 µl	0.04%	0.01%
dATP	100 mM	8 µl	800 µM	200 µM
dTTP	100 mM	8 µl	800 µM	200 µM
dCTP	100 mM	8 µl	800 µM	200 µM
dGTP	100 mM	8 µl	800 µM	200 µM
Triton X100	10%	40 µl	0.4%	0.1%
H$_2$O		640 µl		

[a] The pH of the Tris-HCl should be 8.3 at room temperature. When incubated at 72°C, the pH of the reaction drops to approximately 7.2.

lower the T_m by about 5°C. (A mismatch at the 3' end will not be tolerated by the polymerase.) When using degenerate primers, it is advisable to set the annealing temperature considerably lower than the T_m to allow for several possible mismatches. The destabilizing effect depends on the exact nature of the mismatch. An A-C mismatch has a great destabilizing effect, whereas a G-T mismatch has very little effect.

C. PCR REACTION CONDITIONS

PCR reaction buffer: 10 mM Tris-HCl (pH 8.3 at room temperature), 50 mM KCl, 2 mM MgCl$_2$, 0.01% gelatin, 0.1% Triton X100, 200 µM dATP, 200 µM dCTP, 200 µM dGTP, 200 µM dTTP. This buffer is stored as a 4x stock in aliquots at –20°C (Table 1), and 1 U of *Taq* polymerase (ABI, Foster City, CA) is added per reaction. Primers are added to a concentration of 1 µM and reactions are performed in a 50-µl volume overlaid with 60 µl of mineral oil.

PCR reactions are performed using a thermal cycler, e.g., Hybaid thermal reactor (Hybaid, Teddington, Middlesex, U.K.). Suggested cycling conditions: Denaturation for 2 min at 92°C, primer annealing for 2 min at 47°C, extension for 2 min at 72°C, instrument on "plate" or "block" control, 30 rounds of cycling. Small PCR products can be visualized by electrophoresis in either a 2% agarose gel or an 8% polyacrylamide gel. Metaphor™ agarose (FMC Bioproducts, Rockland, ME) also works well.[6]

This buffer and these reaction conditions work well for most primers used in our laboratory and are good starting points for any PCR reaction. However it may be necessary to adjust conditions to suit particular primers and templates. In particular, the magnesium ion concentration and the annealing temperature may require modification.

D. DIRECT PCR SCREENING OF LAMBDA LIBRARIES

The lambda library is plated out onto a number of 90-mm plates to give a plaque density of about 100 pfu per plate.[1] About ten plates (1000 plaques) are usually sufficient for a bacterial library, but more may be required for eukaryotic libraries (see discussion in Section IV). A plaque lift is performed on each plate using nylon membranes (Amersham International, U.K.). Each filter is placed plaque side up in a sterile petri dish and washed in 3 ml of sterile distilled water. A 10-µl aliquot is taken from each sample for analysis by PCR, and the products are analyzed by gel electrophoresis. If a plate contains one or more phage plaques that have the target sequence, a band of the appropriate size will be seen in the gel.

Once the plate containing the positive lambda plaque has been identified, a further plaque lift is performed on that plate. The filter is marked out into ten segments, and the position of the segments is marked on the petri dish. The filter is then removed to a sterile petri dish and cut into the ten segments. Each segment is placed in a 1.5-ml Eppendorf tube containing 0.3 ml water, and the tube vortexed briefly. PCR is performed on 10-µl aliquots. In this way the search for the positive plaque is narrowed down to one tenth of a plate (or to about ten individual plaques) by only two sets of PCR reactions. The final round of PCR consists of doing reactions on the individual plaques (about ten) in the segment that gave a positive PCR result. The plaques are picked into 200 µl of SM buffer[1] using either a sterile glass Pasteur pipette or a sterile toothpick, and 10-µl aliquots analyzed by PCR. The positive plaque is plaque purified, and DNA can be isolated using a Qiagen lambda isolation kit (Diagen, Germany) or other standard methods.[1]

E. DIRECT PCR SCREENING OF COSMID OR PLASMID LIBRARIES

The techniques for identifying a positive lambda plaque using PCR can equally well be applied to plasmid or cosmid libraries. The colonies are lifted onto a nylon membrane in a manner similar to that used for plaques. It is important to leave sufficient material from each colony behind on the plate to enable the positive colony to be subsequently picked and purified. Alternatively, duplicate or replica plates can be made and stored at 4°C. The membranes are placed colony side up in a sterile petri dish and washed in 3 ml of sterile distilled water. PCR reactions are performed on 10-µl aliquots as described above.

III. RESULTS

A number of genes have been isolated from λ EMBL-3 and λ ZAP II libraries of *Lactococcus lactis* using the method described in this chapter.[6] These include the genes encoding lactate dehydrogenase (*ldh*), 5-enolpyruvylshikimate 3-phosphate synthase (*aroA*), acetolactate decarboxylase (*aldB*), and glutamyl amino peptidase (*pepA*, *gapII*). Due to the AT-rich nature of lactococcal DNA it was necessary to design particularly long primers to ensure an adequate T_m.

IV. DISCUSSION

In the final round of screening, the individual plaques are picked into SM buffer and PCR reactions are performed on 10-µl aliquots. The reason for using SM buffer at this stage is to enable long-term storage of the phage.[1] However, the high $MgCl_2$ concentration in SM buffer causes an almost twofold increase in the final Mg^{2+} concentration of the PCR reaction buffer. This did not adversely affect the PCR reactions described here but could, if necessary, be counteracted by employing a PCR reaction buffer which does not contain $MgCl_2$. The Mg^{2+} ions required would be supplied by the SM buffer. Alternatively, the phage could be resuspended in water, and later transferred to SM buffer for long-term storage.

It is possible to calculate the theoretical number of plaques in a particular library which it is necessary to screen to have a probability of finding any given DNA sequence.[1] This number is calculated by the following formula:

$$N = \frac{\ln(1-P)}{\ln(1-I/G)}$$

where N is the number of plaques that should be screened, P is the probability of having a given sequence represented in this number of clones, I is the insert size of the vector, and G

is the genome size of the organism. For λ EMBL-3 with an insert size between 9 and 23 kbp, the number of plaques that should be screened in a *L. lactis* library ($G = 2,500,000$) is between 500 and 1275 plaques with a 99% probability ($P = 0.99$). Because of the smaller insert size of λ ZAP-II, it is necessary to screen up to 1900 plaques. However, by reducing the P value to 0.9, i.e., assuming a 90% probability, these figures reduce to 250 to 650 plaques for λ EMBL-3, and to 600 to 1000 for λ ZAP-II. In practice, we found that screening approximately 1000 plaques was sufficient to identify a positive plaque in each case. Because of the larger genome size of other organisms compared to *L. lactis*, it may be necessary to screen a greater number of plaques. With a genome size of 4800 kbp (*E. coli*), the theoretical number of plaques it is necessary to screen with a 99% chance of isolating a particular fragment is in the range 1000 to 2500 for λ EMBL-3.

The strategy described in this chapter is a rapid, simple, and effective means of identifying clones containing specific genes in a library. The strategy can be further enhanced by using multiple pairs of primers in the same PCR reaction, thus enabling the isolation of several genes at the same time. Direct screening of lambda libraries by PCR was performed successfully using either homologous primers, degenerate pools of primers designed from a multiple alignment of analogous genes, or using degenerate primers designed from N terminal protein sequences.[6] The method is applicable to lambda, cosmid, or plasmid libraries of any species.

ACKNOWLEDGMENT

The help and advice of M. J. Gasson and K. J. I'Anson is gratefully acknowledged.

REFERENCES

1. **Sambrook, J., Fritsch, E. F., and Maniatis, T.,** *Molecular Cloning: A Laboratory Manual*, 2nd ed., Cold Spring Harbor Laboratory Press, Cold Spring Harbor, NY, 1989.
2. **Griffin, H.G. and Griffin, A.M.,** Cloning and DNA sequence analysis of the *serC-aroA* operon from *Salmonella gallinarum*; evolutionary relationships between the prokaryotic and eukaryotic *aroA*-encoded enzymes, *J. Gen. Microbiol.*, 137, 113, 1991.
3. **Griffin, H. G., Swindell, S. R., and Gasson, M. J.,** Cloning and sequence analysis of the gene encoding L-lactate dehydrogenase from *Lactococcus lactis*: evolutionary relationships between 21 different LDH enzymes, *Gene*, 122, 193, 1992.
4. **Gussow, D. and Clackson, T.,** Direct clone characterization from plaques and colonies by the polymerase chain reaction, *Nucleic Acids Res.*, 17, 4000, 1989.
5. **Bloem, L. J. and Lei, Y.,** A time-saving method for screening cDNA or genomic libraries, *Nucleic Acids Res.*, 18, 2830, 1990.
6. **Griffin, H. G., I'Anson, K. J., and Gasson, M. J.,** Rapid isolation of genes from bacterial lambda libraries by direct PCR screening, *FEMS Microbiol. Letts.*, 112, 49-54.
7. **Casadaban, M. J. and Cohen, S. N.,** Analysis of gene control signals by DNA fusion and cloning in *Escherichia coli*, *J. Molec. Biol.*, 138, 179, 1980.
8. **Miller, J. H.,** *Experiments in Molecular Genetics*, Cold Spring Harbor Laboratory Press, Cold Spring Harbor, NY, 1972.
9. **Norrander, J., Kempe, T., and Messing, J.,** Construction of improved M13 vectors using oligodeoxynucleotide-directed mutagenesis, *Gene*, 26, 101, 1983.
10. **Griffin, A. M. and Griffin, H. G.,** *Computer Analysis of Sequence Data*, Humana Press, Totowa, NJ, 1994.

Chapter 9

METHODS FOR GENERATING MULTIPLE SITE-DIRECTED MUTATIONS *IN VITRO*

Jörg Stappert

TABLE OF CONTENTS

I. Introduction 59
   A. Strategies 59
   B. Site-Directed Mutagenesis by Amplification of the
       Entire Plasmid 60
   C. Site-Directed Mutagenesis by Amplification of
       Short DNA Fragments 62

II. Materials and Methods 64
   A. Materials 64
   B. Recombinant Circle PCR (RCPCR) 64
      1. Primer Design 64
      2. PCR Amplification 64
      3. Purification of the PCR Product 64
      4. Annealing of the PCR Products 64
      5. Transformation of *E. coli* and Screening of Colonies 64
   C. Tagged PCR Mutagenesis Method 65
      1. Primer design 65
      2. 5′ Phosphorylation of the Mutating Primers 65
      3. Denaturation of the ds Template DNA 65
      4. Annealing and Extension Reaction 65
      5. PCR Amplification 65

III. Discussion 66

References 66

I. INTRODUCTION

A. STRATEGIES

Site-directed mutagenesis of cloned genes as well as the mutation of cis-regulatory elements is a powerful and rapid technique for their functional analysis. Since PCR was first described, a number of PCR-based methods have been developed for introducing directed mutations in virtually any position into DNA or for joining unrelated sequences together.

In the simplest cases, point mutations, deletions or insertions can be engineered at the priming sites, using mismatched primers also covering appropriate restriction sites to facilitate recloning of the PCR product. In many cases however, restriction sites are not available, and many mutations cannot be introduced into various sites along a relatively long DNA fragment. To overcome the problem of having to have suitable restriction sites in the vicinity of the desired mutation, several methods have been developed, based on one of the following strategies:

1. In the first strategy the entire plasmid is amplified using a pair of primers located back to back (e.g., inverse PCR, IPCR; enzymatic inverse PCR, EIPCR), or two sets of primers are used to amplify two overlapping parts of the plasmid which, after annealing via their short complementary DNA stretches, build up the entire plasmid again (e.g., recombinant circle PCR, RCPCR).
2. The second strategy leads to the amplification of a defined mutagenized fragment, which then has to be recloned. The desired mutations are generated by using two primer sets resulting in overlapping PCR fragments or by joining the mutated primers to a "tagged" primer, which enables the selective amplification of the mutated DNA strand.

Following the concept guidelines of this book, an illustrative summary of some of the currently published techniques will be given, their advantages and disadvantages will be pointed out, and finally, in Section II, protocols for RCPCR and tagged PCR will be described, representing each of the strategies mentioned. For a detailed description of the other methods summarized in this chapter, the reader is referred to the references cited for each procedure.

B. SITE-DIRECTED MUTAGENESIS BY AMPLIFICATION OF THE ENTIRE PLASMID

As mentioned, earlier, to do site-directed mutagenesis of cloned DNA by amplifying the whole plasmid rather than just a fragment, four different methods have been described:

1. Inverse PCR mutagenesis (IPCR)[1]
2. Enzymatic inverse PCR mutagenesis (EIPCR)[2]
3. Recombinant circle PCR (RCPCR)[3]
4. Recombination PCR (RPCR)[4]

These four methods have been derived from the originally published inverse PCR.[5] In this technique, two primers that are located back to back on the opposing DNA strands of a plasmid drive the PCR. The resultant PCR product is a linear DNA molecule identical in length to the starting plasmid. Because all these procedures rely on PCR amplification of the entire plasmid, it is not necessary to prepare an appropriate vector fragment or a single-stranded DNA template.

EIPCR is a technique that combines strategies of inverse PCR with the class 2s restriction site approach of Tomic et al.[6] (see Figure 1A). The key step to EIPCR is the incorporation of identical class 2s restriction sites in the primer set used for PCR. Class 2s restriction enzymes have a recognition site located 5' of the cut site (e.g., Bsp MI ACCTGC NNNN*NNNN*). Thus, after completing PCR, the ends of the full-length linearized plasmid are digested with the class 2s enzyme incorporated into the primers. Due to the distance between recognition and cut site, all sequences upstream of the cut site will be lost. Thus in the ligation the only part that becomes part of the plasmid is the *NNNN* overhang, which can be made to be the native DNA sequence. Mutations can be placed into one or both primers and at any location between the enzyme cut site and the exact 3' match, which should be of a magnitude >15 bp. Since several 2s restriction enzymes have been described, it should generally be possible to design primers containing an appropriate 2s restriction site. The number of positive clones carrying the desired mutation is more than 95%, which makes EIPCR the most efficient method.

In summary, EIPCR has two major advantages: (1) the high percentage of correct clones and (2) the requirement of only one pair of primers to generate a mutation.

In contrast to EIPCR, two different sets of primers are needed for RCPCR and RPCR.[7] In both methods, the product of one inverse PCR is mixed with the product of a second inverse PCR, which is primed at a different location on the same template. Using RCPCR, these products are combined, denatured, and reannealed *in vitro* before competent *Escherichia coli*

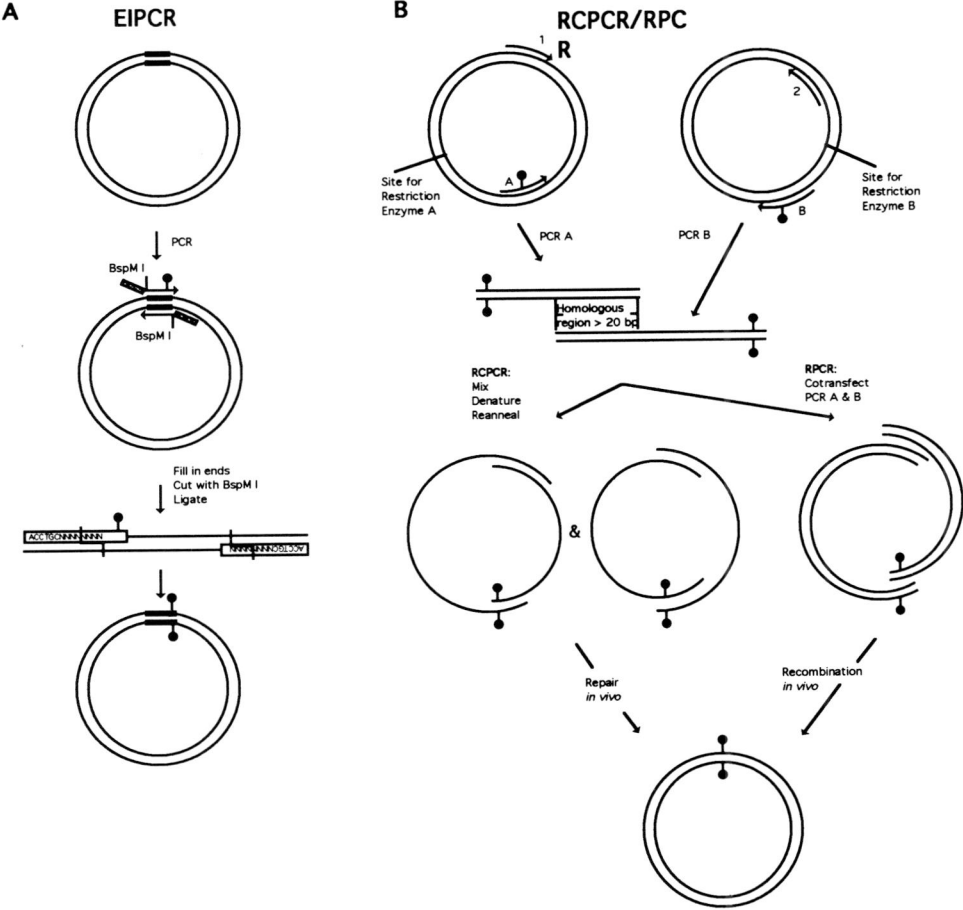

Figure 1. The general scheme for EIPCR (A) and RCPCR/RPCR (B). Double-stranded circles represent double-stranded plasmids; hatched boxes mark 5' add-on sequences; Solid circles indicate mismatches in the primers and resulting mutations in the PCR products.

are transformed with this construct, as shown in Figure 1B. In contrast to RCPCR, this cross-annealing step is omitted when using RPCR. An equal amount of both PCR reactions is transformed into competent *E. coli*. Upon transformation, the DNA ends undergo homologous recombination *in vivo*, resulting in a circular plasmid. The cross-annealing of the two amplified fragments is mediated by homologous DNA sequences at the 5' and 3' and respectively. Whereas the length of homology between mutated ends should be at least 30 bp, the overlapping homology between nonmutated ends depend on the location of primer 1 and 2. Interestingly, it was shown that the number of mutated clones is nearly 100% when using a 22-bp overlap at the nonmutated ends in combination with RPCR, but drops down to 50% when generating the mutation with RCPCR. An inverse correlation was found when increasing the length of homology between nonmutated ends.

All three methods have certain advantages and limitations. All three are rapid and efficient. Nevertheless, the size limit for the template to be amplified is in the range of 5 kb, although mutagenesis of a 7.1-kb construct has been described.[8] A second disadvantage, is the necessity of using relatively long primers; the primers used in these different approaches have to be at least 25 to 30 bp in length. Third, one has to consider the fidelity of the enzyme used for PCR.

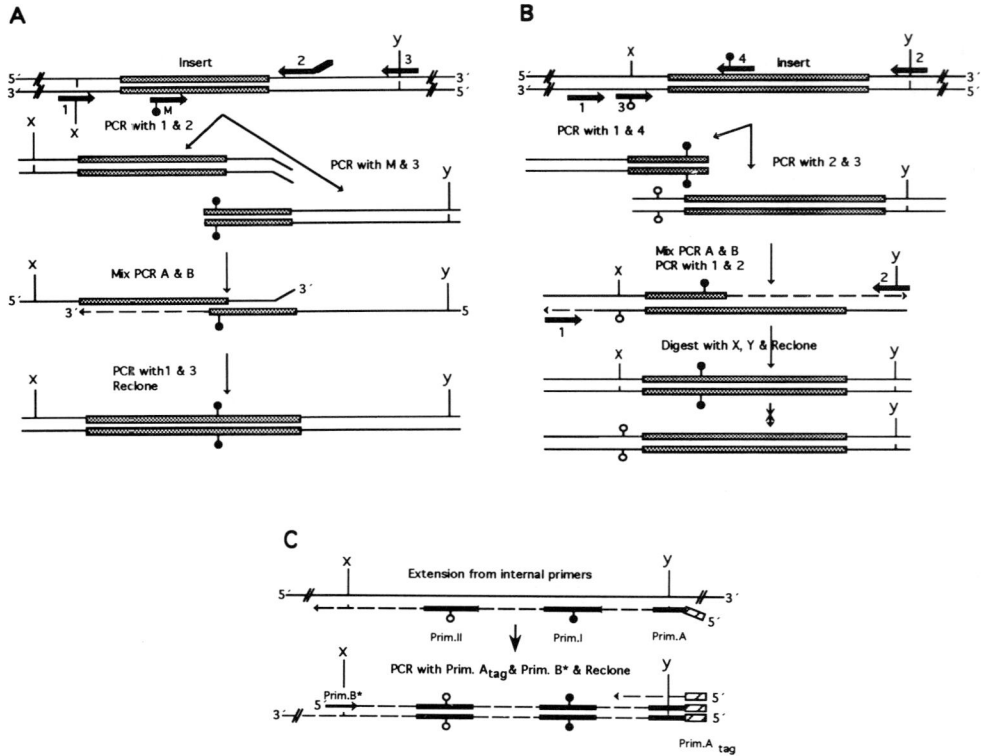

Figure 2. Schematic diagram of three methods used to mutate fragments of a defined length. Primers are represented by black bars; corresponding mismatches are indicated by open and solid circles; Restriction sites are designated X and Y. A detailed description of each method is given in the text.

Since PCR has a low but detectable rate of mutagenesis, polymerases should be used that have a 3′ to 5′ proofreading capability, as this lowers the rate of unwanted mutation.

C. SITE-DIRECTED MUTAGENESIS BY AMPLIFICATION OF SHORT DNA FRAGMENTS

The problem of unwanted PCR errors will be diminished when amplifying only small fragments; this can be achieved by using one of the methods shown in Figure 2.

The two procedures schematically depicted in Figure 2A and 2B represent modifications of the overlap extension method originally published by two groups.[9,10] This method is based on the amplification of two fragments with overlapping ends in which the same mutations are introduced. These fragments are combined and reannealed to each other, and the 3′ overlap of each DNA strand serves as a primer for the 3′ extension of the complementary strand. One disadvantage of the original method is that it requires two new primers for each mutation, limiting one in making an extensive mutagenesis in the same DNA sequence.

The problem has been circumvented by a method described by Mikaelian and Sergeant,[11] as shown in Figure 2A. It requires three universal primers chosen in the vector and only one specific primer for each mutation. As shown, the first step consists of two different rounds of PCR. The two reactions are done using the primer combinations 1,2 and M,3. Primer 1,2 and 3 are homologous to the vector sequence, but primer 2 also contains a mismatched 3′ end; primer M contains the mutation to be introduced into the template DNA. The two amplified fragments are purified, mixed, and subjected to a second PCR with external primers 1 and 3.

During this second PCR, only the mutated DNA strand is amplified, since the 3' add-on end of primer 2 inhibits the extension of the nonmutated DNA strand. The amplified fragment can be then digested with the appropriate restriction enzymes. The efficiency of this procedure reaches 90%.

A similar approach was used by Ito et al.[12] (Figure 2B). To reclone only the mutated DNA strand, they developed a method called MR (an abbreviation of "modification of a restriction site"). As in the previous technique, three primers are selected as common primers so that each mutation requires only one additional primer. Due to the design of the commonly used primers, it is necessary to reclone the insert to be mutated into a polylinker site of an appropriate plasmid. Even if a suitable restriction site does not exist in the target DNA, it is very easy to create a proper site at both ends of the target DNA by using primers carrying 5' add-on sequences. As shown in the diagram, primers 1 and 2 are complementary to the neighboring sequences of a polylinker site. Primer 3 is complementary to the polylinker sequence located downstream of primer 1 and has a mismatched nucleotide (nt) to destroy a restriction site. These three primers can be commonly used in a series of different mutagenesis. The fourth primer, however, carries the mutation of interest. By using the primer combinations 1,4 and 2,3, two fragments are amplified, mixed, denatured, and annealed. The resulting products are further amplified by PCR using the external primers 1 and 2, and digested with two different restriction enzymes, X and Y. Although two kinds of DNA are amplified in the second PCR, only the mutated DNA fragment will be recloned, since the restriction site X has been deleted in the nonmutated DNA fragment. Depending on the oligonucleotides used in this technique, the efficiency of getting positive clones can be as high as 100%.

Due to this high efficiency, the described methods are very useful for introducing several mutations into various sites of the inserted DNA. Nevertheless, each new mutation makes it necessary to run a further PCR. Additionally, each newly mutated DNA fragment has to be sequenced in order to detect nucleotide misincorporation generated during DNA amplification.

To avoid this problem, we have developed a method that allows the introduction of several mutations in only one step.[13] To do this we combined extension of mutated internal primers by T4 DNA polymerase with selective amplification of the mutated DNA strand by PCR from added-on external primers. The principle of this method is summarized in Figure 2C.

In the first step, two or even more mutagenic primers (I, II) and Primer A are annealed to the template. Primer A has a suitable restriction site and a 5' add-on sequence that is not complementary to the template. After primer extension and ligation reaction, the mutated DNA strand is selectively amplified by PCR in the second step by using the two outer primers, A_{tag} and B*, as common primers. Each new mutation thus requires only one additional primer. The percentage of positive clones carrying both mutations depends on whether ss or ds DNA is used as a template. When using ss DNA, up to 80% of all clones will be positive for two mutations, in contrast to 40 to 50% when using ds plasmid DNA. For this method, bacteriophage T4 DNA polymerase or Sequenase should be used in the polymerase extension reaction. Unlike the Klenow fragment of *E. coli* DNA polymerase I, neither of these enzymes is able to displace the mutagenic oligonucleotide from its template. Therefore there should be no limit to the number of different mutagenic primers that can be used in a single *in vitro* site-directed mutagenic reaction. Indeed, as has been reported by Perlack,[15] using the classical procedure of Kunkel,[14] 14% of the clones tested were positive for seven of seven different mutations generated in one single step. This is the same efficiency one would expect when using the PCR-based method.

In the protocols described below, practical approaches concerning the generation of mutations are given exclusively for the mutational step itself. Commonly used techniques like plasmid preparation, ligation, transformation of competent *E. coli*, etc., should be done using accepted protocols.[16,17]

II. MATERIALS AND METHODS

A. MATERIALS

Plasmids used as PCR templates as well as for DNA sequencing were prepared with Qiagen mini- or midi-columns (Qiagen, Düsseldorf). Sequence analysis of the mutants was done on double-stranded DNA, using the T7 Sequenase Kit (U.S. Biochemical) according to the manufacturer's description. All oligonucleotides were synthesized on an Applied Biosystems DNA synthesizer and then purified using a C-18 reversed-phase high-pressure liquid chromatography (HPLC) column. Except the *Taq* DNA polymerase, which was purchased from Amersham Buchler (Braunschweig), all restriction and modification enzymes were purchased from GIBCO BRL (Eggenstein) or from Boehringer Mannheim.

B. RECOMBINANT CIRCLE PCR (RCPCR)

1. Primer Design

RCPCR requires a total of four primers per mutagenesis reaction. If the insert to be mutated has been cloned into a polylinker site of a vector, only two new primers need to be synthesized per mutagenesis reaction. Since the nonmutating primers will be located outside the mutagenesis region, they can be reused to mutate any given region of the insert. The nonmutating primers should be at least 15 nt in length, where the mutating primers have an overall length of >25 nt carrying at least 15 nt of exact complementarity to each other at their 5' ends. It is also possible to use mutating primers with an exact complementarity. The homology length between nonmutated ends will greatly influence the efficiency of this method.

2. PCR Amplification

Before PCR, each plasmid has to be linearized by restriction enzyme digestion outside the region to be amplified. In each of the two separate PCR amplifications for a given mutagenesis, we set up the following reaction mixture in 100 µl: 10 ng of pBlueskript SKII⁺ DNA containing the insert to be mutated, each primer at 1 µM, 100 µM of each dNTP, 2.5 U *Taq* polymerase in Taq-Buffer (10 × Taq-Buffer is 100 mM Tris-HCl, pH 8.4; 500 mM NaCl; 20 mM MgCl$_2$; 1 mg/ml gelatin). Prior to amplification, 60 µl of mineral oil is placed on top of each reaction mixture. For the amplification, reactants are subjected to 30 c of PCR with the following parameters: denaturation at 95°C for 0.5 min; annealing at 50°C for 0.5 min; and extension at 72°C for 2 min.

3. Purification of the PCR Product

As published by Jones and Winistorfer,[7] there is no need to further purify the PCR products. To remove mineral oil, we recommend freezing the probe. Alternatively, each PCR product may be carefully removed with a thin micropipet and transferred into a new reaction tube.

4. Annealing of the PCR Products

In order to generate recombinant circles *in vitro*, equal amounts of the two PCR products have to be combined, denatured at 95°C for 5 min, reannealed at 50°C for 2 h, and returned to room temperature prior to transfection.

5. Transformation of *E. coli* and Screening of Colonies

Competent *E. coli* should be transformed with 2 to 5 µl of combined PCR product. In cases of restriction site deletions or insertions, positive clones can be identified either by preparation of the plasmid DNA or by using a PCR-based procedure described by Güssow and Clackson.[18] Briefly: Colonies are resuspended in 0.5 ml of water (live bacteria can be rescued at this stage) and boiled in a water bath for 5 min. After centrifugation for 2 min at 13 to 16,000 g, 5 µl of the supernatant is subjected to 30 c of PCR amplification with primers that flank the mutagenesis site. The orientation of the insert may be screened using a third primer within the insert. Alternatively 1/10 of the unpurified PCR product may be digested by the appropriate restriction enzyme

C. TAGGED PCR MUTAGENESIS METHOD
1. Primer Design

In order to use Primers A and B* together, they should be complementary to vector sequences flanking the insert to be mutated. Primer A should have a length of >30 nt and contain 15 nt at the 5′ end, which are not homologous to the template. A suitable restriction site should be available within the homologous region. Primer A_{tag}, which is identical to the 5′ add-on sequence of primer A, should have a GC content of >50%. Primer B* has to cover a second restriction site and should have a minimum length of 15 nt. The mutagenic primers we use are, respectively, 18 to 21 nt in length and contain up to four mismatches relative to the template DNA. A total of 6 nt exact homology to the template is included on each end to ensure proper hybridization.

2. 5′ Phosphorylation of the Mutating Primers

Before starting primer extension, the mutating primers have to be phosphorylated with Kinase in order to allow ligation of its 5′ ends. Set up the following reaction mixture in 20 µl: 200 pmol primer, 1 × Kinase Buffer (10 × Kinase Buffer is 0.5 M Tris-HCl, pH 8.0; 0.1 M $MgCl_2$), 10 µM DTT, 2 mM ATP, and 5 U polynucleotide kinase. Incubate at 37°C for 30 min and then at 65°C for 10 min to inactivate the kinase. Dilute the phosphorylated oligonucleotides to a final concentration of 2 pmol/µl in water.

3. Denaturation of the ds Template DNA

In order to facilitate annealing of the primers to the template, it is necessary to denature the ds DNA. This can be done either by heat or alkaline denaturation. In both procedures, use 0.2 to 0.4 pmol ds template DNA. If using alkaline denaturation, incubate the template DNA in a reaction volume of 20 µl, together with 0.2 M NaOH, at room temperature for 5 min. Neutralize the mixture by adding 8 µl 5 M ammonium acetate, pH 7.4. Precipitate the DNA with 100 µl ethanol at –70°C for 5 min. Centrifuge at 16,000 g for 5 min, then wash with 70% ethanol. Dry at room temperature for 10 min.

4. Annealing and Extension Reaction

To anneal primers to the denatured template DNA, incubate 5 to 10 × excess of phosphorylated primers (mutagenized primers and primer A) in 1 × Annealing Buffer (10 × Annealing Buffer is 200 mM Tris-HCl, pH 7.4; 20 mM $MgCl_2$; 500 mM NaCl) in a total volume of 10 µl. Heat to 65°C for 3 min, and allow to cool slowly to room temperature for 30 min.

Synthesis of the complementary DNA strand is done in a volume of 20 µl containing the same annealing mixture plus 2.5 U T4 DNA polymerase, 1 U T4 DNA ligase, and 1 × Synthesis Buffer (10 × Synthesis Buffer is 5 mM of each dNTP; 10 mM ATP; 100 mM Tris-HCl, pH 8.0; 10 mM EDTA). After incubating at 37°C for 90 min, add 10 mM Tris-HCl; 10 mM EDTA, pH 8.0 (final concentration), and stop the reaction by freezing.

5. PCR Amplification

For amplification of the mutated DNA strand use the external primers, A_{tag} and B*, at 1 µM with 1/5 volume of the crude "extension" reaction, 100 µM of each dNTP, and 2.5 U *Taq* polymerase in Taq-Buffer (for 10 × Taq-Buffer, see RCPCR) in a reaction volume of 100 µl. Before amplification, overlay the mixture with 60 µl of mineral oil. Then subject the PCR mixture to 30 c of amplification: 0.5 min at 95°C; 0.5 min at 55°C; 1 min at 72°C. After removing the mineral oil, 5 µl of the PCR mixture can be analyzed on a 1% TBE agarose gel and stained with EtBr. To determine the efficiency of incorporation of the mutated primers, 5 μl of each reaction should be digested with the appropriate restriction enzymes. Before recloning, we recommend purifying the PCR products. Clones can be analyzed as described above.

III. DISCUSSION

As mentioned above, the efficiency of the RCPCR method is influenced by the length of homology between nonmutated ends. Whereas 50% of all clones are positive when containing a 25-bp stretch of homology, up to 90% of the clones will have the mutation if the length of homology is increased to 2800 bp. In order to reduce background transformations by the original nonmutated PCR template, one must use linearized PCR templates, which have been cut outside the region to be amplified.

One big advantage of RCPCR, as well as of all other methods using the mutated primers to amplify the entire plasmid for amplification, is the fact that cloning steps are omitted. Nevertheless, for each mutation it is necessary to do a new round of amplification. Site-directed mutagenesis using the tagged PCR method is therefore the method of choice if several mutations have to be generated.

The possibility of simultaneously introducing several mutations in a single step makes this method much faster than the others as long as suitable "markers" can be introduced in each mutation. For that reason it is necessary to use mutagenic oligonucleotides carrying the desired mutation together with a second nt mismatch. This second, "silent" mutation is necessary to generate or delete a restriction site within the primer sequence in order to facilitate screening for positive clones. If such markers are absent, the method becomes more elaborate because all colonies have to be hybridized using the primers as probes. On the other hand, this problem will still occur in all methods not approaching a mutation efficiency of 100%.

For this reason, it is necessary in each case to choose the *in vitro* mutagenesis method appropriate for a given problem.

REFERENCES

1. **Hemsley, A., Arnheim, N., Toney, M. D., Cortopassi, G., and Galas, D. J.**, A simple method for site-directed mutagenesis using the polymerase chain reaction, *Nucleic Acids Res.*, 17, 6545, 1989.
2. **Stemmer, W. and Morris, K. M.**, Enzymatic inverse PCR: a restriction site independent, single-fragment method for high-efficiency, site-directed mutagenesis, *BioTechniques*, 13, 215, 1992.
3. **Jones, D. H. and Howard, B. H.**, A rapid method for site-specific mutagenesis and directional subcloning by using the polymerase chain reaction to generate recombinant circles, *BioTechniques*, 8, 178, 1990.
4. **Jones, D. H. and Howard, B. H.**, A rapid method for recombination and site-specific mutagenesis by placing homologous ends on DNA using polymerase chain reaction, *BioTechniques*, 10, 62, 1991.
5. **Ochman, H., Gerber, A. S., and Hartl, D. L.**, Genetic applications of an inverse polymerase chain reaction, *Genetics*, 120, 621, 1988.
6. **Tomic, M., Sunjevaric, I., Savtchenko, E. S., and Blumenberg, M.**, A rapid and simple method for introducing specific mutations into any position of DNA leaving all other positions unaltered, *Nucleic Acids Res.*, 18, 1656, 1990.
7. **Jones, D. H. and Winistorfer, S. C.**, Recombinant circle PCR and recombination PCR for site-specific mutagenesis without PCR product purification, *BioTechniques*, 12, 528, 1992.
8. **Yao, Z., Jones, D. H., and Grose, C.**, Site-directed mutagenesis of herpesvirus glycoprotein phosphorylation sites by recombination polymerase chain reaction, *Methods Applic.*, 1, 205, 1992.
9. **Higuchi, R., Krummel, B., and Saiki, R. K.**, A general method for *in vitro* preparation and specific mutagenesis of DNA fragments: study of protein and DNA interactions, *Nucleic Acids Res.*, 16, 7351, 1988.
10. **Ho, S. N., Hunt, H. D., Horton, R. M., Pullen, J. K., and Pease, L. R.**, Site-directed mutagenesis by overlap extension using the polymerase chain reaction, *Gene*, 77, 51, 1989.
11. **Mikaelian, I. and Sergeant, A.**, A general and fast method to generate multiple site directed mutations, *Nucleic Acids Res.*, 20, 376, 1992.

12. **Ito, W., Ishiguro, H., and Kurosawa, Y.,** A general method for introducing a series of mutations into cloned DNA using the polymerase chain reaction, *Gene,* 102, 67, 1991.
13. **Stappert, J., Wirsching, J., and Kemler, R.,** A PCR method for introducing mutations into cloned DNA by joining internal primer to a tagged flanking primer, *Nucleic Acids Res.,* 20, 624, 1992.
14. **Kunke, T. A.,** Rapid and efficient site-specific mutagenesis without phenotype selection, *Proc. Natl. Acad. Sci. U.S.A.,* 82, 488, 1985.
15. **Perlack, F. J.,** Single step large scale site-directed *in vitro* mutagenesis using multiple oligonucleotides, *Nucleic Acids Res.,* 18, 7457, 1990.
16. **Maniatis, T.,** *Molecular Cloning: A Laboratory Manual,* 2nd ed., Cold Spring Harbor Laboratory Press, Cold Spring Harbor, NY, 1989.
17. **Ansubel, F. M.,** *Current Protocols in Molecular Biology,* John Wiley & Sons, New York.
18. **Güssow, D. and Clackson, T.,** Direct clone characterization from plaques and colonies by the polymerase chain reaction, *Nucleic Acids Res.,* 10, 4000, 1989.

Chapter 10

MUTAGENESIS BY PCR

Bernard Y. Tao and K. C. Patrick Lee

TABLE OF CONTENTS

I. Introduction 69
   A. PCR 69
   B. Mutagenesis by PCR 70

II. Primer-Template Mismatch 70
   A. Homology Requirements 70
   B. Introduction of Mutations 71

III. Linear DNA Mutagenesis 71
   A. Overlap Extension PCR 71
   B. Modified Restriction Site PCR 72
   C. New Restriction Site PCR 73
   D. Megaprimer PCR 74

IV. Circular DNA Mutagenesis 75
   A. Inverted PCR 76
   B. Recombinant Circle PCR 76
   C. Recombination PCR 76
   D. Marker-Coupled PCR 78

V. Other PCR-Based Mutation Methods 79
   A. Non-Template Gene Synthesis 79
   B. Deoxyuracil DNA PCR 79

VI. Summary 80

References 81

I. INTRODUCTION

Advances in recombinant genetic engineering tools, such as polymerase chain reaction (PCR) technology, have dramatically facilitated the process of exploring the structural-functional relationships of proteins. Using PCR to create precise mutations in DNA has provided a means of creating site-directed mutations (SDM) in a variety of structural, catalytic, and physiological proteins. Many innovative SDM techniques have been developed using PCR in a variety of conditions and situations.[1] This chapter reviews contemporary methods of SDM-PCR and discusses their various advantages and limitations.

A. PCR

The polymerase chain reaction was first reported by Saiki and co-workers.[2,3] By annealing synthetic DNA oliogmers (primers) to complementary sections of the gene of interest

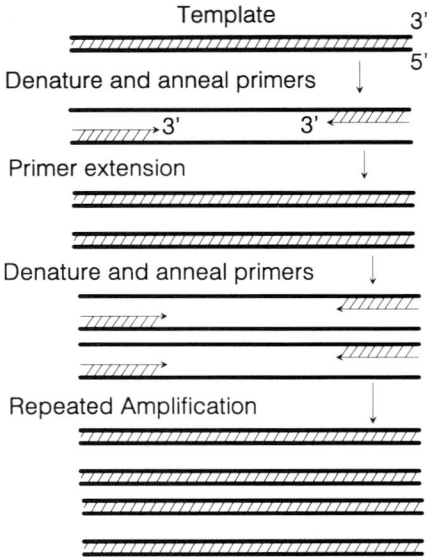

Figure 1. PCR amplification of DNA. (Adapted from Reikofski, J. and Tao, B. Y., *Biotech. Adv.*, 10, 535, 1982. With permission.)

(template DNA), polymerase enzymes (e.g., *Taq* DNA polymerase) can extend the primers to duplicate the intervening template DNA sequence. Using both the original template DNA and the newly created strands as templates, repeated priming and extension amplifies the desired gene (see Figure 1). Since duplicated DNA strands serve as templates, after several rounds of amplification only the DNA between the original primers will be obtained. Therefore, PCR allows both isolation and amplification of specific DNA sequences.

B. MUTAGENESIS BY PCR

The fundamental basis for all PCR mutation is primer mismatching. PCR primers must be sufficiently long sequences to allow unique binding to desired template locations. If long primers in excess of the length needed to locate and bind are used, it is possible to introduce new base pairs into the primer. This process is called mismatching (see Figure 2).

II. PRIMER-TEMPLATE MISMATCH

A. HOMOLOGY REQUIREMENTS

Homology requirements for PCR primer-template mismatching has been a subject of significant attention. Studies by Sommer and Tautz[4] have suggested that primers of 17 to 20 bp in length require at least three homologous base pairs at their 3' end for successful priming. The remainder of the primer needed very little homology with the template. Kwok et al.[5] found that simple internal mismatches had no significant effect on PCR product yield. However, mismatches at the 3' terminus have varying significant effects. A 100-fold decrease in PCR product yield occurred with A:G, G:A, and C:C mismatches, and a 20-fold decrease occurred with an A:A mismatch. They also found that mismatches of T with either G, C, or T had a minimal effect on PCR product yield. Double mismatches within the last 4 bases of a primer-template duplex at the 3' end also reduced PCR product yield dramatically. However, the presence of a mismatched T at the 3' end allowed significant amplification even when coupled with an adjacent mismatch. Nassal et al.[6] also reported that the nature of the mismatch affected the efficiency of amplification, with the efficiency of C:G = T:G > G:G > A:G for 3' terminal mismatches.

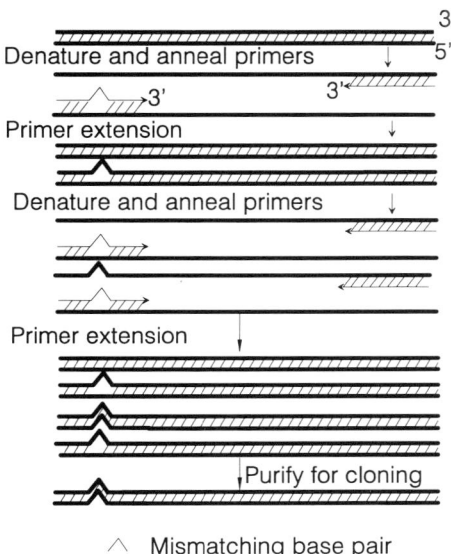

Figure 2. Terminal end PCR mutation. (Adapted from Reikofski, J. and Tao, B. Y., *Biotech. Adv.*, 10, 535, 1992. With permission.)

B. INTRODUCTION OF MUTATIONS

By designing primers that have mismatched sequences to the template DNA, precise mutations can be introduced through the process of PCR amplification. Most early PCR work involved primers flanking the gene of interest, so the location of mutagenesis was initially limited to the terminal ends of the amplified DNA. One of the most useful functions of terminal mutations has been to introduce new restriction sites to simplify ligation and cloning of amplified products. This terminal mutational technique has been used by several researchers to introduce mutations into various proteins.[7-16] However, SDM-PCR is not limited to terminal mutations. While amplification of the entire gene of interest by PCR requires flanking primers, the use of internal mismatch primers can create internal sequence mutations. A number of methods employing SDM-PCR for internal mutation sites have been devised involving both linear and circular DNA.

III. LINEAR DNA MUTAGENESIS

Simple denaturation and control of DNA fragment size are important advantages of using linear DNA templates for PCR reactions. These properties allow for quick, repeated PCR cycles and ease of purification by size, both important processes in SDM-PCR methods. Most linear DNA SDM-PCR procedures are relatively simple and have very high mutational efficiencies. The major limitation of linear DNA SDM-PCR methods is the need to re-ligate/clone the amplified mutants into suitable vectors prior to transformation and expression.

A. OVERLAP EXTENSION PCR

Overlap extension PCR (OE-PCR) utilizes four primers in two sequential PCR reactions to introduce an internal mutation into the target sequence.[17,18] Two separate PCR reactions are run, using primers 1 and 3 and primers 2 and 4 to amplify separate, overlapping sequences of the template DNA (see Figure 3). Primers 2 and 3 have overlapping homologous regions containing the mutation of interest. Following amplification, the PCR product DNA is denatured and re-annealed. Overlapped duplex formations can occur and be extended by DNA

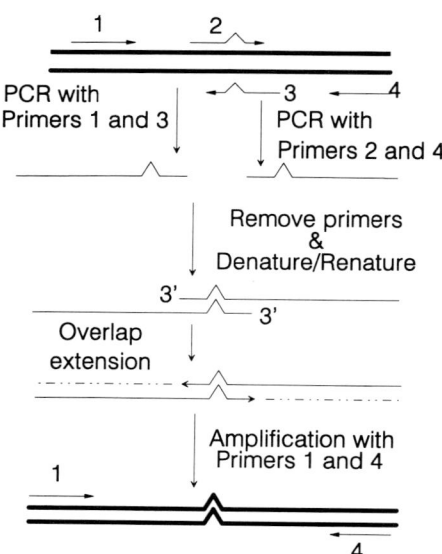

Figure 3. Overlap extension PCR mutation. Primers 2 and 3 contain the desired mutation. (Adapted from Reikofski, J. and Tao, B. Y., *Biotech. Adv.*, 10, 535, 1992. With permission.)

polymerase to give the full-length target sequence containing the mutation. This mutation can be further amplified by PCR using primers 1 and 4, the outer, conserved primers. Modified OE-PCR methods have been used to create fusion genes,[19-23] insert/replace protein domains,[24,25] and create deletion mutations.[26]

Although conceptually appealing, OE-PCR is relatively inefficient at producing mutations. In most cases, the overlap sequences are relatively short, resulting in very few duplexes of the desired type being formed.[27] Therefore, yields of mutations are low and must be selectively identified from a mixture of products. Modified OE-PCR methods have been developed to increase the efficiency, such as exonuclease digestion of PCR products[27] and the annealing of restriction fragments to PCR products.[28] For each new mutation two new primers are needed (increased cost) and primer removal is required between steps. Biotinylated universal primers and streptavidin-coated magnetic beads have also been used to selectively purify mutant PCR products.[29]

B. MODIFIED RESTRICTION SITE PCR

Modified restriction site PCR (MR-PCR) uses terminal restriction sites to increase the efficiency of OE-PCR type mutations. The DNA of interest is inserted into a polylinker site that provides unique restriction sites on both ends of the template DNA (see Figure 4). Primer 2 covers one of these sites and contains a mutation that inactivates this restriction site. Primers 1 and 3 flank the DNA insert, with primer 3 containing the other restriction site. Primer 4 contains the desired mutation. Primers 1 and 4 and primers 2 and 3 are used to create two DNA fragments with extensive overlapping regions. The fragments are combined, denatured, reannealed, and amplified using primers 1 and 3. Restriction enzymes for the external restriction sites are added to the PCR product to create DNA with sticky ends. Since only DNA segments with both functional restriction sites contain the mutation, appropriate re-cloning ligations to vectors should select for the mutation.

This method is highly efficient[30] and is advantageous over the overlap extension method for several reasons. Since the overlap region of the two DNA fragments is relatively long, a greater number of duplexes are formed by this method. Direct sticky-end ligation into the final vector as a selection tool is an effective selective technique. Another advantage of this method

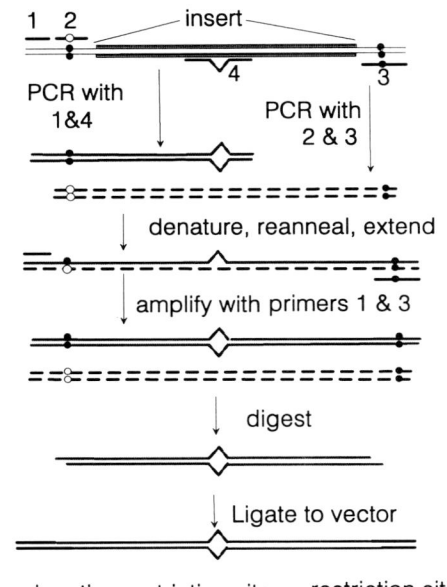

Figure 4. Modified restriction site PCR mutation. Primer 4 contains the desired mutation; primer 2 contains an inactivation mutation for the corresponding restriction site. (Adapted from Jones, D. H. and Winistorfer, S. C., *Methods: Companion to Methods Enzymol.*, With permission.)

is that subsequent mutations require only one additional primer to be produced (primer 4) while primers 1, 2, and 3 are conserved for all reactions. The main disadvantages are that the gene must have been previously cloned and that two unique restriction sites must exist flanking the gene.

C. NEW RESTRICTION SITE PCR

New restriction site PCR (NR-PCR) inserts restriction site sequences into the target template, allowing mutations to be introduced by traditional ligation techniques.[31-35] Four primers are used in this technique: two terminal conserved primers (primers 1 and 4) and two mutant primers each containing the restriction site mutation (primers 2 and 3) (see Figure 5). Primers 1 and 3 and primers 2 and 4 are used in separate PCR reactions to create overlapping sequences, each containing the restriction site mutation. Following digestion of the amplification products with the appropriate restriction endonuclease and removal of small restriction fragments, the target gene fragments are ligated to form the complete target sequence. By repeating this process, new restriction sites can be introduced that flank the region where mutations are desired. Mutation then becomes the relatively simple process of cassette-type mutagenesis, with insertion and removal of mutant sequences via restriction/ligation enzymatic reactions.

To eliminate the problem of introducing new restriction sites internally to genes, Tomic et al.[32] designed mutagenic primers that contained Bsp MI sites that would be eliminated with digestion of Bsp MI. This restriction enzyme cut left a 4-bp 3' overhang. The primers were designed so that after digestion these cohesive ends were complementary to each other and corresponded to the original template DNA. NR-PCR is a highly efficient process and has an advantage over both previously described methods in that after introduction of the restriction sites, mutations can be introduced by simple restriction/ligation methods. The major limitation of NR-PCR is that mutations are restricted to sequences where unique new restriction sites can be introduced without disrupting the original gene sequence (e.g., via codon wobble).

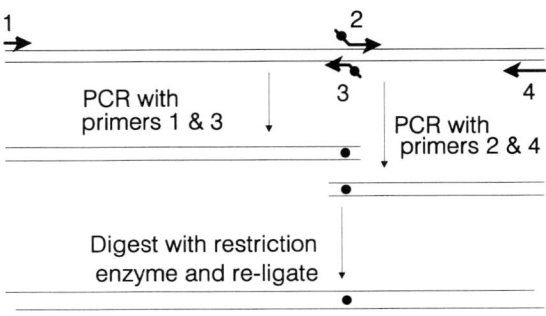

Figure 5. Introduction of a new restriction site by PCR mutation. Primers 2 and 3 contain the restriction site mutation. (Adapted from Reikofski, J. and Tao, B. Y., *Biotech. Adv.*, 10, 535, 1992. With permission.)

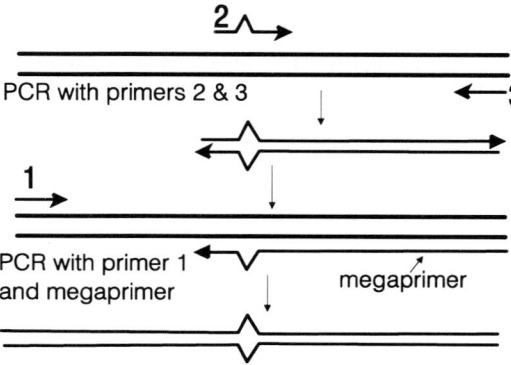

Figure 6. Megaprimer PCR mutation. Primer 2 contains the desired mutation. (Adapted from Reikofski, J. and Tao, B. Y., *Biotech. Adv.*, 10, 535, 1992. With permission.)

D. MEGAPRIMER PCR

The megaprimer method uses only three primers in two rounds of PCR.[36-38] The first round uses primers 2 and 3 to amplify a short segment of the DNA, with primer 2 containing the desired mutation (see Figure 6). This PCR product is then purified to remove the original primers[39,40] and used as a "megaprimer" for the second round of PCR in conjunction with primer 1 to produce the desired mutation. The simplicity and high mutational efficiency of this method make it a very widely used SDM technique.

In addition to its relative simplicity, the megaprimer method has several appealing advantages. Due to the size of the megaprimer, with appropriate selection of primer to template DNA ratios the efficiency of megaprimer SDM can approach 100%.[41] Polymerase fidelity is less likely to be a problem than with OE-PCR or MR-PCR, since the first PCR reaction usually amplifies a relatively short segment of the total sequence. If additional mutations are needed, only one new primer is required per mutation, since the flanking primers are conserved.

Variations of this method have been used with a single PCR reaction,[42,43] single-stranded DNA,[44] and multiple, non-adjacent mutations.[45] Production efficiency of difficult mutations can also be increased by combining OE-PCR and megaprimer methods.[46]

A major constraint of this method is the non-specific, non-template addition of nucleotides at the 3' terminus of the megaprimer DNA by *Taq* DNA polymerase.[47] Use of megaprimers

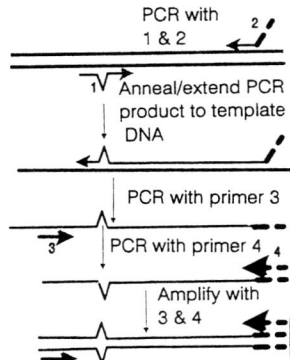

Figure 7. Non-template hybrid PCR mutation. Primer 1 contains the desired mutation; primers 2 and 4 contain complementary non-template sequences for selective binding and amplification. (Adapted from Reikofski, J. and Tao, B. Y., *Biotech. Adv.*, 10, 535, 1992. With permission.)

containing added nucleotides generates unwanted mutations in the final product gene. Several solutions have been suggested. First, sequencing has shown that most of the additions were single adenine residues. Designing the internal mutant primer (primer 2) so that the first 5′ nucleotide of the primer follows a thymine residue in the same strand of the template sequence can obviate this error. When this was done, 100% of the clones sequenced were error free.[41,48] Although this method is applicable in many cases, it may be difficult to find a T residue if a mutation involves a G-C-rich stretch of DNA. In this case, it has been suggested that the primer be designed so that the base immediately upstream of the primer is the third base of a degenerate codon.[41,49] Thus, even if a mutation occurs in the genetic sequence, the amino acid sequence will not vary. This method also approaches 100% mutational efficiency.

A modification of the megaprimer method involving four primers, including a non-template hybrid primer, can be used to increase mutational efficiency (see Figure 7). Primer 1 contains the desired mutation, and primer 3 is the outer primer used for amplification of the entire segment. Primer 2 is a long hybrid primer, with its 3′ sequence complementary to the template and its 5′ sequence non-complementary. Primer 4 is identical to the non-complementary 5′ sequence of primer 2, which is the key to this method. The first PCR amplification is done with primers 1 and 2, producing a megaprimer. This product is then annealed to template DNA and extended by DNA polymerase. Primers 3 and 4 are then added for a second PCR amplification. Since primer 4 is identical to the non-template portion of primer 2, it will only bind to and amplify the mutant strands produced in the first PCR reaction, thus increasing the efficiency of this method.[38]

There are several advantages to this method. Since the mutant is amplified exclusively, the purification steps can be simplified.[38] Also, as with the megaprimer method, only one new primer is needed for each additional mutation, and polymerization errors are reduced since only a short sequence is amplified in the first PCR, not the entire segment. The disadvantages of this method are the same as the megaprimer method. Additionally, the 5′ portion of primers 2 and 4 must be carefully designed so as not to be complementary to either strand anywhere in the gene sequence.

IV. CIRCULAR DNA MUTAGENESIS

SDM-PCR methods have also been developed to mutate genes previously cloned into circular vectors. The use of cloned genes obviates non-template addition problems inherent in the use of Taq DNA polymerase[47] and simplifies ligation/transformation procedures. Since both the gene and vector DNA are amplified, such products can be directly transformed into host cells for expression.

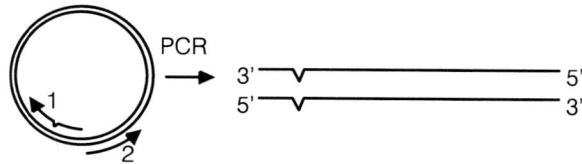

Figure 8. Inverted/counter PCR mutation. Primer 1 contains the desired mutation. (Adapted from Reikofski, J. and Tao, B. Y., *Biotech. Adv.*, 10, 535, 1992. With permission.)

A. INVERTED PCR

Inverted, or "counter", PCR (IPCR) involves using back-to-back primers, one of which contains the desired mutation, to amplify circular DNA (see Figure 8).[50,51] The resulting linear mutation products are end polished with the Klenow fragment and phosphorylated prior to blunt-end ligation. IPCR has a good efficiency, with 82% overall efficiency,[52] and has been used to create fusion genes[53] and gene deletions by positioning a gap between primers.[54] Linear DNA can also be mutated using IPCR by concatamerizing PCR products via blunt-end ligation.[55] A major advantage of ICPR is that only a single SDM-PCR reaction is required.

IPCR techniques do have several limitations, however. First, segments and domains of the protein cannot be individually produced, since both the vector and gene must be completely reproduced. Because the entire vector and gene are amplified, PCR extension cycle times may need to be increased for maximum efficiency,[56] although standard PCR cycle times have been used with acceptable results.[57] The formation of circular mutants for transformation requires post-PCR ligation processing, which is tedious and can reduce efficiency. However, the introduction of restriction sites using IPCR can eliminate post-reaction processing.[58] Finally, since there are no conserved primers used in IPCR, each mutation requires two new primers to be synthesized.

B. RECOMBINANT CIRCLE PCR

Recombinant circle PCR (RCPCR) is a combination of OE-PCR and IPCR, using *in vivo* repair to form circular mutations. RCPCR employs four primers in two separate PCR reactions (see Figure 9). Primers 1 and 3 contain the desired mutation complementary to opposing template strands. Primer 2 is in a back-to-back position with primer 1, and primer 4 is back to back with primer 3. The products of these PCR reactions are purified, combined, denatured, and re-annealed to form gapped, mutant circular DNA (see Figure 9). Transfection into *Escherichia coli* and *in vivo* repair of these gapped recombinant circles produces the desired mutations without requiring *in vitro* Klenow fragment end polishing, restriction enzyme digestion, phosphorylation, or ligation. Gap sizes up to 19 bp have produced mutational efficiencies ranging between 83 and 100%.[59] Circles with gaps of over 1000 nucleotides have been produced with reduced efficiency.[60] Plasmids up to 6.1 kb in size have been used in this technique with success.[61] Simultaneous mutation at distal sites is also possible.[62] RCPCR eliminates most of the post-PCR processing disadvantages of IPCR, but requires twice as many PCR reactions and primers as IPCR.

C. RECOMBINATION PCR

Recombination PCR (RPCR) uses *in vivo* recombination to reduce the number of primers and steps required for creating circular DNA mutations. Using overlapping mutant primers, only one PCR reaction is needed to produce linear mutant DNA (see Figure 10). The linear PCR product is transfected directly into *E. coli*.[63] Since the DNA segment has homologous ends, recombination occurs *in vivo* to produce the mutated plasmid. With this method, only two primers are required, and the denaturation/re-annealing steps are not necessary. The need to form staggered ends for *in vitro* ligation is also eliminated.

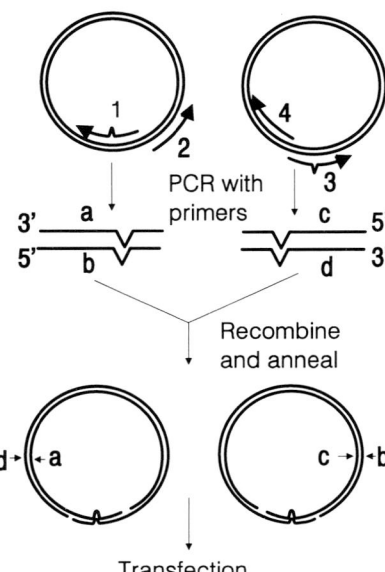

Figure 9. Recombinant circle PCR mutation. Primers 1 and 3 contain the desired mutation; a, b, c, and d indicate the corresponding PCR products in linear and circular configurations; gapped, circular duplexes are repaired *in vivo*. (Adapted from Jones, D. H. and Winistorfer, S. C., *Methods: Companion to Methods Enzymol.*, 2, 2, 1991. With permission.)

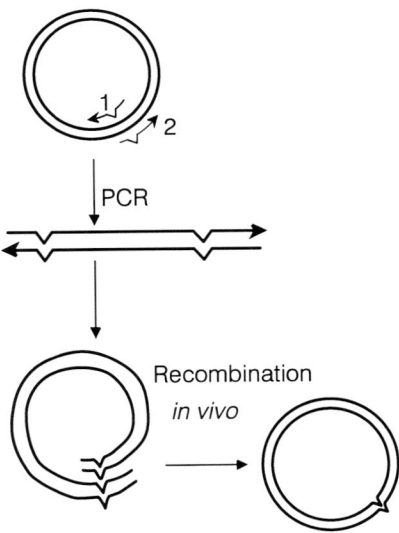

Figure 10. Recombination PCR mutation. Re-circularized duplex is ligated *in vivo* (Adapted from Reikofski, J. and Tao, B. Y., *Biotech. Adv.*, 10, 535, 1992. With permission.)

RPCR has several disadvantages, however. The overall efficiency is quite low; only about 50%, with ligation errors were found in approximately 25% of all clones.[59] There is also a lower yield of clones with RPCR than RCPCR. As with all methods that amplify the entire plasmid in PCR, only the entire protein can be produced, and new primers are required for each additional mutation.

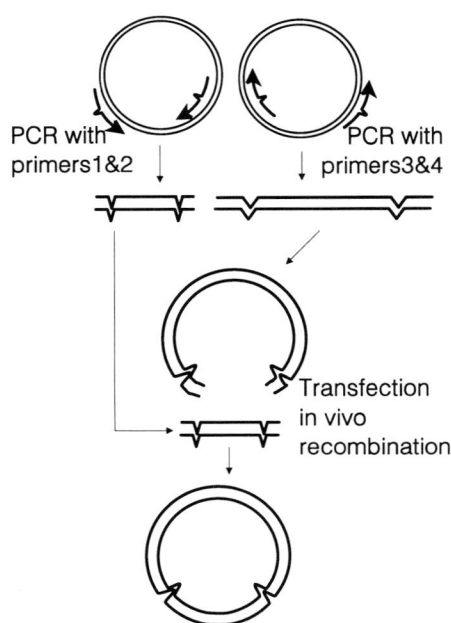

Figure 11. Simultaneous multiple mutation using recombination PCR mutation. All primers contain desired mutations; re-circularized duplexes are ligated *in vivo*. (Adapted from Reikofski, J. and Tao, B. Y., *Biotech. Adv.*, 10, 535, 1992. With permission.)

Figure 11 shows site-directed mutagenesis of two distal sites using RPCR (double mutation RPCR). The procedure is basically the same except that four primers are used and the plasmid is cut with a restriction enzyme before amplification. Two separate PCR amplifications are performed, the products are mixed, transfected into *E. coli*, and recombined *in vivo*. While suffering from the same limitations as single-mutation RPCR, this technique is particularly useful for the introduction of restriction sites for cassette mutagenesis.

D. MARKER-COUPLED PCR

Marker-coupled PCR (MC-PCR) uses a genetic marker activation mutation on a primer to re-activate a disabled marker gene, allowing preferential selection of mutations. Two mutant primers are needed, one that contains the desired mutation and another that restores function to an inactivated marker gene (such as *tet* in pUT 18[64]). Mutagenesis is accomplished in five steps (see Figure 12). In step 1, mutant and marker primers are used in PCR with target DNA previously inserted into the vector. The PCR product fragments contain both the mutation and repaired marker. In step 2, the vector is cut with different restriction enzymes to produce a nicked vector and a large fragment that does not contain either the mutation or marker sites. The nicked vector and large fragment are re-annealed. This large fragment is also used as a primer in later steps. Step 3 combines the PCR product with the annealed nicked fragment to form a gapped duplex plasmid/target DNA template, which is extended and ligated *in vitro* to form a closed, circular plasmid with one of the strands mutated. In step 4, this anyway-strand mutant is nicked and combined with the previous large restriction fragment to produce gapped duplexes, separating the mutant and non-mutant strands. In step 5, the gapped duplexes are transformed and repaired *in vivo*. Transformed cells are spread on plates screening for the marker. Since the marker and the mutant are closely coupled, the only cells that grow contain both the active marker and the desired mutation. Thus both mutation and selection are obtained

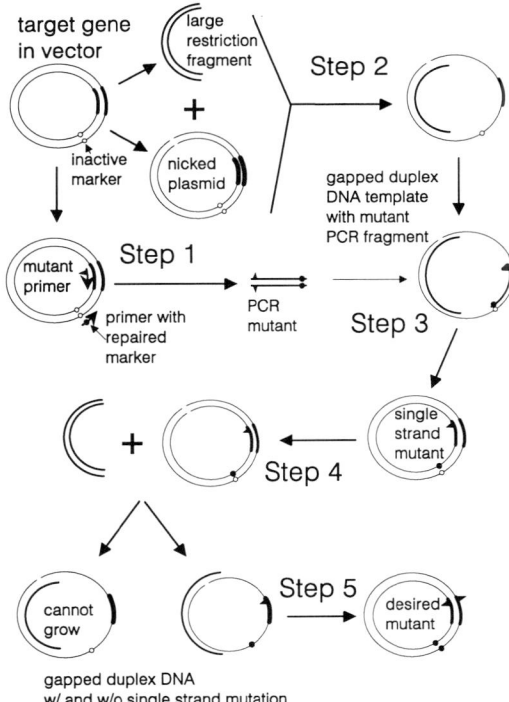

Figure 12. Marker-coupled PCR mutation. Step 1 — binding and amplification of desired mutant primer and repaired marker primer; Step 2 — binding of large restriction fragment with nicked DNA template; Step 3 — binding of mutant PCR product containing repaired marker to gapped partial duplex and extension and ligation to form circular duplex with single strand mutation; Step 4 — binding of large restriction fragment with nicked circular duplex; Step 5 — transformation and *in vivo* repair of mutant gapped duplexes in marker selective media. (Adapted from Shen, T.-J., Zhu, L.-Q., and Sun, X., *Gene*, 103, 1991. With permission.)

in a single process. The major disadvantages to MC-PCR are low efficiency and process complexity. While only mutant colonies are obtained by this technique, relatively few colonies are produced.[64] The obvious complexity of this method, involving several restriction cuts and repeated low-efficiency duplex formation, significantly decrease overall efficiency and make for very tedious work.

V. OTHER PCR-BASED MUTATION METHODS

A. NON-TEMPLATE GENE SYNTHESIS

For genes of known sequence, Majumder[65] describes the synthesis of gene mutations in the absence of template DNA. Overlapping synthetic oligonucleotides spanning the known gene sequences are used in repeated PCR reactions to synthesize the entire gene. The use of mutagenic primers results in SDM of the gene involved (see Figure 13). This method was termed "stepwise elongation of sequence PCR" (SES-PCR). While not practical for most mutagenesis studies because of the number of primers and PCR reactions, if the desired wildtype DNA is a relatively short sequence and high numbers of specific mutations are needed, this method may be advantageous. Note that the SES-PCR method is particularly sensitive to polymerase fidelity and terminal, non-template base pair additions, since it uses multiple internal primers/reactions.[47]

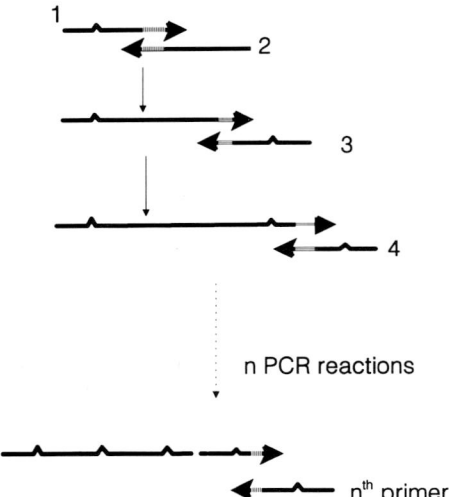

Figure 13. Non-template gene synthesis and mutation. The n multiple mutant primers spanning the entire gene sequence with overlapping complementary ends (hatched areas) are amplified in sequential order to obtain mutated gene.

B. DEOXYURACIL DNA PCR

Rashtchian and co-workers[66,67] have developed a novel PCR method involving the use of deoxyuracil residues, applicable to both IPCR and OE-PCR methods. Their idea was to use a primer in which the thymine residues in the 5' end were replaced with deoxyuracil. After the PCR amplification process, the PCR products were digested with uracil DNA glycosylase (UDG). This destabilized base pairing at the ends of the DNA molecules and generated 3' protruding ends in the opposite strand. Since the 3' protruding ends are complementary to each other, they can anneal with each other. This circularized DNA could be used to transform *E. coli* directly. This method can be used for the cloning of PCR products independent of the non-template addition of nucleotides by DNA polymerase.[47]

VI. SUMMARY

SDM-PCR techniques provide effective methods to precisely control changes of DNA. Relying on fundamental mismatching of base pairs between complementary synthetic primers and DNA templates, mutations can be introduced anywhere within either linear or circular DNA. Primer length, the number/position of mismatched base pairs, and primer homology are all critical factors in successful SDM-PCR.

Various SDM-PCR methods differ primarily by the number of primers needed, the number of PCR reactions needed, the need for duplex formation, the use of restriction endonucleases, and the need for product purification.

Linear DNA template-based methods are somewhat simpler to perform and have higher mutational efficiencies than whole plasmid methods. Most linear SDM-PCR techniques use an isolated gene as the DNA template, although simultaneous isolation and amplification can be performed. The advantages of linear methods are that the efficiency of mutation may be very high (> 90%), DNA segment length is precisely controlled, simple purification methods are available, and few by-products are created. One significant problem with linear methods is non-template terminal base pair addition during extension using *Taq* DNA polymerase.[47] These additions can cause difficulties with undesired internal mutations or problems with subsequent ligation of SDM-PCR products to vectors. Careful selection of primers that terminate next to an A residue can minimize the effects of this problem. While blunt end ligation may be used for inserting SDM-PCR mutations into host vectors, PCR ligation kits, such as Invitrogen's TA Cloning™ kit (Invitrogen, San Diego, CA), may be used with much higher efficiency. Linear methods may also require vector ligation, transformation, and selection procedures to obtain productive mutants. Linear SDM-PCR methods also permit the mutation of short sequences of DNA, which can be inserted into larger genes by restriction cloning methods. This is the basis for highly efficient techniques such as MR-PCR methods.

Terminal mutations can be easily produced by simple mismatch PCR amplification techniques with sufficiently long mutant primers. For internal mutations, the megaprimer method is widely used due to its simplicity, economy, and high efficiency. If multiple mutations are needed, the MR-PCR and NR-PCR can be very effective, since the introduction of endonuclease restriction sites within the gene flanking the desired mutation site, provides a rapid means to directly insert mutations at high efficiency. While this is a highly effective technique, it is dependent upon the existence of native sequences near the desired mutational site that can be easily modified to become restriction sites by using codon wobble.

Circular SDM-DNA mutation methods have the advantages of rapid transformation into host cells and *in vivo* circularization/repair. Problems with non-template terminal additions inherent in linear SDM-PCR methods are thus avoided. However, whole plasmid SDM-PCR techniques are significantly more complex than linear methods. Most circular methods are low-efficiency processes due to requirements of partial overlap duplex formation, amplification of both vector and gene, and post-amplification processing. Some methods involve extensive, multiple restriction cuttings, mutant selection, and *in vivo* gap repair. Since both the entire vector and gene must be amplified, polymerase fidelity is an issue if multiple PCR reactions are needed. Other commercial polymerases, such as thermostable VENT™ polymerases (New England Biolabs, Beverly, MA), have claimed higher levels of fidelity than *Taq* polymerase and may minimize this problem.

REFERENCES

1. **Reikofski, J. and Tao, B. Y.,** Polymerase chain reaction (PCR) techniques for site-directed mutagenesis, *Biotech. Adv.,* 10, 535, 1992.
2. **Saiki, R. K., Gelfand, D. H., Stoffel, S., Scharf, S. J., Higuchi, R., Horn, G. T., Mullis, K. B., and Erlich, H. A.,** Primer-directed enzymatic amplification of DNA with a thermostable DNA polymerase, *Science,* 239, 487, 1988.
3. **Saiki, R. K., Scharf, S., Faloona, F., Mullis, K. B., Horn, G. T., Erlich, H. A., and Arnheim, N.,** Enzymatic amplification of β-globin genomic sequences and restriction site analysis for diagnosis of sickle cell anemia, *Science,* 230, 1350, 1985.
4. **Sommer, R. and Tautz, D.,** Minimal homology requirements for PCR primers, *Nucleic Acids Res.,* 17, 6749, 1989.
5. **Kwok, S., Kellogg, D. E., McKinney, N., Spasic, D., Goda, L., Levenson, C., and Sninsky, J.,** Effects of primer-template mismatches on PCR: human immunodeficiency virus type 1 model studies, *Nucleic Acids Res.,* 18, 999, 1990.
6. **Nassal, M. and Rieger, A.,** PCR-based site-directed mutagenesis using primers with mismatch 3'-ends, *Nucleic Acids Res.,* 18, 3077, 1990.
7. **Kadowaki, H., Kadowaki, T., Wondisford, F. E., and Taylor, S. I.,** Use of polymerase chain reaction catalyzed by Taq DNA polymerase for site-specific mutagenesis, *Gene,* 76, 161, 1989.
8. **Friedman, K. J., Highsmith, W. E., and Silverman, L. M.,** Detecting multiple cystic fibrosis mutations by polymerase chain reaction-mediated site-directed mutagenesis, *Clin. Chem.,* 37, 753, 1991.
9. **Li, X. and Rhode, S. L.,** Mutation of lysine 405 to serine in the parvo virus H-1 NS1 abolishes its functions for viral DNA replication, late promoter trans-activation, and cytoxicity, *J. Virol.,* 64, 4654, 1990.
10. **Scharf, S. J., Horn, G. T., and Erlich, H. A.,** Direct cloning and sequence analysis of enzymatically amplified genomic sequences, *Science,* 233, 1076, 1986.
11. **Hoffman, L. M. and Hundt, H.,** Use of a gas chromatograph oven for DNA amplification by PCR, *BioTechniques,* 6, 932, 1988.
12. **Vallette, F., Mege, E., Reiss, A., and Adesnik, M.,** Construction of mutant and chimeric genes using the PCR, *Nucleic Acids Res.,* 17, 6545, 1989.

13. **Mole, S. E., Iggo, R. D., and Lane, D. P.,** Using the PCR to modify expression plasmids for epitope mapping, *Nucleic Acids Res.,* 17, 3319, 1989.
14. **Tannicha, E., Tummler, M., Arnold, H. H., and Lingelbach, K.,** Deletion mutagenesis in M13 by PCR using universal sequencing primers, *Anal. Biochem.,* 188, 255, 1990.
15. **Kammann, M., Laufs, J., Schel, J., and Gronenborn, B.,** Rapid insertional mutagenesis of DNA by PCR, *Nucleic Acids Res.,* 17, 5404, 1989.
16. **Higuchi, R., Krummel, B., and Saiki, R. K.,** A general method of *in vitro* preparation and specific mutagenesis of DNA fragments study of protein and DNA interactions, *Nucleic Acids Res.,* 16, 7351, 1988.
17. **Higuchi, R.,** in *PCR Protocols: A Guide to Methods and Applications,* Innis, M. A., Gelfand, D.H., Sninsky, J. J., and White, T. J., Eds., Academic Press, San Diego, CA, 1990, 177.
18. **Ho, S. N., Hunt, H. D., Horton, R. M., Pullen, J. K., and Pease, L. R.,** Site-directed mutagenesis by overlap extension using the PCR, *Gene,* 77, 51, 1989.
19. **Horton, R. M., Hunt, H. D., Ho, S. N., Pullen, J. K., and Pease, L. R.,** Engineering hybrid genes without the use of restriction enzymes: gene splicing by overlap extension, *Gene,* 77, 61, 1989.
20. **Horton, R. M., Cai, Z., Ho, S. N., and Pease, L. R.,** Gene splicing by overlap extension: tailor-made genes using PCR, *BioTechniques,* 8, 528, 1990.
21. **Horton, R. M. and Pease, L. R.,** Recombination and mutagenesis of DNA sequences using PCR, in *Directed Mutagenesis: Practical Approach,* Pherson, M. J., Ed., IRL Press, Oxford, 1991, 217.
22. **Yon, J. and Fried, M.,** Precise gene fusion by PCR, *Nucleic Acids Res.,* 17, 4895, 1989.
23. **Cao, Y.,** Direct cloning of a chimeric gene fused by the PCR, *Technique, J. Meth. Cell Molec. Biol.,* 2, 109, 1990.
24. **Clackson, T. and Winter, G.,** "Sticky feet"-directed mutagenesis and its application to swapping antibody domains, *Nucleic Acids Res.,* 17, 10163, 1989.
25. **Near, R. I.,** Gene conversion of immunoglobulin variable regions in mutagenesis cassettes by replacement PCR mutagenesis, *BioTechniques,* 12, 88, 1992.
26. **Kahn, S. M., Jiang, W., Borner, C., O'Driscoll, K., and Weinstein, I. B.,** Construction of defined deletion mutants by thermal cycled fusion: applications to protein kinase C, *Technique J. Meth. Cell Molec. Biol.,* 2, 27–30, 1990.
27. **Shyamala, V. and Ames, F.-L. G.,** Use of exonuclease for rapid polymerase-chain-reaction-based *in vitro* mutagenesis, *Gene,* 97, 1, 1991.
28. **Herlitze, S. and Koenen, M.,** A general and rapid mutagenesis method using PCR, *Gene,* 91, 143, 1990.
29. **Hall, L. and Emery, D.,** A rapid and efficient method for site-directed mutagenesis by PCR, using biotinylated universal primers and streptavidin-coated magnetic beads, *Protein Eng.,* 4, 601, 1991.
30. **Ito, W., Ishiguro, H., and Kurosawa, Y.,** A general method for introducing a series of mutations into cloned DNA using the polymerase chain reaction, *Gene,* 102, 67, 1991.
31. **Dulau, L., Cheyrou, A., and Aigle, M.,** Directed mutagenesis using PCR, *Nucleic Acids Res.,* 17, 2873, 1989.
32. **Tomic, M., Sunjevaric, I., Savtchenko, E. S., and Blumenberg, M.,** A rapid and simple method for introducing specific mutations into any position of DNA leaving all other positions unaltered, *Nucleic Acids Res.,* 18, 1656, 1990.
33. **Margis R., Viry, M., Pinck, M., and Pinck, L.,** Cloning and *in vitro* characterization of the grapevine fanleaf virus proteinase ciston, *Virology,* 185, 779, 1991.
34. **Yang, X.-J., Chen, C.-Q., and Wang, D.-B.,** An efficient site-directed mutagenesis using PCR, *Sci. China, Ser. B,* 34, 712, 1991.
35. **Caffrey, P., Green, B., Packman, L. C., Rawlings, B. J., Staunton, J., and Leadlay, P. F.,** An acyl carrier protein thioesterase domain from the 6-deoxyerythronolide β synthase of *Saccharopolyspora erythraea, Eur. J. Biochem.,* 195, 823, 1991.
36. **Sarkar, G. and Sommer, S. S.,** The "megaprimer" method of site-directed mutagenesis, *BioTechniques,* 8, 404, 1990.
37. **Landt, O., Grunet, H.-P., and Hahn, U.,** A general method for rapid site-directed mutagenesis using the polymerase chain reaction, *Gene,* 96, 125, 1990.
38. **Nelson, R. M. and Long, G. L.,** A general method of site-specific mutagenesis using a modification of the *Thermus aquaticus* PCR, *Anal. Biochem.,* 180, 147, 1989.
39. **Tautz, D. and Renz, M.,** An optimized freeze-squeeze method for the recovery of DNA fragments from agarose gels, *Anal. Biochem.,* 132, 1983.
40. GeneClean™ product, Bio101, La Jolla, CA.
41. **Bowman, S., Tischfield, J. A., and Stambrook, P. J.,** An efficient and simplified method for producing site-directed mutations by PCR, *Technique J. Meth. Cell Molec. Biol.,* 2, 254, 1990.
42. **Young-Sharp, D., Thomson, N., and Kumar, R.,** Site-directed mutagenesis using three primers and diagnostic RFLPs in a single-step PCR, *Technique J. Meth. Cell Molec. Biol.,* 2, 155, 1990.
43. **Kumar, R.,** Method for Site-Directed Mutagenesis. European Patent Appl. 91,111,360.3, 1991.

44. **Perrin, S. and Gilliland, G.**, Site-specific mutagenesis using asymmetric polymerase chain reaction and a single mutant primer, *Nucleic Acids Res.*, 18, 7433, 1990.
45. **Merino, E., Osuna, J., Bolivat, F., and Soberon, X.**, A general, PCR-based method for single or combinatorial oligonucleotide-directed mutagenesis on pUC/M13 vectors, *BioTechniques*, 12, 508, 1992.
46. **Aiyar, A. and Leis, J.**, Modification of the megaprimer method of PCR mutagenesis: improved amplification of the final product, *BioTechniques*, 14, 366, 1993.
47. **Clark, J. M.**, Novel non-templated nucleotide addition reactions catalyzed by procaryotic and eucaryotic DNA polymerases, *Nucleic Acids Res.*, 16, 9677, 1988.
48. **Kuipers, O. P., Boot, H. J., and DeVos, W. M.**, Improved site-directed mutagenesis method using PCR, *Nucleic Acids Res.*, 19, 4558, 1991.
49. **Sharrocks, A. D. and Shaw, P. E.**, Improved primer design for PCR-based, site-directed mutagenesis, *Nucleic Acids Res.*, 20, 1147, 1992.
50. **Ochman, H., Gerber, A. S., and Hartl, D. L.**, Genetic applications of inverse PCR, *Genetics*, 120, 621, 1988.
51. **Ochman, H., Ajioka, J. W., Garza, D., and Hartl, D. L.**, Inverse polymerase chain reaction, in *PCR Technology: Principles and Applications for DNA Amplification*, Erlich, H. A., Ed., Macmillan, New York, 1989, 105.
52. **Hemsley, A., Arnheim, N., Toney, M. D., Cortopassi, G., and Galas, D. J.**, A simple method for site-directed mutagenesis using the PCR, *Nucleic Acids Res.*, 17, 6545, 1989.
53. **Atreya, C. D., Atreya, P. L., and Pirone, T. P.**, Construction of in-frame chimeric plant viral genes by simplified PCR strategies, *Plant Mol. Biol.*, 19, 517, 1992.
54. **Imai, Y., Matshshima, Y., Sugimura, T., and Terada, M.**, A simple and rapid method for generating a deletion by PCR, *Nucleic Acids Res.*, 19, 2785, 1991.
55. **Heda, G. D., Henion, T. R., and Galili, U.**, A simple *in vitro* site-directed mutagenesis of concatamerized cDNA by inverse PCR, *Nucleic Acids Res.*, 20, 5241, 1992.
56. **Street, I. P., Coffman, H. R., and Poulter, C. D.**, Isopentenyl Diphosphate Isomerase. Site-directed mutagenesis of Cys139 using "Counter" PCR amplification of an expression plasmid, *Tetrahedron*, 47, 5919, 1991.
57. **Stemmer, W. P. and Morris, S. K.**, Enzymatic inverse PCR: restriction site independent, single-fragment method for high-efficiency, site-directed mutagenesis, *BioTechniques*, 13, 215, 1992.
58. **Stemmer, W. P., Morris, S. K., Kautzer, C. R., and Wilson, B. S.**, Increased antibody expression from *Escherichia coli* wobble-base library mutagenesis by enzymatic inverse PCR, *Gene*, 123, 1, 1993.
59. **Jones, D. H. and Howard, B. H.**, A rapid method for site-specific mutagenesis and directed subcloning by using the PCR to generate recombinant circles, *BioTechniques*, 8, 178, 1990.
60. **Jones, D. H. and Winistorfer, S. C.**, Recombinant circle PCR and recombination PCR for site-specific mutagenesis without PCR product purification, *BioTechniques*, 12, 528, 1992.
61. **Jones, D. H., Sakamoto, K., Vorce, R. L., and Bruce, B. H.**, DNA mutagenesis and recombination, *Nature*, 344, 793, 1990.
62. **Jones, D. H. and Winistorfer, S. C.**, Simultaneous site-specific mutagenesis of two distal sites by *in vivo* recombination of PCR products, *Technique J. Meth. Cell Molec. Biol.*, 2, 273, 1990.
63. **Jones, D. H. and Winistorfer, S. C.**, Site-specific mutagenesis and DNA recombination by using PCR to generate recombinant circles in vitro or by recombination of linear PCR products *in vivo*, *Methods: Companion to Methods Enzymol.*, 2, 2, 1991.
64. **Shen, T.-J., Zhu, L.-Q., and Sun, X.**, A marker-coupled method for site-directed mutagenesis, *Gene*, 103, 1991.
65. **Majumder, K.**, Ligation-free gene synthesis by PCR: synthesis and mutagenesis at multiple loci of a chimeric gene encoding OmpA signal peptide and hirudin, *Gene*, 110, 89–94, 1992.
66. **Rashtchian, A., Thornton, C. G., and Heidecker, G.**, A novel method for site-directed mutagenesis using PCR and uracil DNA glycosylase, *PCR Methods Appl.*, 2, 124, 1992.
67. **Rashtchian, A., Buchman, G. W., Schuster, D. M., and Berninger, M. S.**, Uracil DNA glycosylase-mediated cloning of polymerase chain reaction-amplified DNA: application to genomic and cDNA cloning, *Anal. Biochem.*, 206, 1992.

Chapter 11

DIRECT AUTOMATED DNA SEQUENCING OF DS AND SS PCR PRODUCTS

Vikki M. Marshall and Andrew M. Lew

TABLE OF CONTENTS

I. Introduction 86

II. Materials and Methods 86
    A. Reagents 86
        1. Commercially Available Products 87
    B. PCR Amplification of AMA-1 87
        1. Sequence of PCR Primers 87
    C. Removal of Unincorporated PCR Primers and Nucleotides 87
        1. Size-Exclusion Chromatography (Method 1) 88
            a. Preparation of Sephacryl® S400 and Sepharose®
                CL-6B Spin Columns 88
            b. Use of Sephacryl® S400 and Sepharose CL-6B
                Spin Columns 89
        2. Electrophoretic Separation in Agarose and Subsequent
            Purification (Method 2) 89
            a. Purification on Glassmilk 90
            b. Freeze/Spin Method 90
        3. Selective Precipitation of ds DNA by PEG (Method 3) 91
        4. Selective Adsorption onto Glassmilk (Method 4) 91
    D. Preparation of Single-Stranded PCR Products for
        DNA Sequencing 91
        1. Digestion of ds PCR Product with λ Exonuclease
            (Method 1) 91
        2. Digestion of ds PCR Product with T7 Gene 6 Exonuclease
            (Method 2) 92
        3. Asymmetric PCR (Method 3) 93
    E. DNA Sequencing 93
        1. Automated DNA Sequencing 93
            a. Quantitation of PCR Template 93
            b. Quantitation of Sequencing Primer 93
            c. Sequencing Reaction 93
        2. Manual DNA Sequencing 94

III. Results and Discussion 94
    A. Relative Yield of PCR Products Using Four Different
        Purification Methods 95
    B. Generation of ss Templates 95
    C. Comparison of Sequence Quality Using Eight
        Template-Preparation Protocols 96

D. Importance of Template Quantitation for Automated
   DNA Sequencing 97
E. Purification of Sequencing Primer 98

IV. Summary 99

References 99

I. INTRODUCTION

A reliable and universal method to sequence PCR products directly is in great demand but has not been forthcoming.[1] Many investigators have resorted to cloning the PCR product into suitable sequencing vectors, but this procedure is labor intensive and time consuming. Another problem is that the cloning of PCR amplification products selects single molecules which may harbor nucleotide misincorporations. Several clones from a transfection experiment must therefore be sequenced in order to generate a consensus sequence, further increasing the labor and cost involved. Since PCR products containing misincorporated bases usually represent a small proportion of the total reaction, direct sequencing of the product "en masse" allows one to determine accurately the target sequence by "masking" the misincorporated nucleotides. However, direct sequencing of PCR templates is still regarded as technically difficult and unreliable. It requires, first, the removal of unincorporated primers and nucleotides. Of the several purification methods possible, none can be used to purify every PCR product due to variation in the size, yield, and specificity of the product.

This chapter will focus on a selection of current methods used to purify PCR products for DNA sequencing, suggesting two universally applicable strategies. We have optimized and streamlined most of the purification methods such that the number of manipulations required is kept to an absolute minimum and can be carried out very rapidly. Some are amenable to automation using robotic pipettors.

It is possible to sequence double-stranded (ds) PCR products directly by cycle sequencing, also known as linear amplification sequencing, without the need for chemical denaturation of the DNA.[2] Cycle sequencing relies on the repeated thermal denaturation of the ds PCR product, and is a simple procedure requiring only small amounts of DNA. Alternatively, single-stranded (ss) PCR products can be directly sequenced following some form of PCR modification. They may be generated during the PCR reaction itself, as in asymmetric PCR[3,4] or in a post-amplification procedure, such as the hydrolysis of one or both DNA strands with a 5' exonuclease.[5-8]

We will describe in detail most of the methods available, and use them to sequence directly a PCR product amplified from a region of the gene encoding the Apical Membrane Antigen I (AMA-1) isolated from the malaria parasite *Plasmodium falciparum*.[9,10] The gene encoding AMA-1 varies amongst isolates of *P. falciparum* at single nucleotide positions. Therefore, accurate sequence identification is essential.

II. MATERIALS AND METHODS

A. REAGENTS
Phenol saturated with TE
Phenol mix: 68% phenol, 18% H_2O, and 14% chloroform
Absolute ethanol (analytical grade)
70% ethanol
80% isopropanol

10 M ammonium acetate
3 M sodium acetate
4 M LiCl
10 mM ATP
1x TE buffer (10 mM Tris-HCl; 1 mM EDTA, pH 8.0)
0.1 x TE buffer
10 X PCR buffer (100 mM Tris-HCl, pH 8.3; 500 mM KCl; 25 mM MgCl$_2$; 1 mg/ml gelatin)
PEG solution (20% polyethylene glycol$_{6000}$; 5 mM MgCl$_2$; 600 mM sodium acetate, pH 5.3)
10X λ Exonuclease Buffer (670 mM Glycine-KOH, pH 9.4; 25 mM MgCl$_2$; 500 mg/ml BSA)
10X T7 Gene 6 Exonuclease Buffer (500 mM Tris-HCl, pH 8.1; 50 mM MgCl$_2$; 200 mM KCl; 50 mM 2-mercaptoethanol)
10X Kinase Buffer (500 mM Tris-HCl, pH 7.5; 100 mM MgCl$_2$; 50 mM DTT)

1. Commercially Available Products

Sephacryl®* S400 (Pharmacia, #17-0609-01)
Sepharose® CL-6B (Pharmacia, #17-0160-01)
T4 polynucleotide kinase (Promega, #M4101)
λ Exonuclease (BRL, #8023SA)
T7 Gene 6 exonuclease (United States Biochemicals, #70025)
PRISM™** DyeDeoxy™** Terminator Cycle Sequencing Kit (Applied Biosystems, #401384)
β-Agarase I (BioLabs, #392S)
Magic™*** PCR Preps DNA Purification System (Promega, #A7170)
DNA ladder 50 to 2000 bp (Amersham, #T3405)
Geneclean™**** II DNA Purification Kit (Bio101)

B. PCR AMPLIFICATION OF AMA-1

A cocktail mix was prepared containing 200 µM each of dATP, dCTP, dGTP, and dTTP (Pharmacia, Uppsala, Sweden), 0.3 µM sense and antisense primers (A and B), 1X PCR Buffer, and 2.5 U *Taq* polymerase (Perkin Elmer, Melbourne, Australia). The cocktail mix was aliquoted into ten 100-µl volumes, and 0.1 mg of *P. falciparum* genomic DNA (strain NF7) was added to each tube. Reactions were subjected to 35 cycles at 94, 50, and 72°C for 60 s at each temperature in the robotic arm Gene Machine (Innovonics, Melbourne, Australia).

1. Sequence of PCR Primers

Primer A (sense): 5' TGG AAC TCA ATA TAG ACT TCC 3'
Primer B (antisense): 5' GGT CTA AAA CAA AAC ATG CTG 3'

Primers A and B correspond to positions 414 to 434 and 813 to 833 of the *P. falciparum* AMA-1 gene, respectively.

C. REMOVAL OF UNINCORPORATED PCR PRIMERS AND NUCLEOTIDES

Four rapid methods for the removal of unincorporated primers and deoxynucleotide triphosphates are outlined below. Each is based on an initial PCR reaction volume of 100 µl and assumes a yield of 50 to 100 ng/µl of specific product. Analyze 5% of the total PCR product by analytical agarose gel electrophoresis. If the predominant band is the desired

---

*   Sephacryl and Sepharose are registered trademarks of Pharmacia, Uppsala, Sweden.
**   PRISM and DyeDeoxy are trademarks of Applied Biosystems, Inc., Foster City, CA.
***   Magic is a trademark of the Promega Corporation, Madison, WI.
**** Geneclean is a registered trademark of Bio101, Inc., La Jolla, CA.

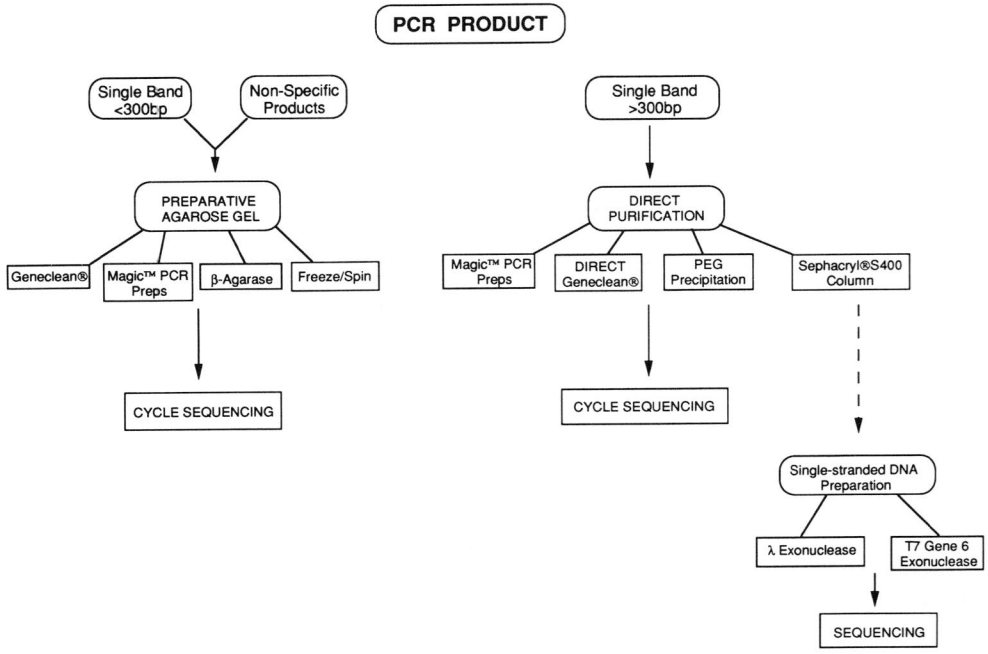

Figure 1. Flow chart summarizing strategies for direct sequencing of PCR products.

product, and there are minimal (<5%) non-specific amplification products, proceed to either method 1, 3, or 4. If background bands are clearly visible, they must be separated from the major PCR product by preparative agarose gel electrophoresis (method 2). If the level of non-specific amplification products is high, it is essential to re-optimize the PCR amplification conditions. Parameters for optimization are discussed in detail elsewhere.[11] An overview of the sequencing strategies to be described is given in Figure 1.

Methods for removing unincorporated primers and nucleotides are as follows:

1. Size exclusion chromatography (spin columns)
2. Electrophoretic separation (agarose gel electrophoresis)
3. Selective precipitation of ds DNA
4. Selective adsorption of ds DNA onto glassmilk

1. Size Exclusion Chromatography (Method 1)

Spin columns which exclude DNA larger than oligonucleotides and primer dimers may be used to purify PCR products. The PCR product should be essentially free of non-specific amplification products, and its size must exceed the exclusion limit of the column.

a. Preparation of Sephacryl® S400 and Sepharose® CL-6B Spin Columns

Sephacryl® S400 and Sepharose® CL-6B pre-made spin columns are available from several commercial sources, but the cost is prohibitive for routine use. The columns can be prepared easily in house at a small cost. Several (20 to 30) spin columns may be prepared simultaneously and stored at +4°C for months. The following protocol applies to Sephacryl® S400 spin columns. For Sepharose® CL-6B columns, the exclusion limit is 194 bp and the maximum speed is 1100 g. For Sephacryl® S400 columns, the exclusion limit is 271 bp and the maximum speed is 400 g.

Prepare the column as follows:

1. Discard plunger of a 1-ml disposable syringe and plug with a very small amount of sterile cotton wool. Place the syringe in a 10-ml centrifuge tube.
2. Fill the syringe to the 1-ml mark with Sephacryl® S400 slurry using a Pasteur pipette and spin to pack the matrix at a maximum of 400 g.
3. Refill the columns to 0.8 ml and wash once with 1X TE Buffer.
4. Store upright in a sealed container at +4°C, with the columns in contact with TE buffer to prevent dessication of the matrix. Use columns exactly as described below. It is very important to adhere to the recommended centrifugation speeds given above.

b. Use of Sephacryl® S400 and Sepharose® CL-6B Spin Columns

We generally achieve higher recoveries using Sephacryl® S400 columns than Sepharose® CL-6B. However, Sepharose® CL-6B columns must be used for purification of small PCR products between 194 and 280 bp. The higher exclusion limit of Sephacryl® S400 (271 bp) ensures that large primer-dimers and small non-specific amplification products are removed.

1. Wash the column 1 to 2 times with 1X TE or purified water and remove liquid from collection tube. Spin Sephacryl® S400 columns at 400 g (maximum) for 5 min in a benchtop centrifuge.
2. Re-centrifuge the column until the void volume has been completely removed.
3. Transfer the spin column to a fresh 10-ml tube and carefully apply 80 µl of the PCR reaction directly to the *center* of the column, i.e., apply 10 µl sample for every 0.1-ml packed column volume.
4. Spin the column(s) for 5 min at the recommended speed.
5. Collect the purified PCR product from the collection tube and transfer to a microfuge tube.
6. Electrophorese 5 µl of the initial PCR product against 5 µl of the purified product to check recovery. Include a DNA size standard of known concentration, e.g., the Amersham 50 to 2000-bp DNA ladder. The recovery should be at least 70%.
7. Estimate the concentration of the purified PCR product and adjust to approximately 0.1 µg/µl with H_2O. If necessary, the DNA may be concentrated by ethanol precipitation and resuspension in a smaller volume. Alternatively, reduce the volume by evaporation in a vacuum centrifuge.
8. Use a total of 0.3 to 0.5 µg directly in a cycle sequencing reaction (see below).

Alternatively, Centricon™* 100 columns may be used to remove residual PCR primers and nucleotides from the PCR product by ultrafiltration. However, the cost is prohibitive for routine use.

2. Electrophoretic Separation in Agarose and Subsequent Purification (Method 2)

Electrophoretic separation of the amplification products is essential if spurious or background bands make up a significant proportion (>5%) of the total amplification product. Electrophorese the PCR reaction on a preparative agarose gel as follows:

1. Remove as much oil overlay as possible from the PCR product.
2. Extract once with an equal volume of chloroform ($CHCl_3$).
3. Add 1/10 vol 3 *M* sodium acetate (NaAc), pH 5.3, and 2.5 vol ethanol (EtOH) to the reaction.

---

* Centricon is a trademark of Amicon®, W. R. Grace and Company, Danvers, MA.

4. Leave on wet ice for 10 min, then spin for 10 to 15 min to collect the DNA pellet.
5. Resuspend in gel loading buffer and electrophorese on a low-melting-point agarose gel of the appropriate percentage (depending on the size of the product). Excise the major band.

Note that gels prepared with NuSieve™* GTG agarose or combination NuSieve™/agarose gels may be used to resolve smaller PCR products. NuSieve™ is also of high purity.

The PCR product can be purified from agarose by one of several methods.

a. Purification on Glassmilk

Glassmilk purification kits may be obtained from several commercial suppliers. Three volumes of sodium iodide (NaI) or sodium perchlorate ($NaClO_4$) are used to solubilize the excised agarose block. Glassmilk is added, which binds DNA in the high-salt conditions provided by the NaI or $NaClO_4$. After washing the matrix in high-salt buffer, elute the DNA under low-salt conditions by adding purified water or TE. High DNA recoveries have been achieved using the Geneclean II® kit supplied by Bio101 (La Jolla, CA), where the purification is carried out in microcentrifuge tubes. More recently, excellent results have been achieved using the Magic™ PCR Prep DNA Purification System™ (Promega, Madison, WI). This kit utilizes 2.5-ml syringes as columns to wash the glassmilk matrix, minimizing tube handling. Several PCR templates (usually four) can be purified simultaneously within 15 min. Using a vacuum manifold for the column-wash steps, it is possible to purify up to 20 PCR templates simultaneously in approximately 20 min.

DNA should be eluted in the smallest possible volume (typically, 20 µl for the Geneclean II® Kit and 30 to 40 µl for the Magic™ PCR Prep Kit). Increased recoveries of DNA using the Geneclean II® Kit can be achieved using 10 µl of glassmilk matrix per purification instead of the recommended 1 µl/1 µg DNA. Note that DNA purified using the Magic™ PCR Prep Kit must be excised from a low-melting-point preparative agarose gel, while either low-gelling-temperature or normal agarose gels may be used with the Bio101 Geneclean II® Kit.

b. Freeze/Spin Method

An inexpensive method for recovering DNA from agarose gels is provided by a slight modification of the Freeze/Spin method.[12] This protocol utilizes readily available laboratory reagents and consumables and can be completed in approximately 30 min.

1. Place the excised agarose block in a blue tip plugged with a small amount of sterile cotton wool. Leave in dry ice for 10 to 15 min.
2. Cut the end off a microfuge tube and place the blue tip inside. Place this assembly in a second microfuge tube (with lid removed), and place the whole assembly in a 10-ml disposable centrifuge tube and spin at about 1000 g for 5 min in a benchtop centrifuge. The agarose matrix remains in the tip while the aqueous solution containing DNA passes into the collection tube.
3. Transfer the aqueous solution to a fresh eppendorf tube and add 1/10 vol 4-M LiCl. Extract with an equal volume of phenol saturated with TE and add 3 vol ethanol.
4. Leave at –70°C for 10 to 15 min; then collect DNA pellet by centrifugation at 13,000 g for 10 to 15 min.
5. Wash pellet once with 70% ethanol, air dry, and resuspend in 20 µl purified water or 0.1XTE.

---

* NuSieve is a registered trademark of FMC Bioproducts, Rockland, ME.

This method generally gives low yields relative to the other methods described, but results in very high-purity PCR templates.

A further alternative for generating PCR templates from agarose is to treat the excised band with β-Agarase I, following the manufacturer's instructions. β-Agarase I cleaves the carbohydrate bonds of agarose to produce neoagaro-oligosaccharides, releasing DNA into solution. These oligosaccharides do not appear to interfere with subsequent procedures.

3. Selective Precipitation of ds DNA by PEG (Method 3)

Double-stranded PCR products can be selectively precipitated from PCR primers and unincorporated dNTPs by the addition of polyethylene glycol (PEG):

1. Transfer the PCR reaction to a fresh Eppendorf tube and extract x1 with an equal volume of $CHCl_3$.
2. Add an equal volume of the 20% PEG solution to the PCR product. Leave at room temperature for 10 min.
3. Spin at 13,000 g for 10 to 15 min. Remove supernatant carefully as the DNA pellet may not be readily visible. Wash the pellet twice with 70% ethanol.
4. Air dry and resuspend in a small volume (15 to 20 µl) of purified water or 0.1X TE

PEG is inhibitory to many polymerases; therefore adequate washing of the DNA pellet is essential for optimal DNA sequencing. This protocol should precipitate ds PCR products greater than 100 bp, while primers, primer-dimers and unincorporated dNTPs remain in the supernatant.

4. Selective Adsorption onto Glassmilk (Method 4)

Purification can be achieved directly using glassmilk, without the need for electrophoretic separation if PCR amplification side-products represent <5% of the total amplification product. The glassmilk matrix binds small DNA and single-stranded DNA strongly but elutes them inefficiently. Both the Geneclean II® Kit and the Promega Magic™ PCR Prep Kit may be used for this purpose, following manufacturer's instructions. We have observed higher recoveries from PCR products purified using the Magic™ PCR purification kit than those prepared using the Geneclean II® kit, and the costs are comparable.

PCR templates purified by any of the above methods can be directly sequenced in a cycle-sequencing reaction using *Taq* DNA polymerase. Cycle sequencing is recommended due to its simplicity, the requirement for only small amounts of template, and the fact that no chemical denaturation of the double-stranded DNA is required. Cycle-sequencing protocols using fluorescently labeled nucleotides or primers for subsequent analysis on an automated DNA sequencer are now well established. However, analogous cycle-sequencing protocols for DNA sequencing with radioisotopes are not as well established. Several kits are commercially available, but they have the disadvantage that ^{32}P-labeled sequencing primers are required. Recently, a cycle-sequencing kit has been released that allows one to sequence PCR products using radiolabeled nucleotides such as $\alpha^{35}S$-dATP (Cyclist DNA Sequencing Kit, Stratagene, La Jolla, CA). This kit utilizes a 3'-5' exonuclease-deficient form of *Pfu* DNA polymerase which, unlike *Taq* polymerase, shows no discrimination against α-thionucleotides, and therefore incorporates radiolabeled nucleotides efficiently.

D. PREPARATION OF SINGLE-STRANDED PCR PRODUCTS FOR DNA SEQUENCING

1. Digestion of ds PCR Product with λ Exonuclease (Method 1)

λ Exonuclease is a highly processive 5' exonuclease that digests ds DNA in the 5'-3' direction. This enzyme digests ds DNA approximately ten times more rapidly from a phosphorylated 5' terminus than from a non-phosphorylated 5' terminus, leaving a 3'-5' single strand which is 90% full length.[5,6,8] This property can be exploited to obtain ss sequencing templates from PCR products.

One of the PCR primers is phosphorylated prior to PCR using T4 polynucleotide kinase, and the PCR reaction is set up in a standard manner. Unincorporated oligonucleotides and dNTPs are removed by one of the methods described above, and the product is digested with λ exonuclease and then sequenced.

1. Phosphorylate both PCR primers (in separate reactions) as follows:

    Set up in an Eppendorf tube:
    5 µl 10 x kinase buffer
    1 to 15 µg oligonucleotide
    5 µl 10-mM ATP
    20 U T4 polynucleotide kinase
    Make up to 50 µl with purified water.
    Incubate at 37°C for 30 to 60 min.

2. Inactivate the kinase at 70°C for 10 min.
3. If sequence from both DNA strands is required, set up two PCR reactions per sample, one with only the sense primer phosphorylated and the other with only the antisense primer phosphorylated. Use 10 to 30 pmol of each oligonucleotide directly without purification in a 100-µl PCR reaction.
4. Run 5% of the PCR product on a 1% agarose gel to check the yield and purity.
5. Remove the residual amplification primers by size exclusion chromatography. Other methods described in Section I.C may also be used. Keep 10% of the product aside as a control.
6. To the 80-µl PCR product recovered, add 10-µl 10X λ exonuclease buffer, 6 U λ exonuclease and purified H_2O to 100 µl.
7. Incubate at 37°C for 20 min and inactivate the λ exonuclease at 65°C for 10 min.
8. Run 5 µl undigested DNA against 6 µl of λ exonuclease digested DNA in a mini-sub agarose gel (10 µl well-loading capacity) and stain with ethidium bromide. A small amount of undigested ds DNA will be present even after λ exonuclease digestion. Since ss DNA binds ethidium bromide approximately ten times less efficiently than ds DNA, estimate the extent of digestion by comparing the intensity of the ds band in each sample pair. Typically, 90 to 95% of the PCR product is hydrolyzed to ss DNA.
9. Use approximately 0.2 µg ss (typically 5 µl) template directly in a sequencing reaction.

Note that the DNA may be concentrated by ammonium acetate/ethanol precipitation prior to sequencing if necessary. Always use as a sequencing primer the same oligonucleotide that was kinased in step 1, or an internal primer with sequence derived from the same strand.

2. Digestion of ds PCR Product with T7 Gene 6 Exonuclease (Method 2)

T7 gene 6 exonuclease is another 5' exonuclease, which, unlike λ exonuclease, hydrolyzes ds DNA in a stepwise non-processive reaction, at an equal rate from both 5'-hydroxyl and 5'-phosphoryl termini.[7,13,14] Digestion from both ends of the ds PCR product continues until both strands have been hydrolyzed to approximately 50% of their full length, at which point they essentially fall apart. This generates two single-stranded templates, one from the sense and the other from the antisense strand.

1. Remove excess PCR primers and dNTPs using spin column chromatography or one of the methods described in Section I.C, depending on PCR product size and specificity.

2. If spin columns are used, DNA is recovered in a volume of 80 µl. Set aside 10% of the purified product as a control. Add 10 µl 10X T7 Gene 6 exonuclease Buffer and 120 U of T7 Gene 6 exonuclease. Add H_2O to a final volume of 100 µl.
3. Incubate at 37°C for 30 min.
4. Inactivate the enzyme at 65°C for 10 to 15 min.
5. Use 2 to 3 µl directly in a sequencing reaction, depending on yield (see discussion in Section III).

The extent of digestion of ds DNA with T7 Gene 6 exonuclease may be estimated by checking an aliquot of the DNA against undigested material. The ss template may be concentrated by ammonium acetate precipitation prior to DNA sequencing.

3. Asymmetric PCR (Method 3)

Asymmetric PCR is discussed in detail elsewhere[3,4] and will not be included here. While excellent sequencing results can be obtained using this technique, in our hands it requires optimization with each new primer pair, and is critically dependent on the accurate quantitation of PCR primers, which in turn is dependent on oligonucleotide purity. Such factors demand extra work, which is not necessary given the success of other techniques available.

E. DNA SEQUENCING

1. Automated DNA Sequencing

AMA-1 PCR products purified using each of the methods described above were cycle sequenced using fluorescently labeled dideoxynucleotide triphosphates and *Taq* polymerase (PRISM™ Ready Reaction DyeDeoxy™ Terminator Cycle sequencing kit, Applied Biosystems).

Cycle sequencing was carried out on the Perkin Elmer 9600 Thermocycler. This type of thermocycler is highly recommended for cycle-sequencing reactions since it requires no oil overlay.

Note that ss PCR products do not need to be cycle sequenced, but can be used in a single annealing/extension-type sequencing reaction, with heat-labile DNA polymerases (e.g., Sequenase®) (see discussion in Section III).

a. Quantitation of PCR Template

Electrophorese 10 to 20% of purified ds PCR product on an analytical agarose gel along with DNA markers of known concentration, and estimate template concentration by comparison of relative fluoresence intensity after ethidium bromide staining. Adjust the template to approximately 0.1 µg/µl. We recommend the Amersham 50- to 2000-bp DNA ladder as a standard. Each band in this DNA ladder contains 10 ng of DNA per microliter of marker loaded.

b. Quantitation of Sequencing Primer

Calculate the primer concentration in pmol/µl using the formula pmol/µl = $61.63 \times [O.D._{260} (U/ml)]$/no. bases. Dilute to 1 pmol/µl.

c. Sequencing Reaction

1. In a PCR tube (eg) MicroAmp™* reaction tube for the Perkin Elmer 9600 thermal cycler, add:

---

* MicroAmp is a trademark of the Cetus Corporation, Norwalk, CT.

9.5 µl Dye-Deoxy Terminator Pre-mix
4 µl Primer A or B (1 pmol/µl)
x µl Purified PCR template (~0.2 to 0.3 µg)
x µl H_2O, to a final volume of 20 µl.

2. For the Perkin Elmer 9600, cycle under the following conditions:

96°C for 10 s
50°C for 5 s
60°C for 4 min, 25 cycles total
Rapid thermal ramp to 4°C and hold (total cycle time approximately 2.1 h)

Cycle-sequencing conditions need to be optimized for each different type of thermal cycler.

3. Add 80 µl H_2O to each 20-µl reaction, and transfer to an Eppendorf tube containing 100 µl phenol mix.
4. Vortex thoroughly (30 s each tube) and spin 2 to 3 min at 13,000 g in a microfuge. Remove the lower organic phase (which now appears pink) and discard. Repeat phenol extraction with another 100 µl phenol mix and spin. Thorough phenol extraction is essential for accurate sequence determination since residual dye-terminators co-migrate with the sequencing products.
5. Transfer the upper aqueous phase to a fresh tube and precipitate the sequencing reaction product by adding 10 µl 3M NaAc, pH 5.3, and 3 vols absolute ethanol. Leave on ice for 10 to 15 min; then spin 15 min in a microfuge.
6. Wash the pellet with 70% EtOH, dry in a vacuum centrifuge, and resuspend in 4 µl of 5:1 deionized formamide/50 mM EDTA mix, prior to heat denaturation at 95°C for 2 to 5 min. Snap-cool on ice. Load on to a pre-electrophoresed 6% acrylamide/8M urea sequencing gel on the automated DNA sequencer (refer to ABI Model 373A DNA Sequencer User's Manual).

2. Manual DNA Sequencing

ds PCR products may be sequenced by conventional radioisotopic DNA sequencing methods, using a single chemical and/or heat denaturation followed by extension with a heat labile polymerase, instead of the repeated thermal denaturation achieved in cycle sequencing.[15,16] While some success has been achieved using these protocols, re-association of the ds template strands during the sequencing reaction is often a problem. This association is favored over template-primer associations and can lead to background bands and inaccurate sequence determination. Cycle sequencing of ds PCR products using radiolabeled nucleotides is preferred for the generation of primary sequence data from ds PCR templates.

Alternatively, single-stranded PCR products generated using the methods described in Section I.D can be sequenced in the same manner as ssM13 templates in conventional DNA sequencing reactions. We have found the Sequenase® Kit (USB) gives reliable results using α^{35}S-dATP as the radiolabel. Detailed protocols on manual DNA sequencing are given elsewhere.[17]

III. RESULTS AND DISCUSSION

We have employed eight different PCR template-preparation methods to sequence directly a 419-bp PCR product amplified from the *P. falciparum* AMA-1 gene by automated DNA sequencing. All purification methods were successful except for PEG precipitation, in which the DNA recovery was inadequate for accurate sequence determination. A flow chart summarizing the strategies used is shown in Figure 1.

Initial analysis of the PCR product by analytical agarose gel electrophoresis is important in deciding the most appropriate method for purification. In addition, prior optimization of the PCR reaction components and cycling conditions is essential for each different template-primer pair, so that extensive purification can be avoided.[11]

Differences in the quality of sequence data generated using each of the ds purification methods were essentially negligible. Therefore, it is logical to choose the most rapid and inexpensive method of purification, and several examples have been presented in this chapter. We routinely purify ds PCR templates by size-exclusion chromatography or electrophoretic separation followed by glassmilk purification if there are non-specific amplification products. In this way, the same template preparation can be used for generating DNA sequence from both strands.

In this study a relatively short PCR fragment (400 bp) was used to demonstrate the usefulness of the various purification methods possible. However, we have amplified and sequenced products up to 900 bp and can routinely obtain 480-bp sequence from each strand. As much as 520 bp of sequence has been achieved from one PCR product using a single primer and the λ exonuclease preparation method. However, the most appropriate strategy is dependent on a number of factors, which will be discussed.

A. RELATIVE YIELD OF PCR PRODUCT USING FOUR DIFFERENT PURIFICATION METHODS

Agarose gel electrophoresis of a 20% fraction of the AMA-1 PCR purified using each of the methods described is shown in Figure 2. Recoveries vary significantly depending on the method used to purify the ds PCR product. As expected, lower yields were obtained by methods involving gel purification of the DNA. Of these, DNA purified by the Freeze/Spin method resulted in the lowest overall recovery (approximately 10%). However, sequence data generated from DNA purified by this purification method was superior to that generated after purification by all other methods. The highest yield (~1 µg) was obtained by direct purification using the Magic™ PCR preps.

The most appropriate purification method is largely dependent on the yield and specificity of the PCR product, since each purification protocol has its own set of constraints. For instance, direct purification without agarose gel electrophoresis requires that the PCR product be >95% free of non-specific amplification products. Using size-exclusion chromatography with Sephacryl® S400 columns, the PCR product must also be >271 bp (the exclusion limit of the matrix). Similarly, PCR products 700 to 800 bp long can be PEG precipitated efficiently with typical yields in the range of 70 to 80%, whereas we have recently experienced very poor recoveries (10% or less) with PCR products <400 bp long. Recoveries from glassmilk preparations such as Geneclean® or Magic PCR™ Resin decrease considerably with DNA less than 200 bp.

B. GENERATION OF SS TEMPLATES

The λ exonuclease method produces ss templates that are 90% full length and provide excellent sequence data in both automated and manual DNA sequencing systems. However, this method has the disadvantage that PCR primers must be kinased prior to the PCR reaction, and two separate PCR reactions must be set up in order to obtain DNA sequence from both strands, effectively doubling the work required. If cycle sequencing is employed, it is more convenient to simply purify the ds PCR template and carry out a cycle-sequencing reaction. If a non-heat stable sequencing enzyme such as Sequenase® is used, however, generation of ss PCR templates using λ exonuclease is the method of choice.

The preparation of ss templates using T7 gene 6 exonuclease results in ss DNA that is 50% the length of the ds starting DNA. Therefore, this method should be used only in the preparation of larger PCR templates >600 bp for sequencing, so that a reasonable length of the PCR product can be sequenced from the 300-bp single strands generated. Further, by using

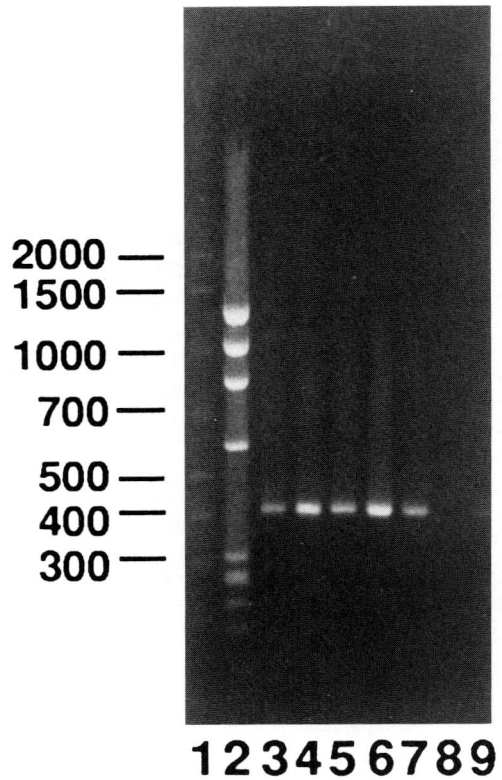

Figure 2. Typical relative yield of AMA-1 PCR product following different methods of purification. PCR products were electrophoresed in a 1.2% agarose gel and stained with ethidium bromide. Each track contains 20% v/v of the total purified material, except for track 2, which contains 10%. Lane 1, Amersham DNA ladder (50 ng of DNA per band); lane 2, φX174 DNA digested with HaeIII; lane 3, unpurified PCR product; lane 4, Magic™ PCR Prep from agarose gel; lane 5, Genecleaned DNA from agarose gel; lane 6, Direct Magic™ PCR prep; lane 7, Sephacryl S400 column purification; lane 8, PEG precipitation; lane 9, Freeze/Spin method.

T7 gene 6 exonuclease it is not possible to sequence all of one single-strand to completion, since only half of any one DNA strand is intact. However, T7 gene 6 exonuclease has an advantage over the λ exonuclease method for generation of ss templates in that no pretreatment of PCR primers is required.

C. COMPARISON OF SEQUENCE QUALITY USING EIGHT TEMPLATE PREPARATION PROTOCOLS

Using each of the methods described it was possible to obtain 380 of a possible 391 bp between the PCR primers used to amplify *P. falciparum* AMA-1. Each protocol resulted in a sequence that was 100% accurate compared with the published AMA-1 sequence.

Sequence data generated from gel-purified PCR products contained slightly fewer ambiguities than those generated from directly purified PCR fragments. A portion of a typical sequence chromatogram obtained after cycle sequencing the gel purified AMA-1 PCR is compared with that obtained following direct purification by size-exclusion chromatography in Figure 3. Less background was observed after gel purification over the 1- to 18-bp range. A comparison of results obtained using all methods described is summarized in Table 1.

Each of the few sequencing errors (unresolvable ambiguities or miscalls) occurred very close to the sequencing primer or at nucleotides that co-migrate with unincorporated dye-

Direct Automated DNA Sequencing of ds and ss PCR Products

A

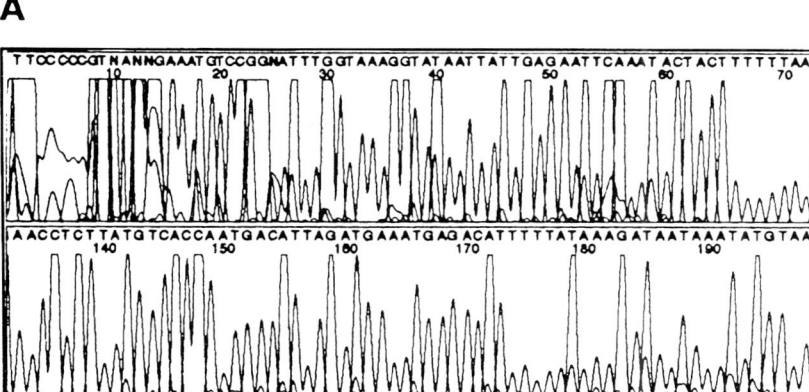

B

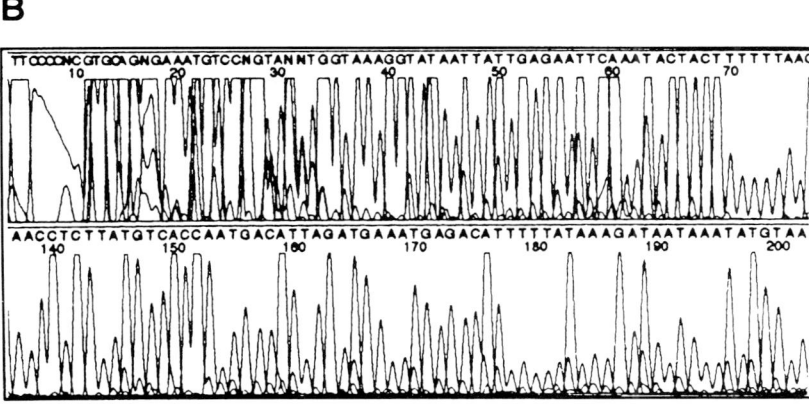

Figure 3. Sequence chromatograms obtained (A) following agarose gel electrophoresis, Magic™ PCR purification and cycle sequencing with dye-labeled terminators, and (B) following direct purification by size-exclusion chromatography. Only a portion of each chromatogram is shown for comparison.

deoxy terminators. If these data are excluded from calculations of percentage accuracy, the sequence readout is essentially identical using each of the protocols described. The comparison of template-preparation methods was repeated with AMA-1 PCR products amplified from two other strains of *P. falciparum*, D10 and HB3 (data not shown). Comparable results were obtained.

D. IMPORTANCE OF TEMPLATE QUANTITATION FOR AUTOMATED DNA SEQUENCING

In order to maximize the amount of sequence data generated in an automated DNA sequencing system, it is essential to conduct some sort of quantitation of the purified PCR template by comparing its relative fluorescence intensity following staining with ethidium bromide against DNA markers of known concentration. If insufficient template DNA is used in the sequencing reaction, the signal will be low, the background high, and the sequence readout inaccurate. If excess DNA template is used, reaction components are exhausted early. This causes high-intensity fluorescent peaks over the first 150 to 200 bp, followed by low signal with high background and inaccurate sequence after this point, effectively reducing the readable sequence length. In theory a total of ~0.3 pmol purified ds PCR template (~0.2 µg

TABLE 1
Comparison of DNA Sequence Accuracy Using Eight Template Preparation Methods

Method	Sequence Length (bp)	No. Ambiguities	No. Resolvable Ambiguities	Miscalls[b]	Quantity DNA used[c] (ng)	% Accuracy
Magic-Gel	381	3	3	1	160	99.7
Geneclean®	380	2	2	1	150	99.7
Freeze/Spin	381	6	6	1	120	99.7
Sephacryl® S400	382	7	6	1	250	99.5
Magic-Direct	379	11	7	1	120	98.6
T7 Gene 6[a]	175	2	2	3	200	98.2
λ Exo[a]	380	6	4	3	200	98.7
PEG Precipitation	—	—	—	—	—	—

Note: The table compares sequence accuracy achieved using each of the methods described. An ambiguity is a base that could not be accurately assigned by the Model 373A Data Analysis Software (Version 1.1.1, Applied Biosystems). A resolvable ambiguity is an ambiguity that could be assigned after manual inspection of the sequence chromatogram. A miscall is an incorrect basecall, insertion, or deletion in the sequence readout. Percent accuracy was calculated according to the formula:

$$\% \text{ Accuracy} = \frac{\text{Total sequence length (bp)} - \text{No. unresolvable ambiguities} - \text{No. miscalls}}{\text{total sequence length (bp)}} \times 100$$

[a] Single-stranded templates were generated.
[b] Miscalls occur mainly within 12 bp of the sequencing primer.
[c] Values given in this column are approximate, based on comparison of relative fluorescent intensity against a DNA marker of known concentration.

of a 1-kbp ds fragment) is required for dye-deoxy terminator cycle sequencing. However, the relationship between template quantity and fluoresence intensity upon automated sequencing is not always linear. Presumably, this is due to differences in the purity of the PCR templates. For instance, excellent results have been achieved using as little as 120 ng DNA prepared by the Freeze/Spin method, while larger quantities of DNA are required for equivalent results using PCR products purified by other methods. As a general rule we aim for 0.2 to 0.3 µg total DNA, but it may be necessary to determine the optimal template quantity in an empirical manner, depending on the method of purification used. Smaller amounts (0.1 pmol) are required for manual DNA sequencing, and accurate quantitation is generally not necessary.

E. PURIFICATION OF SEQUENCING PRIMER

PCR primer A was used in this study to prime the sequencing reactions. We have not found it necessary to synthesize internally nested oligonucleotides in order to obtain clean sequence data using the PCR purification methods described above. Further, it is possible to use relatively crude oligonucleotide preparations. After cleavage from the column support with concentrated ammonium hydroxide, oligonucleotides are deprotected by heating to 55°C for 8 h. Subsequently, they are dried by evaporation in a vacuum centrifuge, resuspended in 2 ml purified H_2O and quantitated by UV spectrophotometry at 260 nm. Generally, no further purification is necessary. Occasionally we have observed that the quality of sequence data achieved by dye-terminator automated sequencing can be somewhat "primer-specific". Two different primers used to sequence the same PCR template can vary greatly with respect to the quality of sequence data generated, even though they may have both been used successfully in PCR. The problem is sometimes rectified by some form of oligonucleotide purification. Several rapid and cost-effective purification methods are available, one involving a simple desalting step on a Bio-Gel P-6 column (BioRad, Richmond, CA) and the other, extraction with *n*-butanol.[18,19] Another alternative is to use oligonucleotide purification cartridges (OPC cartridges, Applied Biosystems, CA), but these require trityl-on synthesis and are expensive for routine use. Failed sequences due to problems with the oligonucleotide, however, are rare.

IV. SUMMARY

We have presented several methods for the direct automated DNA sequencing of PCR products. All protocols can be carried out using standard laboratory reagents and equipment, and are rapid and simple. None of the methods requires any special pre-treatment of PCR primers, except for the λ exonuclease method, which requires a simple phosphorylation reaction. The protocols presented can be used to accurately sequence PCR products by automated DNA sequencing using dye-labeled terminators, or using dye-labeled primers if a universal primer sequence is incorporated into the PCR oligonucleotide. They can also be used for conventional manual DNA sequencing using radioisotopic detection of reaction products. Although the comparison of sequence accuracy using different protocols given here is derived from one PCR product, we have achieved equal success with numerous other PCR products of different size and base composition and with target DNA from different organisms. By direct automated DNA sequencing, it is possible to obtain sequence data within 24 h from start to finish.

REFERENCES

1. **Saiki, R. K., Gelfand, D. H., Stoffel, S., Scharf, S. J., Higuchi, R., Horn, G. T., Mullis, K. B., and Erlich, H. A.,** Primer-directed enzymatic amplification of DNA with a thermostable DNA polymerase, *Science,* 239, 487, 1988.
2. **Murray, V.,** Improved double-stranded DNA sequencing using the linear polymerase chain reaction, *Nucleic Acids Res.,* 17, 8889, 1989.
3. **Innis, M. A., Myambo, K. B., Gelfand, D. H., and Brow, M. D.,** DNA sequencing with Thermus Aquaticus DNA polymerase and direct sequencing of polymerase chain reaction-amplified DNA, *Proc. Natl. Acad. Sci. U.S.A.,* 85, 9436, 1988.
4. **Wilson, R. K., Chen, C., and Hood, L.,** Optimization of asymmetric polymerase chain reaction for rapid fluorescent DNA sequencing, *BioTechniques,* 8, 184, 1990.
5. **Little, J. W.,** Lambda exonuclease, in *Gene Amplification and Analysis,* Chirikjian, J. G. and Papas, T. S., Eds., Elsevier/North-Holland, Amsterdam, 1981, 135.
6. **Higuchi, R. G. and Ochman, H.,** Production of single-stranded DNA templates by exonuclease digestion following the polymerase chain reaction, *Nucleic Acids Res.,* 17, 5865, 1989.
7. **Lee, L. G., Connell, C. R., Woo, S. L., Cheng, R. D., McArdle, B. F., Fuller, C. W., Halloran, N. D., and Wilson, R. K.,** DNA sequencing with dye-labeled terminators and T7 DNA polymerase: effect of dyes and dNTP's on incorporation of dye-terminators and probability analysis of termination fragments, *Nucleic Acids Res.,* 20, 2471, 1992.
8. **Thomas, K. R. and Olivera, B. M.,** Processivity of DNA exonucleases, *J. Biol. Chem.,* 253, 424, 1978.
9. **Peterson, M. G., Marshall, V. M., Smythe, J. A., Crewther, P. E., Lew, A., Silva, A., Anders, R. F., and Kemp, D.,** Integral membrane protein located in the apical complex of *Plasmodium falciparum, Mol. Cell. Biol.,* 9, 3151, 1989.
10. **Thomas, A. W., Waters, A. P., and Carr, D.,** Analysis of variation in Pf83, an erythrocytic merozoite vaccine candidate antigen of *Plasmodium falciparum, Mol. Biochem. Parasitol.,* 42, 285, 1990.
11. **Saiki, R. K.,** The design and optimization of the PCR, in *PCR Technology: Principles and Applications for DNA Amplification,* Erlich, H. A., Ed., Stockton Press, New York, 1989, 7.
12. **Koenen, M.,** Recovery of DNA from agarose gels using liquid nitrogen, *Trends Genet.,* 5, 137, 1989.
13. **Kerr, C. and Sadowski, P. D.,** Gene 6 exonuclease of bacteriophage T7. I. Purification and properties of the enzyme, *J. Biol. Chem.,* 247, 305, 1972.
14. **Kerr, C. and Sadowski, P. D.,** Gene 6 exonuclease of bacteriophage T7. II. Mechanism of the reaction, *J. Biol. Chem.,* 247, 311, 1972.
15. **Winship, P. R.,** An improved method for directly sequencing PCR amplified material using dimethyl sulphoxide, *Nucleic Acids Res.,* 17, 1266, 1989.
16. **Kusukawa, N., Uemori, T., Asada, K., and Kato, I.,** Rapid and reliable protocol for direct sequencing of material amplified by the polymerase chain reaction, *BioTechniques,* 9, 66, 1990.

17. **Slatko, B. E., Albright, L. M., and Tabor, S.,** DNA sequencing, in *Current Protocols in Molecular Biology,* Suppl. 19, Ausubel, F. M., Brent, R., Kingston, R. E., Moore, D. D., Seidman, J. G., Smith, J. A., and Struhl, K., Eds., John Wiley & Sons, New York, 1992, sec. 7.
18. **Sawadogo, M. and Van Dyke, M. W.,** A rapid method for the purification of deprotected oligodeoxynucleotides, *Nucleic Acids Res.,* 19, 674, 1990.
19. **Fischer, I., Magil, S., and Kopczynski, M.,** A rapid and inexpensive procedure for desalting synthetic oligonucleotides, *BioTechniques,* 9, 300, 1990.

Chapter 12

DISTINCTION BETWEEN ALMOST-IDENTICAL DNA SEQUENCES BY POLYMERASE CHAIN REACTION

Jean-Paul Charlieu

TABLE OF CONTENTS

I. Introduction 101

II. Materials and Methods 102
    A. Primers 102
    B. Buffers and Reagents 102
    C. DNA Templates 102
    D. PCR Conditions 103
    E. Analysis of PCR Products 103

III. Results 103

IV. Discussion 104

Acknowledgments 106

References 106

I. INTRODUCTION

Two DNA fragments sometimes present very strong sequence homology. For example, the alleles of a gene can differ at only one position, when a point mutation occurs. Several techniques, such as denaturing gradient gel electrophoresis (DGGE), single-strand conformation polymorphism (SSCP), and hybridization with allele-specific oligonucleotides (ASO) allow discrimination between almost-identical DNA sequences.

It has been shown recently[1] that PCR can also be used to distinguish DNA variants in the HIV genome. Based on this study, PCR conditions were developed to distinguish the alpha-satellite subfamilies of chromosomes 13 and 21.[2] Alpha-satellite is a family of tandemly repeated DNA sequences present at the centromeric region of all human chromosomes.[3] Slight variations in the nucleotide sequence of the ~171-bp basic motif and in the distribution of restriction sites define higher order repeats that are specific for one or a few chromosomes.[4] The subfamilies from chromosomes 13 and 21 are almost identical, however, as they share 99.7% homology (2 out of 680 bp).[5] Hybridization experiments performed with a probe from one of these chromosomes reveal the alpha-satellite fragments from both.[6]

A membrane-bound PCR[7] approach allowing the chromosomal origin (13 or 21) of alpha-satellite fragments detected by Southern blot hybridization is presented here. The experimental parameters that influence the PCR specificity also have implications for other purposes such as the detection of point mutations and the discrimination between the alleles of these models.

II. MATERIALS AND METHODS

A. PRIMERS
Based on the nucleotide sequence of alpha-satellite subfamilies from chromosomes 13 and 21,[5] two pairs of primers (13A + 13B and 21A + 21B) were designed:

13A: 5' TGATGTGTGTACCCAGCT 3'
13B: 5' GCTATCCAAATATCCACT 3'
21A: 5' TGATGTGTGTACCCAGCC 3'
21B: 5' GCTATCCAAATATCCACC 3'

Each primer carries a chromosome-specific nucleotide at its 3' end. They were resuspended in sterile distilled water to obtain stock solutions at 20 µM.

B. BUFFERS AND REAGENTS
Distilled water was filtered through 0.22-µm filters and autoclaved at 120°C for 20 min before use. Small aliquots (1 ml) were reserved for PCR assays. Bovine Serum Albumine (BSA) was purchased from Boehringer Mannheim, diluted with water to 0.5 mg · ml^{-1}, and stored as 0.5-ml aliquots at –20°C. Taq DNA polymerase buffer 10 × concentrated was supplied by Promega: 500 mM KCl, 100 mM Tris-HCl, pH 8.8 at 25°C, 15 mM MgCl$_2$, 1% Triton X-100. Nucleotides, from Boehringer Mannheim, were separately resuspended in water to obtain stock solutions at 50 mM. Working solution (10 × concentrated) consists of a mix of the four dNTPs at 0.025 mM each. Diluted dNTPs are quite unstable and should be kept as small aliquots (20 µl) at –20°C. Taq DNA polymerase was from Promega, and Perfect Match polymerase enhancer was from Stratagene.

C. DNA TEMPLATES
The following somatic hybrids containing the specified human chromosome(s) in parentheses in a rodent genomic background were used: WA17 (human chromosome 21) was obtained from Dr. Devine;[8] HY124VT4, HY73DMT3 (both with human chromosome 21 in a hamster genomic background), RJ387.58T1, HY25T1 (both containing human chromosome 13), RJ387.91CT8 (chromosome 2), and HY129T14 (chromosome 14) were kindly provided by Dr. M. Rocchi; BCHE (human chromosomes 3, 4, 6, 8, 9, 10, 13, 14, 15, 16, 18, 19, Xp$^+$) and C35B2 (chromosome 11) were from Dr. N'Guyen Van Cong.

Cells were collected and washed twice with phosphate saline buffer (PBS) and lysed with 200 µg·ml^{-1} proteinase K (Appligène) in 0.5 M EDTA; 1% L-lauryl sarcosyl, pH 8.5 for 48 h at 50°C. The DNA was then purified by phenol extraction and ethanol precipitation according to standard methods.[9]

DNA was denatured in 0.5 M NaOH, 1.5 M NaCl for 5 min at room temperature and dotted onto nylon membrane (hybond N, Amersham). When dried, DNA was cross-linked to the membrane with UV light for 5 min. Small pieces (~1 × 1 mm) carrying ~100 ng of DNA were cut to obtain membrane-bound DNA templates for PCR.

Human genomic DNA was prepared embedded in low-melting agarose from blood or cultured cells as previously described.[6] Restriction enzyme hydrolysis, separation of DNA fragments by pulsed field gel electrophoresis (PFGE), Southern blotting and hybridization with the alpha-satellite probe α-RI 680, 368[5] from human chromosome 21 were performed as described[6] except that hybond N (Amersham) nylon membrane was used. After blotting, the filter was dried in an oven at 80°C and the DNA was cross-linked to the membrane with UV light (5 min on a UV table).

Pieces of membrane carrying the alpha-satellite fragments detected by hybridization were used as templates in PCR experiments.

D. PCR CONDITIONS

PCR experiments were performed in a PREM thermocycler as follows:

1. Prepare the sample mix (for 1 sample):

H$_2$O	9 µl
Taq DNA polymerase buffer 10 × concentrated	2 µl
Primer A, 20 µM	1 µl
Primer B, 20 µM	1 µl
BSA, 0.5 mg·ml^{-1}	2 µl

    Multiply these quantities by the number of PCR samples.

2. Distribute 15 µl of the sample mix in 0.5-ml reaction tubes, and add the DNA template (100 ng) and one drop of mineral oil to prevent evaporation.

3. Prepare the "enzyme mix" (for one sample):

H$_2$O	2.3 µl
Taq DNA polymerase (Promega), 5 U·µl^{-1}	0.5 µl
dNTPs 10 × concentrated	2.0 µl
Perfect Match polymerase enhancer, 1 U·µl^{-1}	0.2 µl

    Multiply by the number of samples and keep on ice.

4. Program the thermocycler as follows:

    1. 95°C, 5 min
    2. 59°C, 1 s
    3. Suspend the program
    4. 92°C, 5 s
    5. 59°C, 30 s
    6. Repeat from step 4, 40 times
    7. 72°C, 5 min

5. Place the tubes in the thermocycler and run the program. When the program is suspended at the annealing temperature after the initial denaturation, add 5 µl of "enzyme mix" to each reaction tube and continue the program.

For membrane-bound PCR, each piece of nylon filter carrying the DNA template was washed in 10 ml of H$_2$O for 30 min at 65°C and saturated in 200 µl of 1 × *Taq* DNA polymerase buffer containing 0.5 mg·ml^{-1} of BSA for 15 min at 65°C. The PCR conditions were the same except that 60 cycles were performed.

E. ANALYSIS OF PCR PRODUCTS

PCR products were analyzed in a 10% polyacrylamide gel prepared in TBE 0.5 × from a stock solution 38% acrylamide to 2% bisacrylamide. After 3 h of electrophoresis at 150 V, the DNA was stained by soaking the gel in an ethidium bromide solution (0.5 µg·ml^{-1}) for 30 min and visualized with UV light.

III. RESULTS

The PCR conditions allowing discrimination between the alpha-satellite subfamilies from chromosomes 13 and 21 were developed using the DNA from somatic hybrids containing one of these two chromosomes. A DNA fragment of the expected size (98 bp) was amplified with the two pairs of primers 13A + 13B and 21A + 21B. The nature of this DNA fragment was checked by Southern blot hybridization with the alpha-satellite probe used in this study (not shown).

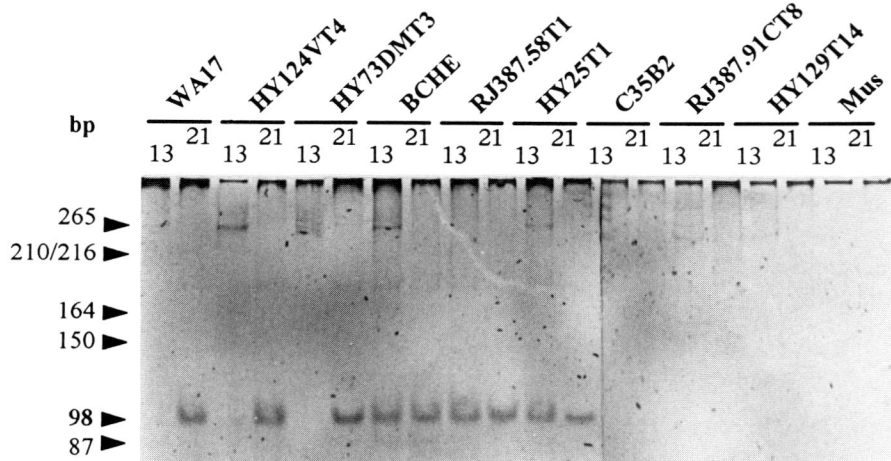

Figure 1. Membrane bound PCR with dot blots carrying the DNA from somatic hybrids noted at the top of the lanes. Numbers 13 and 21 indicate which pair of primers was used. 10 µl of the PCR products were loaded in a 10% polyacrylamide gel. The DNA was revealed by ethidium bromide staining and UV detection. (From Charlieu, J.-P., Murgue, B., Laurent, A.-M., Marçais, B., Bellis, M., and Roizès, G., *Genomics*, 14, 515, 1992. With permission.)

Membrane-bound PCR conditions were tested with dot blots carrying the DNA from the same somatic hybrids (Figure 1).

As for the experiment described above, the 98-bp amplified fragment was obtained with the chromosome 13-specific pair of primers (13A + 13B) only when using somatic hybrids containing chromosome 13, whereas the chromosome 21-specific pair of primers (21A + 21B) allowed the amplification of this DNA fragment in both chromosome 21- and 13-containing hybrids. The negative controls showed that this PCR product does not originate from the rodent genomic background or human chromosomes other than 13 and 21.

In order to test the possibility of determining the chromosomal origin of alpha-satellite fragments revealed by Southern blot hybridization, a CEPH (Centre d'Étude du Polymorphisme Humain, Paris) was used. Genomic DNA prepared in agarose plugs was digested with Bam HI, and the fragments were separated by PFGE. After hybridization and autoradiography, one band characterizing each allele determined by segregation analysis (A1, A2, B1, and B2 from the mother and C1, C2, D1, and D2 from the father)[6] was picked up and used in membrane-bound PCR (Figure 2). Alleles A1, A2, D1, and D2 were found to originate from chromosome 13, and alleles B1, B2, C1, and C2 from chromosome 21.

When analyzed by PFGE, the alpha-satellite Bam HI fragments were found to be quite variable in size. In addition, these alleles were found to segregate in a Mendelian fashion.[6] Alpha-satellite can therefore define a very informative centromeric marker when it is possible to distinguish by PCR the fragments from chromosomes 13 and 21.

IV. DISCUSSION

Annealing temperature is generally the main parameter considered for PCR specificity. In fact, several other parameters should be taken into account to obtain a specific PCR amplification. These are presented in this discussion, based on the example of PCR amplification of alpha-satellite subfamilies from chromosomes 13 and 21. The given values should be considered only as indications for the study of other models for which very stringent PCR conditions are needed.

It has been shown[1,10] that PCR amplification can be performed using primers containing up to 50% of mismatching nucleotides, but there is an absolute requirement for the correct base

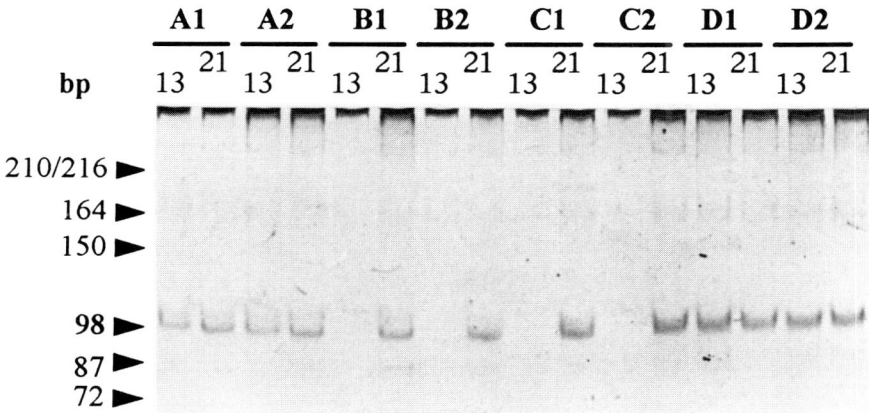

Figure 2. Membrane-bound PCR was performed with alpha-satellite primers on pieces of a Southern blot carrying the alpha-satellite Bam HI fragments revealed by hybridization. The PCR products, obtained with PCR conditions identical to those described in Figure 1, were analyzed in a 10% polyacrylamide gel. The CEPH family K1418 was used in this experiment. (From Charlieu, J.-P., Murgue, B., Laurent, A.-M., Marçais, B., Bellis, M., and Roizès, G., *Genomics*, 14, 515, 1992. With permission.)

pairing of their 3′ end. Thus, the main rule for the design of PCR primers allowing the discrimination between homologous DNA sequences is the *3′ position* of the mismatching nucleotide. In addition, Kwok et al.[1] have shown that all base mispairings do not work with the same efficiency. In particular, they have found that the T:G mismatch has no effect on PCR specificity. In the primary sequence of alpha-satellite DNA from chromosomes 13 and 21, the only differences are T-to-C changes, however.[5] The T:G mismatch (T on 13A and 13B primers and G on the chromosome 21 alpha-satellite DNA template) was found in this study to allow the correct discrimination between these two alpha-satellite DNA sequences: the 98-bp fragment was amplified with the pair of primers 13A + 13B only with the alpha-satellite template from chromosome 13. Thus, there might not be absolute rules in the nature of the mispairing nucleotide but this may depend on the model studied. Amplification of the 98-bp fragment from chromosome 13 with the chromosome 21-specific pair of primers could indicate either that the C:A mismatch (C on the 21A and 21B primers and A on the chromosome 13 alpha-satellite DNA sequence) does not affect PCR specificity or that the alpha-satellite subfamily of chromosome 21 is also present on chromosome 13.

The nucleotide concentration can also determine PCR specificity. In our laboratory, a "standard" PCR sample contains 0.2 mM of each dNTP. In these conditions, however, a 98-bp fragment was amplified in all cases with both chromosome 13 and 21 alpha-satellite-specific pairs of primers. Correct specificity was obtained by decreasing the nucleotide concentration to 2.5×10^{-3} mM.

MgCl$_2$ concentration also influences the PCR. We found that 1.5 mM was a good concentration for our system. Promega now provides *Taq* DNA polymerase buffer without MgCl$_2$, which can be added to give the desired concentration of this salt.

Another parameter affecting PCR specificity is the *Taq* DNA polymerase concentration. With the enzyme used, 2.5 U was the limit required to obtain reproducible and specific PCR amplification. We have also tested the thermostable DNA polymerase from *Thermus flavis* (Tfl polymerase), and we have found that the 98-bp alpha-satellite DNA fragment can be obtained with 1 or 0.5 U of enzyme, but results were not reproducible. If no specific amplification occurs in a given model, it is advisable to decrease the enzyme concentration or to test other enzymes.

The addition of chemicals such as tetramethylammonium chloride (TMAC)[11] or Perfect Match polymerase enhancer (Strategene) can also make it possible to obtain a specific amplification. TMAC is generally used at 10^{-4} to 10^{-5} M. We have found that these conditions are too stringent, and we could not find the correct concentration for our system. The manufacturer recommends addition of 1 U of Perfect Match to the PCR sample. Again, we found the conditions too stringent for our case, and 0.2 U was sufficient. We recommend addition of Perfect Match with the enzyme and the dNTPs *after* the initial denaturation step in hot-start PCR. This product seems to be inactivated when submitted to a high temperature for a long time. We have also observed that non-specific amplification is reduced in these conditions. The mix of TMAC and Perfect Match can also be tried, but we found that they have a very strong effect on PCR when used together.

The stringency of the PCR is defined by a combination of these factors. We have described the conditions that work for our primers and in our hands. The study of other models or of our model in another laboratory and/or with another PCR machine will probably need some adjustments of the indicated values.

For membrane-bound PCR, we have tested several nylon supports. We have found that positively charged membranes are not convenient for our purpose. It seems that they have an extremely high capacity to capture the DNA and that, if amplification occurs, the PCR products are not released in the medium and cannot be detected. All uncharged nylon membranes we have tested present good qualities for membrane-bound PCR.

ACKNOWLEDGMENTS

Dr. E. Devine, Dr. N'Guyen Van Cong, and Dr. M. Rocchi are acknowledged for providing the somatic cell lines. I am grateful to Dr. B. Murgue, who performed some of the experiments described here. This work was carried out under the direction of Dr. G. Roizès in his laboratory at CNRS UPR 9008 and INSERM U 249, Institut de Biologie, Bd Henri IV, F-34060 Montpellier cedex, France.

REFERENCES

1. **Kwok, S., Kellog, D. E., McKinney, N., Spasic, D., Goda, L., and Sninsky, J. J.**, Effect of primer-template mismatches on the polymerase chain reaction: human immunodeficiency virus type 1 model studies, *Nucleic Acids Res.*, 18, 999, 1991.
2. **Charlieu, J.-P., Murgue, B., Laurent, A.-M., Marçais, B., Bellis, M., and Roizès, G.**, Discrimination between alpha-satellite DNA sequences from chromosomes 21 and 13 by using polymerase chain reaction, *Genomics*, 14, 515, 1992.
3. **Manuelidis, L.**, Chromosomal localization of complex and simple repeated human DNAs, *Chromosoma*, 66, 1, 1978.
4. **Willard, H. F. and Waye, J. S.**, Hierarchical order in chromosome-specific human alpha-satellite DNA, *Trends Genet.*, 3, 192, 1987.
5. **Jørgensen, A. L., Bostock, C. J., and Bak, A. L.**, Homologous subfamilies of human alphoid repetitive DNA on different nucleolus organizing chromosomes, *Proc. Natl. Acad. Sci. U.S.A.*, 84, 1075, 1987.
6. **Marçais, B., Bellis, M., Gérard, A., Pagès, M., Boublik, Y., and Roizès, G.**, Structural organization and polymorphism of the alpha satellite DNA sequences of chromosomes 13 and 21 as revealed by pulsed field gel electrophoresis, *Hum. Genet.*, 86, 311, 1991.
7. **Kadokami, Y. and Lewis, R. V.**, Membrane-bound PCR, *Nucleic Acids Res.*, 18, 3082, 1990.
8. **Witek, M. P., Devine, E., Dutkowski, R., Blume, A., Mullikin-Kilpatrick, D., Miller, D. L., Brown, T., and Wrayne, W.**, Isolation of specific chromosomes and their DNA, *Gene Anal. Technol.*, 2, 16, 1985.
9. **Sambrook, J., Frisch, E. F., and Maniatis, T.**, *Molecular Cloning: A Laboratory Manual*, 2nd ed., Cold Spring Harbor Laboratory Press, Cold Spring Harbor, NY, 1989.
10. **Parker, J. D., Rabinovitch, P. S., and Burner, G. C.**, Targeted genewalking polymerase chain reaction, *Nucleic Acids Res.*, 9, 3055, 1991.
11. **Hung, T., Mak, K., and Fong, K.**, A specific enhancer for polymerase chain reaction, *Nucleic Acids Res.*, 18, 4953, 1990.

Chapter 13

DETECTION OF MUTATIONS BY PCR

E. P. H. Yap and J. O'D. McGee

TABLE OF CONTENTS

I. Introduction 107
   A. DNA Labeling and Detection 108

II. Detection of Known Mutant Sequences 108
   A. Principles 108
   B. PCR-Based Methods 110
      1. Product Absence/Presence 110
      2. Product Size 110
      3. Allele-Specific PCR 111
   C. Hybridization-Based Methods 112
   D. Enzyme-Based Methods 112
      1. Restriction Cleavage 112
      2. Methylation 113
      3. Ligation 113
      4. Primer Extension 113

III. Detection of Unknown Mutation 113
   A. Principles 113
   B. Double-Stranded Mismatch Detection 114
      1. Heteroduplex Detection by Electrophoresis 114
      2. Mismatch Modification 114
         a. Enzymatic Cleavage 114
         b. Chemical Cleavage 115
         c. Carbodiimide Labeling 115
      3. Thermal Instability 115
   C. Single-Stranded DNA Analysis 116
      1. Single-Stranded Conformation Polymorphism 116
      2. Nucleotide Analogue Incorporation 116
      3. Sequencing 116

IV. Conclusion 117

References 117

I. INTRODUCTION

The identification and characterization of mutations have been historically important in every branch of biology and medicine. Gregor Mendel developed the field of classical genetics in the last century by studying phenotypic polymorphisms, while Archibald Garrod first published studies on inherited enzyme mutations in his landmark tome *Inborn Errors of Metabolism* (1909). Since then, the study of mutations has contributed significantly to our understanding of the mechanisms and pathways of both normal physiological processes as well as disease pathogenesis. It has now moved on to the study of the relationships between protein structure and function, and correlation between genotype and disease phenotype. Various developments in molecular genetics and biology have revolutionized our ability to analyze genes at a nucleotide sequence level, not least the advent of DNA hybridization/manipulation techniques and *in vitro* DNA amplification. The current productiveness of efforts to map the human genome has yielded exponentially increasing amounts of sequence data deposited in computer databases such as *Genbank* and *EMBL*. Therefore the emerging constraint on advances in molecular pathology appears to be the ability to correlate mutant genotype with disease phenotype. To these ends, a growing number of methods have evolved to enable the rapid analysis of specific sequences in genomes, with decreasing requirements on sample quality and quantity, time, and manual effort.

This review presents a conceptual and mechanistic description of the methods, using the polymerase chain reaction (PCR), to detect small mutations or polymorphisms, involving alterations to one or several bases. For a discussion of the comparative effectiveness of these methods, the reader is referred to other recent reviews. [1,2] The application of these methods to human pathology is also described elsewhere.[3]

Methods for detecting small and point mutations may be grouped into

- Those useful for detecting specific mutations at specific sites which have previously been characterized and sequenced (diagnostic methods) (Table 1)
- Those useful for detecting the presence of unknown sequence differences in a given length of DNA (screening methods) (Table 2).

The latter methods may also be used for detecting known mutations, and are particularly useful for rapidly detecting complex multi-allelic loci such as the HLA complex.

A. DNA LABELING AND DETECTION

Several developments in the labeling and detection of nucleic acids are applicable to most of the methods described in the following sections. Isotopic labeling is traditionally performed with ^{32}P. However since the sensitivity of detecting amplified DNA is not usually limiting, the short half-life and concerns of safety and health may be overcome by using ^{35}S-labeled DNA. This provides the added advantage of a better resolution on autoradiography. Recently, ^{33}P has been used for sequencing and DNA detection, offering features between those of ^{32}P and ^{35}S. Tritium (^{3}H) is used for the detection of DNA in solution by liquid scintillation. This allows rapid quantitative analysis of large numbers of samples.

Isotopes have been replaced, to a large extent, by non-isotopic labels such as biotin and digoxigenin. These molecules may be detected with immunochemistry to give a chromogenic, fluorescent, or chemiluminescent signal. Oligonucleotides may also be labeled directly with enzymes or fluorochromes; fluorescent detection can now be performed on automated gel scanners.

II. DETECTION OF KNOWN MUTANT SEQUENCES

A. PRINCIPLES

The methods for detecting known single base substitutions (e.g., c-Ras mutations in human cancer) or small deletions (e.g., 3-bp deletion in cystic fibrosis) can be developed once the mutation-disease relationship has been characterized. Diagnostic tests are designed to specifically detect the presence or absence of the particular mutated sequence. The desirable features

TABLE 1.
Detection of Known Mutations (Diagnosis)

Oligonucleotide hybridization
    Solution phase hybridization: PCR primers
        Absence/presence of product
            Multiplex PCR
        Size difference
        Allele-specific PCR
            Homoduplex-dependent extension: ASPCR, ARMS, ASA, PASA
                Double ARMS, PAMSA, ARMS-SSCP
            Competitive annealing of primers: color complementation assay
    Immobilized phase hybridization: allele-specific oligonucleotide (ASO)
      probes
        Southern blot
        Dot blot
        Reverse dot blot
        Microtiter plate (sandwich assay)
Enzyme-based recognition of sequences
    Recognition of specific homoduplexes
        Restriction endonucleases
        Methylases
    Modification of perfectly matched homoduplexes
        Polymerases
        Ligases

TABLE 2.
Detection of Unknown Mutations (Screening)

Detection of mismatched double-stranded DNA (heteroduplexes)
    Electrophoretic mobility change
        Agarose, polyacrylamide, Hydrolink, etc.
    Modification of mismatch
        Cleavage
            Chemical
                Hydroxylamine, osmium tetroxide
                Hydroxylamine, potassium permanganate
            Enzymatic
                S1 Nuclease
                RNase Cleavage
        Labeling
            Carbodiimide
Detection of differences in single-stranded DNA
    Secondary conformation: single-strand conformation polymorphism (SSCP)
    Nucleotide analogue incorporation
    Sequencing
        Direct
        Indirect

of such tests are that they should be rapid, capable of being automated, and suitable for handling large numbers of samples. Furthermore, the ability to quantitate the ratio of mutant to wild-type DNA is an advantage. Techniques in routine use for diagnosis must, in addition, have low false-negative and false-positive rates.

These methods (Table 1) are based on the recognition of a short length (4 to 20 bases) of specific sequence in amplification products ranging from 40 bp to 2 kb. This is achieved either by specific annealing/hybridization of an oligonucleotide or by enzyme-based recognition.

The thermal stability and melting temperature of duplex DNA is affected by the type and number of mismatches between the two strands. The shorter the hybrid, the greater will be the instability due to mismatches and hence the more easily is this detected by different hybridization/annealing conditions. However, a DNA sequence has to be longer than a certain size to be statistically unique in a genome of particular size; for the human genome, this lower limit is 17 bases.[4] Oligonucleotides used for detection of specific human sequences in genomic DNA are therefore typically about 20 bases in length. Thus, separate oligonucleotides can be designed to be complementary to and specific for the normal (wild-type) sequence and for the mutant sequence(s). Hybridization conditions have therefore to be carefully optimized and controlled such that only completely homologous sequences hybridize. Sequence-specific oligonucleotide hybridization may be performed in one of two ways:

- In solution as primers to specifically select sequences for amplification during PCR
- As labeled probes for hybridization to PCR products immobilized on a solid support

Enzyme-based recognition may take one of two forms:

- The recognition of a specific (often palindromic) 4- to 10-bp sequence of double-stranded DNA by restriction endonucleases and methylases
- The fidelity of ligases and DNA polymerases to bind to or extend a primer in a DNA template-dependent manner.

Detection of the products of these enzymatic reactions may involve either size determination for gel electrophoresis, or incorporation of a labeled substrate.

B. PCR-BASED METHODS

The PCR accomplishes the dual functions of specifically selecting a DNA sequence and exponentially amplifying it. Thus, PCR may itself be used for mutation detection without any post-PCR manipulation, apart from some form of product quantification. This is typically done by gel electrophoresis, or by capture or immobilization onto a solid support (facilitating automation).

1. Product Absence/Presence

In its simplest form, PCR is used directly to show the absence or presence of normal or mutated sequences respectively. One of the first applications of PCR using a thermostable polymerase was for the prenatal diagnosis of α-thalassemia caused by a 23-kb deletion.[5] The absence of the α-globin PCR product in the presence of a control fragment (β-globin) co-amplified simultaneously denotes the lethal homozygous state. Similarly, chromosomal translocations (which occur frequently in a variety of hematological cancers) can be demonstrated by PCR of the breakpoint or junction site.

If the same sample is to be screened for several potential mutations, co-amplification of multiple DNA segments in the same reaction can be performed. This is provided that the amplification conditions for each primer pair are similar and that the different fragments can be adequately separated by electrophoresis. Such a *multiplex PCR* approach[6] has been employed for the detection of dystrophin gene mutations in Duchenne muscular dystrophy, where at least nine hot spots[7] are amplified and screened for deletional mutations in the same tube. Absence of a band denotes a deletion that involves at least one of the two primer-annealing sites.

2. Product Size

Smaller deletion/insertion mutations can be detected as changes in the expected size of PCR products. A common mutation in the CFTR gene, causing 80% of cystic fibrosis cases in Caucasian populations, is a 3-bp deletion in codon 508. Normal and mutated PCR DNA can be resolved by electrophoresis on polyacrylamide gel electrophoresis (PAGE);[8] heterozygous individuals can further be identified by presence of heteroduplexes (see later). A recently discovered class of mutations are the trinucleotide repeats causing a variety of neuro-psychiatric diseases such as the fragile-X syndrome; somatic amplification of these unstable repeats results in a spectrum of disease phenotype depending on number of repeats.[9] These can readily be studied by PCR amplification across the unstable region followed by agarose gel electrophoresis.

3. Allele-Specific PCR

Conventional gel electrophoresis is not sensitive enough to discriminate single base substitutions in PCR products. However primers may be designed to anneal to the region being studied. If the PCR is optimized so that only perfectly complementary sequences are amplified, different alleles can be preferentially amplified, depending on the primer used. This has variously been termed *allele-specific PCR* (ASPCR),[10] *allele-specific amplification* (ASA),[11] *PCR amplification of specific alleles* (PASA),[12] and *amplification refractory mutation system* (ARMS).[13] These methods are based on either of two principles:

- Mismatches near the middle of the primer result in decreased thermal stability, i.e., annealing will not occur under stringent hybridization conditions.
- Mismatches at the 3' end will not allow primer elongation, since *Taq* polymerase and some other thermostable polymerases do not have 3'-to-5' exonuclease activity.

A step that enhances selectivity is the deliberate introduction of another mismatch near the 3' end, which destabilizes it; false-positive extension of the double-mismatched primer is thus eliminated.[13] These methods are also contingent on the maintenance of a temperature sufficiently stringent to prevent non-specific annealing during the course of the reaction,[13] a requirement that is met by the Hot-Start protocol.[14] Thus, each sample is analyzed in separate reactions for normal or mutant sequences. Homozygous and heterozygous genotypes can therefore be easily identified.

One modification of allele-specific amplification is to use the allele-specific primers together, allowing competitive hybridization to template DNA; the perfectly matched rather than the mismatched primer would tend to be extended.[15] Thus if the two (or more) competing primers were labeled differently, genotyping could be performed in one tube by simply identifying the label that had been incorporated into the product. An elegant application of this principle was to label primers for normal hemoglobin and the sickle-cell mutant with the fluorochromes, flourescein, and rhodamine, respectively.[16] This *color complementation assay* therefore gave normal products that fluoresced green and mutant products that fluoresced red, thus allowing quick visual identification, fluoroscopy, and automation.[17]

Various other modifications and applications of allele-specific PCR have been described. In *double ARMS*[18] or *PCR amplification of multiple specific alleles* (PAMSA),[19] allele-specific primers are used at both ends of the template so that larger numbers of alleles may be detected. When combined with inverse PCR, haplotypes involving loci more than several kilobases apart may be detected.[18] To facilitate the analysis of poly-allelic loci such as the HLA complex, a two-step approach (*ARMS-SSCP*) involving, first, identifying groups of similar alleles using ARMS, followed by their specific typing using another screening method; single-strand conformation polymorphism (see later) was used.[20]

A specialized application of allele-specific PCR is for "digital" DNA typing or fingerprinting.[21] The sequence polymorphisms within each repeat unit of a minisatellite allele can be analyzed to give a unique binary code, while a ternary code may be derived from a diploid genome. These extremely variable digital DNA profiles are appropriate for forensic investigations, including computer databasing.

C. HYBRIDIZATION-BASED METHODS

Hybridization of target DNA to labeled allele-specific oligonucleotide (ASO) probes is a well-used method for studying small mutations in genomic DNA,[22] and is particularly well suited for amplified DNA. The principles are similar to those for allele-specific PCR; under sufficiently stringent conditions, only perfectly matched probes will bind to the target. However, in contrast to PCR, formamide may be added to reduce hybridization temperature and trimethylammonium chloride helps to prevent non-specific hybridization. The post-hybridization washing step is often the critical stringency-determining step.

There are a large variety of different formats for hybridizing PCR products to ASO probes. Generally hybridization is performed with the target DNA immobilized onto a solid support either by baking/UV cross-linking on a nylon/nitrocellulose membrane or by ligand-mediated (e.g., biotin-avidin) binding to a plastic surface.

Classically, products are size fractionated by gel electrophoresis before Southern transfer onto filter and isotopic ASO hybridization (*Southern blot*). This additional selection process allows the elimination of non-specific amplification products of incorrect size. Furthermore, background signal caused by non-specific binding of probe to nonhomologous DNA and other constituents in the sample or membrane may be readily detected.

If the PCR is well optimized, it may be sufficient to directly immobilize PCR products onto filter without electrophoresis either as *dot blots* or *slot blots*, depending on whether a slot vacuum manifold is used. If a large number of alleles are to be studied, it may be more convenient to prepare, beforehand, filters or plastic strips containing immobilized ASO probes; these are hybridized to labeled PCR products in solution (*reverse dot blot*).[23]

A *microtiter plate* format may be employed by coating it with avidin and biotinylating one of the primers. Biotin-containing DNA product may then be captured, denatured, and probed with oligonucleotides in 96-well plates, and the process may be automated.

D. ENZYME-BASED METHODS
1. Restriction Cleavage

Bacterial DNA-modifying enzymes that recognize specific DNA sequences may be used to detect the presence or absence of point mutations. *Restriction endonucleases* have been widely used for this purpose. For instance, the 7-base recognition site of OxaNI, present in normal β-globin, is abolished by the A-to-T mutation causing sickle-cell anemia.[5] Digestion of PCR products spanning this site therefore results in restriction fragment length polymorphisms (RFLP) in a population of normal and affected individuals. Samples from heterozygote carriers exhibit both digested as well as resistant DNA. However, because of the formation of digestion-resistant heteroduplexes (by template re-annealing) during the later cycles of PCR, heterozygotes may appear to have disproportionately more of one allele than the other. This may be exploited to detect low levels of mutant (resistant) DNA, or it may require computation using the binomial equation for quantification of DNA.[24] Furthermore, it is often necessary for the digestion to be performed to completion, particularly when quantification or high sensitivity are required. An invariable restriction site present in another part of the amplicon may be used to ensure this.

The normal or mutant sequence forms known endonuclease restriction sites only in a minority of instances. However, such sites can be artificially introduced into most sequences by using a mismatched primer (site-directed mutagenesis, *artificial RFLP*).[25] Furthermore, the sample may be enriched for a mutant sequence that exists as a minority species by performing wild-type sequence-specific restriction. The digestion may be carried out between two rounds of nested PCR analysis[26] or even during PCR cycles if a heat-stable endonuclease is used (RFLP/PCR).[27] Such enrichment allows the detection of minority mutations present in as few as 1 cell per 200,000 normal cells.

2. Methylation

A recently developed method is the use of another class of DNA-modifying enzymes, *DNA methylases*, which recognize specific sequences analagous to endonucleases. These methylation sites may be originally present in the mutant/wild-type sequence, or be introduced by primer-mediated mutagenesis.[28] Methylated DNA can be detected if labeled S-adenosylmethionine is used as substrate. Large numbers of samples may be studied by tritium-labeling and scintillation counting. As with restriction analysis, heteroduplex formation has to be taken into account.

3. Ligation

In contrast to the nicking or methylating specific sequences of double-stranded DNA described above, *ligases* may be used to bind two abutting primers in a template-dependent fashion. Such activity occurs only when the adjacent ends of both primers are perfectly matched to the complementary strand. Thus using appropriate mutant or wild-type specific primers, DNA sequences may be inferred by the presence or absence of ligation (*oligonucleotide ligation assay*).[29] Ligated products may be assayed either by gel electrophoresis, by binding to avidin beads, or by capture onto 96-well plates. A variety of detection formats may therefore be envisaged.

Sequential cycles of ligation may be used to amplify specific sequences from genomic DNA in a linear or exponential manner, depending on the number of oligonucleotides used.[30] The *ligase amplification reaction* (LAR) or *ligase chain reaction* (LCR) employing a thermostable ligase can therefore be used to specifically amplify mutant or wild-type sequences in a manner analagous to PCR.[31] However due to non-specific blunt-end ligation of oligoduplexes, mutant sequences present at 1% of wild-type amounts cannot be detected.[32]

4. Primer Extension

Another strategy for enzymatic detection of point mutations in amplified DNA is to extend a primer having its 3' end abutting the base substitution. DNA polymerase is used in the presence of a single labeled nucleotide substrate. The identity of the template sequence can be determined by whether the labeled base has been added onto the primer, as determined by gel electrophoresis,[33] on a solid support (*solid-phase mini-sequencing*),[34] or by liquid scintillation.[35]

The primer may also be extended by the use of a chain-termination sequencing reaction. If a dideoxynucleotide corresponding to the mutated nucleotide is used, mutated samples will not be extended beyond this nucleotide whereas normal sequences will. Products from such single-tube mini-sequencing are electrophoresed on conventional sequencing gels.[36]

III. DETECTION OF UNKNOWN MUTATIONS

A. PRINCIPLES

Screening a PCR product for a base change anywhere along its length requires a different approach from the diagnostic methods described above. A much longer sequence has to be analyzed, typically 100 bp to 3 kb. The trade-off for this is that the actual sequence and precise location of the mutation cannot usually be determined accurately, since screening is often followed by definitive sequencing. Furthermore, the majority of these screening methods are not able to detect all possible mutations. Tolerance for such false-negatives is often allowable in the research applications in which these methods are employed.

Sequence differences in DNA can be detected on the basis of various physicochemical properties (Table 2):

- **Double-stranded DNA** — Denaturation and annealing of approximately equimolar mixtures of the wild-type and mutant DNA results in the formation of two possible mismatched duplex strands (heteroduplexes), in addition to the two perfectly matched duplexes. Heteroduplexes may be detected by changes in electrophoretic mobility, changes in thermal stability, or modification of mismatch site.
- **Single-stranded DNA** — Sequence-dependent properties of ssDNA include its secondary conformation and base composition.

B. DOUBLE-STRANDED MISMATCH DETECTION

The use of dsDNA mismatches to detect mutations is performed by the pair-wise comparison of a sample DNA with "reference" DNA of known sequence or of known origin, i.e., wild type. Therefore these methods do not readily distinguish heterozygotes from mutant homozygotes.

1. Heteroduplex Detection by Electrophoresis

The most straightforward method for detecting mismatches is by electrophoresis on non-denaturing gels; this method is often simply termed "heteroduplex detection". Heteroduplexes migrate more slowly than their respective homoduplexes on gels. Homoduplexes of the same length cannot be electrophoretically distinguished from each other unless there are major sequence differences of more than several percent.[37] On the other hand, the two heteroduplexes formed from any mixed hybridization often migrate as separate bands on gels.

The retardation in electrophoretic mobility is assumed to be due to a deviation from the super-helical conformation rather than local steric changes to the double helix; this is well discussed by Sorrentino et al.[38] Hence, deletions or larger mismatches are more easily detected, whereas point mutations require gels with high resolving power. Furthermore, the degree of gel retardation has been empirically shown to be dependent on both the number and types of mismatches, with pyrimidine-pyrimidine mispairing exerting the largest influence.

Deletions of several bases have been detected as heteroduplexes on sieving (high-percentage) agarose gels, and they often appear as non-specific shadow bands when several heterologous loci are amplified by the same primer set during PCR.[39] This is of little diagnostic importance because the homoduplexes can themselves be resolved on agarose. However, a non-denaturing polyacrylamide gel is able to detect smaller deletional and insertional mutations of 3 to 4 bases.[8,40] Point mutations can be detected on proprietary acrylamide-substitute gels, which are purported to have better resolving ability than polyacrylamide, for example, Hydrolink (AT Biochem, U.S.; Hoeffer Scientific, U.K.).[41]

This method does not detect all potential point mutations, and the understanding of the mechanisms involved is largely empirical. Heteroduplex detection is suitable for discriminating between sequences with multiple mismatches and for rapid genotyping of large numbers of samples. One particular application suited to this method is genetic profiling; the multiple heteroduplexes formed from amplification of HLA alleles and pseudogenes give a unique banding pattern.[42]

2. Mismatch Modification

Mismatched base pairs may be modified either by cleavage with enzymes/chemicals or by labeling with a chemical marker. The advantage of these methods is that the site of the mismatch can generally be determined; this is useful to aid sequencing that follows.

a. Enzymatic Cleavage

Single-strand-specific *S1 nuclease* was originally used to map deletions in DNA; heteroduplexes are cleaved at the sites of mismatch.[43] However, the method was useful only for mapping large deletions/insertions. More recently, ribonuclease A (*RNase A*), which similarly cuts RNA at sites where it is single stranded (i.e., mispaired with its complementary DNA or RNA strand), has been used.[44] A labeled single-stranded RNA probe of defined length and sequence is hybridized to genomic or amplified DNA and digested with RNase A. The cleaved probe is detected by denaturing gel electrophoresis and autoradiography. The template DNA may also be cleaved by simultaneous digestion with S1 nuclease.[45] The disadvantages of this method are the need to synthesize the RNA probe and the fact that mismatched purines are not cleaved by RNase A.

b. Chemical cleavage

Cleavage of the mismatched bases can also be effected by chemicals. Mismatched T and C bases are preferentially susceptible to modification by osmium tetroxide and hydroxylamine, respectively. The modified bases are then cleaved with piperidine. Both complementary strands of a labeled reference probe are used to detect all possible base mismatches.[46] This method has been variously termed *chemical mismatch cleavage* and *amplification and mismatch detection* (AMD).[47] Potassium permanganate, in the presence of tetramethylammonium chloride, has also been used to cleave mismatched thymines.[48] The advantage of this method is that long sequences can be screened and a 100% detection rate is theoretically possible. However, the reactions require care to perform since multiple manipulations are required and the chemicals used are toxic.

c. Carbodiimide Labeling

A third method for mismatch modification is the use of carbodiimide, which reacts preferentially with unpaired bases. This large molecule, therefore, labels heteroduplexes formed between mutant and wild-type DNA. The carbodiimide label may be detected by one of several methods:

- Gel retardation (which does not provide the location of mutation)[49]
- Cleavage by ABC excinuclease[50]
- Immuno-electronmicroscopy[51]
- Inhibition of primer extension[52]

3. Thermal Instability

The decreased thermal stability of a heteroduplex with base mismatches is utilized in *denaturing gradient gel electrophoresis* (DGGE).[53] Double-stranded DNA is electrophoresed in polyacrylamide gels containing an increasing gradient of denaturant (urea or formamide) at a fixed temperature. The strands are progressively denatured by domains of similar stability; as the duplex becomes partially single stranded, its mobility is retarded. The conditions at which each domain denatures is dependent on its sequence as well as the presence of mismatches; differences in these are reflected by different positions of bands after electrophoresis. Heteroduplexes are more easily compared with homoduplexes than differences between pairs of homoduplexes; hence a reference sequence is usually hybridized to each sample. The denaturation conditions at which mismatched domains will separate may be predicted by calculation of theoretical melting profiles. Mutations in the last domain to denature are not detected because complete strand separation results in much slower migrating ssDNA. The addition of a GC-rich sequence (GC-clamp), which separates last, during PCR allows the whole sequence studied to be screened.[54]

Owing to the technical inconvenience of pouring gradient gels, modified approaches have been used. A special apparatus may be used to create a temperature gradient across the gel during electrophoresis (*temperature gradient gel electrophoresis*, TGGE).[55] Alternatively, the temperature of electrophoresis can be steadily increased during the period of electrophoresis (*temperature sweep gel electrophoresis*, TSGE).[56] A further simplification has also been used to detect mutations within a specific region of the DNA amplified; a constant gel denaturation condition, optimized for the specific segment studied, is used (*constant denaturant gel electrophoresis*, CDGE).[57]

The various forms of DGGE allow single-step separation of the alleles comprising the heteroduplexes for purification and sequencing. However, despite the use of GC-clamps that are costly to synthesize, not all mutants are routinely detected.

C. SINGLE-STRANDED DNA ANALYSIS
1. Single-Stranded Conformation Polymorphism

Single-stranded conformation polymorphism (SSCP) analysis is performed by denaturing double-stranded DNA and fractionating the single strands on a non-denaturing polyacrylamide gel.[58,59] Under the appropriate conditions, the electrophoretic mobility of the ssDNA is dependent not only on its length and molecular weight, but also on its overall conformation. This secondary structure is determined by the balance between destabilizing thermal forces and weak local stabilizing forces such as intra-strand base pairings and stackings, which are in turn determined by the primary sequence. In practice, not only do the complimentary strands migrate as separate bands on the gel, but small differences in sequence can also be detected.

This method has been adapted for use on both genomic and amplified DNA, and a variety of formats and modifications have been described.[60] Technical details of this method are reviewed and described in Chapter 20. SSCP fragments are restricted to 100–500 bp, and detection of mutations is seldom complete. In order to increase the scope of mutation detection, SSCP may be combined with other detection methods such as allele-specific PCR (ARMS-SSCP)[20] and heteroduplex/double-stranded detection (double-and single-strand conformation polymorphism, DSSCP).[37]

2. Nucleotide Analogue Incorporation

A method related to labeling with carbodiimide used for detecting differing base compositions of single-stranded sequences is the incorporation of specific nucleotide analogues (such as biotinylated deoxynucleotide) during PCR.[61] A difference in the number of labeled nucleotides incorporated manifests as a mobility shift compared to normal DNA during gel electrophoresis.

3. Sequencing

Recent developments in nucleotide sequencing of PCR products have made it suitable both for confirmation and characterization of mutants detected by one of the screening methods described above, and for direct mutational screening of DNA samples. Two classical methods have been described: the chemical method involves partial degradation of DNA at all occurrences of each base using different chemicals,[62] while the chain-termination method involves partial synthesis of DNA by enzymatic extension of a primer in the presence of dideoxynucleotides.[63] The latter method is more commonly used for sequencing products of PCR; direct sequencing of purified amplified DNA obviates the need for intermediate and time-consuming cloning steps. Some of the important modifications to this method include the following:

- Use of thermostable polymerases for sequencing reactions,[64] allowing use of double-stranded templates, linear amplification by thermal cycling, and avoidance of sequence compressions at GC-rich regions
- Coupled amplification and sequencing (CAS)[65] performed in a two-stage reaction

- Improvements in gel analysis of sequencing products, including the use of automated fluorescent gel readers,[66] capillary electrophoresis,[67] and ultra-thin gels[68]

Another sequencing approach under development, with particular relevance to short PCR products, is the hybridization approach, where template DNA is hybridized to arrays of oligonucleotides immobilized on a solid support.[69] This has potential advances of speed and automation.

IV. CONCLUSION

The introduction of PCR into the molecular geneticist's armamentarium has enabled the identification and characterization of gene mutations in many novel applications and situations. In the field of molecular pathology, for instance, miniscule amounts of tissue and samples (e.g., smears, swabs, aspirates, needle biopsies) now provide sufficient DNA for multiple analyses, as does the degraded DNA present in archival tissue sections that have been routinely stored for decades. Multifactorial diseases such as cancer, caused by multiple somatic mutations, can now be studied at the single cell level.

The development many mutation screening methods over the past few years is a response to the need to rapidly correlate genotypic differences with phenotypic variation. While there has been an exponential accumulation of data at the gene level, the understanding of function at the protein or cellular level still requires lengthy investigation. Similarly, rapid diagnostic methods for mutation detection will see further development and usage as disease genes are identified. Whether these methods will be used for routine genetic screening of human populations and high-risk individuals is dependent on a host of scientific, sociological, and ethical issues.

REFERENCES

1. **Rossiter, B. and Caskey, C.**, Molecular scanning methods of mutation detection, *J. Biol. Chem.*, 265, 12753, 1990.
2. **Cotton, R. G. H.**, Current methods of mutation detection, *Mut. Res.*, 285, 125, 1993.
3. **Yap, E. P. H. and McGee, J. O'D.**, Future directions in molecular biology, in *Oxford Textbook of Pathology*, Vol. 2b, McGee, J. O'D., Wright, N., and Isaacson, P., Eds., Oxford University Press, Oxford, 1992, 2337.
4. **Thein, S. and Wallace, R.**, The use of synthetic oligonucleotides as specific hybridization probes in the diagnosis of genetic disorders, *Human Genetic Diseases: A Practical Approach*, Davies, K., Ed., IRL Press, Oxford, 1986, 33.
5. **Chehab, F. F., Doherty, M., Cai, S., Kan, Y. W., Cooper, S., and Rubin, E. M.**, Detection of sickle cell anaemia and thalassaemias, *Nature*, 329, 293, 1987.
6. **Chamberlain, J. S., Gibbs, R. A., Ranier, J. E., Nguyen, P. N., and Caskey, C. T.**, Deletion screening of the Duchenne muscular dystrophy locus via multiplex DNA amplification, *Nucleic Acids Res.*, 16, 11141, 1988.
7. **Chamberlain, J. S., Gibbs, R. A., Ranier, J. E., and Caskey, C. T.**, Multiplex PCR for the diagnosis of Duchenne muscular dystrophy, in *PCR Protocols: A Guide to Methods and Applications*, Innis, M. A., Gelfand, D. H., Sninsky, J. J., and White, T. J., Eds., Academic Press, San Diego, CA, 1990, 272.
8. **Rommens, J., Kerem, B.-T., Greer, W., Chang, P., Tsui, L.-C., and Ray, P.**, Rapid nonradioactive detection of the major cystic fibrosis mutation, *Am. J. Hum. Genet.*, 46, 395, 1990.
9. **Hirst, M., Knight, S., Bell, M., Super, M., and Davies, K.**, The fragile X syndrome, *Clin. Sci.*, 83, 255, 1992.
10. **Wu, D. Y., Ugozzoli, L., Pal, B. K., and Wallace, B.**, Allele-specific enzymatic amplification of β-globin genomic DNA for diagnosis of sickle cell anemia, *Proc. Natl. Acad. Sci. U.S.A.*, 86, 2757, 1989.
11. **Okayama, H., Curiel, D. T., Brantly, M. L., Holmes, M. D., and Crystal, R. G.**, Rapid, nonradioactive detection of mutations in the human genome by allele-specific amplification, *J. Lab. Clin. Med.*, 114, 105, 1989.

12. **Sommer, S., Cassady, J., Sobell, J., and Bottema, C.**, A novel method for detecting point mutations or polymorphisms and its application to population screening for carriers of phenylketonuria, *Mayo Clin. Proc.*, 64, 1361, 1989.
13. **Newton, C. R., Graham, A., Hepinstall, L. E., Powell, S. J., Summers, C., Kalsheker, N., Smith, J. C., and Markham, A. F.**, Analysis of any mutation in DNA: the amplification refractory mutation system, *Nucleic Acids Res.*, 17, 2503, 1989.
14. **Chou, Q., Russel, M., Birch, D. E., Raymond, J. and Bloch, W.**, Prevention of pre-PCR mis-priming and primer dimerization improves low-copy-number amplifications, *Nucleic Acids Res.*, 20, 1717, 1992.
15. **Gibbs, R., Nguyen, P.-N., and Caskey, C. T.**, Detection of single DNA base differences by competitive oligonucleotide priming, *Nucleic Acids Res.*, 17, 2437, 1989.
16. **Chehab, F. F. and Kan, Y. W.**, Detection of sickle cell anaemia mutation by colour DNA amplification, *Lancet*, 335, 15, 1990.
17. **Chehab, F. and Kan, Y.**, Detection of specific DNA sequences by fluorescence amplificaiton: a color complementation assay, *Proc. Natl. Acad. Sci. U.S.A.*, 86, 9178, 1989.
18. **Lo, Y.-M. D., Lo, E. S.-F., Patel, P., Tse, C. H., and Fleming, K. A.**, Heteroduplex formation as a means to exclude contamination in virus detection using PCR, *Nucleic Acids Res.*, 19, 6653, 1991.
19. **Dutton, C. and Sommer, S.**, Simultaneous detection of multiple single-base alleles at a polymorphic site, *Biotechniques*, 11, 700, 1991.
20. **Lo, Y.-M. D., Patel, P., Mehal, W. Z., Fleming, K. A., Bell, J. I., and Wainscoat, J. S.**, Analysis of complex genetic systems by ARMS-SSCP: application to HLA genotyping, *Nucleic Acids Res.*, 20, 1005, 1992.
21. **Jeffreys, A. J., Macleod, A., Tamaki, K., Neil, D. L., and Monckton, D. G.**, Minisatellite repeat coding as a digital approach to DNA typing, *Nature*, 354, 204, 1991.
22. **Wallace, R., Johnson, M., Hirose, T., Miyake, T., Kawashima, E., and Itakura, K.**, The use of synthetic oligonucleotides as hybridization probes. II. Hybridization of oligonucleotides of mixed sequence to rabbit β-globin DNA, *Nucleic Acids Res.*, 9, 879, 1981.
23. **Saiki, R. K., Walsh, P. S., Levenson, C. H., and Erlich, H. A.**, Genetic analysis of amplified DNA with immobilized sequence-specific oligonucleotide probes, *Proc. Natl. Acad. Sci. U.S.A.*, 86, 6230, 1989.
24. **Iland, H. J. and Todd, A. V.**, Estimation of the proportions of mutant and normal N-ras alleles by allele specific restriction analysis, *Nucleic Acids Res.*, 20, 620, 1992.
25. **Haliassos, A., Chomel, J. C., L., T., Baucis, M., Kruh, J., Kaplan, J. C., and Kitzis, A.**, Modification of enzymatically amplified DNA for the detection of point mutations, *Nucleic Acids Res.*, 17, 3606, 1989.
26. **Chen, J. and Viola, M. V.**, A method to detect ras point mutation in small subpopulations of cells, *Proc. Natl. Acad. Sci. U.S.A.*, 86, 2617, 1991.
27. **Sandy, M., Chiocca, S., and Cerutti, P.**, Genotypic analysis of mutations in Taq I restriction recognition sites by restriction fragment length polymorphism/polymerase chain reaction, *Proc. Natl. Acad. Sci. U.S.A.*, 89, 890, 1992.
28. **Petty, E. M., Gold, E., and Bale, A. E.**, DNA diagnosis with mutation-specific artificial methylation sites: application to rapid screening of ΔF508, *Clin. Chem.*, 38, 2422, 1992.
29. **Landegren, U., Kaiser, R., Sanders, J., and Hood, L.**, A ligase-mediated gene detection technique, *Science*, 241, 1077, 1988.
30. **Wu, D. Y. and Wallace, R. B.**, The ligation amplification reaction (LAR) — duplication of specific DNA sequences using sequential rounds of template-dependent ligation, *Genomics*, 4, 560, 1989.
31. **Barany, F.**, Genetic disease detection and DNA amplification using cloned thermostable ligase, *Proc. Natl. Acad. Sci. U.S.A.*, 88, 189, 1991.
32. **Kalin, I., Shephard, S., and Cancrian, U.**, Evaluation of the ligase chain reaction (LCR) for the detection of point mutations, *Mut. Res.*, 283, 119, 1992.
33. **Sokolov, B. P.**, Primer extension technique for the detection of single nucleotide in genomic DNA, *Nucleic Acids Res.*, 18, 3671, 1989.
34. **Syvanen, A.-C., Ikonen, E., Manninen, T., Bengstrom, M., Soderlund, H., Aula, P., and Peltonen, L.**, Convenient and quantitative determination of the frequency of a mutant allele using solid-phase minisequencing: application to aspartylglucoasminuria in Finland, *Genomics*, 12, 590, 1992.
35. **Kaye, S., Loveday, C., and Tedder, R. S.**, A microtitre format point mutation assay: application to the detection of drug resistance in human immunodeficiency virus type-I infected patients treated with zidovudine, *J. Med. Virol.*, 37, 241, 1992.
36. **Lee, J.-S. and Anvret, M.**, Identification of the most common mutation within the porphobilinogen deaminase gene in Swedish patients with acute intermittent porphyria, *Proc. Natl. Acad. Sci. U.S.A.*, 88, 10912, 1991.
37. **Yap, E. P. H. and McGee, J. O'D.**, Rapid nonisotopic demonstration of genetic heterogeneity in viruses by double and single stranded conformation polymorphism (DSSCP), *J. Pathol.*, 168, 123A, 1992.
38. **Sorrentino, R., Iannicola, C., Costanzi, S., Chersi, A., and Tosi, R.**, Detection of complex alleles by direct analysis of DNA heteroduplexes, *Immunogenetics*, 33, 118, 1991.

39. **Wenger, R. H. and Nielsen, P. J.,** Reannealing of artificial heteroduplexes generated during PCR-mediated genetic isotyping, *Trends Genet.*, 7, 178, 1991.
40. **Nagamine, C. M., Chan, K., and Lau, Y.-F. C.,** A PCR artifact: generation of heteroduplexes, *Am. J. Hum. Genet.*, 45, 337, 1989.
41. **Keen, J., Lester, D., C., I., Curtis, A., and Bhattacharya, S.,** Rapid detection of single base mismatches as heteroduplexes on Hydrolink gels, *Trends Genet.*, 7, 5, 1991.
42. **Clay, T. M., Bidwell, J. L., Howard, M. R., and Bradley, B. A.,** PCR-fingerprinting for selection of HLA matched unrelated marrow donors, *Lancet*, 337, 1049, 1991.
43. **Shenk, T. E., Rhodes, C., Rigby, P. W. J., and Berg, P.,** Biochemical method for mapping mutational alterations in DNA with S1 nuclease: the location of deletions and temperature-sensitive mutations in simian virus 40, *Proc. Natl. Acad. Sci. U.S.A.*, 72, 989, 1975.
44. **Myers, R. M., Lumelsky, N., Lerman, L., and Maniatis, T.,** Detection of single base substitutions in total genomic DNA, *Nature*, 313, 495, 1985.
45. **Atweh, G. F., Baserga, S. J., and Brickner, H. E.,** Detecting small mutations in expressed genes by a combination of S1 nuclease and RNase A, *Nucleic Acids Res.*, 16, 8709, 1988.
46. **Cotton, R. G. H., Rodrigues, N. R., and Campbell, R. D.,** Reactivity of cytosine and thymine in single-base-pair mismatches with hydroxylamine and osmium tetroxide and its application to the study of mutations, *Proc. Natl. Acad. Sci. U.S.A.*, 85, 4397, 1988.
47. **Montandon, A., Green, P., Giannelli, F., and Bentley, D.,** Direct detection of point mutations by mismatch analysis: application to haemophilia B, *Nucleic Acids Res.*, 17, 3347, 1989.
48. **Gogos, J. A., Karayiorgou, M., Aburatani, H., and Kafatos, F. C.,** Detection of single base mismatches of thymine and cytosine residues by potassium permanganate and hydroxylamine in the presence of tetralkylammonium salts, *Nucleic Acids Res.*, 18, 6807, 1990.
49. **Novack, D. F., Casna, N. J., Fischer, S. G., and Ford, J. P.,** Detection of single base-pair mismatches in DNA by chemical modification followed by gel electrophoresis in 15% polyacrylamide gel, *Proc. Natl. Acad. Sci. U.S.A.*, 83, 586, 1986.
50. **Thomas, D. C., Kunkel, T. A., Casna, N. J., Ford, J. P., and Sancar, A.,** Activities and incision patterns of ABC excinuclease on modified DNA containing single-base mismatches and extrahelical bases, *J. Biol. Chem.*, 261, 14496, 1986.
51. **Ganguly, A., Rooney, J. E., Hosomi, S., Zeiger, A. R., and Prockop, D. J.,** Detection and location of single-base mutations in large DNA fragments by immunomicroscopy, *Genomics*, 4, 530, 1989.
52. **Ganguly, A. and Prockop, D. J.,** Detection of single-base mutations by reaction of DNA heteroduplexes with a water-soluble carbodiimide followed by primer extension: application to products from the polymerase chain reaction, *Nucleic Acids Res.*, 18, 3933, 1990.
53. **Myers, R. M., Larin, Z., and Maniatis, T.,** Detection of single base substitutions by ribonuclease cleavage at mismatches in RNA-DNA duplexes, *Science*, 230, 1242, 1985.
54. **Sheffield, V. C., Cox, D. R., Lerman, L. S., and Myers, R. M.,** Attachment of a 40-base-pair G+C-rich sequence (GC-clamp) to genomic DNA fragments by the polymerase chain reaction results in improved detection of single-base changes, *Proc. Natl. Acad. Sci. U.S.A.,*, 86, 232, 1989.
55. **Po, T., Steger, G., Rosenbaum, V., Kaper, J. and Riesner, D.,** Double-stranded cucumovirus associated RNA 5: experimental analysis of necrogenic and non-necrogenic variants by temperature-gradient gel electrophoresis, *Nucleic Acids Res.*, 15, 5069, 1987.
56. **Yoshino, K., Nishigaki, K., and Husimi, Y.,** Temperature sweep gel electrophoresis: a simple method to detect point mutations, *Nucleic Acids Res.*, 19, 3153, 1991.
57. **Borresen, A.-L., Hovig, E., Smith-Sorensen, B., Malkin, D., Lystad, S., Andersen, T., Nesland, J., Isselbacher, K., and Friend, S.,** Constant denaturation gel electrohoresis as a rapid screening technique for p53 mutations, *Proc. Natl. Acad. Sci. U.S.A.*, 88, 8405, 1991.
58. **Orita, M., Iwahana, H., Kanazawa, H., Hayashi, K., and Sekiya, T.,** Detection of polymorphisms of human DNA by gel electrophoresis as single-strand conformation polymorphisms, *Proc. Natl. Acad. Sci. U.S.A.*, 86, 2766, 1989.
59. **Orita, M., Suzuki, Y., Sekiya, T., and Hayashi, K.,** Rapid and sensitive detection of point mutations and DNA polymorphisms using the polymerase chain reaction, *Genomics*, 5, 874, 1989.
60. **Iizuka, M., Mashiyama, S., Oshimura, M., Sekiya, T., and Hayashi, K.,** Cloning and polymerase chain reactionsingle-strand conformation polymorphism analysis of anonymous Alu repeats on chromosome 11, *Genomics*, 12, 139, 1992.
61. **Kornher, J. S. and Livak, K. J.,** Mutation detection using nucleotide analogs that alter electrophoretic mobility, *Nucleic Acids Res.*, 17, 7779, 1989.
62. **Maxam, A. M. and Gilbert, W.,** A new method for sequencing DNA, *Proc. Natl. Acad. Sci. U.S.A.*, 74, 560, 1977.
63. **Sanger, F., Nicklen, S., and Coulson, A.,** DNA sequencing with chain terminating inhibitors, *Proc. Natl. Acad. Sci. U.S.A.*, 74, 5463, 1977.

64. **Murray, V.,** Improved double-stranded DNA sequencing using linear polymerase chain reaction, *Nucleic Acids Res.,* 17, 8889, 1989.
65. **Ruano, U. and Kidd, K. K.,** Coupled amplification and sequencing of genomic DNA, *Proc. Natl. Acad. Sci. U.S.A.,* 88, 2815, 1991.
66. **Voss, H., Schwager, C., Wirkner, U., Sproat, B., Zimmermann, J., Rosenthal, A., Erfle, H., Stegemann, J., and Ansorge, W.,** Direct genomic fluorescent on-line sequencing and analysis using in vitro amplification of DNA, *Nucleic Acids Res.,* 17, 2517, 1989.
67. **Swerdlaw, H. and Gesteland, R.,** Capillary gel electrophoresis for rapid, high resolution DNA sequencing, *Nucleic Acids Res.,* 18, 1415, 1990.
68. **Brumley, R. L. and Smith, L. M.,** Rapid DNA sequencing by horizontal ultrathin gel electrophoresis, *Nucleic Acids Res.,* 19, 4121, 1991.
69. **Bain, W.,** Hybridization methods for DNA sequencing, *Genomics,* 11, 294, 1991.

Chapter 14

MAPPED RESTRICTION SITE POLYMORPHISMS (MRSPS) IN PCR PRODUCTS FOR RAPID IDENTIFICATION AND CLASSIFICATION OF GENETICALLY DISTINCT ORGANISMS

David Ralph and Michael McClelland

TABLE OF CONTENTS

I. Introduction 121

II. Materials and Methods 124
    A. Choosing Primers for Amplification 124
    B. PCR Protocols 125
    C. Restriction Digests of PCR Products 125
    D. Polyacrylamide Gels 126

III. Results 126

IV. Discussion 127

References 129

I. INTRODUCTION

The polymerase chain reaction (PCR) provides a means by which extremely minute amounts of DNA can be amplified to near macroscopic amounts within a small number of hours.[1] Sequences thus amplified can be further analyzed without the need to propagate a microbe *in vitro* or to purify DNA from large volumes of primary material. Propagating pathogens to high titers is never desirable for safety reasons, and it is often impossible or time consuming. Eliminating the need to propagate an organism for analysis may be particularly advantageous in studies of microbial ecology where only a fraction, perhaps a very small fraction, of all bacteria encountered in nature can be or have been cultivated on standard bacteriological media.[2-6]

PCR is specific, permitting the amplification of a rare sequence within an extremely complex mixture. The characteristics of vast amplification, coupled with great specificity, suggest that PCR can be an extremely sensitive tool for microbial identification and/or diagnosis. In this later aspect, PCR has an additional advantage over traditional immunological techniques. Immunology can detect only what an immune system has been exposed to in the past, but PCR can detect what microbes are present at the time of sampling. In addition to the ecological applications described above, PCR can be used to detect microbes in clinical samples such as blood and sera,[7-11] urine,[12-14] milk,[15] cerebral spinal fluid,[16-18] lung biopsies and sputum,[19-23] tumor biopsies,[24] fecal samples,[25,26] plant tissue,[27] and museum specimens.[28] It is difficult to imagine a need for microbial identification for which PCR would not be an appropriate solution.

One possible shortcoming of PCR is that the sequence at the ends of the piece of DNA to be amplified must be known. The primers must anneal specifically to the ends of the target sequence in order to obtain the desired amplification. If there is sequence heterogeneity at the sites of primer annealing between individuals from which an amplified product is desired, inefficient priming and insufficient PCR yields will result in some cases. Insufficient yields could be incorrectly scored as a false-negative in a diagnostic assay or as reduced diversity in an environmental survey.

Two strategies have been commonly employed to overcome this problem. The first is to decrease the stringency at which the primers are annealed to the target DNA (e.g., Reference 29). While this strategy decreases the probability that a desired target will not be sufficiently amplified, it increases the possibility that an undesired sequence will be amplified. The more complex the DNA in which the desired target is found, the more likely it is that reducing the stringency of annealing will result in undesired amplification. Actually, this type of artifact has been used to great advantage. If the stringency of primer annealing is sufficiently lowered, any arbitrarily chosen oligonucleotide will act as a primer that directs the synthesis of a specific set of PCR products.[30] These PCR products will be reproducible and both primer and template specific. After electrophoresis, these PCR products can be viewed as a fingerprint that unambiguously identifies the source of the template, distinguishing even between strains of the same species.[30a,30b] This technology is called Arbitrarily Primed PCR (AP-PCR), and its many useful applications are reviewed in Chapters 22 and 23. The only limitations of AP-PCR toward microbial identification are that it is less sensitive, requiring much more microbial DNA, and that it is very sensitive to contamination by other DNAs, such as host cells. AP-PCR does, however, vividly illustrate that synthesis of a PCR product in and of itself is insufficient to demonstrate the presence of a target sequence. Some additional assay must be performed on the PCR product to confirm its authenticity.

A second commonly used strategy to overcome the problems associated with target sequence heterogeneity is to design primers that anneal to sequences that are conserved between different phylogenetic taxa. Such primers can be designed to amplify sequences from an extremely diverse group of phylogenetic taxa.[31-33] While this strategy significantly diminishes the possibility that no PCR product will be produced when the desired target is present, it decreases the information derived from the PCR product itself by amplifying sequences from a wide variety of organisms. As with the previously described strategy, further characterization of the PCR product is required to identify or diagnose a microbe or exclude artifactual amplification.

Three techniques have been used to acquire more taxonomic information from a PCR product and to exclude artifactual amplifications. The first of these is to hybridize a probe to the portion of the PCR product not included in the primers.[8,16,18-21,25,27,34-39] Typically, these probes are oligonucleotides designed to hybridize to regions in the amplified product that show sequence variability between the desired amplification product and those with which it is likely to be mistaken. Under conditions of high stringency, these oligos will hybridize only to sequences with perfect or near perfect complimentarity. This assay, as well as the two described below, eliminate confusion due to artifactual amplification. It is extremely unlikely that any artifactual amplification product would have any sequence similarity to the desired PCR product other than the primers. A major drawback of this approach is that it assumes that the sequence variability of organisms with the desired targets for hybridization are known. This variation can be estimated only by statistical sampling. The uncertainty of such estimates leads to the possibility of false-negatives. Another drawback of this approach is that it requires a different hybridizing oligonucleotide for every sequence variation. Failure to meet this criterion requires that a failure to hybridize be scored as an informative taxonomic character. Even with the best of controls, interpreting a negative result is not optimal.

The second approach used to confirm a PCR product and extract more taxonomic information from it is to determine its primary sequence.[24,31,33,36,40-43] This approach yields the maximal amount of information that can be obtained from any PCR product. Each nucleotide position becomes a potentially informative taxonomic character. Not only are artifactual products easily determined, but phylogenetically related organisms can be easily differentiated. This approach is also instantly expandable to include previously unrecognized genetic variants. Frequently, this approach yields sufficient information for the phylogenetic relationships of the organisms being studied to be estimated. While great advances have been made toward improvement of DNA sequencing technologies, DNA sequencing still tends to be sufficiently time consuming and technically difficult to render it unsatisfactory for routine microbial identification.

The third approach used to confirm a PCR product and extract more information from it is to analyze these products using mapped restriction site polymorphisms (MRSPs). In MRSP analysis, PCR products are digested with various restriction endonucleases. The resulting fragments are separated by electrophoresis, and examined for restriction site polymorphisms.[7,10,15,25,26,32,39,44-57] This approach can be useful in differentiating between members of any group of organisms for which a region of sequence heterogeneity that is flanked by conserved sequence domains can be identified. Sequence heterogeneity results, upon electrophoresis, in restriction fragment pattern heterogeneity. These different restriction fragment patterns can be viewed as fingerprints that can unambiguously identify any organism from which the PCR products were derived. Like the DNA sequencing approach, MRSP analysis will recognize previously unencountered genetic variants that produce novel restriction fragment patterns. Like the hybridization approach, every restriction site contains roughly the same information content as an oligonucleotide hybridizing to a PCR product.

This strategy has several important advantages over the other two. One advantage is that every restriction site becomes an internal positive control for every other restriction site. Restriction endonucleases should be chosen for which several recognition sites are conserved between PCR products while others are variable. Under these conditions, misinterpretation of this assay is excluded except in cases of cross-contamination of samples because the conserved restriction sites act as controls for oligonucleotide specificity during the PCR, while the variable restriction sites act as taxonomic characters that can unambiguously identify any strain. Absence or presence of a restriction site is much more believable if several adjacent restriction sites were successfully digested in the same reaction.

Another important advantage is that relatively large PCR products can be examined over their entire length. For example, an enzyme that cuts, on average, once every 256 nucleotides can be expected to digest a 1536-nucleotide-log PCR product at roughly six sites. These six recognition sites are small sequence samples scattered over the entire length of the PCR product. By repeating these digestions with a number of different restriction enzymes, several dozen or more restriction sites can be surveyed from a single PCR amplification. If these restriction site are mapped, each of these mapped restriction site polymorphisms (MRSPs) can be used as a phylogenetic character that can be further used to predict phylogenetic relationships between individuals (see Section III on results). If enough polymorphisms can be mapped, MRSP analysis can be as phylogenetically informative as DNA sequence data.

The last major advantage of this method is that it is relatively fast and efficient. Typically, the time elapsed between taking the template DNAs out of the freezer to set up the PCR reactions until reading the MRSPs is roughly 36 h; however, this time could be cut to as little as 15 h with some minor technical improvements. Furthermore, many samples can be processed simultaneously. Typically, 48 samples are amplified and analyzed at one time. It is this characteristic of speed and efficiency that suggests that MRSP analysis of PCR products could be a useful approach for analyzing the many samples implied in environmental surveys and clinical diagnostics.

II. MATERIALS AND METHODS

A. CHOOSING PRIMERS FOR AMPLIFICATION

In order for PCR-linked MRSP analysis to yield taxonomic information most efficiently, it is imperative that the oligonucleotide primers used in the PCRs be carefully designed. These primers must anneal to sequences that are highly conserved or identical between all the organisms for which taxonomic differentiation is desired. In addition, the sequences to which these primers anneal must flank a region that contains sequence variation among all these same organisms.

We have found that it is useful if the degree of sequence variation in this primer-flanked domain is great enough that some, but not all, of the restriction enzyme recognition sites are polymorphic between the various organisms to be differentiated. It is most advantageous if the sequence of the domain to be amplified is known for one or more of the organisms to be differentiated. If these two conditions are met, it becomes relatively easy to know the position of every polymorphic site simply by comparing the sizes of the resulting restriction fragments with those predicted by the known sequence. New recognition sites are also easy to identify since they almost always occur at positions at which a single nucleotide change results in creating the previously unobserved site. In this way all the polymorphic sites can be mapped for future phylogenetic analysis.

For the phylogenetic groups that we have been most interested in, which are all eubacterial, we have found that the genes encoding the ribosomal RNAs (rRNAs) have been most useful. There are several reasons for this choice. One is that all organisms possess genomes that encode functionally and structurally homologous rRNAs. Another reason is that the rRNAs have been sequenced from a large array of organisms.[58] These rRNAs contain various sequence domains that vary dramatically in their degree of conservation between species or taxonomic groups.[59,60] Some of these sequence domains are extremely similar even when compared between organisms that are only distantly related phylogenetically. These domains make ideal sites for primer design. For our studies of the 16S rRNA genes we have used primers that are very similar to those used by Weisburg et al.[33] For studying 23S rRNA genes, we have designed a novel set of primers.[44] Other domains in the rRNA genes are known to be less conserved and contain the variability needed to generate the required MRSPs.

The PCR products that result from using these primers on eubacterial genomes range in size from 1.2 to 2.0 kb. When these products were digested with a wide variety of restriction enzymes, it was observed that most of the restriction sites are conserved between species. However, for many enzymes, one or two restriction sites were found to be polymorphic between any two species. These polymorphisms result upon electrophoretic separation in species- or strain-specific patterns of DNA fragments (see below). Some of the fragments (restriction sites) were also found to be conserved between any pair of compared organisms. These conserved restriction fragments are important controls in these MRSP analyses because they show that there is sequence similarity and homology between the compared PCR products. Confirmation of this similarity confirms that no artifactual amplification occurred during the PCR.

Many other investigators have also used restriction digests of PCR-amplified rRNA genes to differentiate between species and/or strains.[32,46,48-50,56,61] There are, however, many other kinds of sequence domains that might be superior for different phylogenetic applications. One example of such an application might be an investigation to distinguish between very closely related organisms. Ribosomal RNA genes evolve more slowly than do the rest of the genome as a whole. This could lead to situations in which considerable genetic variation could occur within a group of organisms while there is little variation in the restriction profiles of their respective rRNA genes. The obvious solution to this problem is to design primers that amplify some other sequence that is expected to diverge more rapidly than coding domains. The primers would anneal to sequences that would be evolutionarily conserved while they would amplify regions that would be expected to diverge much more rapidly. Another example of an application for which amplification of rRNAs is inappropriate is the investigation of genetic diversity in viruses.[7,9,12,39,53] Since viral genomes do not encode rRNAs, other domains must be utilized to investigate restriction site polymorphisms.

B. PCR PROTOCOL

Typical PCR reaction conditions for amplifying 1.2- to 2.0-kb portions of rRNA genes were as follows:

0.5 µM	Primers
200 nM	Each dNTP
1.5 mM	MgCl$_2$
10 mM	Tris Cl (pH 8.3)
50 mM	KCl
1.25 U	*Taq* polymerase
2.5 µCi	α^{32}P-dCTP 3000 Ci/mmol (optional)
1.0 ng	Bacterial template DNA
total reaction volume = 50 µl	

Reactions were subjected to 30 cycles of 1.0 min at 94°C, 1.0 min at 50°C, and 2.0 min at 72°C in a Perkin Elmer 9600 thermocycler. In our investigations, radiolabeled nucleotides are used in the reaction mixtures. This results in uniformly radiolabeled PCR products. For several reasons discussed below, the use of radiolabeled nucleotides greatly increased the efficiency of our investigations. Isotopic labeling is, however, not required for MRSP analysis of PCR products (see discussion in Section IV).

One of the keys to making this type of analysis maximally efficient is to array the various template DNAs in 96-well microtiter plates. These various template DNAs should be at roughly 0.2 ng/µl. Once in an array of this nature, these master plates of template DNAs can be stored at −20°C until needed. For the PCR reactions 5.0 µl of each template DNA is transferred from the master plate to a separate 400-µl PCR reaction tube or other series of vessels for PCR. We use 96-tube arrays (Perkin Elmer) but other thermocyclers that use conventional microtiter plates (e.g., as manufactured by M. J. Research, Inc.) are acceptable. The feature of these two styles that makes them attractive is that neither of them requires the investigator to manipulate individual tubes with individual lids. Individual tubes are easy to confuse and must be individually labeled. The need to open and close these tubes makes them cumbersome and tedious to handle. Positional information inherent in fixed arrays makes all of this unnecessary. It is highly advisable to use an eight-channel multipipetter to transfer solutions into and between these 96-well arrays (e.g., as manufactured by Finnpipette).

After the template DNAs (50 µl) are added to the individual tubes of the 96-well reaction array, 45 µl of a solution containing all the other reagents required for the PCR reactions are added to each tube. An eight-channel multipipetter makes this much easier and is highly recommended. The solution containing all the other PCR reagents, including the α^{32}P-dCTP, are mixed together behind a Plexiglas™ shield in such quantities as to complete all the desired PCR reactions plus a little extra for ease of manipulation. This solution is made at 1.11X so that it becomes 1X when added to the template DNA. For ease of manipulation, this solution is placed in a trough (CoStar) before being transferred with the eight-channel multipipetter. Using the solutions and equipment described here, it takes roughly 15 min to set up and start each array of PCR reactions. The PCR reactions themselves take about 3 h to complete.

C. RESTRICTION DIGESTS OF PCR PRODUCTS

After the PCR reactions have been completed, 4 to 5 µl of each reaction are transferred to a separate well of a fresh 96-well microtiter plate. Fifteen microliters of a solution containing

2.0 to 10 U of a restriction enzyme are added to each well and mixed by pipetting with the eight-channel multipipetter. Like the PCR reagent solution, this restriction-enzyme-containing solution is made at a concentration such that when it is added to the PCR reactions, the concentration of the buffer for each restriction enzyme is close to that recommended by the manufacturer. These restriction digests of PCR products are sealed by simply placing a piece of plastic adhesive tape over the top of the microtiter plate. Digestions were normally done for 2 or 3 h but could be left overnight without drying out. Digestions were incubated at the temperature recommended by the manufacturer (usually in a 37°C incubator). After incubation, 3 to 5 µl of a solution of 25% glycerol, 0.25% bromophenol blue, 25 mM Tris Cl (pH = 7.5) were added to each reaction. These reactions could then be stored at –20°C for several days or loaded directly onto a polyacrylamide gel. Typically, about 2 to 4 ml of this final restriction digested solution were added onto a single lane.

D. POLYACRYLAMIDE GELS

We found that thin (0.35 to 0.40 mm) 40-cm-long polyacrylamide gels were best for these MRSP experiments. The glass plates, spacers, and combs used for these gels were previously used for DNA sequencing. The most important difference between the gels used for our MRSP experiments and those used for DNA sequencing was that the gels for the MRSP experiments were nondenaturing gels that did not contain urea. Eliminating urea in these gels means that the restricted PCR products to not have to be diluted in formamide or denatured at 95°C, which decreases sample manipulation. It should be noted that nondenaturing gels have much lower electrical resistances, so that they run at relatively high currents and low voltages (around 1000 V or less) as compared to urea-containing gels. If possible, run these gels at a constant power setting of between 40 and 50 W/hr. Usually, these gels are formulated as 5.0% polyacrylamide (19:1 acrylamide to bisacrylamide) and are 1X for TBE electrophoresis buffer.[63] Typical run times vary from 2 to 4 h, depending on the desired resolution.

After electrophoresis, the glass plates are separated and the gels are picked up on large sheets of filter paper. The gels stick to the filter paper and can be easily peeled off the remaining glass plate. The gels are then covered with a plastic wrap on the side not facing the filter paper and dried on a standard sequencing gel dryer (BioRad). It is not necessary to fix or stain the gels before drying. If imperfections occur while the glass plates are being separated, fixing the gels in a solution of 12% methanol and 10% acetic acid provides an opportunity to remedy these problems. Gels without urea or fixers dry fairly rapidly, usually less than 45 min. After drying, the gels are used to expose X-ray film 16 to 24 h. An intensifying screen is sometimes used (optional).

III. RESULTS

In one example, a *Bfa*I digest of PCR products derived from the 16S rRNA genes of spirochetes in the genus *Borrelia* were predicted to be polymorphic in this genus by comparing published sequences of their 16S rRNAs.[64,65] The *Bfa*I sites at bases 86 and 244 and a third site at either base 150 or 495 in the standard numbering system distinguish the three groups most closely related to *B. burgdorferi*.[66] *B. burgdorferi* has a site at 244, *B. garinii* has sites at 86 and 244, and group VS461 has sites at 86 and either 150 or 495. The three groups could be reliably distinguished from each other by this single digest.[63a] All three groups were distinguishable from the closely related *B. anserina* and *B. hermsii* by multiple restriction site polymorphisms.

If many such experiments are performed, it is possible to map many polymorphisms that differentiate between closely related species. Because these polymorphisms are mapped, they can be used in parsimony analysis to reconstruct the phylogenetic relationships between the various taxa.[67-69] In one study,[44] 21 polymorphisms that differentiated between 48 strains

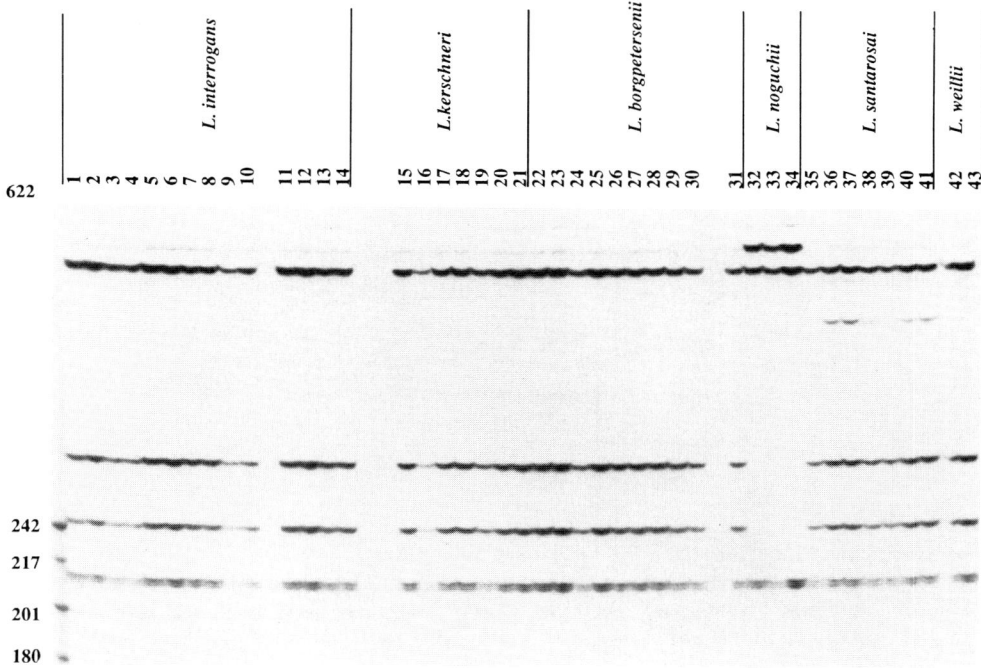

Figure 1. *Hin*f1 site polymorphisms in PCR-amplified 23S rRNA genes from *Leptospira*. A *Hin*f1 site at 1517 is absent in *L. noguchii* but is present in five other species. Products were separated by polyacrylamide gel electrophoresis and visualized by autoradiography. Numbers indicate fragment sizes in nucleotides.

representing seven species in the spirochete genus *Leptospira* were mapped. An example is shown in Figure 1. These species were easily differentiated by this approach. In addition, some intraspecific variation was observed. A phylogenetic tree resulting from the parsimony analysis of these data plus 11 polymorphisms we have subsequently mapped is shown in Figure 2. Such a phylogenetic tree can form the basis of many interesting physiological, ecological, and epidemiological studies of this important group of pathogenic and nonpathogenic bacteria. It is expected that this type of analysis could be extended to virtually any group of organisms.

IV. DISCUSSION

Mapped restriction site polymorphism (MRSP) analysis of PCR products represents a sensitive, efficient, and widely applicable approach to many problems inherent in microbial strain differentiation and identification. While certainly not limited to clinical applications, the use of this technology in diagnostics is facile. Certainly most of the published applications of this technology have so far been directed towards clinical diagnostics (see introduction in Section I).

One of the great advantages of PCR as a means of detection is its great sensitivity. Typically, the sensitivity of PCR to detect target sequences in DNA derived from clinical specimens is somewhat lower than its sensitivity using DNA derived from more abundant sources. This is unfortunate because the greatest sensitivity is desired when analyzing clinical specimens. One strategy to increase the sensitivity of PCR is to perform two consecutive PCRs using nested oligonucleotide primers.[9,12,15,16,18,24] This second PCR is associated with a significant increase in sensitivity and specificity. Kaltenboeck et al.[15] reported that nested PCR could detect one bacterial genome in a background composed of the DNA extracted from 21,000 human cells.

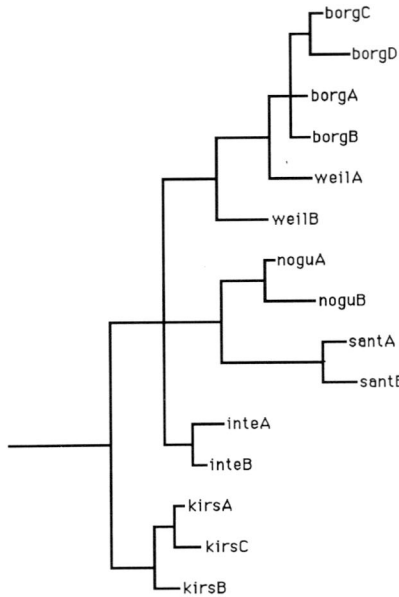

Figure 2. Phylogenetic tree for *Leptospira* species constructed by parsimony analysis. A cladogram constructed using the PAUP phylogenetic package of Swofford.[44,67] *Leptospira biflexa* was defined as the out group. This cladogram was constructed using a complete search by the Branch and Bound method. Capital letters indicate different intraspecific variants detected within individual species. The seven *Leptospira* species described are *L. borgpetersinii* (borg), *L. weillii* (weil), *L. kirschneri* (kirs), *L. interrogans* (inte), *L. santarosai* (sant), *L. noguchii* (nogu), and *L. biflexa* (biflexa).

A system could be imagined in which nested PCR was performed using primers with different phylogenetic specificities. In this case, the primers used in the second PCR anneal to only a subset of the templates to which the first primers annealed. The primers used in the first PCR could have very broad phylogenetic specificity (e.g., Reference 33) while the internal primers used in the second PCR could direct the amplification of templates derived from only a specific taxonomic group. A series of secondary PCRs could be performed using separate aliquots of the primary PCR. Each of these secondary PCRs would be directed by a different primer pair, each of which amplifies DNA derived from different taxonomic groups. The resulting secondary PCR products could be examined by MRSP analysis to identify the specific strains or species. This scheme combines the sensitivity of nested PCR with the taxonomically broad requirements of clinical diagnostics in an efficient and timely assay.

In our experiments, we used thin sequencing-style polyacrylamide gels without urea to separate the radiolabeled PCR-derived restriction fragments generated in our MRSP analyses. These individual DNA fragments are then detected by autoradiography. Most other investigators (e.g., Reference 32) use agarose gels or thicker polyacrylamide gels and ethidium bromide staining to separate and detect the various DNA fragments. For many applications, agarose gels and ethidium bromide staining is perfectly acceptable. However, polyacrylamide gels and autoradiography have several important advantages. One is that many more restriction digests can be performed on each PCR product if the products are radiolabeled. Typically, 10 to 12 restriction digests were performed on every 50-μl PCR product, but this number could easily be extended to 20 or more by simply extending the exposure times for the X-ray film. Another advantage in using polyacrylamide gels is that small DNA fragments that are usually difficult to separate on agarose gels or stain with ethidium bromide are easily resolved and visualized with polyacrylamide gels and autoradiography. Visualization of small DNA fragments helps significantly in mapping the observed polymorphisms. The last advantage of polyacrylamide gels is that length polymorphisms of as small as one or a few nucleotides are easily visualized on these gels. Frequently these small length polymorphisms can be mapped and used as taxonomic characters themselves. With all these considerations taken into account, it might be best to use polyacrylamide gels and autoradiography when initiating an MRSP for phylogenetic analysis but convert to agarose gels or thicker polyacrylamide gels and ethidium bromide staining for routine strain identifications.

REFERENCES

1. **Saiki, R. K., D. H. Gelfand, S. Stoffel, S. J. Scharf, R. Higuchi, G. T. Horn, K. B. Mullis, and H. A. Ehrlich.** 1988. Primer directed enzymatic amplification of DNA with a thermostable DNA polymerase. *Science* 239:487–491.
2. **Fuhrman, J. A., K. McCallum, and A. A. Davis.** 1992. Novel major archaebacterial group from marine plankton. *Nature* 356:148–148.
3. **Ward, D. M., R. Weller, and M. M. Bateson.** 1990. 16S rRNA sequences reveal uncultured microorganism in a natural community. *Nature* 345:63–65.
4. **Schmidt, T. M., E. F. DeLong, and N. R. Pace.** 1991. Analysis of a marine picoplankton community by 16S rRNA gene cloning and sequencing. *J. Bacteriol.* 173:4371–4378.
5. **Liesack, W., and E. Stackebrandt.** 1992. Occurrence of novel groups of the domain *Bacteria* as revealed by analysis of genetic material isolated from an Australian terrestrial environment. *J. Bacteriol.* 174:5072–5078.
6. **Giovannoni, S. J., T. B. Britschgi, G. L. Moyer, and K. G. Field.** 1990. Genetic diversity in Sargasso Sea bacteriolplankton. *Nature* 345:60–63.
7. **Shih, J. W.-K., L. C. Cheung, H. J. Alter, L. M. Lee, and J. R. Gu.** 1991. Strain analysis of hepatitis B virus on the basis of restriction endonuclease analysis of polymerase chain reaction products. *J. Clin. Microbiol.* 29:1640–1644.
8. **Anderson, B. E., J. W. Sumner, J. E. Dawson, T. Tzisanabos, C. R. Greene, J. G. Olson, D. B. Fishbein, M. Olsen-Rasmussen, B. P. Holloway, E. H. George, and A. F. Azad.** 1992. Detection of the etiologic agent of human ehrlichiosis by polymerase chain reaction. *J. Clin. Microbiol.* 30:775–780.
9. **Brytting, M., W. Xu, B. Wahren, and V.-A. Sundqvist.** 1992. Cytomegalovirus DNA detection in sera from patients with active cytomegalovirus infections. *J. Clin. Microbiol.* 30:1937–1941.
10. **Ballinger, S. W., T. G. Schurr, A. Torroni, Y. Y. Gan, J. A. Hodge, K. Hassan, K.-H. Chen, and D. C. Wallace.** 1992. Southeast Asian mitochondrial DNA analysis reveals genetic continuity of ancient mongoloid migrations. *Genetics* 130:139–152.
11. **Long, G. W., J. J. Oprandy, R. B. Narayanan, A. H. Fortier, K. R. Porter, and C. A. Nacy.** 1993. Detection of *Francisella tularensis* in blood by polymerase chain reaction. *J. Clin. Microbiol.* 31:152–154.
12. **Liou, T.-C., T.-T. Chang, K.-C. Young, X.-Z. Lin, C.-Y. Lin, and H.-L. Wu.** 1992. Detection of HCV RNA in saliva, urine, seminal fluid, and ascites. *J. Med. Virol.* 37:197–202.
13. **Gerritsen, M. J., T. Olyhoek, M. A. Smits, and B. A. Bokhout.** Sample preparation method for polymerase chain reaction-based semiquantitative detection of *Leptospira interrogans* serovar harjo subtype hardjobovis in bovine urine. *J. Clin. Microbiol.* 29:2805–2808.
14. **Van Eys, G. J. J. M., C. Gravekamp, M. J. Gerritsen, W. Quint, M. T. E. Cornelissen, J. Ter Schegget, and W. J. Terpstra.** 1989. Detection of leptospires in urine by polymerase chain reaction. *J. Clin. Microbiol.* 27:2258–2262.
15. **Kaltenboeck, B., K. G. Kousoulas, and J. Storz.** 1991. Detection and strain differentiation of *Chlamydia psittaci* mediated by a two-step polymerase chain reaction. *J. Clin. Microbiol.* 29:1969–1975.
16. **Jaton, K., R. Sahli, and J. Bille.** 1992. Development of polymerase chain reaction assays for detection of *Listeria monocytogenes* in clinical cerebrospinal fluid samples. *J. Clin. Microbiol.* 30:1931–1936.
17. **Muir, P., F. Nicholson, M. Jhetam, S. Neogi, and J. E. Banatvala.** 1993. Rapid diagnosis of enterovirus infection by magnetic bead extraction and polymerase chain reaction detection of enterovirus RNA in clinical specimens. *J. Clin. Microbiol.* 31+:31–38.
18. **Noordhoek, G. T., E. C. Wolters, M. E. J. DeJonge, and J. D. A. Van Embden.** 1991. Detection by polymerase chain reaction of *Treponema pallidum* DNA in cerebrospinal fluid from neurosyphilus patients before and after antibiotic treatment. *J. Clin. Microbiol.* 29:1976–1984.
19. **Shawar, R. M., F. A. K. El-Zaatari, A. Nataraj, and J. E. Clarridge.** 1993. Detection of *Mycobacterium tuberculosis* in clinical samples by two-step polymerase chain reaction and nonisotopic hybridization methods. *J. Clin. Microbiol.* 31:61–65.
20. **Altamirano, M., M. T. Kelly, A. Wong, El. T. Bessuille, W. A. Black, and J. A. Smith.** 1992. Characterization of a DNA probe for detection of *Mycobacterium tuberculosis* complex in clinical samples by polymerase chain reaction. *J. Clin. Microbiol.* 30:2173–2176.

21. **Fauville-Dufaux, M., B. Vanfleteren, L. De Wit, J. P. Van Vooren, M. D. Yates, E. Serruys, and J. Content.** Rapid detection of tuberculous and non-tuberculous mycobacteria by polymerase chain reaction amplification of a 162 bp DNA fragment from antigen 85. *Eur. J. Microbiol. Infect. Dis.* 11:797–803.
22. **Sjobring, U., M. Mecklenburg, Andersen A. Bengard, and H. Miorner.** 1990. Polymerase chain reaction for detection of *Mycobacterium tuberculosis*. *J. Clin. Microbiol.* 28:2200–2204.
23. **Buck, G. E., L. C. O'Hara, and J. T. Summersgill.** 1992. Rapid, simple method for treating clinical specimens containing *Mycobacterium tuberculosis* to remove DNA for polymerase chain reaction. *J. Clin. Microbiol.* 30:1331–1334.
24. **Tyan, Y.-S., S.-T. Liu, W.-R. Ong, M.-L. Chen, C.-H. Shu, and Y.-S. Chang.** 1993. Detection of Epstein-Barr virus and human papillomavirus in head and neck tumors. *J. Clin. Microbiol.* 31:53–56.
25. **Ibrahim, A., W. Liesack, and E. Stackebrandt.** 1992. Polymerase chain reaction-gene probe detection system specific for parthogenic strains of *Yersinia enterocolitica*. *J. Clin. Microbiol.* 30:1942–1947.
26. **Tachibana, H., S. Kobayashi, K. C. Paz, I. S. Aca, S. Tateno, and S. Ihara.** 1992. Analysis of pathogenicity by restriction-endonuclease digestion of amplified genomic DNA of *Entamoeba histolytica*. *Parasitol. Res.* 78:433–436.
27. **Deng, S. and C. Hiruki.** 1991. Genetic relatedness between two nonculturable mycoplasmalike organisms revealed by nucleic acid hybridization and polymerase chain reaction. *Mol. Plant Path.* 81:1475–1479.
28. **Persing, D. H., S. R. Telford III, P. N. Rys, D. E. Dodge, T. J. White, S. E. Malawista, and A. Speilman.** 1990. Detection of *Borrelia burgdorferi* DNA in museum specimens of *Ixodes dammini* ticks. *Science* 249:1420–1423.
29. **Welsh, J. and M. McClelland.** 1992. PCR-amplified length polymorphisms in tRNA intergenic spacers for categorizing staphylococci. *Mol. Microbiol.* 6:1673–1680.
30. **Welsh, J., C. Petersen, and M. McClelland.** 1991. Polymorphisms generated by arbitrarily primed PCR in the mouse: application to strain identification and genetic mapping. *Nucleic Acids Res.* 19:303–306.
30a. **Welsh, J. and M. McClelland.** 1990. Fingerprinting genomes using PCR with arbitrary primers. *Nucleic Acids Res.* 18:7213–7218.
30b. **Williams, J. G. K., A. R. Kubelik, K. J. Livak, J. A. Rafalski, and S. V. Tingey.** 1990. DNA polymorphisms amplified by arbitrary primers are useful as genetic markers. *Nucleic Acids Res.* 18:6531–6535.
31. **Medlin, L., H. J. Elwood, S. Stickel, and M. L. Sogin.** 1988. The characterization of enzymatically amplified eukaryotic 16S-like rRNA-coding regions. *Gene* 71:491–499.
32. **Vilgalys, R. and M. Hester.** 1990. Rapid genetic identification and mapping of enzymatically amplified ribosomal DNA from several *Cryptococcus* species. *J. Bacteriol.* 172:4238–4246.
33. **Weisburg, W. G., S. M. Barns, D. A. Pelletier, and D. J. Lane.** 1991. 16S ribosomal DNA amplification for phylogenetic study. *J. Bacteriol.* 173:697–703.
34. **Simonet, P., P. Normand, A. Moiroud, and R. Bardin.** 1990. Identification of *Frankia* strains in nodules by hybridization of polymerase chain reaction products with strain-specific oligonucleotide probes. *Arch. Microbiol.* 153:235–240.
35. **Morita, K., M. Tanaka, and A. Igarashi.** 1991. Rapid identification of dengue virus serotypes by using polymerase chain reaction. *J. Clin. Microbiol.* 29:2107–2110.
36. **Adam, T., G. S. Gassmann, C. Rasiah, and U. B. Gobel.** 1991. Phenotypic and genotypic analysis of *Borrelia burgdorferi* isolates from various sources. *Infect. Immun.* 59:2579–2585.
37. **Hookey, J. V.** 1992. Detection of Leptospiraceae by amplification of 16S ribosomal DNA. *FEMS Microbiol. Lett.* 90:267–274.
38. **Kron, M. A., A. Gupta, and C. Mackenzie.** 1991. Identification of related DNA sequences in *Borrelia burgdorferi* and two strains of *Leptospira interrogans* by using polymerase chain reaction. *J. Clin. Microbiol.* 29:2338–2340.
39. **Todd, D., K. A. Mawhinney, and M. S. McNulty.** 1992. Detection and differentiation of chicken anemia virus isolates by using the polymerase chain reaction. *J. Clin. Microbiol.* 30:1661–1666.
40. **Bottger, E. C.** 1989. Rapid determination of bacterial ribosomal RNA sequences by direct sequencing of enzymatically amplified DNA. *FEMS Microbiol. Lett.* 65:171–179.
41. **Kitada, K., S. Oka, S. Kimura, K. Shimada, T. Serikawa, J. Yamada, H. Tsunoo, K. Egawa, and Y. Nakamura.** 1991. Detection of *Pneumocystis carinii* sequences by polymerase chain reaction: Animal models and clinical application to noninvasive specimens. *J. Clin. Microbiol.* 29:1895–1990.
42. **Rainey, F. A., M. Dorsch, H. W. Morgan, and E. Stackebrandt.** 1992. 16S rDNA analysis of *Spirochaeta thermophila*: Its Phylogenetic position and implications for the systematics of the order Spirochaetales. *Syst. Appl. Microbiol.* 15:197–202.
43. **Dame, J. B., S. M. Mahan, and C. A. Yowell.** 1992. Phylogenetic relationship of *Cowdria ruminantium*, agent of heartwater, to *Anaplasma marginale* and other members of the order Rickettsiales determined on the basis of 16S rRNA sequence. *Int. J. Syst. Bacteriol.* 42:270–274.
44. **Ralph, D., M. McClelland, J. Welsh, G. Baranton, and P. Perolat.** 1993. *Leptospira* categorized by arbitrarily primed PCR and by mapped restriction site polymorphisms in PCR-amplified rDNA. *J. Bacteriol.*, 175:973–981..

45. **Liu, Z. L. and J. B. Sinclair.** 1992. Genetic diversity of *Rhizoctonia solani* Anastomosis Group 2. *Mol. Plant Pathol.* 82:778–787.
46. **Clarke, C. G. and L. S. Diamond.** 1991. Ribosomal RNA genes of "pathogenic" and "nonpathogenic" *Entamoeba histolytica* are distinct. *Mol. Biochem. Parasitol.* 49:297–302.
47. **Rodriguez, P., A. Verkis, B. De Barbeyrac, B. Dutilh, J. Bonnet, and C. Bebear.** Tyoing of *Chlamydia trachomatis* by restriction endonuclease analysis of the amplified major outer membrane protein gene. *J. Clin. Microbiol.* 29:1132–1136.
48. **Vaneechoutte, M., R. Rossau, P. De Vos, M. Gillis, D. Janssens, N. Paepe, A. De Rouck, T. Fiers, G. Claeys, and K. Kersters.** 1992. Rapid identification of bacteria of the Comamonadaceae with amplified ribosomal DNA-restriction analysis (ARDRA). *FEMS Microbiol. Lett.* 93:227–234.
49. **Jayarao, B. M., J. J. E. Dore, Jr., G. A. Baumbach, K. R. Matthews, and S. P. Oliver.** 1991. Differentiation of *Streptococcus uberis* from *Streptococcus parauberis* by polymerase chain reaction and restriction fragment length polymorphism analysis of 16S ribosomal DNA. *J. Clin. Microbiol.* 29:2774–2778.
50. **Gurtler, V., V. A. Wilson, and B. C. Mayall.** 1991. Classification of medically important clostridia using restriction endonuclease site differences of PCR-amplified 16S rRNA. *J. Gen. Microbiol.* 137:2673–2679.
51. **Regnery, R. L., C. L. Spruill, and B. D. Plikaytis.** 1991. Genotypic identification of Rickettsiae and estimation of intraspecific sequence divergence from portions of two rickettsial genes. *J. Clin. Microbiol.* 173:1576–1589.
52. **Manor, E., J. Ighbarieh, B. Sarov, I. Kassis, and R. Regnery.** 1992. Human and tick spotted fever group rickettsia isolates from Israel: a genotypic analysis. *J. Clin. Microbiol.* 30:2653–2656.
53. **Carman, W. F., C. Williamson, B. A. Cunliffe, and A. H. Kidd.** 1989. Reverse transcription and subsequent DNA amplification of rubella virus RNA. *J. Virol. Methods.* 25:21–39.
54. **Beati, L., J.-P. Finidori, B. Gilot, and D. Raoult.** 1992. Comparison of serologic typing, sodium dodecyl sulfate-polyacrylamide gel electrophoresis protein analysis, and genetic restriction fragment length polymorphism analysis for identification of Rickettsiae: characterization of two new Rickettsial strains. *J. Clin. Microbiol.* 30:1992–1930.
55. **Gutekunst, K. A., B. P. Holloway, and G. M. Carlone.** 1992. DNA sequence heterogeneity in the gene encoding a 60-kilodalton extracellular protein of *Listeria monocytogenes* and other *Listeria* species. *Can. J. Microbiol.* 38:865–870.
56. **Ahrens, U. and E. Seemuller.** 1992. Detection of DNA of plant pathogenic mycoplasmalike organisms by a polymerase chain reaction that amplifies a sequence of the 16S rRNA gene. *Phytopathology* 82:828–832.
57. **Akopyanz, N., N. O. Bukanov, T. U. Westblom, and D. E. Berg.** 1993. PCR-based RFLP analysis of DNA sequence diversity in the gastric pathogen *Helicobacter pylori*. *Nucleic Acids Res.*, in press.
58. **Dams, E., L. Hendriks, Y. Van de Peer, J.-M. Neefes, G. Smits, I. Vandenbempt, and R. De Wachter.** 1988. Compilation of small ribosomal subunit RNA sequences. *Nucleic Acids Res.*, Suppl. 16:r87–r173.
59. **Lane, D. J., B. Pace, G. J. Olsen, D. A. Stahl, M. L. Sogin, and N. R. Pace.** 1985. Rapid determination of 16S ribosomal RNA sequences for phylogenetic analysis. *Proc. Natl. Acad. Sci. U.S.A.*, 82:6955–6959.
60. **Woese, C. R.** 1987. Bacterial evolution. *Microbiol. Rev.* 51:221–271.
61. **Gaydos, C. A., T. C. Quinn, and J. J. Eiden.** 1992. Identification of *Chlamydia pneumoniae* by DNA amplification of the 16S rRNA gene. *J. Clin. Microbiol.* 30:796–800.
62. **Gotoda, T., N. Yamada, T. Murase, H. Shimano, M. Shimada, K. Harada, M. Kawamura, K. Kozaki, and Y. Yazaki.** 1992. Detection of three separate DNA polymorphisms in the human lipoprotein lipase gene by gene amplification and restriction endonuclease digestion. *J. Lipid Res.* 33:1067–1072.
63. **Sambrook, J., E. F. Fritsch, and T. Maniatis.** 1989. *Molecular Cloning*, ed. C. Nolan, Cold Spring Harbor Laboratory Press, Cold Spring Harbor, NY.
63a. **Ralph, D., D. Postic, G. Baranton, C. Pretzman, and M. McClelland.** 1993. Species of *Borrelia* distinguished by restriction site polymorphisms in 16S rRNA genes. *FEMS Micro. Lett.* 111:239–244.
64. **Marconi, R. T. and C. F. Garon.** 1992. Phylogenetic analysis of the genus *Borrelia*: a comparison of North American and European isolates of *Borrelia burgdorferi*. *J. Bacteriol.* 174:241–244.
65. **Paster, B. J., E. Stackebrandt, R. B. Hespell, C. M. Hahn, and C. R. Woese.** 1984. The phylogeny of the spirochetes. *Syst. Appl. Microbiol.* 5:337–351.
66. **Welsh, J., C. Pretzman, D. Postic, I. Saint Girons, G. Baranton, and M. McClelland.** 1992. Genomic Fingerprinting by arbitrarily primed polymerase chain reaction resolves *Borrelia burgdorferi* into three distinct phyletic groups. *Int. J. Syst. Bact.* 42:370–377.
67. **Swofford, D. L.** 1991. PAUP: Phylogenetic Analysis Using Parsimony. Version 3.0. Illinois Natural History Survey, Champaign.
68. **Carr, S. M., A. J. Brothers, and A. C. Wilson.** 1987. Evolutionary inferences from restriction maps of mitochondrial DNA from nine taxa of *Xenopus* frogs. *Evolution* 41:176–188.
69. **Felsenstein, J.** 1988. PHYLIP: Phylogenetic Inference Package. Version 3.0. University of Washington. Seattle.

Chapter 15

DETECTION OF POLYMORPHIC DNA SEQUENCES AT THE 3' END OF ALU REPEATS BY PCR

Jean-Paul Charlieu

TABLE OF CONTENTS

I. Introduction 133

II. Materials and Methods 134
    A. Primers 134
    B. DNA Templates 134
    C. 5' End Labeling of Primers 134
    D. PCR Conditions 134
    E. Analysis of PCR Products 136

III. Results 136
    A. Choice of Primers 136
    B. Detection of Polymorphic Microsatellite at the 3' End of
       Alu Repeats 136
    C. Characterization of Detected Markers 137

IV. Discussion 137

Acknowledgments 139

References 140

I. INTRODUCTION

The localization of genes or loci involved in genetic diseases requires the construction of genetic maps consisting of an alignment and ordering of polymorphic markers with two markers being separated by a genetic distance expressed in the centiMorgan (cM). To create such maps, pluriallelic "variable number of tandem repeats" (VNTRs) are increasingly searched today because they generally provide more information than the biallelic "restriction fragment length polymorphisms" (RFLPs). Based on the length of the basic motif, VNTRs can be classified as "minisatellites" or "microsatellites". Among the microsatellites, the dinucleotide repeats such as (TG)n are very popular because of their frequency in the human genome.[1,2] It has been shown recently that the 3' end of Alu repeats may also contain polymorphic microsatellites.[3-5] Alu repeats are dimeric sequences found in high copy numbers (~10⁶) interspersed throughout the human genome.[6] In 6.7 to 10% of cases,[3,7] the adenine-rich extension of these retrotransposon-like elements is a microsatellite which may be polymorphic. The detection of such polymorphisms, as for dinucleotide repeat microsatellites, can be achieved by PCR amplifying a DNA fragment containing the microsatellite, and by determining the size of the amplified fragment(s), which reflects the number of the basic motif of each allele of the VNTR. However, such a study is often time consuming as the nucleotide sequence flanking the microsatellite of interest has to be determined, in order to design sequence-specific PCR primers.

We have developed another strategy to detect polymorphic microsatellites present at the 3'-end of Alu repeats.[8] The method, called "3' Alu PCR" and consisting of the PCR amplification of a DNA fragment between a primer directed toward the 3' end of an Alu repeat and a randomly chosen primer, presents several advantages over the way presented above: (1) Knowledge of the nucleotide sequence downstream to the Alu repeat is not necessary. The low-stringency PCR conditions used allow *in vitro* amplification of the DNA fragment containing the 3' extension of the Alu repeats even with an imperfectly matching primer.[9] (2) For this reason, no specific PCR primer has to be designed. If a particular region of the human genome is studied, the presence of an Alu repeat containing a polymorphic microsatellite can be tested in the vicinity of other genetic markers or of sequence tagged sites (STSs) with previously designed primers. (3) Several polymorphic markers are generally detected at different loci with a single pair of primers. Linkage analysis could therefore be performed with a reduced number of experiments.

II. MATERIALS AND METHODS

A. PRIMERS

The primer Alu 2, designed from a consensus Alu sequence[10] and directed toward the 3' extension, is common to all pairs of primers. The second primer of each pair was selected from existing PCR primers previously designed for other PCR purposes. The primers for which results are presented in this chapter are listed in Table 1.

All primers were resuspended in sterile distilled water to obtain a stock solution at 100 μM.

B. DNA TEMPLATES

The following rodent/human somatic hybrid cell lines were used: WA17 containing chromosome 21 as the sole human DNA source was kindly provided by Dr. E. Devine; C35B2 (human chromosome 2) and BCHE (human chromosomes 3, 4, 6, 8, 9, 10, 13, 14, 15, 16, 18, 19, Xp⁺) were from Dr. N'Guyen Van Cong.

The DNA was prepared by lysing the cells with 200 µg/ml of proteinase K (Appligène) in 0.5 M EDTA, 1% L-lauryl sarcosyl, pH 8.5, for 12 h at 50°C, then purified by phenol extraction and ethanol precipitation using standard methods.[11]

DNA from unrelated individuals and from complete pedigrees was provided by CEPH (Centre d'Étude du Polymorphisme Humain, Paris).

C. 5' END LABELING OF PRIMERS

In order to detect the amplified DNA fragments in a sequencing gel, one of the PCR primers was labeled with [$\gamma^{35}S$] ATP by kination, as described in the following paragraph.

In a 1.5-ml Eppendorf tube, mix the following components: 1 nmol of 5' OH synthetic primer, 1/10 of the final volume of kinase buffer 10 × concentrated (0.5 M Tris-HCl, 0.1 M MgCl$_2$, 0.05 M dithiotreitol, pH 7.5), 10 U of polynucleotide kinase (Boehringer), 50 µCi (50 nmols) of [$\gamma^{35}S$] ATP (Amersham) and with sterile distilled water added to give a final volume of 50 or 500 µl. This makes it possible to have the labeled primer at a final concentration of 20 or 2 μM. The reaction is carried out at 37°C in a water bath for 30 min. The labeled primer can then be used directly without any further purification.

D. PCR CONDITIONS

PCR experiments were performed in a PREM thermocycler. In order to obtain amplification with partially matched primers at several loci, an annealing temperature of 15 to 10°C under the melting temperature (T_m) of primers was used. The T_m of each primer was estimated

TABLE 1.

Name	Nucleotide sequence
Alu 2	5'-TTGCAGTGAGCCGAGATCGCGCC-3'
D21S172(92)	5'-TACAGTGGCAAATGTCATTG-3'
D21S172(93)	5'-GAATATGTGTTAGGTCCTGC-3'
D21S13E	5'-TATGCCATTCACTGTGATA-3'
APP17A	5'-CCTCATCCAAATGTCCCCTGCAAT-3'
APP17B	5'-GCCTAATTCTCTATAGTCTTAATTCCCAC-3'
TPA(7227)	5'-TCAGACCTTGTCTCTAAA-3'
TPA(34028)	5'-CCAGCTAATTGATATAAA-3'
ApoB1	5'-GGACAGTGAAACGAGGGC-3'
ApoB2	5'-CGCACATGAAGACACCAGAGG-3'
5'CAA	5'-GGAGTTCAGAGAGCTGGCGA-3'
Alu 30-2	5'-AAACTGGGTGTCCCTGAAGGTCC-3'

according to Lathe[12]: $T_m = 2°C \times (A + T) + 4°C \times (C + G)$. Five hundred nanograms to 1 µg of template DNA were used in a 50-µl PCR sample containing the *Taq* DNA polymerase buffer supplied by Promega (10 × buffer = 500 mM KCl, 100 mM Tris-HCl, pH 8.8 at 25°C, 15 mM MgCl$_2$, 1% Triton X-100), 50 µg·µl^{-1} BSA (Boehringer), 2 pmols of Alu 2 primer, 20 pmols of the primer to be tested, one of the primers being 5' end labeled, 0.2 mM each dNTP (Boehringer), and 2.5 U of *Taq* DNA polymerase (Promega).

In practice, testing a primer for 3' Alu PCR was performed as follows:

1. Prepare the sample mix:

Sterile distilled water	32.5 µl
10 × *Taq* DNA polymerase buffer	5.0 µl
BSA, 1 mg·ml^{-1}	2.5 µl
Alu 2 primer, 2 pmol·µl^{-1}	1.0 µl
Second PCR primer, 20 pmol·µl^{-1}	1.0 µl

   These quantities are for one tube. Multiply them by the number of samples
2. Distribute 42 µl of sample mix per 0.5-µl tube. Add 2.5 µl of template DNA (200 to 400 ng·µl^{-1}) and one drop of mineral oil. Keep on ice.
3. Prepare the nucleotide/*Taq* DNA polymerase mix:

dNTP, 8 mM (total)	5.0 µl
Taq DNA polymerase, 5 U·µl^{-1}	0.5 µl

   Multiply by the number of samples. Keep on ice.
4. Program the thermocycler as follows:

   1. 95°C, 5 min (initial denaturation)
   2. Annealing temperature, 1 s
   3. Suspend the program
   4. 72°C, 1 min (elongation)
   5. 92°C, 5 s (denaturation)
   6. Annealing temperature, 30 s (annealing)
   7. Repeat from step 4, 30 times
   8. 72°C, 10 min (final elongation)

5. Place the sample tubes in the thermocycler, and denature the template DNA and primers. When the program is suspended, add 5.5 µl of dNTP/*Taq* polymerase mix to the samples and continue the program.

E. ANALYSIS OF PCR PRODUCTS

PCR products were analyzed either in a 1.5% agarose gel for 16 h at 30 mA or in a 10% polyacrylamide (9.5% acrylamide, 0.5% bisacrylamide) gel for 3 h at 100 V. The DNA fragments were detected by ethidium bromide staining, and λ DNA digested by Pst I was used as a size marker.

In order to detect polymorphic bands, the PCR products (10 µl), obtained with one of the primers labeled at its 5' end with [$\gamma^{35}S$] ATP and polynucleotide kinase, were mixed with 5 µl of formamide-loading buffer (95% deionized formamide, 0.5% bromophenol blue, 0.5% Xylene cyanol), denatured in a water bath at 80°C for 10 min and analyzed in a sequencing gel (5% polyacrylamide, 7 M urea). After electrophoresis (3 to 5 h at 55 W), the gel was soaked in 10% acetic acid, 20% ethanol for 20 min, then dried under vacuum at 80°C for 30 min. The DNA fragments were detected by autoradiography with the film (Hyperfilm MP, Amersham) directly in contact with the gel, at room temperature, for two days. M13 mp8 DNA was used in sequencing reactions with the "sequenase" kit (USB corporation) to provide a size maker.

III. RESULTS

A. CHOICE OF PRIMERS

Primers can be selected for their ability to generate numerous polymorphic markers by 3' Alu PCR or because they map in a particular region of the human genome. We first tested a series of primers (Table 1) among those previously designed for other PCR purposes in our laboratory. The PCR products were analyzed in an agarose gel (Figure 1) and in a polyacrylamide gel (not shown). The number of amplified fragments clearly depends on the annealing temperature. For example, no amplification product was obtained with the TPA(34028) primer at 57°C, which corresponds to $T_m + 11°C$, but a few bands were obtained at 50°C ($T_m + 4°C$). For all primers, the number of bands tends to increase when the annealing temperature is lowered. All these DNA fragments should correspond to amplification from partially matched primers at several loci.

The search of polymorphic markers in a particular region of the human genome can be achieved in two ways:

1. YAC clones or somatic hybrid DNAs can be used as templates for testing primers (Figure 2). Those primers allowing amplification in both somatic hybrid DNA and total genomic DNA can be tested directly for detecting polymorphism. If DNA fragments are amplified from a YAC clone or from a somatic hybrid but not in total genomic DNA, PCR conditions can be improved or amplified fragments can be isolated and sequenced in order to design specific PCR primers (see Section III.C on the characterization of detected markers).
2. The PCR products of a randomly chosen primer used in combination with Alu 2 primer can be used as probes for *in situ* hybridization (not shown). Primers giving a signal of interest can then be tested for polymorphism. Alternatively, primers can be tested in primer *in situ* elongation (PRINS) reaction[13,14] at low stringency (not shown) before being tested in PCR experiments.

In both cases, the detected markers need to be localized either by linkage analysis or by *in situ* hybridization with the purified product.

B. DETECTION OF POLYMORPHIC MICROSATELLITE AT THE 3' END OF ALU REPEATS

When a primer is found to generate DNA fragments within the length resolvable in a DNA sequencing gel, the PCR products can be tested for polymorphism using unrelated individuals.

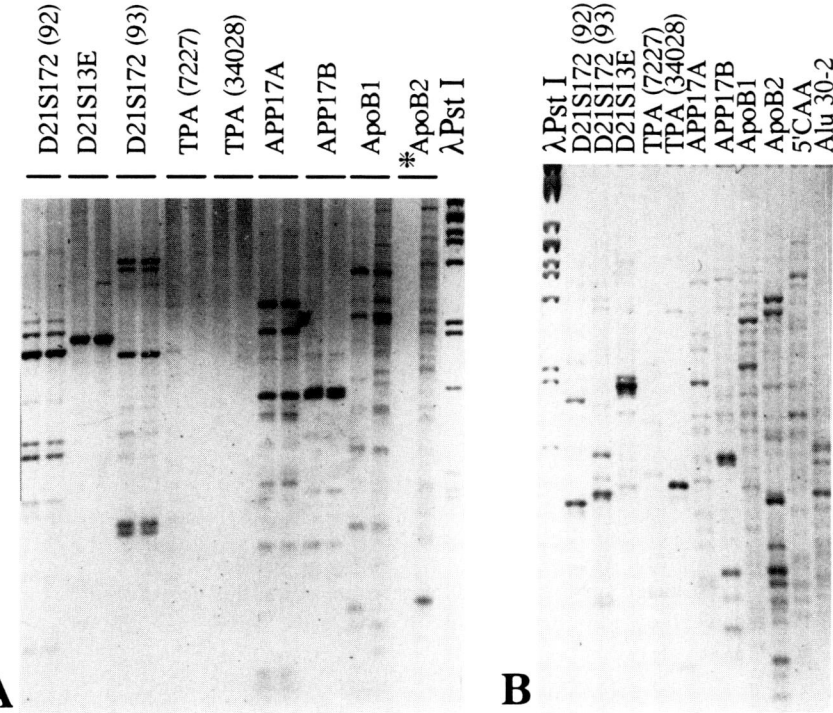

Figure 1. Amplification products from total human DNA with Alu 2 primer in combination with the primers listed in Table 1. The annealing temperature was (A) 57°C or (B) 50°C. The primers used are listed above each lane. PCR products were electrophoresed in 1.5% agarose gels. In (A), each amplification was performed in duplicate to test for reproducibility. The asterisk indicates a track that could not be loaded. (From Charlieu, J. P., Laurent, A.-M., Carter, D. A., Bellis, M., and Roizès, G., *Nucleic Acids Res.*, 20, 1333, 1992. With permission.)

All the primers we have tested in combination with Alu 2 allowed the detection of at least two polymorphic fragments (not shown). Mendelian inheritance was checked using CEPH pedigrees (one example is shown in Figure 3).

C. CHARACTERIZATION OF DETECTED MARKERS

As the 3′ Alu PCR method generates anonymous markers when the primers are used directly in total genomic DNA, they must be localized. CEPH pedigrees can be used in order to genetically locate the detected polymorphisms by linkage analysis. A way to determine their physical location is to pick up each polymorphic fragment from a sequencing gel and use it as an *in situ* hybridization probe. A greater quantity of DNA can be obtained by a second round of PCR amplification on the gel slice, with the same primers. In this case, the annealing temperature can be increased up to the T_m, and the primers should be used in a molar ratio (1 pmol µl^{-1} each).

PCR products purified in this way can also be sequenced in order to determine the nature of the polymorphic microsatellite or to design a sequence-specific primer. This can be useful if PCR products are obtained from a YAC clone or a somatic hybrid cell line but not in total genomic DNA.

IV. DISCUSSION

A general method for the detection of polymorphic microsatellites present at the 3′ end of Alu repeats in the human genome is described here. However, a polymorphic marker detected

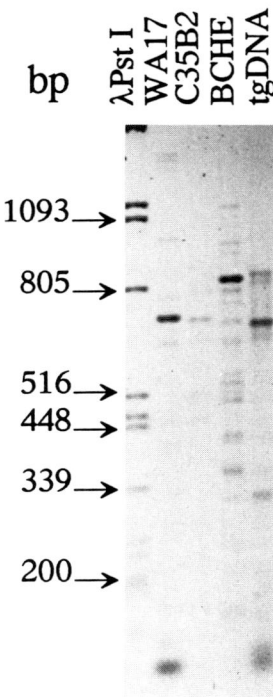

Figure 2. PCR products obtained with different mouse/human somatic hybrid DNAs and with total genomic DNA (tgDNA), using Alu 2 and Alu 30-2 primers analyzed in a 1.5% agarose gel. Annealing temperature was 58°C. (From Charlieu, J. P., Laurent, A.-M., Carter, D. A., Bellis, M., and Roizès, G., *Nucleic Acids Res.*, 20, 1333, 1992. With permission.)

in one laboratory could be difficult to amplify in another because PCR conditions can change slightly from one machine to another. The description of a genetic marker developed by 3′ Alu PCR should include the sequence of the primer, the thermocycler used, and the exact PCR conditions. To reproduce the results, some changes in the PCR conditions may be necessary. Only experience will say if the nucleotide sequence of the polymorphic DNA fragment should be included. These data could be used for the design of a PCR primer specific of the nucleotide sequence downstream to the Alu repeat of the studied marker.

Because one of the primers corresponds to an Alu repeat, the 3′ Alu PCR method also reveals loci polymorphic for the presence or absence of the Alu element. If the Alu repeat is absent, no amplification product is obtained by PCR. When the size of the amplified fragment is constant from individuals to individuals, only homozygotes for the absence of the Alu repeat can be detected, however.[8] These polymorphisms will therefore not be useful as genetic markers.

Two technical points of the 3′ Alu PCR method should also be detailed:

1. A 1:10 (Alu 2 to second primer) molar ratio of the primers was used in the experiments described here to amplify a series of DNA fragments instead of a smear. This ratio was empirically determined, however, and the 3′ Alu PCR method may be improved by testing this parameter, in order to obtain a greater number of DNA fragments in PCR experiments.
2. The T_m of oligonucleotides was estimated according to Lathe.[12] Several methods can be used, based on the nearest-neighbor frequencies, to determine the T_m of the primers used. For example the computer program *OLIGO*, for the design of primers, also calculates the T_m; however, a more precise value is obtained if the nucleotide sequence of the DNA fragment to be amplified is known.

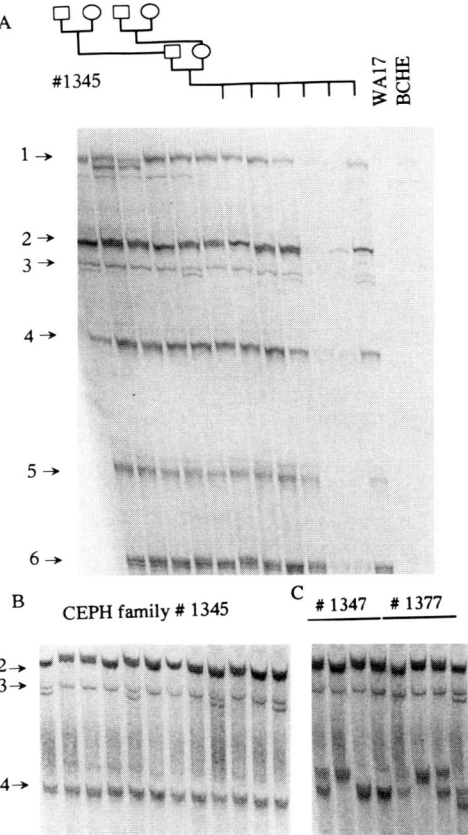

Figure 3. Detection of polymorphic DNA fragments by 3′ Alu PCR in a sequencing gel. DNA from individuals from the CEPH family #1345 were used in PCR-amplification analysis, using Alu 2 and D21S13E as primers, with the former being 5′ end labeled. The annealing temperature was 50°C. (A) Three polymorphic bands are visible (arrows #1, 2, and 3). The amplified fragment #1 can be assigned to one of the chromosomes of the somatic hybrid BCHE as the same band occurs in the DNA from this hybrid. The results were reproducible as several gels from independent PCR have been performed and gave the same pattern with only slight changes in the faintest bands. This can be illustrated in (B), where the DNA from the same family was resolved to show polymorphism #2 more clearly. (C) Eight grandparents from two other CEPH pedigrees were analyzed, allowing the detection of a fourth polymorphic (triallelic) band (#4), which was monomorphic in pedigree #1345. (From Charlieu, J. P., Laurent, A.-M., Carter, D. A., Bellis, M., and Roizès, G., *Nucleic Acids Res.*, 20, 1333, 1992. With permission.)

The same type of approach as the 3′ Alu PCR described here could also be performed using the large number of L1 sequences present in the human genome. Oligonucleotides have already been designed that could serve for this purpose.[15]

AKNOWLEDGMENTS

Dr. E. Devine and D. N'Guyen Van Cong are acknowledged for providing the somatic cell lines. R. Orti is thanked for providing technical assistance, and Dr. A.-M. Laurent and Dr. G. Roizès for performing some of the experiments described here. This work was carried out under the direction of Dr. G. Roizès in his laboratory at CNRS UPR 9008 and INSERM U 249, Institut de Biologie, Bd Henri IV, F-34060 Montpellier cedex, France.

REFERENCES

1. **Weber, J. L. and May, P. E.**, Abundant class of human DNA polymorphisms which can be typed using the polymerase chain reaction, *Am. J. Hum. Genet.*, 44, 388, 1989.
2. **Manor, H., Rao, B. S., and Martin, R. G.**, Abundance and degree of dispersion of genomic d(GA)n.d(TC)n sequences, *J. Mol. Evol.*, 27, 96, 1988.
3. **Economou, E. P., Bergen, A. W., Warren, A. C., and Antonorakis, S. E.**, The polydeoxyadenylate tract of Alu repetitive elements is polymorphic in the human genome, *Proc. Natl. Acad. Sci. U.S.A.*, 87, 2951, 1990.
4. **Zuliani, G. and Hobbs, H. H.**, A high frequency of length polymorphisms in repeated sequences adjacent to Alu sequences, *Am. J. Hum. Genet.*, 46, 963, 1990.
5. **Epstein, N., Nahor, O., and Silver, J.**, The 3' end of Alu repeats are highly polymorphic, *Nucleic Acids Res.*, 18, 4634, 1990.
6. **Deininger, P. L.**, SINEs short interspersed repeated DNA elements in higher eucaryotes, in *Mobile DNA*, Howe, M. and Berg, D., Eds., ASP Press, Washington DC, 619–636.
7. **Beckman, J. S. and Weber, J. L.**, Survey of human and rat microsatellites, *Genomics*, 12, 627, 1992.
8. **Charlieu, J.-P., Laurent, A.-M., Carter, D. A., Bellis, M., and Roizès, G.**, 3' Alu PCR: a simple and rapid method to isolate human polymorphic markers, *Nucleic Acids Res.*, 20, 1333, 1992.
9. **Parker, J. D., Rabinovitch, P. S., and Burmer, G. C.**, Targeted gene walking polymerase chain reaction, *Nucleic Acids Res.*, 19, 3055, 1991.
10. **Bains, W.**, The multiple origins of human Alu sequences, *J. Mol. Evol.*, 23, 189, 1986.
11. **Sambrook, J., Fritsch, E. F., and Maniatis, T.**, *Molecular Cloning: A Laboratory Manual*, 2nd ed., Cold Spring Harbor Laboratory Press, Cold Spring Harbor, NY, 1989.
12. **Lathe, R.**, Synthetic oligonucleotide probes deduced from amino acid sequence data; theoretical and practical considerations, *J. Mol. Evol.*, 183, 1, 1985.
13. **Koch, J. E., Kelvraa, S., Petersen, K. B., Gregersen, N., and Bolund, L.**, Oligonucleotide-priming methods for the chromosome-specific labeling of alpha satellite DNA *in situ*, *Chromosoma*, 98, 259, 1989.
14. **Godsen, J. R., Hanratti, D., Starling, J., Fantes, J., Mitchell, A., and Parteous, D.**, Oligonucleotide primed *in situ* DNA synthesis (PRINS): a method for chromosome mapping, banding, and investigation of sequence organization.
15. **Ledbetter, S. A., Nelson, D. L., Warren, S. T., and Ledbetter, D. H.**, Rapid isolation of DNA probes within specific chromosome regions by interspersed repetitive elements sequence polymerase chain reaction, *Genomics*, 6, 475, 1990.

Chapter 16

LIGATION-ANCHORED PCR

Anthony B. Troutt

TABLE OF CONTENTS

I. Introduction 141

II. Materials and Methods 142
    A. Preparation of cDNA 142
    B. Preparation of the Anchor Oligonucleotide 143
    C. Anchor-cDNA Ligation 143
    D. PCR Amplification 143
    E. Product Analysis 143

III. Results 144

IV. Discussion 144

References 145

I. INTRODUCTION

Ligation-anchored PCR (LA-PCR), or single-strand ligation to ss-cDNA (SLIC), is a simple, efficient, and sensitive method for amplification of cDNAs encoding molecules with 5' regions of unknown sequence.[1,2] The success of LA-PCR relies on the ability of T4 RNA ligase to catalyze efficiently the covalent linking of single-stranded DNA molecules bearing 5'-phosphate and 3'-hydroxyl groups.[3] In brief (Figure 1), T4 RNA ligase was used to covalently join an "anchor" oligonucleotide of defined sequence to reverse-transcribed first-strand cDNA. The cDNA has hydroxyl groups at both termini, making only the 3' end a potential substrate for ligation, while the anchor oligonucleotide is blocked at the 3' end by addition of a dideoxynucleotide, leaving only the 5'-phosphorylated terminus as a potential substrate. Thus, ligation of a single anchor molecule to the 3' end of the cDNA is the favored reaction. The resultant anchored product was then subjected to PCR amplification using one primer specific for a sequence within the anchor and another specific for a sequence within the cDNA of interest. The anchor oligonucleotide has been designed to allow multiple uses of the amplified PCR product (Figure 2). It includes a T3 promoter sequence for direct *in vitro* transcription, and it allows sequential use of two nested sets of PCR primers as a means of increasing sensitivity and product yield and of enabling direct sequencing of PCR product without prior cloning.[4] Not I and Hha I sites have been included in the anchor oligonucleotide to facilitate cloning of the PCR product, and sequences for convenient restriction sites can be included at the 5' end of the primer specific for the cDNA of interest for optimal cloning efficiency.

This three-step procedure does not require product purification between steps and avoids many of the technical difficulties associated with established anchored PCR protocols. Moreover, the efficacy of LA-PCR has been demonstrated by amplification of a specific immuno-

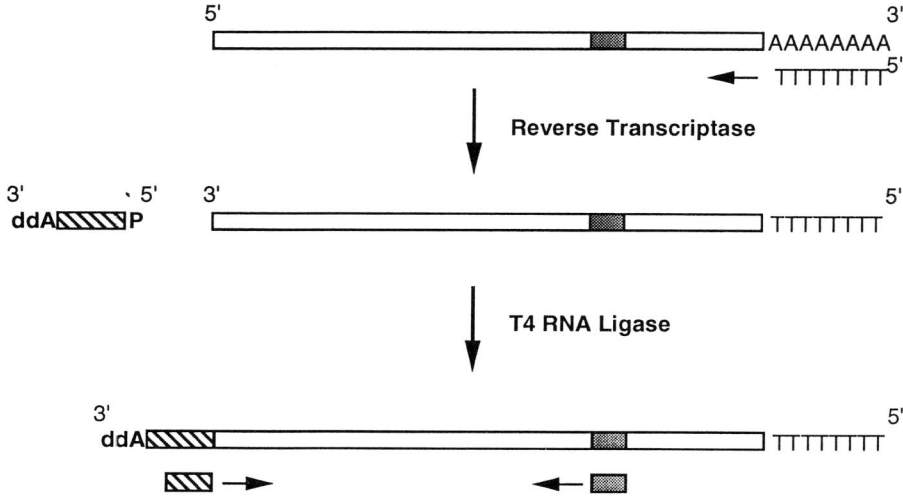

FIGURE 1. LA-PCR strategy. Oligo-(dT) was used as primer for cDNA synthesis by reverse transcriptase. The 5'-phosphorylated, 3'-end blocked anchor oligonucleotide was then ligated to the first-strand cDNA. PCR amplification of the anchored cDNA was carried out using oligonucleotide primers specific for a known sequence within the cDNA of interest (gray box) and for a sequence within the anchor (hatched box).

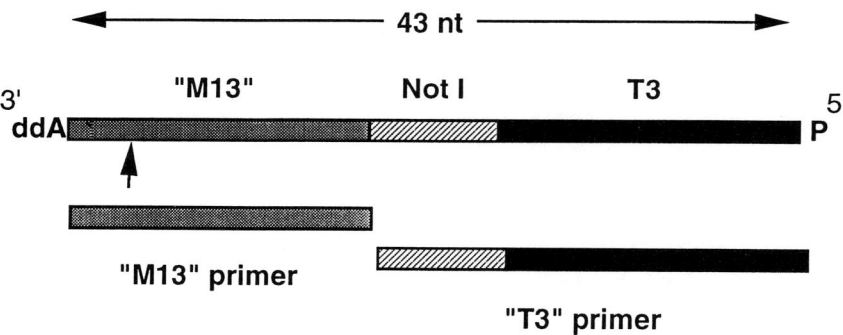

FIGURE 2. Diagram of anchor oligonucleotide. The 43-nucleotide (nt) anchor contains an altered M13 sequence including an Hha I restriction site (arrow), a Not I restriction site, and a T3 promoter sequence. Nested "M13" and "T3" primers suitable for PCR amplification or direct sequencing of product are illustrated.

globulin cDNA using total RNA equivalent to as few as 100 cells.[2] The simplicity and sensitivity of this LA-PCR suggest that it could have widespread utility.

II. MATERIALS AND METHODS

A. PREPARATION OF CDNA

Total RNA was prepared by lysis in guanidinium-thiocyanate and acid phenol-chloroform extraction.[5] For samples containing fewer than 10^6 cells, or less than 1 mg of tissue, 1 µg of *Escherichia coli* tRNA was included in the lysis buffer to minimize sample losses and to aid in subsequent precipitation. The precipitated RNA was recovered by centrifugation, and each sample was resuspended in diethyl pyrocarbonate-treated, distilled deionized water at a maximum concentration of 0.5 µg/ml. First-strand cDNA synthesis was performed by incubating 5 µl of each RNA sample in a total reaction volume of 20 µl containing 50 mM KCl,

10 m*M* Tris-HCl, pH 8.3, 8 m*M* MgCl$_2$, 1 m*M* dithiothreitol, 500 µ*M* each dNTP, 10 µg/ml oligo-(dT)$_{15}$, 8 U rRNAsin, and 2 U AMV reverse transcriptase (Promega, Madison, WI) at 42°C for 1 to 2 h.

Since T4 RNA ligase can use both single-stranded DNA and single-stranded RNA as substrates, the sensitivity of the LA-PCR was improved by removing residual RNA from the cDNA preparation. This was achieved by combining 3 µl of cDNA with 3 µl 300 m*M* NaOH, incubating in a boiling water bath for 5 min, immediately chilling on ice, and neutralizing by addition of 3 µl 300 m*M* HCl. This treatment results in selective hydrolysis of the ribonucleic acid and production of fragments bearing 5'-hydroxyl and 3'-or 2'-phosphate groups,[6] which do not serve as substrates for T4 RNA ligase.

B. PREPARATION OF THE ANCHOR OLIGONUCLEOTIDE

The sequence of the anchor oligonucleotide was TTTAGTGAGGGTTAATAAGCGGCCG-CGTCGTGACTGGGAGCGC. The 5' end of the anchor was phosphorylated by incubating 2 nmol of the oligonucleotide and 100 nmol of ATP in a 40-µl reaction containing 70 m*M* Tris-HCl, pH 7.4, 10 m*M* MgCl$_2$, 5 m*M* 2-mercaptoethanol, and 20 U T4 polynucleotide kinase (Promega, Madison, WI). After incubation at 37°C for 60 min, the phosphorylated anchor was precipitated with 0.4 *M* NaCl and 2.5 vol of 95% v/v ethanol and redissolved in 20 µl 10 m*M* Tris-HCl, pH 7.4, 1 m*M* EDTA. The 3' end of the anchor was blocked by incubating 1 nmol of the phosphorylated oligonucleotide with 14 nmol of ddATP in a total volume of 70 µl containing 100 m*M* potassium cacodylate pH 7.2, 2 m*M* CoCl$_2$, 0.2 m*M* dithiothreitol, and 10 U terminal deoxynucleotide transferase (TdT) (GIBCO BRL, Gaithersburg, MD). The reaction was incubated for 60 min at 37°C and terminated by adding 3 vol of ice-cold 16 m*M* EDTA. The 5'-phosphorylated, 3'-blocked anchor oligonucleotide was precipitated in 0.4 *M* NaCl and 2.5 vol of 95% v/v ethanol and resuspended in 10 m*M* Tris-HCl, pH 7.4, 1 m*M* EDTA at a final concentration of 10 µ*M*.

C. ANCHOR-CDNA LIGATION

Ligation reactions were performed at 22°C with 1 µl of cDNA and 10 pmol of the phosphorylated, blocked anchor in 10 µl vol containing 50 m*M* Tris-HCl, pH 8, 10 m*M* MgCl$_2$, 10 µg/ml bovine serum albumin, 25% w/v polyethylene glycol 8000, 1 m*M* hexamine cobalt chloride, 20 µ*M* ATP and 1 U T4 RNA ligase (New England Biolabs, Beverly, MA). Ligations were terminated after 12 to 18 h by adding 32 µl of 0.5 *M* NaCl, 10 m*M* Tris-HCl, pH 7.4, 1 m*M* EDTA, and were stored at –20°C.

D. PCR AMPLIFICATION

PCR was performed using 2 µl of the terminated ligation reactions in 25 µl vol containing 50 m*M* KCl, 10 m*M* Tris-HCl, pH 8.3, 2 m*M* MgCl$_2$ 200 µ*M* each dNTP, 400 n*M* each specific primer, and 1 U *Taq* polymerase (Perkin Elmer Cetus, Norwalk, CT). Oligonucleotide primers specific for the anchor were as follows: "M13", GCGCTCCCAGTCACGAC; "T3", GCGGCCGCTTATTAACCCTCACTAAA. Primers specific for the cDNA of interest will, of course, vary, but should be designed to have an annealing temperature compatible with that of at least one of the anchor-specific primers. The cDNA-specific primers used in the present application were as described.[2] The reactions were run using an immersion thermocycler (Bartelt Instruments, Melbourne, Australia) as follows: cycle 1, 5 min at 94°C, 60 s at 55°C, and 90 s at 70°C; cycles 2 to 40, 45 s at 94°C, 60 s at 55°C, and 90 s at 70°C.

E. PRODUCT ANALYSIS

The specificity and sensitivity of the LA-PCR was assessed by Southern blot analysis. A 5-µl aliquot of each PCR reaction was electrophoresed through a 1.8% agarose gel, transferred to Zeta-Probe membrane by alkaline blotting, and hybridized with a random-decamer radiolabeled probe specific for the cDNA of interest according to the manufacturer's instructions (Bio-Rad, Richmond, CA). As detailed above, the anchor oligonucleotide has been designed to facilitate sequencing, expression, and cloning of the amplified product.

III. RESULTS

The efficacy of the LA-PCR protocol was tested using total RNA from a B-cell hybridoma expressing an IgG1 gene of known sequence.[2] Amplifications conducted using the "T3" anchor-specific primer and a primer specific for the constant region of the IgG1 molecule yielded a PCR product of the predicted size that hybridized with a radiolabeled probe specific for the cDNA of interest. Control studies showed that generation of this specific PCR product required the presence of both the anchor oligonucleotide and T4 RNA ligase in the ligation reaction, and required that both anchor-specific and cDNA-specific primers were present in the PCR. When LA-PCR was performed after NaOH treatment of cDNA samples to remove residual RNA as a potential competitive inhibitor of anchor-cDNA ligation, a >100-fold increase in sensitivity was achieved; specific PCR product was detected when as little as 1 ng of total RNA was used as starting material.

IV. DISCUSSION

The established protocols for anchored PCR rely on homopolymer tailing of cDNA using TdT, subsequent amplification using one primer specific for the molecule of interest and a second primer containing a defined "anchor" sequence attached to a homopolymer sequence complementary to the tail, and, finally, re-amplification using one primer specific for the cDNA of interest and one specific for the anchor.[7,8] LA-PCR has a number of advantages over these procedures, particularly its avoidance of TdT tailing, which has often proved technically demanding.[9] Furthermore, direct ligation of the anchor to first-strand cDNA eliminates the use of homopolymer-containing primers with limited specificity and the necessity of performing two rounds of PCR. Since the LA-PCR technique involves only three simple enzymatic reactions and does not require purification of products between steps, it may be possible to perform the entire set of reactions in a single tube by using oligo-(dT) coupled to a solid substrate during cDNA synthesis. Finally, LA-PCR can yield specific product from as little as 1 ng of total RNA, whereas conventional approaches have generally used much higher amounts of starting material.

Although LA-PCR is quite simple both in theory and in practice, there are a number of potential difficulties that should be taken into account when using this method. If a primer specific for the molecule of interest is to be used in cDNA synthesis, this primer must not be complementary to the cDNA-specific primer to be used in subsequent PCR. If such a primer were used, ligation of the anchor to residual reverse-transcription primer would yield a product that could act as a competing substrate in PCR, decreasing the efficiency of cDNA amplification. Under some circumstances, NaOH-treatment of the cDNA reaction to remove residual RNA may result in the formation of an insoluble magnesium hydroxide-DNA complex. This can be prevented by addition of sufficient EDTA to chelate all magnesium ions prior to base treatment. It is also important that the amount of NaCl transferred into the subsequent ligation reaction be kept to a minimum, since concentrations as low as 20 mM have been reported to inhibit T4 RNA ligase activity.[10] Although the ligation conditions reported here are likely to be acceptable for most applications, optimization of reaction parameters may be required in some instances. Alterations of polyethylene glycol, ATP, and hexamine cobalt chloride concentrations have all been reported to have major effects on ligation efficiency,[3] and would be likely candidates for optimization. Finally, the optimal conditions of the PCR itself will be expected to vary between different substrates.[7,8]

The simplicity and sensitivity of the LA-PCR should allow a more widespread application of anchored-PCR technology than has been possible with standard protocols, and it could conceivably lend itself to semi-automated analysis of multiple samples, perhaps even of samples derived from single cells.

REFERENCES

1. **Edwards, J. B. D. M., Delort, J., and Mallet, J.**, Oligodeoxyribonucleotide ligation to single-stranded cDNAs: a new tool for cloning 5' ends of mRNAs and for constructing cDNA libraries by *in vitro* amplification, *Nucleic Acids Res.*, 19, 5227, 1991.
2. **Troutt, A. B., McHeyzer-Williams, M. G., Pulendran, B., and Nossal, G. J. V.**, Ligation-anchored PCR: a simple amplification technique with single-sided specificity, *Proc. Natl. Acad. Sci. U.S.A.*, 89, 9823, 1992.
3. **Tessier, D. C., Brousseau, R., and Vernet, T.**, Ligation of single-stranded oligodeoxyribonucleotides by T4 RNA ligase, *Anal. Biocem.*, 158, 171, 1986.
4. **McHeyzer-Williams, M. G., Nossal, G. J. V., and Lalor, P. A.**, Molecular characterization of single memory B cells, *Nature*, 350, 502, 1991.
5. **Chomczynski, P. and Sacchi, N.**, Single-step method of RNA isolation by acid guanidinium thiocyanate-phenol-chloroform extraction, *Anal. Biochem.*, 162, 156, 1987.
6. **Brown, D. M. and Todd, A. R.**, Chemical bonds in nucleic acids, in *The Nucleic Acids:Chemistry and Biology*, Vol. 1, Chargaff, E. and Davidson, J. N., Eds., Academic Press, New York, 1955, 409.
7. **Frohman, M. A., Dush, M. K., and Martin, G. R.**, Rapid production of full-length cDNAs from rare transcripts: amplification using a single gene-specific oligonucleotide primer, *Proc. Natl. Acad. Sci. U.S.A.*, 85, 8998, 1988.
8. **Loh, E. Y., Elliot, J. F., Cwirla, S., Lanier, L. L., and Davis, M. M.**, Polymerase chain reaction with single-sided specificity: analysis of T cell receptor δ chain, *Science*, 243, 217, 1989.
9. **Okayama, H., Kawaichi, M., Brownstein, M., Lee, F., Yokota, T., and Arai, K.**, High efficiency cloning of full-length cDNA: construction and screening of cDNA expression libraries for mammalian cells, *Methods Enzymol.*, 159, 3, 1987.
10. **Harrison, B., and Zimmerman, S. B.**, Polymer-stimulated ligation: enhanced ligation of oligo- and polynucleotides by T4 RNA ligase in polymer solutions, *Nucleic Acids Res.*, 12, 8235, 1984.

Chapter 17

PCR-LIMITING DILUTION ANALYSIS

Anthony B. Troutt

TABLE OF CONTENTS

I. Introduction 147

II. Materials and Methods 147
    A. Preparation of RNA 147
    B. PCR Amplification 148
    C. Southern Analysis 148
    D. Frequency Estimations 149

III. Results 149

IV. Discussion 149

References 150

I. INTRODUCTION

The combination of PCR amplification of cDNA with limiting dilution analysis (PCR-LDA) allows quantitation of the number of cells containing a particular mRNA in a mixed population. In this method (Figure 1), total RNA prepared from multiple samples containing graded numbers of cells was reverse-transcribed, and the cDNA was subjected to two rounds of PCR amplification using nested oligonucleotide primers. The resulting PCR products were assessed for the presence of the specific cDNAs of interest by Southern blot hybridization. Estimations of the frequencies of mRNA-containing cells were made by statistical analysis of the relationship between cell number per sample and the percentage of negative samples at each cell input.[1,2] The full details of this statistical analysis are beyond the scope of this chapter, but are considered in full in References 1 and 2. In its simplest form, given that the assay is of sufficient sensitivity to detect mRNA from single positive cells, plotting the percentage of negative samples on a logarithmic scale vs. cell input per sample on a linear scale will yield a straight line passing through the origin, so-called single-hit kinetics. The frequency of positive cells in the population will then be equal to the cell input per sample that yields 37% negative samples. Application of the PCR-LDA approach to a murine-graft-vs.-host reaction (GVHR) demonstrated that it allowed simultaneous analysis of multiple transcripts in the same sample, it was able to detect a single positive cell among >40,000 negative cells, and it was of sufficient sensitivity to determine mRNA production patterns of individual cells.

II. MATERIALS AND METHODS

A. PREPARATION OF RNA
Pooled spleen and lymph node cells were prepared as a single cell suspension and dispensed into 1.5-ml tubes in graded numbers as mid-log serial dilutions ranging from 10^5 cells

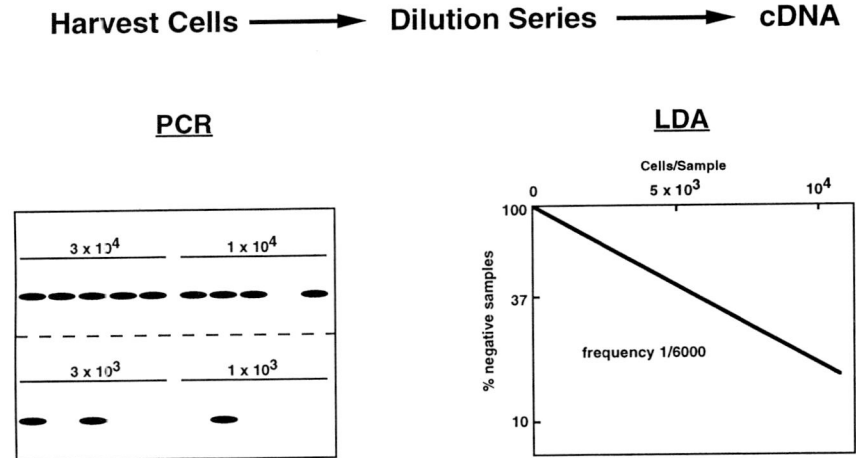

Figure 1. Protocol for PCR-LDA. The cell population of interest was harvested, serial dilutions of the cells were prepared, total RNA was extracted from multiple samples at each cell input, and this RNA was used to synthesize cDNA. The cDNA samples were subjected to PCR amplification and subsequent Southern blot analysis. A schematic representation of the results of such an analysis conducted using five samples at each of four different cell inputs is illustrated. The LDA plot of the percent negative samples vs. the cell input per sample allows estimation of the frequency of cells containing the mRNA of interest in the starting cell population (1/6000 in the illustrated example).

per sample to 10 cells per sample (10 samples per cell number). In some cases, individual cells were micromanipulated into tubes. The cells were immediately lysed in guanidinium-thiocyanate solution with 1 µg *Escherichia coli* tRNA as carrier per sample, and total RNA was purified by acid phenol-chloroform extraction.[3] The precipitated RNA from each sample was resuspended in a total volume of 20 µl diethyl pyrocarbonate-treated, distilled deionized water. First-strand cDNA synthesis was performed using 5 µl of each RNA sample in a total reaction volume of 20 µl containing 50 mM KCl, 10 mM Tris-HCl, pH 8.3, 8 mM MgCl$_2$, 1 mM dithiothreitol, 500 µM each dNTP, 10 µg/ml oligo-(dT)$_{15}$, 8 U rRNAsin, and 2 U AMV reverse transcriptase (Promega, Madison, WI) at 42°C for 1 to 2 h.

B. PCR AMPLIFICATION

Two-microliter aliquots of each cDNA sample were used in 25-µl PCR reactions containing 50 mM KCl, 10 mM Tris-HCl pH 8.3, 2 mM MgCl$_2$, 200 µM each dNTP, 400 nM each specific primer, and 1 U *Taq* polymerase (Perkin Elmer Cetus, Norwalk, CT). The first- and second-round PCR primers used in our application of this technique have been described,[4,5] but of course these will vary, depending on the particular transcripts of interest. All primer pairs were included in each reaction. Following first-round PCR, the products were diluted 1:1000, and 2 µl of each dilution was used in each second-round PCR. In each round of PCR the reactions were run using an immersion thermocycler (Bartelt Instruments, Melbourne, Australia) as follows: cycle 1, 5 min at 94°C, 60 s at 60°C, and 90 s at 70°C; cycles 2 to 40, 45 s at 94°C, 60 s at 60°C, and 90 s at 70°C. At least seven negative control samples were included at each step in the procedure; *E. coli* tRNA and HeLa cell total RNA were included during cDNA synthesis, and reactions containing no cDNA were included in each round of PCR. The expression of β-actin mRNA was analyzed as a positive control.

C. SOUTHERN ANALYSIS

Five microliters of each second-round PCR reaction was electrophoresed through a 1.8% agarose gel, transferred to Zeta-Probe membrane by alkaline blotting, and hybridized with random-hexamer radiolabeled cDNA probes according to the manufacturer's instructions (Bio-Rad, Richmond, CA). Each membrane was sequentially hybridized with each of the cDNA probes of interest by multiple rounds of hybridization and stripping according to the manufacturer's recommendations.

D. FREQUENCY ESTIMATIONS

Frequencies of mRNA-containing cells were obtained from the slope of best fit for the experimental data determined using the maximal likelihood estimator,[1] or by direct enumeration of positive micromanipulated single cells.

III. RESULTS

PCR-LDA was used to analyze and characterize *in vivo* activated lymphokine-producing cells in an acute murine GVHR.[5,6] Each sample was analyzed for interferon-γ, granulocyte-macrophage colony-stimulating factor, interleukin-3, interleukin-4, and β-actin mRNA. Single-hit kinetics indicated that the technique could detect an individual positive cell among >40,000 negative cells, and mRNA was detected in micromanipulated single GVHR cells following *in vitro* restimulation with anti-CD3. Frequency estimates demonstrated that while nearly 70% of pooled spleen and lymph node cells from GVHR mice could be triggered to produce lymphokines following *in vitro* restimulation, at most 3% contained detectable lymphokine transcripts *in vivo*. All frequency estimates for cells from normal mice were 20- to 175-fold lower than equivalent estimates for GVHR cells. Furthermore, frequency estimates of mRNA production determined by PCR-LDA corresponded to those obtained by limiting dilution analysis of cells secreting biologically active protein.

IV. DISCUSSION

PCR-LDA has important advantages over other methods of analyzing the frequency of mRNA-producing cells such as *in situ* hybridization; it is exquisitely sensitive and allows simultaneous analysis of multiple transcripts in a single cell preparation. Control experiments demonstrated that the detection limit in this system was <30 molecules of cDNA, and in other studies we have performed simultaneous analysis of eight different mRNAs in a single sample.[4]

One of the key requirements of PCR-LDA is that the assay be able to detect individual positive cells. If this condition is not fulfilled, limiting dilution analysis will yield multi-hit curves, which cannot be used to obtain frequency estimates.[7] However, appropriate primer design and, when necessary, use of nested primers in two rounds of PCR should enable the requisite sensitivity to be achieved.

The accuracy of frequency estimates obtained by PCR-LDA is directly related to the number of informative samples analyzed in the experiment.[8] While the limited number of samples used in our studies was sufficient to reveal the gross frequency disparities found in the GVHR system, detection of more subtle differences would require analysis of a much larger set of samples. Given the large number of samples that must be analyzed at each cell input for accurate frequency estimations, it may be necessary in practice to perform preliminary experiments in which a wide range of cell numbers is surveyed using only a few samples per cell input. Based on these results, multiple samples at only a few different cell inputs can then be analyzed in more definitive studies.

Although the high sensitivity required by PCR-LDA makes the generation of false-positives a particular worry, the reliance of the technique upon statistical analysis of relatively large numbers of samples minimizes the effect of such artifacts. A false-positive rate of as high as 6% would be expected to lead to, at most, a two- to threefold alteration in frequency estimates determined using the protocol outlined in this chapter.[5] However, as with all procedures involving PCR amplification, great care should be taken to avoid sample contamination with amplifiable template, and appropriate negative controls must be included at each step to detect any false-positives. It is equally important that expression of a constitutively produced transcript such as β-actin be analyzed in all samples as a positive control for the presence of mRNA. It is helpful if the different PCR products to be analyzed by Southern blotting are easily distinguished based on their molecular weight, and if sequential probing of the blots starts with the cDNA expected to show the lowest frequency of expression and ends with that expected to show the highest. This reduces the risk of any ambiguities arising due to incomplete stripping of probe between rounds of hybridization.

In summary, PCR-LDA allows analysis of multiple transcripts in the same sample, even within the same single cell, and can enumerate mRNA-containing cells over a wide frequency range. It should prove useful for many types of analysis where the number of cells expressing a particular transcript, or set of transcripts, is of interest.

REFERENCES

1. **Good, M. F., Boyd, A. W., and Nossal, G. J. V.**, Analysis of true anti-hapten cytotoxic clones in limit dilution microcultures after correction for "anti-self" activity: precursor frequencies, Ly-2 and Thy-1 phenotype, specificity, and statistical methods, *J. Immunol.*, 130, 2046, 1983.
2. **Lefkovits, I. and Waldmann, H.**, Limiting dilution analysis of the cells of immune system. I. The clonal basis of the immune response, *Immunol. Today*, 5, 265, 1984.
3. **Chomczynski, P. and Sacchi, N.**, Single-step method of RNA isolation by acid guanidinium thiocyanate-phenol-chloroform extraction, *Anal. Biochem.*, 162, 156, 1987.
4. **Maraskovsky, E., Troutt, A. B., and Kelso, A.**, Co-engagement of CD3 with LFA-1 or ICAM-1 adhesion molecules enhances the frequency of activation of single CD4$^+$ and CD8$^+$ T cells and induces synthesis of IL-3 and IFN-γ but not IL-4 or IL-6, *Int. Immunol.*, 4, 475, 1992.
5. **Troutt, A. B. and Kelso, A.**, Enumeration of lymphokine mRNA-containing cells *in vivo* in a murine graft-versus-host reaction using the PCR, *Proc. Natl. Acad. Sci. U.S.A.*, 89, 5276, 1992.
6. **Troutt, A. B., Maraskovsky, E., Rogers, L. A., Pech, M. H., and Kelso, A.**, Quantitative analysis of lymphokine expression *in vivo* and *in vitro*, *Immunol. Cell Biol.*, 70, 51, 1992.
7. **Lefkovits, I. and Waldmann, H.**, *Limiting Dilution Analysis of the Cells of Immune System*, Cambridge University Press, Cambridge, 1979, 64.
8. **Lefkovits, I. and Waldmann, H.**, *Limiting Dilution Analysis of the Cells of Immune System*, Cambridge University Press, Cambridge, 1979, chap. 7.

Chapter 18

DIRECT PCR FROM WHOLE BLOOD USING FORMAMIDE AND LOW TEMPERATURES

Michael Panaccio and Andrew M. Lew

TABLE OF CONTENTS

I. Introduction 151

II. Materials and Methods 152
    A. Standard PCR 152
    B. Folt PCR 152
        1. Method A 153
        2. Method B 153
    C. Primers 154
    D. Anticoagulants 154

III. Results 154

IV. Discussion 155

Acknowledgments 157

References 157

I. INTRODUCTION

Polymerase chain reaction (PCR) is an *in vitro* method for amplifying DNA.[1] One of the major limitations to the rapid adoption of the use of PCR for diagnosis has been the inability to apply PCR directly to clinical material. This is because many substances found in the various types of clinical material inhibit PCR. Many factors have been postulated to give rise to this inhibition, including the presence of iron, hemoglobin, chelation of free magnesium ions. The consequence of the inability to amplify DNA directly from clinical material has been that DNA or target purification step(s) are necessary before the PCR reaction can be applied.[2-5] In a theoretical clinical situation a PCR-based test will proceed through the steps outlined in Figure 1.

The type of DNA-purification method used depends on the nature of the clinical material and the types of inhibitors present within the sample. Obviously it would be better to have rapid DNA-purification methods to allow a faster turnaround time. However, the need to purify DNA has important implications on the economics of the test and its reliability. First, the need to purify DNA increases the labor input of the test or the need to invest in expensive automated DNA-purification equipment. The risk of sample contamination is increased since most DNA-purification methods involve multiple tube transfers and manipulations. These sample transfers also increase the risk of contamination of the operator by infectious material that may be in the sample. The overall effect of DNA or target purification is that the cost of the test is increased as well as the time required for diagnosis. In an ideal PCR test, handling

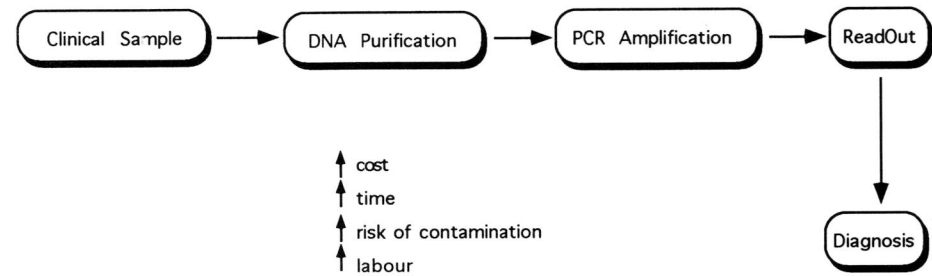

Figure 1. Flow diagram of a theoretical PCR test. The effects of the DNA-purification step are summarized in point form.

would be reduced to a minimum, all steps of the process would be performed in a single tube, and the whole process would be automated.

FoLT (formamide low temperature) PCR, a technique that has recently been developed for PCR from whole blood,[6] has the characteristics of an ideal PCR diagnostic test. It allows direct PCR; all manipulations occur in a single tube and the whole process can be automated. The genesis of this technique was experiments designed to shed light on the inhibition of PCR when whole blood was used as the target. Initial experiments to identify a particular inhibitor of PCR were unsuccessful. Purified hemoglobin added to a PCR reaction were not inhibitory; nor was the presence of iron. The nature of the inhibition seemed elusive. An important finding was that an alternative DNA polymerase, *Tth* polymerase, was less sensitive than *Taq* polymerase to the presence of blood.[7] This is illustrated in Figure 2. Purified target DNA or DNA in whole bacteria in the presence of blood can be amplified by *Tth* polymerase but not by *Taq* polymerase. This was an important difference between *Taq* and *Tth* polymerase.

We have found that the major problem for PCR in blood was due to the inability of the DNA polymerase to access the target DNA.[6] Maybe DNA from leukocytes in whole blood was physically trapped by coagulated organic material that results after the first step (95°C incubation) of a standard PCR protocol. In support of this, both *Taq* and *Tth* polymerase are able to amplify purified DNA added to blood.[7] Therefore, conditions that reduce the amount of protein coagulation may be a superior way of amplifying directly from blood.

FoLT PCR uses formamide (which solubilizes blood cells) and reduced incubation temperatures to reduce protein coagulation. This is possible because a window of formamide concentration (15 to 25% v/v) can be used in combination with reduced temperatures (80–85, 40, and 60°C) to achieve amplification of DNA in whole blood.

II. MATERIALS AND METHODS

A. STANDARD PCR

Fifty-microliter PCR reactions containing 200 μM of dATP, dCTP, dTTP, and dGTP (Pharmacia, Uppsala, Sweden), 0.2 μM of forward and reverse primers, 10 mM Tris, pH 9.0 at 25°C, 50 mM KCl, 1.5 mM MgCl$_2$, 0.01% gelatin, 0.1% Triton X-100, and 2.5 U of *Taq* polymerase (Perkin Elmer Cetus, Melbourne, Australia) or *Tth* polymerase (Toyobo, Japan) were performed in a robotic arm Gene Machine (Innovonics, Melbourne, Australia) for 30 c of 30 s for each of the temperatures 95, 50, and 72°C.

B. FOLT PCR

There are two ways of doing FoLT PCR. In Method A, all reagents are put into a tube at the same time, and in Method B, the blood and formamide are heated in the tube first, before

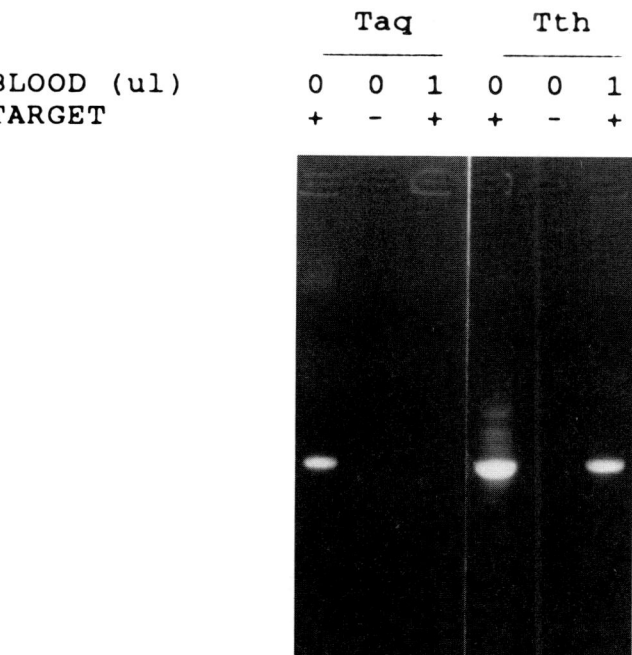

Figure 2. Differences in sensitivity between *Taq* and *Tth* polymerase to blood for PCR. The PCR products were electrophoresed in a 1% agarose gel stained with ethidium bromide. Each track contains 20 of a 100-µl standard PCR (without formamide) using *C. fetus* primers. Reactions were performed with (+) or without (–) target (approximately 10^3 *C. fetus* bacteria) and with (1 µl) or without (0 µl) sodium heparinized whole blood. The samples were subjected to 30 c of 30 s for each of the following temperatures: 95, 50, and 70°C.

the other reagents are added. Method B is more sensitive. The following describes the conditions for a 100-µl reaction.

1. **Method A**
   In the same tube, add

1 µl of whole blood containing anticoagulant (see below)	
18 µl of formamide	
10 µl 10X PCR buffer	100 m*M* Tris, pH 9.0, 500 m*M* KCl, 15 m*M* MgCl$_2$, 0.1% gelatin, 1% Triton X-100
10 µl dNTP stock	(2 m*M* of dATP, dCTP, dTTP, and dGTP Pharmacia, Uppsala, Sweden)
1 µl forward primer	2 µ*M*
1 µl reverse primer	2 µ*M*
57 µl H$_2$O	

   Overlay with liquid paraffin, and amplify for 30 c.

2. **Method B**
   Step 1 — Mix 1 µl of whole blood containing anticoagulant (see below) with 18 µl of deionized formamide, and heat the sample to 95°C for 5 min to solubilize all the blood. For hot-start PCR, a pastille of paraffin wax (#36084, BDH, Poole, U.K.) can be added to the sample at this stage.
   Step 2 — Make up the reaction mix (similar to above):

10 µl 10X PCR buffer
10 µl dNTP stock
1 µl forward primer
1 µl reverse primer
57 µl H$_2$O
1 µl Tth polymerase

Step 3 — Using a pipette, mix 81 µl of the reaction mix with the blood formamide mixture. Do not spin the tubes. Overlay with liquid paraffin if appropriate, and amplify for 30 c.

The optimal temperatures and incubation times vary between models of PCR machines. As a guide, conditions for two different machines are as follows. For the robotic arm Gene Machine (Innovonics, Melbourne, Australia), using three water baths, the following conditions are recommended: 85°C for 30 s, 40°C for 30 s, and 60°C for 30 s. For the Perkin Elmer 9600: 80°C for 30 s, 40°C for 10 s, and 60°C for 10 s.

Notes:

1. FoLT PCR reactions can be done in as little as 20 µl volume.
2. Accurate pipetting is essential since the final percentage of formamide in the reaction mixture is critical. This is more important when small reaction volumes are used.
3. Up to 10% v/v of whole blood can be used in FoLT PCR. However, using more blood may not result in greater sensitivity since the increase in target number may be negated by the increase in organic matter.

C. PRIMERS

HLA-DO primers which correspond to the 5' untranslated region of the human major histocompatibility complex class II gene were 5'TGCAGGCAAACAATGGTTGAG 3' and 5'GGACCCACCCAGAACCCAT 3'. *Campylobacter fetus* primers have been described previously.[7] The PCR product using the HLA-DO primers is 320 bp, whereas that with the *C. fetus* primers is 1.3 kb.

D. ANTICOAGULANTS

Various anticoagulants were used to collect blood from a single individual. These include lithium heparin (60 USP U/10 ml, Becton Dickinson, New Jersey), sodium heparin (60 USP U/10 ml, Becton Dickinson, New Jersey), fluoride heparin, ammonium oxalate (6 mg/5 ml), fluoride oxalate, potassium oxalate (4 mg/5 ml), EDTA (K3) (7.2 mg/5 ml), di-sodium EDTA (10.5 mg/7 ml), sodium citrate (0.105 M), sodium citrate (0.129 M), Alsever's solution, citrate phosphate dextrose (CPD), acid citrate dextrose (ACD) solution A, and ACD solution B.

III. RESULTS

Because the type of anticoagulants used in blood collections may have an effect on the quality of DNA that is recovered from blood samples stored at room temperature, the suitability of different anticoagulants was determined for FoLT PCR. Whole blood was collected from a single individual using different anticoagulants and stored at room temperature over a three-month period. Each week a 1-µl aliquot of blood was taken and the HLA-DO gene was amplified using FoLT PCR, Method B. The results for selected weeks are shown in Figure 3. Of the 14 anticoagulants tested, only two, namely, sodium citrate (0.105 M) and sodium citrate (0.129 M), inhibited FoLT PCR at week 0. Possibly the citrate ions at such high concentration bound all Mg^{2+} ions. Interestingly, after the blood samples were stored at room temperature for six weeks, this inhibition was no longer present. This is presumably due to the cell lysis and release of sequestered Mg^{2+}. All the other anticoagulants tested were still

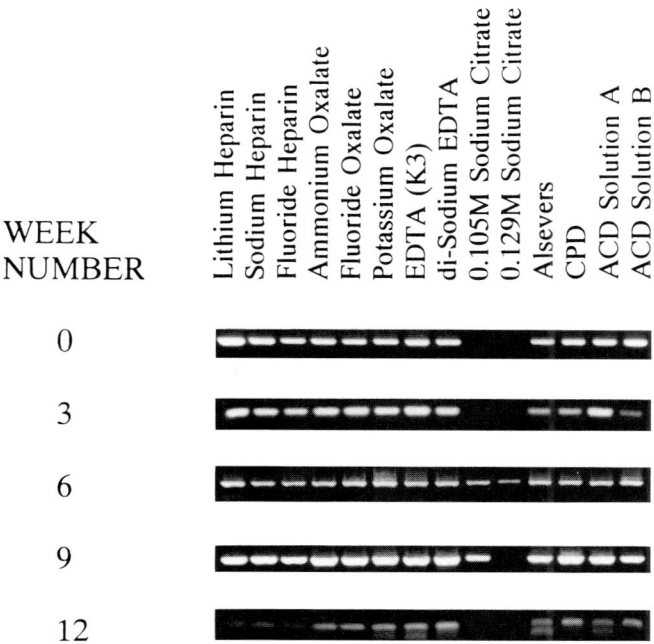

Figure 3. Effect of type of anticoagulant on FoLT PCR. Photograph of a 2% agarose gel stained with ethidium bromide. Each track represents 20 of a 100-µl FoLT PCR reaction containing the HLA-DO primers and 1 µl of whole blood. Blood was collected in the anticoagulants indicated. The samples were subjected to 30 c of 30 s for each of the following temperatures: 85, 40, and 60°C.

successful after storage at room temperature for 12 weeks. However, the signal strength was reduced for samples stored in the various types of heparin salt. This reduction in signal strength is consistent with the DNA breakdown observed in the samples when they were electrophoresed in agarose gels (data not shown).

IV. DISCUSSION

FoLT PCR is a method that allows direct PCR from whole blood without the need to purify DNA. As such, it represents a major advance in the diagnostic uses of PCR where whole blood is the target. FoLT PCR is based on two observations. First, *Tth* polymerase is a more robust enzyme than *Taq* polymerase for use in reactions containing a high degree of organic matter,[7] and second, the failure to amplify DNA from whole blood was at least partially due to the physical entrapment of target DNA by the coagulated organic matter.[6] Given these two observations, a number of protocols could have been designed to allow direct PCR from whole blood. The use of different detergents to inhibit DNA entrapment was not found to be successful. The alternative was to develop a low-temperature PCR protocol involving formamide. PCR reactions containing formamide (18% v/v), utilizing *Tth* polymerase as the DNA polymerase, preformed under reduced temperature conditions were found to allow the direct amplification of DNA from whole blood.

Formamide (2.5% v/v) has been previously used to increase the specificity of PCR. Sarkar et al.[8] found that concentrations of 10% formamide inhibited *Taq* polymerase when used under standard PCR conditions. In contrast *Tth* polymerase can tolerate 10% formamide under standard conditions and 24% formamide under reduced temperature conditions.[6] It is this tolerance of formamide by *Tth* polymerase that makes FoLT PCR possible. *Tth* polymerase from suppliers other than Toyobo, such as TetZ, (Amersham) and Thermalase (Kodak), were found to be successful in FoLT PCR. In contrast alternative thermostable polymerases, such as *Taq* polymerase and *Pfu* polymerase (Stratagene), were not found to be suitable in FoLT PCR.

There may be a need to titrate the formamide concentration for a particular oligonucleotide pair the first time they are used. This is because the behavior of different primer pairs may differ under FoLT PCR conditions due to their different primer lengths and GC content. We would recommend testing over a narrow range of 17 to 22% (v/v) formamide. This will ensure that the optimal formamide concentration is used for a particular primer set; 18% formamide has worked well with the three primer pairs tested.

Preheating formamide and blood is not mandatory (see Method A); however, it increases the sensitivity of FoLT PCR by at least tenfold by ensuring that the target DNA is accessible to *Tth* polymerase. A further advantage of mixing the blood with formamide, in the appropriate ratio (18:1 to 10), is that it allows the samples to be stored at 4°C for a long time (at least months) before PCR amplification. As such, it provides a convenient way of collecting and storing blood samples for subsequent analysis.

The sensitivity of FoLT PCR in whole blood has been determined using various dilutions of human blood in horse blood.[6] This was done so as to keep the total protein, DNA, and red blood cell concentration constant while varying the number of human nucleated leucocytes in each sample. This method of estimating the sensitivity of FoLT PCR provides a far better estimate than determining the total number of *purified* targets that are necessary for amplification because it takes into account the effects of inhibitors, protein, and the excess of nontarget DNA, which are present in PCR amplifications in which whole blood is the source of DNA. The limit of sensitivity of FoLT PCR was approximately 5.5 targets per microliter of whole blood, which corresponds to a target concentration of 5.5×10^6 per liter. Employing a second round of amplification or probing the PCR reaction product should allow even greater sensitivity. With a sensitivity of 5.5 targets per microliter of whole blood, FoLT PCR compares favorably with the sensitivity of rapid DNA-purification methods such as the Chelex 100 protocols.[2-4]

Determination of the minimum target concentration of FoLT PCR in whole blood allows one to estimate the usefulness of FoLT PCR for a particular application. For example, in applications such as HLA typing, where every nucleated cell contains the target sequence, a sensitivity of 100 to 1000 targets per microliter of whole blood will allow the efficient screening of a population. FoLT PCR, therefore, has applications in HLA typing, parentage testing, detection of genotypes linked to known diseases, and the identification of transgenic animals where as few as 10% of cells contain the introduced gene. It is envisaged that a sensitivity greater than 1000 targets per microliter of whole blood would be necessary for the diagnosis of some protozoal, viral, and bacterial infections.

The inhibition of *Taq* polymerase by samples containing heparin has been reported.[9-11] This inhibition of *Taq* polymerase was shown to be caused by heparin itself since incubating target DNA with Heparinase II prior to amplification removed the inhibition.[9,11] In contrast, *Tth* polymerase, when used under FoLT PCR conditions, was not found to be inhibited by heparin. However, in previous work the type of heparin salt used was found to be critical.[6] Fluoride heparin was found to totally inhibit FoLT PCR, lithium heparin had a slightly inhibitory effect, whereas no inhibition was observed with sodium heparin. In our current study inhibition was not observed with fluoride heparin. These seemingly inconsistent results may be due to variability in the amount of blood drawn into vacutaire tubes. Until further information is derived, the use of fluoride heparin and lithium heparin is not recommended. In contrast, no inhibition of FoLT PCR has been observed by sodium heparin, and reactions containing up to 10% v/v sodium-heparinized whole blood have been successful.

ACD solution B has been reported to be the anticoagulant of choice for preserving DNA in blood transported at room temperature.[12] ACD solution B is also suitable for FoLT PCR, and strong PCR signals were still obtained after the blood had been stored at room temperature for three months. Since ACD solution B, EDTA, and sodium heparin are used in general blood collecting, FoLT PCR can be used in population screenings without any modification to existing blood-collecting methods.

FoLT PCR is generally applicable because we have now used it in several examples (including microsatellite DNA) and three different species (sheep, mouse, and human). It should be suitable for all PCR tests (such as disease diagnosis, HLA-typing, family studies, and parentage testing) that use whole blood or clinical material with a high organic content, as the source of target DNA, without compromising sensitivity. FoLT PCR is suitable for blood collected with a range of commonly used anticoagulants and is not affected by transportation or storage of the samples. As such, FoLT PCR offers a convenient method of applying PCR analysis to blood samples. FoLT PCR can be easily automated, and thus it offers a rapid method, with minimal handling of amplifying DNA from whole blood.

ACKNOWLEDGMENTS

The authors wish to thank Michael Georgesz and Dianne Beck for their technical support.

REFERENCES

1. **Saiki, R. K., Gelfand, D. H., Stoffel, S., Scharf, S. J., Higuchi, R., Horn, G. T., Mullis, K. B., and Erlich, H. A.**, Primer-directed enzymatic amplification of DNA with a thermostable DNA polymerase, *Science,* 239, 487, 1988.
2. **Higuchi, R.**, Rapid, efficient DNA extraction for PCR from cells or blood, *Amplifications,* 2, 1 1989.
3. **Singer-Sam, J., Tanguay, R., and Riggs, A. D.**, Use of Chelex to improve the PCR signal from a small number of cells, *Amplifications,* 3, 11, 1989.
4. **Skalnik, D. G. and Orkin, S.**, A rapid method for characterising transgenic mice, *BioTechniques,* 8, 34, 1990.
5. **Winberg, G.**, A rapid method for preparing DNA from blood, suited for PCR screening of transgenes in mice, *PCR Methods Appl.,* 1, 72, 1991.
6. **Panaccio, M., Georgesz, M., and Lew, A.**, FoLT PCR: a simple PCR protocol for amplifying DNA directly from whole blood, *BioTechniques,* 14, 238,1993.
7. **Panaccio, M. and Lew, A.**, PCR in the presence of 8% (v/v) blood, *Nucleic Acids Res.,* 19, 1151, 1991.
8. **Sarkar, G., Kapelner, S. and Sommer, S. S.**, Formamide can dramatically improve the specificity of PCR, *Nucleic Acids Res.,* 18, 7465, 1990.
9. **Beutler, E., Gelbart, T., and Kuhl, W.**, Interference of heparin with the polymerase chain reaction, *BioTechniques,* 9, 166, 1990.
10. **Holodniy, M., Kim, S., Katzenstein, D., Konrad, M., Groves, E., and Merigan, T.**, Inhibition of human immunodeficiency virus gene amplification by heparin, *J. Clin. Microbiol.,* 29, 676, 1991.
11. **Wang, J.-T., Wang, T.-H., Sheu, J.-C., Lin, S.-M., Lin, J.-T., and Chen, D.-S.**, Effects of anticoagulants and storage of blood samples on efficacy of the polymerase chain reaction assay for hepatitis C virus, *J. Clin. Microbiol.,* 30, 750, 1992.
12. **Gustafson, S., Proper, J. A., Bowie, E. J. W., and Sommer, S. S.**, Parameters affecting the yield of DNA from human blood, *Anal. Biochem.,* 175, 294, 1987.

Chapter 19

IMMUNO-PCR: A GENERIC METHOD FOR PURIFYING TARGET FOR PCR

Andrew M. Lew and Michael Panaccio

TABLE OF CONTENTS

I. Introduction 159

II. Materials and Methods 160
    A. Anti-Histone Antibodies 160
    B. Immuno-PCR Using Anti-Histone Antibodies 160
    C. Purification of Anti-Bacteria Antibody 160
    D. Antibody Coating to Beads 160
        1. Glass Beads 160
        2. Magnetic Beads 161
        3. Latex Beads 161
    E. Immuno-PCR Using Anti-Bacteria Antibodies 161
        1. Antibody Coated to Surface of Microcentrifuge Tube 161
        2. Antibody Coated to Beads 161

III. Results 161
    A. Immuno-PCR Using Anti-Histone Antibodies 161
    B. Sensitivity of Immuno-PCR 161
    C. Immuno-PCR Using Anti-Bacteria Antibodies 161

IV. Discussion 162

Acknowledgments 163

References 163

I. INTRODUCTION

One of the limitations to the polymerase chain reaction (PCR) for DNA amplification directly from biological material (especially clinical samples) has been the need to usually treat the source material. Various sources of DNA have been used for PCR. Phage plaques or bacterial colonies carrying plasmid DNA have multiple copies of the DNA of interest, and hence PCR can be performed directly without the need for DNA purification. Some purification (e.g., making a nuclear pellet followed by proteinase K digestion) is, however, generally required for PCR amplification of single-copy genes from genomic DNA, especially if there is a mixed population of cells. This is presumably due to the large amount of other cell constituents interfering with the PCR.

We describe an expedient method for capturing target in a tube by coating the surface with an antibody. We called this technique *immuno-PCR*,[1] but it should not be confused with a subsequent unrelated technique given the same name by other workers.[2]

The two examples we describe are the amplification of chromosomal DNA from whole blood in the same tube as that to be used for PCR and the amplification of bacterial DNA from a mixture. For the former, anti-histone antibodies adsorbed onto a polypropylene tube bind chromatin and its associated chromosomal DNA, thus providing a suitable substrate for PCR. The PCR can be performed in this tube even a week later. Antibodies to surface molecules of bacteria (or indeed probably any cell type) may also be used to capture selectively what is to be amplified.

II. MATERIALS AND METHODS

A. ANTI-HISTONE ANTIBODIES

Human anti-histone antibodies were purified on protein A agarose[3] from serum from a patient treated with hydralazine (such patients often develop autoantibodies to histones). The serum was a gift from Dr. S. Whittingham, The Walter and Eliza Hall Institute, Melbourne, and was previously shown to react against calf and human histones by Western blotting. Alternatively, anti-histone antibodies could be generated in mice as previously described.[4]

B. IMMUNO-PCR USING ANTI-HISTONE ANTIBODIES

Fifty microliter of the immunoglobulin fraction of anti-histone serum were coated onto polypropylene microcentrifuge tubes at 10 µg/ml in 0.01 M phosphate-buffered saline, pH 7.4 (PBS), for 2 h at room temperature or 4°C overnight. After three washings in PBS, 45 µl of 100-µM EDTA, pH 8, containing 0.5% Triton X-100 (Sigma, St Louis) and 5 µl of whole blood were added. This solution was used, because Triton X-100 lyses the cell membrane and the high EDTA concentration destabilizes the nuclear membrane without interfering with the histone-DNA interaction or the antigen-antibody reaction.

Two systems were tried: amplification of a *Plasmodium falciparum* gene from malaria blood and amplification of the pyruvate dehydrogenase gene from normal human blood. For the former, 5 µl of whole blood (heparinized 10 IU/ml) containing *P. falciparum* parasites was added and mixed. For the latter, just 5 µl of whole blood (heparinized 10 IU/ml) were used. After a 2-hr incubation, tubes were washed three times with PBS and PCR as performed in 50-µl volume (50 mM KCl, 2 mM MgCl$_2$, 10 mM Tris, pH 8, 200 nM of each primer, and 1 U *Taq* polymerase), and 30 c of 95°C for 1 min, 50°C for 1 min, and 70°C for 1 min were completed in a robot-arm machine with three water baths (Innovonics, Melbourne). Oligonucleotides (specific for a gene encoding a merozoite surface antigen or pyruvate dehydrogenase) were used for priming.[1,5] The PCR product was electrophoresed in 1% agarose gel containing 0.1 µg/ml ethidium bromide for visualization.

C. PURIFICATION OF ANTI-BACTERIA ANTIBODY

Rabbit anti-*Campylobacter fetus* antibodies were purified from 15 ml of serum by precipitation with caprylic acid and ammonium sulfate.[6]

D. ANTIBODY COATING TO BEADS

1. Glass Beads

One milligram of 75-Å aminopropyl glass beads (G9514, Sigma Chemicals, St Louis) was combined with 300 µg anti-*C. fetus* immunoglobulin, and the solution was made up to 1.5 ml 50 mM 1-ethyl-3-(3-dimethylaminopropyl)-carbodiimide (Sigma) and 100 mM sodium borate, pH 4.5. The beads were rotated for 5 h at room temperature before being washed three times with PBS.

2. Magnetic Beads

One milliliter of undiluted Dynabeads (4×10^8/ml; M-450, Dynal, Oslo) was activated and coupled with 360 µg of anti-*C. fetus* immunoglobulin according to the manufacturer's instructions.

3. Latex Beads

One milliliter of a 10% suspension of latex beads (Styrene divinylbenzene SD91, Sigma Chemicals, St Louis) was incubated with 360 µg anti-*C. fetus* immunoglobulin for 16 h prior to being washed three times with PBS. The beads were then stored in 1 ml of PBS at 4°C.

To ensure that the coupling/coating of antibody to the beads was successful, each bead type was used to immunoprecipitate *C. fetus* proteins from a *C. fetus* lysate. In all cases, immunoprecipitation was successful.

E. IMMUNO-PCR USING ANTI-BACTERIA ANTIBODIES
1. Antibody Coated to Surface of Microcentrifuge Tube

The tube was coated with antibody as outlined above for anti-histone antibodies. 50 µl of a PBS solution containing *C. fetus* bacteria were added, and after a 1-h incubation, the tube was washed three times with PBS before being subjected to PCR amplification as described earlier.

2. Antibody Coated to Beads

Ten microliters of each anti-*C. fetus* coupled/coated bead type were incubated for 1 h with the target solution in a microcentrifuge tube before being washed three times with PBS. The glass and latex beads were allowed to settle between washes, whereas a magnet was used to collect all the magnetic beads on the side of the tube between washes. Following the three washes, the samples were subjected to PCR amplification.

III. RESULTS

A. IMMUNO-PCR USING ANTI-HISTONE ANTIBODIES

Figure 1 illustrates the strategy of immuno-PCR. It has been shown to amplify chromosomal DNA from malaria parasites and from human leukocytes in whole blood directly in the tubes, using the captured chromatin as the source of DNA.[1,4] If normal Ig or non-human blood was used, no visible PCR product was obtained.

B. SENSITIVITY OF IMMUNO-PCR

To determine the sensitivity of the immuno-PCR assay, a *P. falciparum* culture with a parasitemia of 20% (20 parasites per 100 erthrocytes) was diluted in human whole blood to a parasitemia of 1, 0.2, and 0.04%. The parasites were mainly in the ring stage of development to mimic the clinical situation. Five microliters of this were added to the antibody-coated tube containing 0.5% Triton X-100 and 100 mM EDTA, and immuno-PCR was performed as described above. When the tubes were coated with mouse anti-*P. falciparum* histone antibody, a visible DNA fragment was obtained at 1% parasitaemia, but not at 0.2 or 0.04% parasitemia.

C. IMMUNO-PCR USING ANTI-BACTERIA ANTIBODIES

An array of immuno-capture supports were explored for use in immuno-PCR. These included inert supports such as glass, as well as latex and magnetic supports which are now widely used in immunological procedures. Immuno-PCR was successful for all matrices used (surface of tube and glass, magnetic and latex beads), when the bacteria concentration was 10^9 cells per milliliter (Figure 2). At 10^6 cells per milliliter, a stained product could be seen only for the tube capture system.

STRATEGY of IMMUNO-PCR

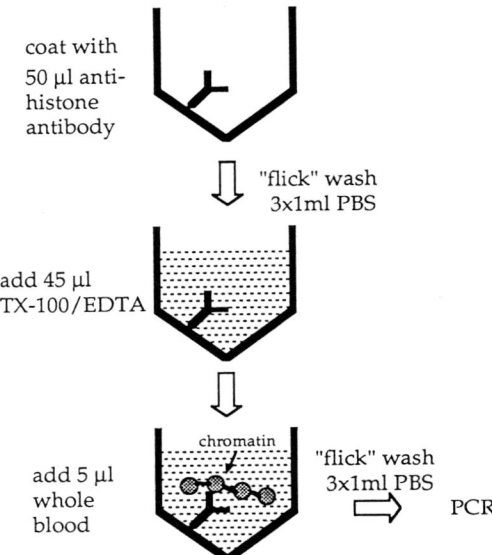

Figure 1. Scheme of immuno-PCR. The example shown is the use of anti-histone antibodies to capture chromatin from whole blood in the one microcentrifuge tube (see Section II on Materials and Methods).

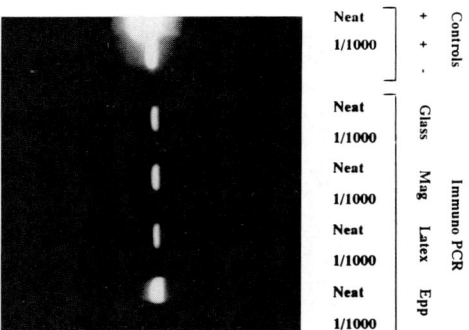

Figure 2. Demonstration of the different types of support structures that can be used in immuno-PCR. Photograph of a 1% agarose gel stained with ethidium bromide. In all reactions the *C. fetus* primers P1f and P1r were used. The target in each reaction was a 50-µl suspension of neat (approximately 10^9 cells per milliliter), 1/1000 (approximately 10^6 cells per milliliter), or no *C. fetus* bacteria, as indicated. In the positive control samples (uncoated tube), 1 µl of the *C. fetus* bacteria suspensions was added directly to the PCR mixture. For the anti-*C. fetus* attached glass, magnetic (mag), latex and microcentrifuge eppendorf tube (Epp) all the antibody capture steps and subsequent PCR amplification were performed as described in Section II on materials and methods.

IV. DISCUSSION

Antibodies can be used to capture target for PCR. We have used anti-histone antibodies to selectively capture chromatin (and hence, chromosomal DNA from eukaryotes) and anti-bacteria antibodies to capture a particular bacterial species. Using anti-histone antibodies in a capture assay in a polypropylene microcentrifuge tube or in microtiter trays, chromosomal DNA can be selectively bound in the presence of a large amount of organic material (e.g., blood), and then PCR can be performed in the same tube. Because chromatin is found in all eukaryotes, this method is generic. The antibody coated tubes can be stored for at least one week (longer times were not tested) in the cold in PBS containing 0.1% azide without significant loss in activity (data not shown). Moreover, this immuno-PCR procedure would seem to be easily adapted for field use, as there are few manipulations and no required centrifugation steps. During the actual collection of material, no freezing facilities are required, and once the chromatin has been captured in the antibody-coated tubes, it can be dried and stored for at least one week before PCR. Since the volume of sample required is small (5 µl), it can also be obtained by scratching the surface of the contents of a vial (containing malaria parasites frozen at −70°C) with a pipette tip; thus a precious sample (e.g., primary isolates) need not be thawed. Where large numbers of samples need to be processed, immuno-PCR could be performed in microtiter trays.[4]

The anti-malaria histone antibody was effective, when the starting material was at 1% parasitemia. The antibody used was obtained from mice that had received only two injections of *P. falciparum* histones. It is probable that a more avid antibody obtained from hyperimmunization might be more efficacious, or a monoclonal antibody against histones specific for a particular organism could be used. Alternatively, there are ways of increasing the amount of functional antibody bound to the tube, e.g., by coating the tube with protein A or protein G (which are polyvalent for the Fc of immunoglobulin) before the addition of affinity-purified antibody; thus, not only might more antibody be bound but all the antibodies would be orientated in the appropriate way.

The use of different bead matrices were explored, because they allow capture of target from larger volumes than can be achieved with the use of microcentrifuge tube surface. This is particularly important in clinical situations where the sample volume may be large, e.g., 10 ml, and the target concentration is low, e.g., 10 targets per milliliter. Bead matrices for immuno-PCR need to be resistant to the high temperatures of PCR. The three bead types described here (glass, magnetic, and latex) were found to be suitable, whereas Sephadex and Sepharose beads were not (data not shown). We have not made any quantitative comparison among the bead types. The use of magnetic beads has one important advantage over the other bead types in that they can be collected on the side of the tube with a magnet, thus separating the heavy debris that can contaminate spun pellets.

We have used the immuno-PCR technique to capture chromatin of malaria parasites from whole blood, human chromatin from whole blood and bacterial cells. We also believe that it could be used for the detection of any target DNA in body fluids or tissue as long as a suitable antibody was available.

ACKNOWLEDGMENTS

Thanks are due to Dianne Beck and Nicole Hunter for technical assistance.

REFERENCES

1. **Lew, A. M. and Kelly, J.,** Capture of chromosomal DNA by anti-histone antibodies for PCR, *Nucleic Acids Res,* 19, 3459, 1991.
2. **Sano, T, Smith, C. L., and Cantor, C. R.,** Immuno-PCR: very sensitive antigen detection by means of specific antibody-DNA conjugates, *Science,* 258, 120, 1992.
3. **Ey, P. L., Prowse, S. J., and Jenkin, C. R.,** Isolation of pure IgG1, IgG2a and IgG2b immunoglobulins from mouse serum using protein A-Sepharose, *Immunochemistry,* 15, 429, 1978.
4. **Lew, A. M. and Panaccio, M.,** Preparation of DNA from blood for polymerase chain reaction in microtitre dish, *Methods Enzymol.,* in press.
5. **Smythe, J. A., Peterson, M. G., Coppel, R. L., Saul, A., Kemp, D. J., and Anders, R. F.,** Structural diversity in the 45-kilodalton merozoite surface of *Plasmodium falciparum, Mol. Biochem. Parasitol.,* 39, 227, 1990.
6. **McKinney, M. M. and Parkinson, A.,** A simple, non-chromatographic procedure to purify immunoglobulins from serum and ascites fluid, *J. Immunol. Meth.,* 96, 271, 1987.

Chapter 20

NON-ISOTOPIC SINGLE-STRAND CONFORMATION POLYMORPHISM (SSCP) ANALYSIS OF PCR PRODUCTS

E. P. H. Yap and J. O'D. McGee

TABLE OF CONTENTS

I. Introduction 165
   A. Definition and Principle 165
   B. Conceptual Development 166
   C. Recent Innovations 168

II. Materials and Methods 168
   A. Equipment 168
   B. Materials 169
   C. Methods 169

III. Results 170
   A. Denaturation 170
   B. Loading Dyes 171
   C. Detection Methods and Sensitivity 172
   D. Gel Composition 172
   E. Electrophoretic Conditions 173

IV. Discussion 173
   A. Future Developments 173
   B. Applications 174
   C. Summary 174

Acknowledgments 174

References 175

I. INTRODUCTION

A. DEFINITION AND PRINCIPLE

Single-strand conformation polymorphism (SSCP) analysis is a simple method for detecting sequence differences between nucleic acid samples.[1] Sequence variations as small as single base point mutations can be identified in PCR products amplified from a variety of genomic or complementary DNA sources.

SSCP is performed by denaturing dsDNA and fractionating the strands on a non-denaturing polyacrylamide gel. Under the appropriate conditions, the electrophoretic mobility of the DNA is dependent not only on its length and molecular weight, but also on its conformation. This secondary structure is determined by the balance between destabilizing thermal forces and weak stabilizing forces such as intra-strand base pairings and stackings, which are in turn dependent on the primary structure of the DNA strand.[2] In practice, not only do the comple-

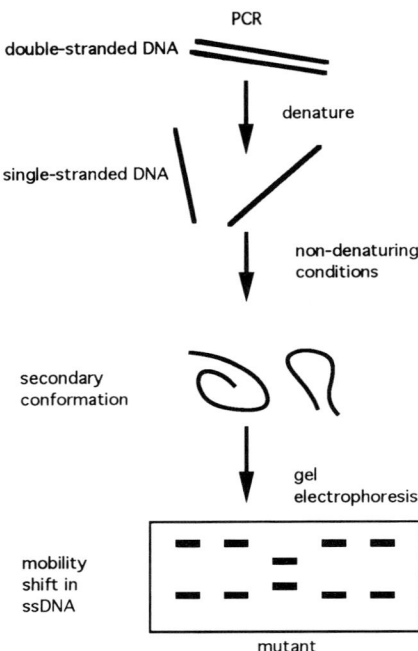

Figure 1. Principle of single-strand conformation polymorphism (SSCP).

TABLE 1.
Summary of Conceptual Development of SSCP

Polyacrylamide gel discriminates between:	Ref.
RNA vs. DNA	3
Linear vs. coiled vs. supercoiled	3′
ssDNA vs. dsDNA	3
dsDNA with different sequence	4
Complementary ssDNA strands	6
ssDNA strands of different sequences	7
ssDNA strands with single base change	8

mentary strands migrate as separate bands on the gel, but small differences in sequence also alter the mobility of these strands (Figure 1).

B. CONCEPTUAL DEVELOPMENT

It has long been recognized that native polyacrylamide gel electrophoresis (PAGE) fractionates nucleic acids not merely on the basis of size but also molecular conformation (Table 1). In a study of physical conditions affecting PAGE, Fisher and Dingman[3] reported that electrophoretic mobility of nucleic acids depended on chemical (whether RNA or DNA) and gross (whether single- or double-stranded, linear or circular) structures, and suggested that the mobility of polynucleotides reflected their relative molecular radii rather than their molecular weights. The fact that molecular weight determination of dsDNA by low-temperature PAGE is inaccurate[4] led to the development of denaturing gels, and the differential migration of complementary ssDNA strands forms the basis for strand-separating gels.[5,6] The anomalous migration of cloned mutant genes in strand-separation gel electrophoresis was subsequently described.[7]

TABLE 2.
Some Applications of SSCP

Application		Ref.
Detection of polymorphisms		
Human		
Rapid demonstration of known polymorphisms		
Biallelic	p53	
Polyallelic	HLA	13
HLA-DQ	Bone marrow transplant	27
Characterization of new polymorphisms		
Anonymous		
D8S8		28
Alu repeat		2
		29
Coding gene		
Dopamine receptor D2	Alcoholism	30
3′ untranslated regions	Detect new polymorphisms	31
dystrophin	Duchenne dystrophy carrier	32
GLUT4/glucose transporter gene		33
CFTR	Cystic fibrosis	34
Detection of mutations		
Human		
Known		
Ras oncogenes	Cancer cell lines	2
Unknown		
Tumor suppressor genes		
APC	Adenomatous polyposis coli	35
Polygenic disease genes		
Insulin receptor	NIDDM	36
Transthyretin	Hereditary polyneuropathy	9
Viral		
Human papillomavirus		18
Human rhinovirus 14		17
Quantification		
Transcripts		
Mouse complement C4-Related genes		37
Gene dosage		
Loss of heterozygosity (deletion) of tumor suppressor genes		
p53	Breast cancer	22
Allele separation		
Sequencing	Lipoprotein lipase mutation	38
Species identification		
Mouse vs. rat esterase-10 gene	Somatic cell hybrid analysis	39
Exclusion of PCR contamination		
Hepatitis B virus	Hepatocellular carcinomas	40

The term SSCP was first used by Hayashi and his group to describe the intentional use of sequence-dependent mobility shifts to detect DNA polymorphisms in Southern blots.[8] Genomic DNA was restricted, alkali denatured, electrophoresed on non-denaturing polyacrylamide gel, transferred onto filter, and hybridized to ^{32}P-labeled probes. Polymorphic and mutant fragments detected with H-ras and anonymous probes varied from 100 to 700 bp. Subsequently, SSCP analysis was applied to small PCR-amplified genomic DNA[2] and cDNA (Table 2). Isotopic ^{32}P label was incorporated either by 5′ end labeling of one or both primers, or by including labeled nucleotide triphosphate in the PCR. The DNA, denatured by heat, was loaded onto non-denaturing polyacrylamide gels with or without glycerol. Electrophoresis was performed at 4°C or room temperature for 3 to 24 h, the gel fixed and dried, and exposed for autoradiography for 0.5 to 72 h.

TABLE 3.
Methodological Developments in SSCP Analysis

SSCP Methodology	Ref.
Format	
Southern blots using ^{32}P	8
PCR products	2
Detection	
Silver staining	10
Chemiluminescence	9
Fluorochrome-labeled primers	13
Ethidium bromide staining	14
Modifications	
Endonuclease predigestion	16
Two-dimensional	17
RNA	19,20

C. RECENT INNOVATIONS

Improvements have been made in non-isotopic DNA detection of SSCP (Table 3), thus avoiding the need to radioactively prelabel PCR product/primers and shortening the duration of the procedure. Chemiluminescent detection of biotinylated probe hybridized to SSCP gels has been used.[9] DNA can be directly detected by silver staining, either manually[10] or using a semi-automated electrophoresis and staining system (Phastsystem, Pharmacia).[11,12] Fluorochrome-labeled primers have also been used,[13] facilitating automated gel reading and comparison of several samples in the same lane. With optimization of the denaturation step to maximize strand separation, ssDNA bands can be detected on SSCP gels using conventional ethidium bromide staining,[14] the detailed protocol for which is presented in the following section.

Modifications of the procedure have also been described to overcome the effective length limitation of about 400 bp. Long PCR products may be restricted with endonucleases, with the resultant fragments being analyzed by SSCP either individually after gel separation and excision,[15] or together.[16] In two-dimensional SSCP, the DNA is similarly digested into smaller fragments, separated on a denaturing polyacrylamide capillary gel, then run under non-denaturing conditions in the perpendicular axis on a conventional slab SSCP gel. Point mutations in 2.7-kb DNA were detected, and this method required only one pair of primers.[17]

The sensitivity of SSCP for detecting point mutations can be improved by optimizing the variables involved[18] or using a variety of SSCP conditions. Factors that affect SSCP migration and detection, including temperature, composition of gel, and electrophoretic conditions, are discussed below. An alternative approach is to perform SSCP on RNA transcribed from PCR-amplified DNA with promoter-primed RNA polymerase.[19,20] The enhanced sensitivity of RNA-SSCP for detecting mutations reported is attributed to the larger repertoire of secondary and tertiary structures (and hence mobilities) of ssRNA because short hairpins can form stable duplexes and the 2' hydroxyl group is available for additional hydrogen bonding.

II. MATERIALS AND METHODS

An optimized protocol for non-isotopic SSCP analysis is described here.

A. EQUIPMENT
1. Vertical electrophoresis gel tank apparatus with cooling compartment, e.g., Biorad Protean II: 20- by 16-cm gel plates
2. Thermostat-controlled cooled water supply (optional), e.g., CU400, Bette-Tech

3. Spacers, well-forming comb 0.4 mm thick
4. Power source of at least 300 V
5. Staining trays
6. UV transilluminator and Polaroid camera

B. MATERIALS

1. Polyacrylamide/bis-acrylamide solution 40% (w/v) in 19:1 and 49:1 ratios (available as premixed powder from Sigma)
2. Ammonium persulfate (AMPS) solution 25% (w/v) —freshly made, or stored at 4°C for up to one week
3. TEMED (N,N,N',N'-Tetramethylethylenediamine) —stored airtight at room temperature
4. Glycerol (reagent grade)
5. 10x Tris-borate EDTA (TBE) solution: Tris-borate, 0.89 M; boric acid, 0.89 M; EDTA, 0.02 M
6. 10x Alkali denaturing buffer (NaOH, 500 mM; EDTA, 10 mM)
7. 10x Loading dye (bromophenol blue 0.5% [w/v], xylene cyanol 0.5% [w/v] in deionized formamide)
8. Ethidium bromide solution (10 mg/ml)
9. Silicon solution in organic solvent: Sigmacote®

C. METHODS

1. Prepare fresh gel solution with the desired acrylamide (A) concentration (6 to 10%), bis-acrylamide cross-linker (C) concentration (2 to 5%), and glycerol (G) content (0 to 5%) in 0.5x TBE solution; 16 ml of the solution is sufficient for a 0.4-mm-thick gel. The viscous glycerol should be pipetted through a large-bore pipette or a disposable tip with the end cut off so that it mixes homogeneously. If a number of gels of a certain composition are to be poured, one can prepare a bottle of gel solution. This pre-mixed stock can be stored for several weeks in the dark at 4°C; it is filtered before use.

Volumes (ml)	10% A	10% A 5% G	6% A	6% A 5% G
40% Acrylamide mix	4.0	4.0	2.4	2.4
10x TBE	0.8	0.8	0.8	0.8
Glycerol	—	0.8	—	0.8
Water	11.2	10.4	12.8	12.8
Total volume/gel	16	16	16	16

2. Clean the glass plates and comb regularly (every 5 to 10 gels) with soap and water. Degrease and dry by wiping with 95% ethanol on a tissue, and siliconize by wiping on a thin layer of Sigmacote (to be done in fume cupboard or well-ventilated area). This will facilitate pouring and removal of gels later, and it needs to be done every 2 to 5 gels. Assemble the glass plates and spacers carefully onto gel-pouring stand to prevent leakage from the bottom. In particular, ensure that both plates and spacers are flush with each other.
3. Add 32 µl each of AMPS 25% and TEMED per 16 ml of gel solution. Mix and pour between the plates, using a 10-ml syringe. If an air bubble is detected, stop adding solution, tilt the plates so that the solution-air surface falls to the level of the bubble. Place the comb into position. Allow at least 20 min for the gel to set. Prewarming the gel solution to 37 to 42°C will allow the gel to set within 5 to 10 min.
4. At this stage, the electrophoresis tank may be set up (without the gel) and precooled to the final running temperature with either running tap water (the inlet temperature should be measured) or preferably a constant-temperature cooled-water recycling unit.

5. Meanwhile prepare the PCR samples. Remove the mineral oil (if used) in each tube with 50- to 100-μl chloroform. The aqueous sample, initially beneath the oil layer, will float on the chloroform-oil mix, which can be pipetted away. Use 10 to 30 μl for each lane, depending on the size of the well and the DNA concentration. Add 1 μl 10x alkali denaturing buffer per 10-μl sample. Denature for 10 min in a 42°C oven.
6. After the gel has polymerized (check the remaining excess gel solution for gelation), remove excess gel above the comb with a scalpel blade. Carefully remove the comb, avoiding air bubbles in wells by displacing comb with buffer. Install the gel onto the cooling unit and electrophoresis tank, ensuring that the gasket forms a good seal by first wetting it. Flush the wells with running buffer using a Pasteur pipette or pipette tip to remove unpolymerized acrylamide, air bubbles, and free glycerol.
7. Add 1 μl of 10x loading dye to each 10μl sample and load onto gel with ultra-fine pipette tip. The formamide-containing loading dye should be added just before the sample is loaded (see later).
8. Perform electrophoresis at 300 to 600 V. If using a constant voltage of 300 V, the current and power should initially be approximately 12 to 15 mA and 4 W per gel, and will decrease with time to 7 mA, 2 W, respectively.
9. On completion of electrophoresis, separate the plates with the help of a blade. Immerse the gel that is adherent to one of the plates into aqueous ethidium bromide solution 0.5 μg/ml. Stain at room temperature for 5 to 10 min. The gel may swell with absorbed water and become uneven or dislodge from the plate. If this occurs or if the gel folds on itself, the use of a gentle stream of running tap water (instead of gloved fingers) may help salvage the situation. Remove gel and plate, deftly invert onto surface of UV transilluminator, and carefully remove plate. Visualize and photograph through UV filter, using exposures approximately twice as long as those for agarose gels.

III. RESULTS

A systematic study of the factors affecting ethidium bromide detection of ssDNA, and SSCP band mobility and resolution was undertaken.[21] A summary of the findings is discussed here. An example of the use of the protocol above was for the quantitative analysis of a single base substitution polymorphism in the p53 gene[22] (Figure 2).

A. DENATURATION

Heat denaturation in 10 to 90% formamide yielded sufficient ssDNA to be detected by ethidium bromide staining. However *alkali* denaturation yielded consistently better results, two- and fourfold higher sensitivity depending on the DNA, and did not affect the mobility and position of bands on the gel. Incubation of the sample at 42°C for 10 min was found to give higher ssDNA yields on some occasions. The optimum sodium hydroxide concentration was determined to be approximately 50 mM final concentration. Concentrations at or above 200 mM altered gel pH and interfered with electrophoresis.

Alkali denaturation results in greater ssDNA yield probably because the alkali prevents reannealing from taking place while the samples are being loaded and during the initial period of electrophoresis before the DNA has entered the gel. It is rapidly neutralized by buffer in the gel, and does not affect the electric field nor the ssDNA conformation.

Other chemicals were added to samples in attempts to denature DNA but did not give the desired result. Formaldehyde (final concentration, 3.7%) reacted with the polyacrylamide gel and prevented DNA separation, and urea (final concentration, 1.5 M) did not denature the DNA sufficiently to give visible ssDNA bands.

Non-Isotopic Single-Strand Conformation Polymorphism (SSCP) Analysis of PCR Products

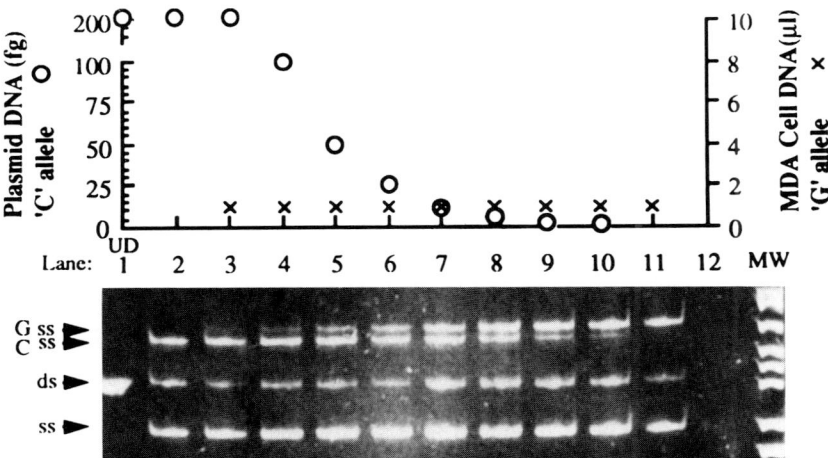

Figure 2. SSCP analysis of p53 gene polymorphism. The single-base substitution polymorphism in codon 72 in exon 4 of the p53 gene was analyzed by PCR amplification of a 185-bp fragment and SSCP analysis. The single-stranded (ss) and double-stranded (ds) DNA of the two alleles, designated C and G, were resolved on 10% polyacrylamide, containing 5% glycerol and 5% cross-linker at 22°C. The slower migrating ssDNA demonstrated a sequence-dependent mobility shift, as indicated in lanes 2 and 11 containing plasmid (C allele) and homozygous MDA cell (G allele) DNA, respectively. Undenatured (UD) DNA (lane 1), blank (lane 12) control, and MspI-digested pBR322 ladder (MW) are indicated. A serial dilution of known concentrations of plasmid DNA was added to fixed amounts of MDA cell DNA and used as template for competitive PCR (lanes 3 to 10), as shown in the graph. The relative intensities of the two upper ssDNA bands in the SSCP gel correlated with the initial proportions of template DNA.

B. LOADING DYES

Formamide allowed more convenient loading of samples through very-fine-pore pipette tips into thin wells, compared to other density agents commonly used in sample loading buffers, such as sucrose, glycerol, and Ficoll. The physical properties of formamide[23] that facilitated sample loading include the following:

1. It is more viscous ($\eta \times 10^5$ at 15°C = 4320) and dense than water (specific density = 1.13 at 20°C compared to water).
2. It has a low surface tension, and it facilitates the displacement of buffer during the loading of thin (0.4-mm) wells.

Surface tension	dyn/cm
Water (20°C)	72.75
Sucrose 10%	73.3
Glycerol 5%	72.9
Formamide	58.35

3. Chloroform, used to remove mineral oil, is immiscible with water unless formamide, which dissolves (in) both, is present.

It was found that the formamide loading dye appeared to interfere with denaturation if added at the same time with alkali denaturant. This could be due to the chemical reaction between formamide and alkali:

$$H.CO.NH_2 + H_2O = H.COOH + NH_3$$

$$H.COOH + NaOH = H.COONa + H_2O$$

Hence the formamide dye was added to the sample after its denaturation and just before loading onto the gel.

Bromophenol blue and *xylene cyanole*, each at a final concentration of 0.05% (w/v) in formamide, were used to assist loading and assess DNA migration. However both quench UV fluorescence of ethidium bromide and could interfere with visualization of low-molecular-weight bands, especially at low gel temperatures when dye migration is inhibited more than DNA migration. Where DNA of below 100 bp was separated, xylene cyanole was omitted from the loading buffer.

C. DETECTION METHODS AND SENSITIVITY

Single-stranded DNA bands were stained preferentially and differentially (compared to dsDNA) with the metachromatic dye *acridine orange*, giving a green fluorescence. The limit of visual sensitivity was about 100 ng total DNA, but was poorer (200 ng total DNA) on photographic film due to the red UV filter commonly used.

Optimum sensitivity (50 ng total DNA) for ssDNA detection was obtained with conventional *ethidium bromide* staining. Ethidium bromide intercalation within DNA and fluorescence is pH dependent, and was extinguished at high pH. However with the above protocol, use of buffer was unnecessary for staining because the alkali denaturant was sufficiently neutralized within the gel during electrophoresis. Double-stranded DNA bands were always seen due to the higher affinity and better fluorescent yield of ethidium bromide for dsDNA than for ssDNA.[5]

Prestaining the gel by adding ethidium bromide to the gel solution and buffer (2 l) was not routinely done because of the dye front migrating to the negative electrode, and the large amount of this potential mutagen required.

D. GEL COMPOSITION

Acrylamide concentration (%A) affected the rate of migration of both single- and double-stranded DNA, as well as the indicator dyes. The relative mobilities of the ssDNA bands were also affected; single base differences were best detected at certain concentrations depending on the length of the DNA. The DNA below 200 bp was best separated on 10% acrylamide, 200 to 300 bp on 8%, and greater than 300 on 6% gels. Use of denser gels did not improve discrimination, while gels below 6% were fragile and physical handling was difficult.

The extent of cross-linking is expressed by the ratio of N,N'-methylbisacrylamide to total acrylamide concentrations (%C) and determines the pore size within the separation matrix.[5] This ratio affected both DNA as well as indicator dye migration, but the former to a greater extent. 5%C gels were firmer and easier to handle, and suitable for separation of DNA smaller than 200 bp. The separation of DNA larger than 300 bp was best performed on gels with large pore size (2%C), which were more sensitive to single base differences. Polyacrylamide gels are not routinely made with %C outside this range.[15]

Glycerol content (%G) of gels affected the relative mobilities of ssDNA, because of their denaturant properties on the secondary conformation of macromolecules.[2] Five percent glycerol was found to be optimal for SSCP for a variety of DNA. While dye migration was retarded in its presence, ssDNA migrated relatively faster or slower depending on its length and sequence. Glycerol also improved ease of gel handling since in its absence, gels (particularly low %A and low %C) tended to swell and distort on being soaked in water, glycerol has therefore been used to fix gels before autoradiography. Use of 10% glycerol did not confer any advantage over 5%.

Agarose separated ssDNA from dsDNA but was not sufficiently discriminative for mutation detection. A proprietary alternative gel matrix to polyacrylamide, Hydrolink D600, with potential advantages of ease of preparation in a horizontal format, better resolving ability, and lack of neurotoxicity, was used for SSCP analysis. Single base changes were detected but the resolution of DNA bands was poorer than that in polyacrylamide. Recently, other polyacrylamide substitutes (e.g., MDEE gel, Hoefer Scientific, U.K.), with purportedly better ssDNA band separation have been marketed. The particular gel constitution and physical factors may have to be separately optimized for this purpose.

E. ELECTROPHORETIC CONDITIONS

The *temperature* of coolant and gel/buffer affected both absolute and relative migration speed of DNA and dyes. At lower temperatures, electrophoretic mobility was retarded. The relative ssDNA mobility was also dependent on temperature, probably reflecting the thermal effect on weak intramolecular bonds and hence secondary structure. It was necessary to ensure that temperature gradients within the same gel were minimized by sufficient cooling for the heat/power generated, to enable comparison between lanes. Failure to do so resulted in "smiling" band pattern artifacts. The use of a thermoregulated circulator allowed use of higher voltages, user-defined temperatures, and inter-seasonal consistency.

Voltage field influenced DNA migration speed. When other factors (gel and temperature) were held constant, the distance migrated was proportional to the product of voltage and duration of electrophoresis (volt-hours). The electrical *power* (proportional to the square of the voltage) was dissipated as heat. The ideal conditions were therefore the highest voltage and power that could be used without overheating the gel or creating intra-gel temperature gradients, so as to give maximum resolution and rapidity. The maximum voltage/power that could be used was determined empirically with the gel apparatus, cooling system, buffer, and gel thickness in use. SSCP gels were run at either constant power or constant voltage, though the former gave the theoretical advantage of a more constant temperature.

Resolution of bands was determined by several factors. While field/charge-dependent band separation increased with time, so did random diffusion, and the overall resolution of closely migrating bands did not increase after prolonged electrophoresis. Hence use of long-sequencing gel apparatus, with its attendant problems of handling, was found to be unnecessary. Distortion of gels during staining, prolonged staining, non-homogeneous or improperly poured gels, inappropriate pore size, and fluctuating gel temperatures were also found to reduce SSCP band resolution.

Half-strength Tris-borate EDTA (TBE) *buffer* provided sufficient buffering capacity with the system used and, compared to 1x concentration, was more economical and had a lower conductivity, thus enabling higher voltages to be used. In agreement with published reports,[18] buffer concentration affected band mobility; the same buffer stock should be used for the gel and for both buffer reservoirs.[24]

The *gel thickness* also influenced the current and hence power/heat generated; the mobilities of DNA were otherwise unaffected. Thin (0.4-mm) polyacrylamide gels could be run at a higher voltage, required a shorter staining period of 5 to 10 min, and were adequate for most analytical purposes; thicker (1.5-mm) gels were used where large amounts of DNA were required for preparative purposes (increased loading capacity), for ease of handling of very-low-density gels (<6%), or for loading a large volume of dilute sample. The detection sensitivity in thick gels was slightly lower than that in thin gels.

IV. DISCUSSION

A. FUTURE DEVELOPMENTS

Bifunctional DNA-intercalating dyes with affinities three orders of magnitude greater than that of ethidium bromide have recently been synthesized. Including ethidium homodimer,[25] thiazole orange dimer (TOTO), and oxazole yellow dimer (YOYO),[26] they have these advantages: they can be added to the samples prior to electrophoresis; free unbound dye is quantitatively removed during electrophoresis resulting in low background; very high sensitivity of dsDNA detection (picogram quantities) can be achieved with a confocal fluorescence imaging system; and the use of several dyes simultaneously allows for automation, multiplexing, and intra-lane controls. These dyes also have a high affinity for ssDNA in polyacrylamide, but their effect on the conformation of undenatured ssDNA has yet to be studied.

Occasionally and sporadically, amplified DNA (e.g., mitochondrial D-loop) re-anneals rapidly and therefore is not amenable to conventional non-isotopic SSCP analysis, there being insufficient ssDNA for detection. A discontinuous dual-phase gel, comprising an upper stacking denaturing gel containing formamide, and a lower SSCP-resolving gel, has been developed to maximize strand separation.[41]

In its present form, non-isotopic SSCP analysis may not be suitable for routine diagnostic use on a large scale, due to the inconvenience of pouring vertical polyacrylamide gels, neurotoxicity of acrylamide monomer, DNA size limitations, and the fact that the molecular conformation of ssDNA sequences cannot currently be theoretically predicted. Efforts to overcome these problems include development of acrylamide gel substitutes, horizontal gel formats, precast gels, and automated gel readers (e.g., GeneScanner, Applied Biosystems, U.S.).*

B. APPLICATIONS

The Human Genome Mapping Project has seen an exponential increase in the number of gene and protein sequences being published and submitted to reference databases. This has resulted in the concomitant growth of biomedical research that aims to elucidate gene function and correlate gene abnormalities with clinical phenotype. Recent developments in mutation screening methods have facilitated such work. Since its invention in 1989, SSCP has found increasing use in research (Figure 3).

SSCP has primarily been used to detect known and novel polymorphisms and mutations in human genes and loci (Table 3). Relevant regions (e.g., conserved exons) of the gene of interest are amplified in short fragments and subjected to SSCP analysis. Aberrant bands are then sequenced to confirm the presence of mutation or polymorphism. Other novel applications are for quantification of gene transcripts, allele separation for direct sequencing, to identify genes from different species in transgenic studies and to exclude contamination in virological diagnosis by PCR (see Chapter 27).

C. SUMMARY

SSCP analysis of PCR products provides a rapid and sensitive means of screening small sequence changes in DNA amplified from various sources. A rapid, simple, and non-isotopic method has been developed, involving denaturation of dsDNA by alkali, use of formamide loading dyes, and detection by ethidium bromide staining. The sequence-dependent conformation and mobility of ssDNA are affected by a number of physical factors, including gel composition (acrylamide concentration, degree of cross-linking, glycerol content, and type of gel matrix) and conditions of electrophoresis (temperature, electric field and power, buffer concentration, and gel thickness). Empirical optimization of these factors for DNA length and sequence may be necessary to increase the likelihood of detecting small mutations.

ACKNOWLEDGMENTS

We thank Dr. Damian McManus for critical review of the manuscript. EPHY was a Rhodes Scholar, a Wellcome Trust Advanced Training Fellow in Tropical Medicine (grant 035401/Z/92/Z/14T), and a Junior Research Fellow at Wolfson College, Oxford. This work was supported by the Cancer Research Campaign, U.K.

---

* The development and optimization of non-isotopic SSCP, using ethidium bromide stained minigels and methylmercury hydroxide as denaturant has also recently been described.[42]

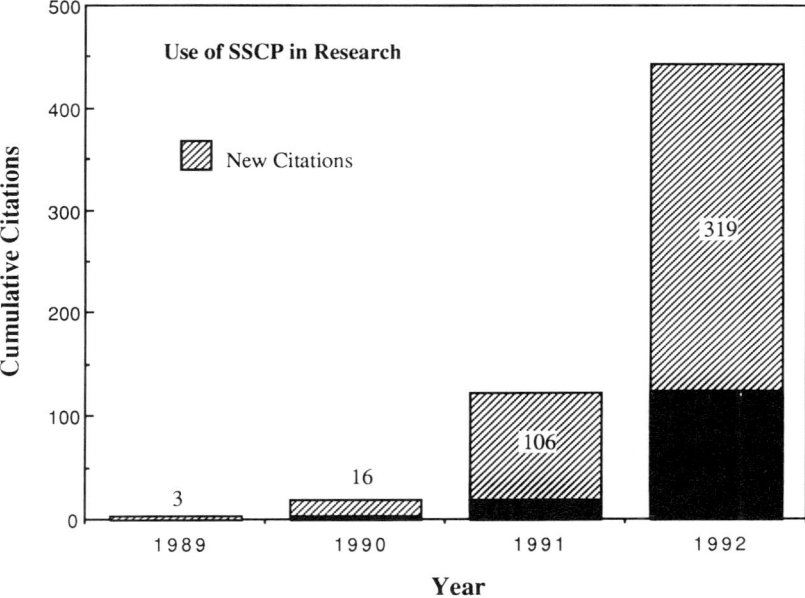

Figure 3. Use of SSCP analysis. Research use of SSCP, based on an online search of the *Science Citation Index* (ISI, U.S.) for publications citing the original SSCP papers.[2,8] Cumulative figures are plotted, and numbers are citations for 12-month periods.

REFERENCES

1. **Hayashi, K.**, PCR-SSCP: a simple and sensitive method for detection of mutations in genomic DNA, *PCR Methods Appl.*, 1, 34, 1991.
2. **Orita, M., Suzuki, Y., Sekiya, T., and Hayashi, K.**, Rapid and sensitive detection of point mutations and DNA polymorphisms using the polymerase chain reaction, *Genomics*, 5, 874, 1989.
3. **Fisher, M. P. and Dingman, C. W.**, Role of molecular conformation in determining the electrophoretic properties of polynucleotides in agarose-acrylamide composite gels, *Biochemistry*, 10, 1895, 1971.
4. **Allet, B., Jeppesen, P. G. N., Katagiri, K. J., and Delius, H.**, Mapping the DNA fragments produced by cleavage of lambda DNA with endonuclease RI, *Nature*, 241, 120, 1973.
5. **Maniatis, T., Fritsch, E. F., and Sambrook, J.**, *Molecular Cloning: A Laboratory Manual*, Cold Spring Harbor Laboratory Press, Cold Spring Harbor, NY, 1982.
6. **Maxam, A. M. and Gilbert, W.**, A new method for sequencing DNA, *Proc. Natl. Acad. Sci. U.S.A.*, 74, 560, 1977.
7. **Kanazawa, H., Noumi, T., and Futai, M.**, Analysis of *Escherichia coli* mutants of the H^+-transporting ATPase: determination of altered sites of the structural gene, *Methods Enzymol.*, 126, 595, 1986.
8. **Orita, M., Iwahana, H., Kanazawa, H., Hayashi, K., and Sekiya, T.**, Detection of polymorphisms of human DNA by gel electrophoresis as single-strand conformation polymorphisms, *Proc. Natl. Acad. Sci. U.S.A.*, 86, 2766, 1989.
9. **Saeki, Y., Ueno, S., Yorifuji, S., Sugiyama, Y., Ide, Y., and Matsuzawa, Y.**, New mutant gene (transthyretin Arg 58), in cases with hereditary polyneuropathy detected by non-isotope method of single-strand conformation polymorphism analysis, *Biochem. Biophys. Res. Commun.*, 180, 380, 1991.
10. **Ainsworth, P. J., Surh, L. C., and Coulter-Mackie, M. B.**, Diagnostic single strand conformation polymorphism (SSCP): a simplified non-radiosotopic method as applied to a Tay-Sachs B1 variant, *Nucleic Acids Res.*, 19, 405, 1991.
11. **Mohabeer, A. J., Hiti, A. L., and Martin, W. J.**, Non-radioactive single strand conformation polymorphism (SSCP) using the Pharmacia 'PhastSystem', *Nucleic Acids Res.*, 19, 3154, 1991.
12. **Dockhorn-Dworniczak, B., Dworniczak, B., Brommelkamp, L., Bulles, J., Horst, J., and Bocker, W. W.**, Non-isotopic detection of single strand conformation polymorphism (PCR-SSCP): a rapid and sensitive technique in diagnosis of phenylketonuria, *Nucleic Acids Res.*, 19, 2500, 1991.
13. **Lo, Y.-M. D., Patel, P., Mehal, W. Z., Fleming, K. A., Bell, J. I., and Wainscoat, J. S.**, Analysis of complex genetic systems by ARMS-SSCP: application to HLA genotyping, *Nucleic Acids Res.*, 20, 1005, 1992.

14. **Yap, E. P. H. and McGee, J. O.,** Nonisotopic SSCP detection in PCR products by ethidium bromide staining, *Trends Genet.*, 8, 49, 1992.
15. **Dean, M. and Gerrard, B.,** Helpful hints for the detection of single-stranded conformation polymorphisms, *BioTechniques*, 10, 332, 1991.
16. **Iwahana, H., Yoshimoto, K., and Itakura, M.,** Detection of point mutations by SSCP of PCR-amplified DNA after endonuclease digestion, *BioTechniques*, 12, 64, 1992.
17. **Kovar, H., Jug, G., Auer, H., Skern, T., and Blaas, D.,** Two dimensional single-strand conformation polymorphism analysis: a useful tool for the detection of mutations in long DNA fragments, *Nucleic Acids Res.*, 19, 3507, 1991.
18. **Spinardi, L., Mazars, R., and Theillet, C.,** Protocols for an improved detection of point mutations by SSCP, *Nucleic Acids Res.*, 19, 4009, 1991.
19. **Danenberg, P. V., Horikoshi, T., Volkenandt, M., Danenberg, K., Lenz, H.-J., Shea, L. C. C., Dicker, A. P., Simoneau, A., Jones, P. A., and Bertino, J. R.,** Detection of point mutations in human DNA by analysis of RNA conformation polymorphism(s), *Nucleic Acids Res.*, 20, 573, 1992.
20. **Sarkar, G., Yoon, H.-S., and Sommer, S. S.,** Screening for mutations by RNA single-strand conformation polymorphism (rSSCP): comparison with DNA-SSCP, *Nucleic Acids Res.*, 20, 871, 1992.
21. **Yap, E. P. H.,** Molecular Genetic Analysis of Human Breast and Cervical Cancers, D. Phil. thesis, University of Oxford, Oxford, U.K., 1992.
22. **Yap, E. P. H. and McGee, J. O'D.,** Nonisotopic SSCP and competitive PCR for DNA quantification: p53 in breast cancer cells, *Nucleic Acids Res.*, 20, 145, 1992.
23. **Budavari, S.,** *The Merck Index*, NH, 1989.
24. **Sambrook, J., Fritsch, E. F., and Maniatis, T.,** *Molecular Cloning: A Laboratory Manual*, 2nd ed., Cold Spring Laboratory Press, Cold Spring Harbor, NY, 1989.
25. **Glazer, A. N., Peck, K., and Mathies, R. A.,** A stable double-stranded DNA-ethidium homodimer complex: application to picogram fluorescence detection of DNA in agarose gels, *Proc. Natl. Acad. Sci. U.S.A.*, 87, 3851, 1990.
26. **Rye, H. S., Yue, S., Wemmer, D. E., Quesada, M. A., Haughland, R. P., Mathies, R. A., and Glazer, A. N.,** Stable fluorescent complexes of double-stranded DNA with bis-intercalating asymmetric cyanine dyes: properties and applications, *Nucleic Acids Res.*, 20, 2803, 1992.
27. **Summers, C., Fergusson, W., Gokhale, D., and Taylor, M.,** Donor-recipient bone-marrow matching by single strand conformation polymorphism analysis, *Lancet*, 339, 621, 1992.
28. **Iizuka, M., Hayashi, K., and Sekiya, T.,** Single-strand conformation polymorphism (SSCP) at the D8S86 locus, *Nucleic Acids Res.*, 19, 6346, 1991.
29. **Iizuka, M., Mashiyama, S., Oshimura, M., Sekiya, T., and Hayashi, K.,** Cloning and polymerase chain reaction-single-strand conformation polymorphism analysis of anonymous Alu repeats on chromosome 11, *Genomics*, 12, 139, 1992.
30. **Bolos, A. M., Dean, M., Lucas-Dorse, S., Ramsburg, M., Brown, G. L., and Goldman, D.,** Population and pedigree studies reveal a lack of association between dopamine D2 receptor gene and alcoholism, *JAMA*, 264, 3156, 1990.
31. **Poduslo, S. E., Dean, M., Kolch, U., and O'Brien, S. J.,** Detecting high-resolution polymorphisms in human coding loci by combining PCR and single-strand conformation polymorphism (SSCP) analysis, *Am. J. Hum. Genet.*, 49, 106, 1991.
32. **Richards, R. I. and Friend, K.,** Determination of Duchenne muscular dystrophy carrier status by single strand conformation polymorphism analysis of deleted regions of the dystrophin locus, *J. Med. Genet.*, 28, 856, 1991.
33. **Muraoka, A., Sakura, H., Kishimoto, M., Akanuma, Y., Buse, J. B., Yasuda, K., Seino, S., Bell, G. I., Yazaki, Y., Kasuga, M., and Kadowaki, T.,** Polymorphism in exon 4A of the human GLUT4 muscle-fat facilitative glucose transport gene detected by SSCP, *Nucleic Acids Res.*, 19, 4313, 1991.
34. **Chillon, M., Nunes, V., and Estivill, X.,** SSCP-polymorphism in intron 12 of the CFTR gene recognized by BclI, *Nucleic Acids Res.*, 19, 6343, 1991.
35. **Groden, J., Thliveris, A., Samowitz, W., Carlson, M., Gelbert, L., Albertsen, H., Joslyn, G., Stevens, J., Spirio, L., Robertson, M., Sargeant, L., Krapcho, K., Wolff, E., Burt, R., Hughes, J. P., Warrington, J., Mcpherson, J., Wasmuth, J., Lepaslier, D., Abderrahim, H., Cohen, D., Leppert, M., and White, R.,** Identification and characterization of the familial adenomatous polyposis coli gene, *Cell*, 66, 589, 1991.
36. **O'Rahilly, S., Choi, W. H., Patel, P., Turner, R. C., Flier, J. S., and Moller, D. E.,** Detection of mutations in insulin-receptor gene in NIDDM patients by analysis of single-stranded conformation polymorphisms, *Diabetes*, 40, 777, 1991.
37. **Huang, Z. M., Takahashi, M., and Nonaka, M.,** Differential expression of the five C4-related genes of H-2w7 mice, *Immunogenetics*, 33, 361, 1991.

38. **Hata, A., Robertson, M., Emi, M., and Lalouel, J.-M.,** Direct detection and automated sequencing of individual alleles after electrophoretic strand separation: identification of a common nonsense mutation in exon 9 of the human lipoprotein lipase gene, *Nucleic Acids Res.,* 18, 5407, 1990.
39. **Pravenec, M., Simonet, L., Kren, V., Lezin, E. S., Levan, G., Szirer, J., Szirer, C., and Kurtz, T.,** Assignment of rat linkage group V to chromosome 19 by single-strand conformation polymorphism analysis of somatic cell hybrids, *Genomics,* 12, 350, 1992.
40. **Yap, E. P. H., Lo, Y.-M. D., Cooper, K., Fleming, K. A., and McGee, J. O.,** Exclusion of false-positive PCR viral diagnosis by single-strand conformation polymorphism, *Lancet,* 340, 726, 1992.
41. **Yap, E. P. H. and McGee, J. S.,** Nonisotopic discontinuous phase single strand conformation polymorphism (DP-SSCP): genetic profiling of D-loop of human mitochondrial (mt) DNA, *Nucleic Acids Res.,* 21, 4153, 1993.
42. **Hongyo, T., Buzard, G. S., Calvert, R. J., and Weghorst, C. M.,** 'Cold SSCP': a simple, rapid and non-radioactive method for optimized single-strand conformation polymorphism analyses, *Nucleic Acids Res.,* 21, 3637, 1993.

Chapter 21

THE USE OF PCR-RAPD ANALYSIS IN PLANT TAXONOMY AND EVOLUTION

Tigst Demeke and Robert P. Adams

TABLE OF CONTENTS

I. Introduction	179
II. Analysis at the Subgeneric (Sectional) Level	180
III. Interspecific Taxonomy and Classification	180
IV. Geographic Variation and Infraspecific Taxonomy	182
V. Cultivar Classification and Identification	184
VI. Genetic Analyses within Clones	186
A. Genetic Linkage	186
B. Analysis of Somaclonal Variation	186
VII. Analyses of Gene Flow between Individuals	186
VIII. Hybridization Studies	186
A. Intergeneric Hybridization	186
B. Interspecific Hybridization and Introgression	186
IX. Analyses of Genetic Diversity	187
X. Analyses of the Evolution of Disease Resistance	187
XI. Studies of Ancient DNA	187
XII. Numerical Analyses of RAPD Data	188
XIII. Conclusion	188
References	189

I. INTRODUCTION

Molecular markers such as restriction fragment length polymorphisms (RFLPs) and isozymes have been extensively used for genetic studies and plant identification.[1-3] Restriction fragments are codominantly and stably inherited in a Mendelian fashion, and they disclose unlimited polymorphic markers. The problem with RFLPs is that the procedure is time consuming, requires a large amount of DNA (2 to 10 µg) and suitable probes, and usually involves the use of radioisotopes. Isozyme analysis is limited by the small number of loci sampled by the technique.

The polymerase chain reaction (PCR) has facilitated genetic studies in plants and animals.[4,5] DNA fingerprinting, forensic analysis, genetic mapping, and phylogenetic studies have tremendously benefited from PCR. One variation of PCR is the random amplified polymorphic DNA (RAPD), which generates DNA fingerprints with a single synthetic oligonucleotide primer. The polymorphisms observed may result from point mutations, insertions, deletions, and inversions.[6] RAPDs are usually dominant markers and are inherited in a simple Mendelian fashion. In comparison with RFLP, the procedure is less expensive, faster, requires a smaller amount of DNA (0.5 to 50.0 ng), does not involve the use of radioisotopes, and requires less skill to operate. Because of these advantages, RAPDs have proven useful in genotype identification and gene mapping. Although details about RAPD analysis and use are presented by Yu and Pauls in Chapters 22 and 23, it is worth mentioning that Penner et al.[7] recently investigated reproducibility in RAPDs using the same target DNA and primers in different laboratories. They found most RAPD markers to be reproducible, with differences between PCR machines accounting for most of the variations. This review will focus on the use of RAPDs for taxonomic and evolutionary analysis.

II. ANALYSIS AT THE SUBGENERIC (SECTIONAL) LEVEL

Although almost all the research to date has focused on taxonomy and classification at the species level or below, one paper (Adams and Demeke[8]) has dealt with taxonomy at the sectional level (subgeneric) in *Juniperus*. The genus *Juniperus* L. is very diverse, with about 75 to 80 taxa worldwide, of which 41 taxa are found in the Western Hemisphere. On the basis of morphology, three sections (*Caryocedrus, Juniperus,* and *Sabina*) have been recognized.[9] The utility of RAPDs at different taxonomic levels was investigated in *Juniperus*.[8] Notice the separation of sections *Caryocedrus, Juniperus,* and *Sabina* (Figure 1) by the principal coordinate analysis (PCO). In this instance, RAPDs were found to be taxonomically useful at the subgeneric level. In addition, species in the two series, entire and serrate, in section *Sabina* were separated on the second principal coordinate (Figure 1). The established relationships based on morphological and terpenoid analysis were confirmed by the RAPD data. Adams and Demeke[8] found that some primers amplify DNA that is highly conserved and will thus help to generate polymorphisms at a high levels of classification whereas other primers amplify DNA that is highly variable and useful for classification and analyses at and below the species level. As more studies are done, it is expected that this observation will be tested and that primers may be classified in this manner and thus be more predictable concerning their utilization.

III. INTERSPECIFIC TAXONOMY AND CLASSIFICATION

RAPDs have been used in several taxonomic studies at the species level. Wilkie et al.[10] used 20 random primers (10-mers) to analyze *Allium* species (and cultivars). The results were in broad agreement with classical classifications. However, *A. roylei* was shown to be the closest relative of *A. cepa*, in contrast to results from previous work.

The evolution and genetics of *Brassica* species has been well documented. The different techniques used include RFLP,[11,12] chloroplast analysis,[13] morphological taxonomy and cytogenetics,[14,15] and analysis of isozymes and rDNA genes.[16] These studies have confirmed the classical U triangle relationship[17] among the diploid and amphidiploid *Brassica* taxa. Demeke et al.[18] examined the potential taxonomic use of RAPDs in *Brassica* species, *Sinapis* and

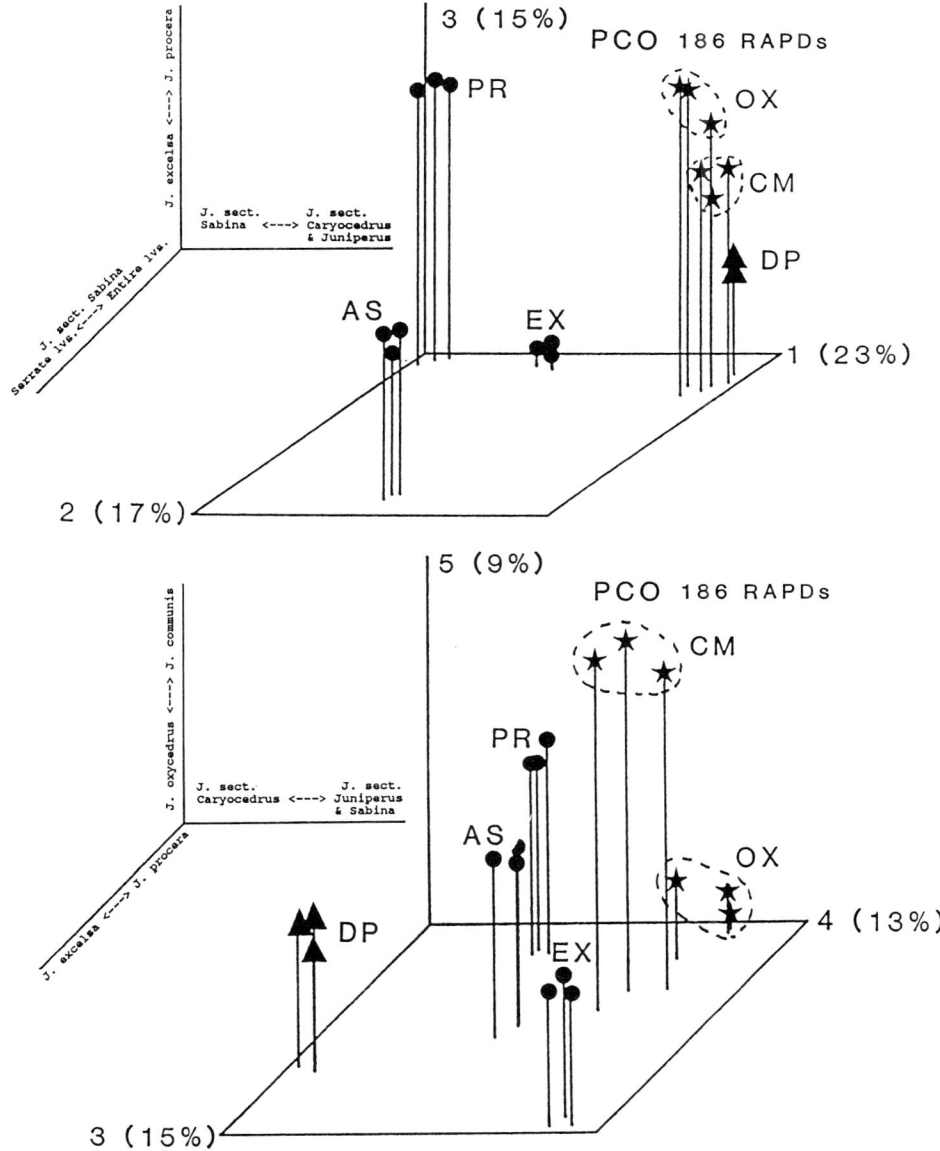

Figure 1. Ordination of *Juniperus* species using PCO3D, based on 186 RAPDs (RAPD bands). Notice (upper) the separation of sections *Sabina* and *Caryocedrus/Juniperus* on the first coordinate (23% of the variance) and section *Caryocedrus* from *Juniperus/Sabina* on the fourth coordinate (lower, 13% of the variance). Each symbol represents a different individual of a taxon. (Adapted from Adams, R. P. and Demeke, T., *Taxon*, 42, 553, 1993, permission of the International Association for Plant Taxonomy from Adams and Demeke.)

Raphanus taxa. Principal coordinate analysis (PCO) of 284 RAPD bands showed the classical U triangle relationships among diploid and amphidiploid *Brassica* taxa (Figure 2). A minimum of ten primers with a total of 100 RAPD bands were needed to explain the genetic relationships. The use of only six primers resulted in a poorer classification. Cultivars of cabbage and cauliflower were also clearly distinguished by the RAPD analysis. Polymorphisms were also obtained among individual seedlings of *B. carinata* cv. *Dodola*. In this study RAPDs proved to be useful at taxonomic levels ranging from individuals to cultivars and species.

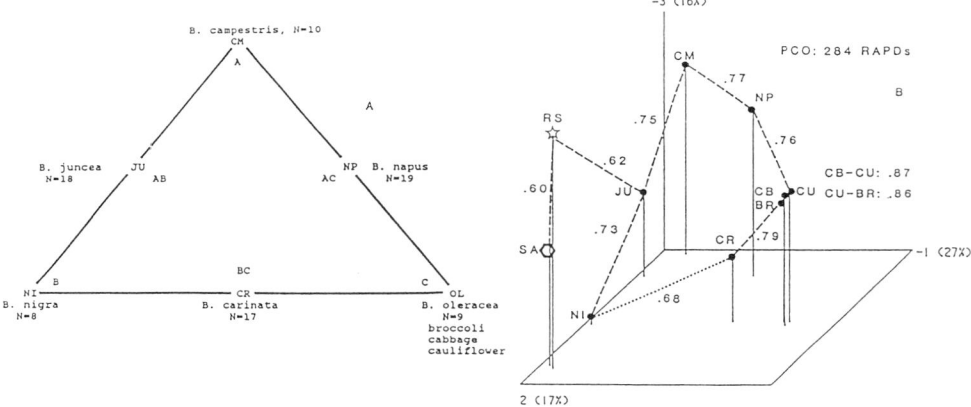

Figure 2. (A) Classically defined relationships among *Brassica* species (U triangle). (B) Results of PCO ordination using 284 RAPDs (markers). Notice the U triangle, with vertices of CM, OL (CU, CB, and BR), and NI, just as in the case of the classical system. Also note the intermediate nature of the amphidiploids, JU, CR, and NP. (From Demeke, T., Adams, R. P., and Chibbar, R., *Theor. Appl. Genet.*, 84, 990, 1992. With permission.)

Halward et al.[19] compared cultivated peanut with 29 wild diploid species of *Arachis* using ten primers. They found no variation in banding patterns among the cultivars but found variation among the wild species. PAUP and HyperRFLP programs were used to generate dendrograms showing genetic relationships among the diploid *Arachis* species.

Four *Stylosanthes* species were analyzed[20] using 22 RAPD primers. This resulted in 200 RAPD markers (RAPDs) that were used to generate clustering. Four main clusters were produced (Figure 3) in which all accessions of each species cluster together. Genetic variation appeared to be greater in *S. guianensis* and lesser in *S. hamata* (Figure 3). The phylogenetic results were in agreement with morphology, cytology, and enzyme electrophoresis.

IV. GEOGRAPHIC VARIATION AND INFRASPECIFIC TAXONOMY

Both RAPDs and leaf volatile terpenoids were used to compare junipers from Abha, Saudi Arabia, with *Juniperus excelsa* from Greece and *J. procera* from Addis Ababa, Ethiopia.[21] *Juniperus procera* Hochst. ex Endl. is the only species of the genus that grows naturally in the Southern Hemisphere. It has been postulated[22] that *J. procera* originated from *J. excelsa* in Asia Minor in the Mio-pliocene as *J. excelsa* expanded its range southward along the Western mountains of the Arabian Peninsula, thence across the Red Sea to Ethiopia and southward along East African rift mountains. Most taxonomic treatments call the juniper from the Saudi Arabian Peninsula *J. excelsa*.[22] The RAPD data clearly show (Figure 4) that the juniper at Abha, Saudi Arabia, is *J. procera*, even though the morphological data were inconclusive. In addition, the volatile leaf terpenoids show the same pattern as the RAPDs. Analyses using RAPDs are clearly shown to be of use in analyses of geographical variation questions.

Other studies using RAPDs for the analysis of geographic variation include that of *Microseris elegans*,[23] where 17 primers were used to generate 134 RAPDs from ten populations from throughout California. Brauner et al.[24] used 16 primers to analyze genetic diversity among populations of *Lactoris fernandeziana*, endemic to the island of Masatierra in the Juan Fernandez Archipelago. They found nearly all variants restricted to single populations.

The only paper depicting geographically contoured variation in RAPD similarities is that of Demeke and Adams[25] on the medicinal plant *Phytolacca dodecandra* in Africa. Geographical variation of *P. dodecandra* in Africa has been investigated in terms of morphology, leaf chemistry, and triterpene aglycones.[26-28] Use of PCO, followed by contour mapping of the

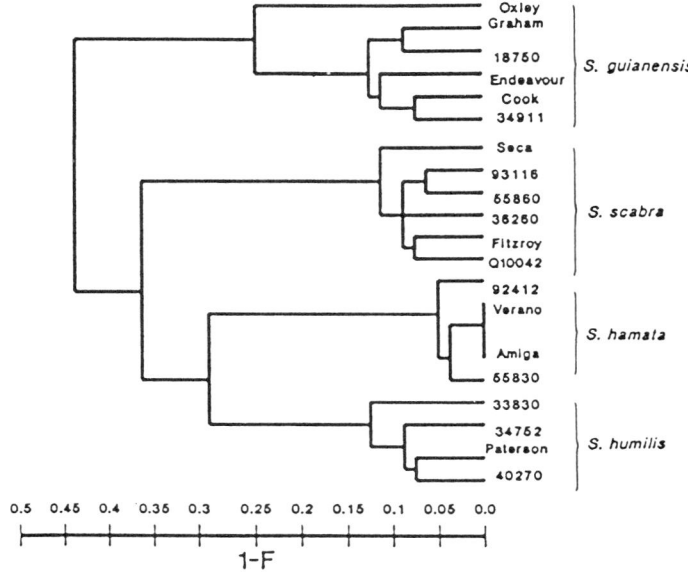

Figure 3. Clustering of four *Stylosanthes* species by UPGMA based on RAPDs data. Notice the clustering of all the accessions by species. (From Kazan, K., Manners, J. M., and Cameron, D. F., *Theor. Appl. Genet.*, 85, 882, 1993. With permission.)

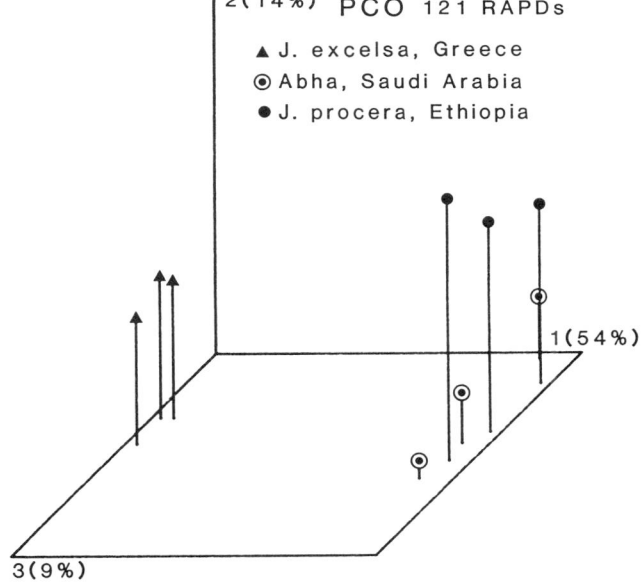

Figure 4. PCO ordination of *J. excelsa*, Greece, *J. procera*, Ethiopia, and junipers from *Saudi Arabia*, based on 121 RAPDs. The Saudi Arabian junipers clearly cluster with *J. procera* (axis 1, 54% of the variance among individuals). (From Adams, R. P., Demeke, T., and Abulfatih, H. A., *Theor. Appl. Genet.*, 87, 22, 1993. With permission.)

population scores on each axis revealed geographic patterns (Figure 5). Several of these trends were the same as those previously found in morphological and chemical data. Four major groups were discovered among *P. dodecandra* populations: Madagascar, Nigeria, Ethiopia, and Southern Africa. The studies suggest these four groups to be the center of focus for future *P. dodecandra* germplasm collection.

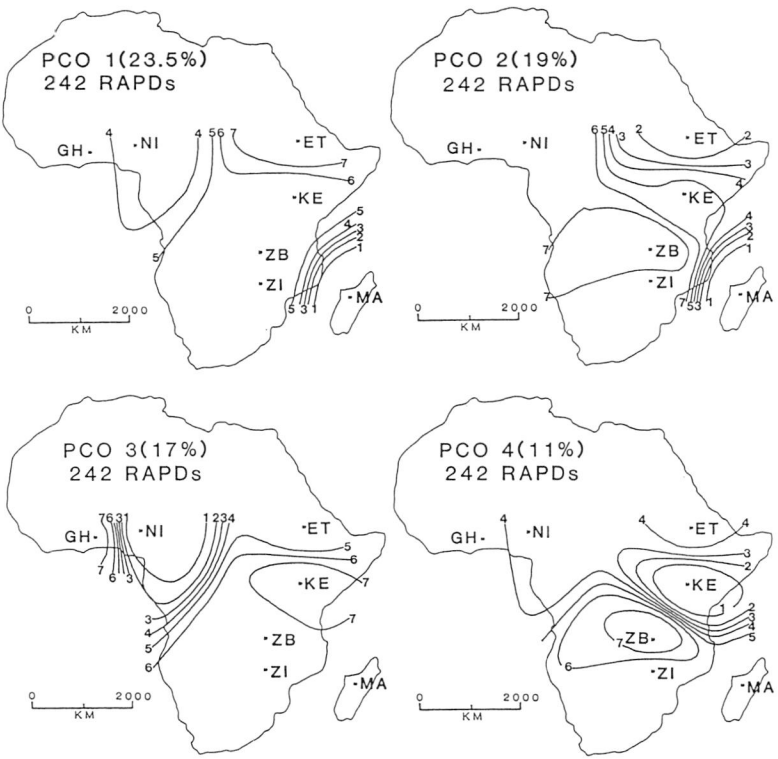

Figure 5. Geographical variation patterns in the RAPDs of *Phytolacca dodecandra*. The population coordinate scores on each principal coordinate (1 to 4) were used at Z values for contouring. (From Demeke, T. and Adams, R. P., in *Conservation of Plant Genes II: Utilization of Ancient and Moder2n DNA*, Adams, R. P., Miller, J. S., Golenberg, E. M., and Adams, J. E., Eds., Missouri Botanical Garden Press, St. Louis, 1994, chap. 11. With permission.)

Finally, it should be mentioned that RAPDs have also been used for the analysis of geographic variation in the fungus *Colletotirchum graminicola*.[29]

V. CULTIVAR CLASSIFICATION AND IDENTIFICATION

Cultivar classification, germplasm collection, and characterization are major undertakings in developing and developed countries. Characterization is usually done with morphological markers, which may not give an accurate picture because of environmental influences. RAPD analysis is a simple, fast, accurate, and relatively inexpensive tool to characterize germplasm collections. The procedure can have wide applicability in developing countries.

A study of the relationships among ten papaya (*Carica papaya* L.) cultivars was made with 11 decamers.[30] A dendrogram was constructed using 102 distinct DNA fragments (Figure 6). The seven Hawaiian types are grouped in the upper branch. Divisions within the Hawaiian types were mostly consistent with the known genetic background of the cultivars. The RAPD procedure was found to be fast, precise, and sensitive for genomic analysis.

Pedigree relationships in spring barley lines were examined by Tinker et al.[31]. Two- and six-row lines were separated by cluster analysis. Within the two- and six row groups, other qualitative similarities were observed. A linear relationship was found between kinship coefficients and genetic distance in the cluster analysis. It was concluded that RAPD markers could be used to gain information about genetic similarities or differences that are not evident from pedigree information.

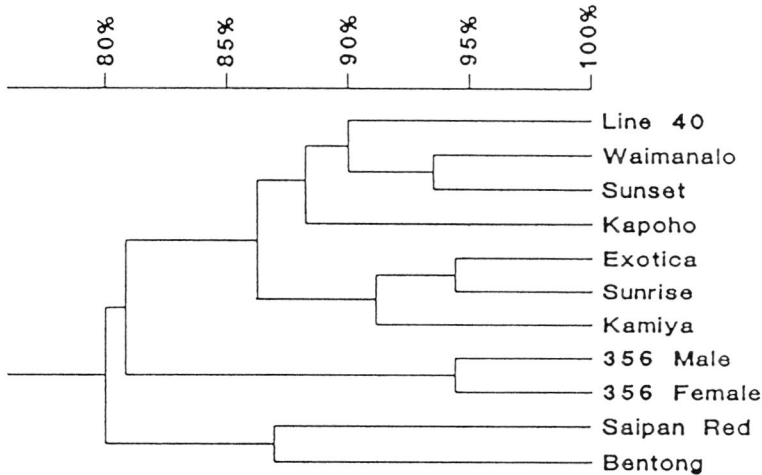

Figure 6. Phenogram of papaya cultivars based on *UPGMA* clustering of 102 RAPDs. The upper seven cultivars are the Hawaiian type and cluster together. Divisions within the Hawaiian branch were mostly consistent with known genetic backgrounds. (From Stiles, J. I., Lemme, C., S., Morshidi, M. B., and Marshardt, R., *Theor. Appl. Genet.*, 85, 697, 1993. With permission.)

Wheat cultivars generally have a low level of intervarietal polymorphisms. According to Gale et al.,[32] RFLP mapping in *Triticum aestivum* ($2n = 6x = 42$) has been hindered by low levels of intervarietal polymorphisms, complexities arising from polyploidy, and large genome size. RAPD analysis often generates more polymorphisms than RFLP procedure. Inter- and intraspecific diversity was found in wheat, barley, rye, and wheat-barley addition lines using conserved, semi-random, and random primers.[33] Vierling and Nguyen[34] investigated the genetic diversity of two diploid wheat species, *T. monococcum* and *T. urartu* ($2n = 2x = 14$) using random primers. A higher rate of polymorphisms were observed in *T. urartu* than in *T. monococcum*.

In order to visualize more markers in cereal crops with low levels of intervarietal polymorphisms such as wheat, denaturing gradient gel-electrophoresis (DGGE) might be used. DDGE has previously been used to analyze polymorphisms.[35,36] DNA fragments differing by single base pair substitutions are separated on denaturing gradient gels. More polymorphisms have been observed by running RAPD products on DGGE gel than on the normal agarose gel.[37] The DGGE method is also reported to generate many more polymorphisms than the RFLP procedure.

One of the problems in using RAPDs for identification of markers in wheat (e.g., disease resistance) is lack of reproducibility.[38] The problem is less pronounced in dicots. The high ploidy level of wheat may contribute to the complexity of the problem. Some primers are more unstable than others. Optimization of the PCR procedure is very important to get reasonably reproducible results. The development of reliable PCR-based markers such as SCARs (sequence characterized amplified regions), explained by Paran and Michelmore,[39] increases the use of RAPDs as genetic markers.

RAPDs were used to identify 16 rice accessions.[40] 28 decamers generated 244 different RAPD bands, and 116 of them were informative in that they differentiated one or more accessions. All accessions were uniquely distinguished by at least one RAPD band, and the accessions clustered into three ecospecies, i.e., *Japonica*, *Javanica*, and *Indica*, supporting previous morphological recognition of the three taxa. The differences among the species were demonstrated with at least 23 polymorphisms between *Japonica* and *Javanica*, 71 between *Japonica* and *Indica*, and 72 between *Indica* and *Javanica*. The results confirmed the previous isozyme and RFLP studies.[41,42] RAPDs are very useful in the identification of rice accessions and are much simpler to use than RFLP analysis.

Very low variability was detected among cultivated peanut cultivars (*Arachis hypogaea* L.) using RFLPs.[43] Peanut cultivars have a narrow genetic base. Genetic polymorphisms were studied in wild peanut plants and cultivated species by RFLP, RAPD, and four-cutter analysis of PCR-amplified fragments.[44] A high level of polymorphism was found among wild species, whereas little polymorphism was found in cultivated peanut in all cases. The utilization of wild germplasm for peanut breeding programs has been emphasized.

Other examples of cultivar and/or accession RAPD analyses include apples,[45] banana,[46] and celery.[47]

VI. GENETIC ANALYSES WITHIN CLONES

A. GENETIC LINKAGE

Tulsieram et al.[48] report on a unique application of RAPD technology for genetic linkage mapping of a single spruce tree using haploid DNA from megagametophyte tissue of individual seeds. Sixty-one segregating loci were analyzed for inheritance and linkage using RAPDs. This tool should be useful in the study of the evolution of linkage among genes.

B. ANALYSIS OF SOMACLONAL VARIATION

Isabel et al.[49] used RAPDs to evaluate the genetic integrity of somatic embyrogensis-derived cell lines of *Picea mariana* (black spruce). No variation was found in the clonal lines in the RAPDs, so the authors propose use of RAPDs to assess genetic stability and certify stability during somatic embryogenesis. This is an interesting application of DNA fingerprinting by RAPDs that may find wider usage in the nursery industry as well as in biotechnology.

VII. ANALYSES OF GENE FLOW BETWEEN INDIVIDUALS

Lewis and Snow[50] show that RAPDs can be of significant use to ecologists in studying mating systems and assigning paternity. The unambiguous assignment of paternity can be extremely valuable in understanding mating patterns of birds, for example. In plants, insect pollination might be more exactly studied because all the adjacent plants (potential pollen sources) could be fingerprinted by RAPDs and the dominant RAPD bands seen in the resulting seeds.

VIII. HYBRIDIZATION STUDIES

A. INTERGENERIC HYBRIDIZATION

Crawford et al.[51] used RAPDs to document the origin of an intergeneric hybrid (*Margyracaena skottsbergii*). One parent, *Acaena argentea* had 18 species-specific bands, the other parent, *M. digynus*, had 27 species-specific bands, and the putative intergeneric hybrid had all 45 bands present. They conclude that RAPDs are particularly attractive for the study of hybridization in rare species because they can provide numerous genetic markers while requiring minimal amounts of DNA.

B. INTERSPECIFIC HYBRIDIZATION AND INTROGRESSION

Arnold et al.,[52] in one of the earliest papers that used RAPDs, analyzed a classical case of hybridization and introgression in irises. They reported intermediate frequencies for RAPD markers in putative hybrid populations, corresponding to cpDNA markers. The study used only three primers and reported on only nine RAPD markers. Additional primers would have been desirable.

Waugh et al.[53] used six RAPD primers for the detection of introgression in the potato. The RAPD data showed that dihaploids, generated after interspecific pollination of a tetraploid

Solanum tuberosum cultivar by *S. phureja*, was the result of gene flow, not of parthenogenic origin. They concluded that RAPDs would be useful for the detection of gene introgression in both natural and cultivated plant populations.

In order to confirm the genetic basis of RAPD markers in conifers, Carlson et al.[54] examined inheritance of RAPDs in both Douglas fir and white spruce. The RAPD markers were found to be inherited as dominants, as originally shown for soybeans by Williams et al.[6] They concluded that RAPDs would be useful genetic markers.

IX. ANALYSES OF GENETIC DIVERSITY

A number of workers have used RAPDs for the analyses of genetic diversity. We have already mentioned the work on wheat by Vierling and Nguyen.[34] Mosseler et al.[55] examined populations of red pine, white spruce, and black spruce by RAPDs and found that disjunct Newfoundland populations of red pine were largely monomorphic for 69 RAPD bands and that they relate to genetic isolation from mainland populations. The RAPD data corresponded with genetic diversity estimates using isozymes for red pine, white spruce, and black spruce.

Genetic diversity in *Arachis* (peanut) germplasm was investigated by use of RAPDs.[56] Sixty 10-mer primers yielded 49 polymorphic loci between a cultivated *A. hypogaea* (TMV-3) and a synthetic amphidiploid. The researchers concluded that RAPDs offer a simple, yet efficient tool for the characterization of *Arachis* germplasm and for use in hybridization research.

Other examples of genetic analysis include alfalfa[57] and wheat.[38]

X. ANALYSES OF THE EVOLUTION OF DISEASE RESISTANCE

The evolution of disease resistance in plants is of great interest to crop scientists as well as botanists. Martin et al.[58] used 144 RAPD primers and obtained seven RAPD bands that were associated with *Pseudomonas* resistance in tomato. Four of these RAPDs were tightly linked to the bacterial-resistance *Pto* gene. This provides both an insight into the evolution of disease resistance and information that should be useful for plant breeding.

In lettuce, RAPD markers have been discovered that are linked to downy mildew resistance genes,[39,59] and bacterial blight has been analyzed by RAPDs.[60]

XI. STUDIES OF ANCIENT DNA

It is not our intention to give a detailed review of ancient DNA studies. However, several papers have already been published and others will follow quickly. PCR has already made a tremendous contribution to the study of archaeological and fossil specimens and will usher in the new field of molecular archaeology.

Nanogram quantities of DNA were extracted from plant tissues ranging in age from 22 to more than 45,000 years old,[61] showing the potential of obtaining usable DNA from ancient samples.

Brown et al.[62] used PCR to amplify a 246-bp DNA fragment from charred wheat seeds 700 to 3300 years old. They considered using RAPDs but are cautious that "jumping PCR"[63] might produce spurious bands. However, they do indicate plans to try RAPDs on their material in the future.

Rollo et al.[64] examined DNA from 1000-year-old maize kernels using PCR to amplify the *Mu* termini and the cytochrome *c* oxydase subunit I gene. Although, RAPDs were not used in the study, their application seems appropriate.

A 790-bp cpDNA fragment of rbcL from a 17 to 20 million year old *Magnolia* leaf compression fossil has been amplified by PCR and sequenced.[65] The fossil specimen clustered with extant members of *Magnolia*. However, results from the study were later questioned, with the emergence of a model that predicted that DNA would be completely degraded after 4 million years.[66] In a later study, *Taxodium* material from the same fossil formation (a 1320 bp portion of the chloroplast rbcL gene from a Miocene *Taxodium* specimen from the Clarkia site) has been successfully amplified and sequenced,[67] validating the *Magnolia* study.

The study of genealogical relationships of extinct species and vanished populations has been reported.[68-70] Mitochondrial DNA from a 7000-year-old human brain were amplified by PCR and sequenced. The sequences showed that the individual belonged to a mitochondrial lineage that is rare in the Old World and not previously known to exist among native Americans.

This area of research will grow very rapidly, and PCR and PCR-RAPD will be increasingly applied to these problems.

XII. NUMERICAL ANALYSES OF RAPD DATA

Several numerical methods have been used to analyze RAPD data. In our work, we have coded the data both as presence/absence and intensity values (e.g., 0 to 6, with 0 being no band, and 6 being a very bright band). We found intensity values to be more informative than presence/absence data.[8,18] Similarity measures were computed using absolute character state differences (Manhattan metric) divided by the maximum observed value for that character over all taxa (= Gower metric, Gower;[71] Adams[72]). Principal coordinate analysis was performed according to Gower.[73] Principal coordinate analysis is ideally suited for use with large numbers of characters and numerous taxa. Each eigenroot extracts a separate component of variance among the taxa. This enables one to progressively examine the relationships, particularly in geographic trend analysis.[25] However, care should be taken in the interpretation of three-dimensional figures when the number of eigenroots extracted is large (see Adams and Demeke[8]). In our RAPD research, we obtained very convincing results for *Brassica*, *Juniperus*, and *Phytolacca dodecandra*, as previously noted. The results supported previous chemical and morphological studies, and the *Brassica* data also confirmed the previous molecular studies. In our experience, data from at least ten primers with a total of about 100 RAPD bands are needed to produce a stable classification. Using more primers may not be necessary, as little additional information was gained by increasing the number of primers from 11 to 17 in our *Brassica* study.[18] The computer software for PCO analysis, PCO3D, is available from R.P. Adams for distribution.

A number of similarity coefficients have been used for phenogram constructions in other studies: Jaccard's similarity coefficients;[74] index of genetic similarity;[75] Jaquard index;[76] and Rogers' distance.[77] Phenograms have been constructed by the *UPGMA* program[78] and parsimony analysis.[79] It might be noted that *UPGMA* uses the average similarity of an operational taxonomic unit (OTU) rather than the nearest neighbor to bring a new member into a group. It would seem that evolution would favor the fewest mutations and thus nearest-neighbor or single linkage clustering would be preferable over *UPGMA*. The programs used include *PAUP*, *PHYLIP*, *CLINCH*, *MaClade*, and *NT-SYS*. Some analyses have also been performed with the *SAS* program (SAS Institute).

XIII. CONCLUSION

Because of their technical ease and cost effectiveness, PCR-RAPDs will become commonplace (perhaps a cornerstone) in taxonomic and evolutionary studies. Good agreement has been achieved between RAPD analysis and morphology, chemical, and molecular (isozyme, RFLP) data. RAPDs will also have a wide applicability in germplasm collection and characterization programs. Analyses of ancient DNA will move toward the use of RAPDs in many cases. However, fossil DNA will probably require sequence data due to problems of homology of RAPD bands with extremely ancient taxa. RAPD protocols will also be refined through time so that some of the current problems such as occasional lack of reproducibility will be overcome.

REFERENCES

1. **Beckman, J. S. and Soller, M.**, Restriction fragment length polymorphisms in plant genetic improvement, *Oxford Surv. Plant Mol. Cell Biol.*, 3, 198, 1986.
2. **Gebhardt, C. and Salamini, F.**, Restriction fragment length polymorphism analysis of plant genomes and its application to plant breeding, *Int. Rev. Cytol.*, 135, 201, 1992.
3. **Quiros, C. F. and McHale, N.**, Genetic analysis of isozyme variants in diploid and tetraploid potatoes, *Genetics*, 111, 131, 1985.
4. **Erlich, A., Gelfand, D., and Sninsky, J. J.**, Recent advances in the polymerase chain reaction, *Science*, 252, 1643, 1991.
5. **Mullis, K. B.**, The unusual origin of the polymerase chain reaction, *Sci. Am.*, 56, 1990.
6. **Williams, J. G. K., Kubelik, A. R., Livak, K. J., Rafalski, J. A., and Tingey, S. V.**, DNA polymorphisms amplified by arbitrary primers are useful as genetic markers, *Nucleic Acids Res.*, 18, 6531, 1990.
7. **Penner, G. A., Bush, A., Wise, R., Kim, W., Domier, L., Kasha, K., Laroche, A., Scoles, G., Molnar, S., and Fedak, G.**, Reproducibility of random amplified polymorphic DNA (RAPD) analysis among laboratories, *PCR Methods Appl.*, 2, 341, 1993.
8. **Adams, R. P. and Demeke, T.**, Systematic relationships in *Juniperus* based on random amplified polymorphic DNA, *Taxon*, 42, 553, 1993.
9. **Zanoni, T. A.**, The American junipers of the section *Sabina* (*Juniperus cupressaceae*) — a century later, *Phytologia*, 38, 433, 1978.
10. **Wilkie, S. E., Isaac, P. G., and Slater, R. J.**, Random amplified polymorphic DNA (RAPD) markers for genetic analysis in *Allium*, *Theor. Appl. Genet.*, 86, 497, 1993.
11. **Hosaka, K., Kianian, S. F., McGrath, J. M., and Quiros, C. F.**, Development and chromosomal localization of genome-specific DNA markers of *Brassica* and the evolution of amphidiploids and $n = 9$ diploid species, *Genome*, 33, 131, 1990.
12. **Song, K., Osborn, T. C., and Williams, P. H.**, *Brassica* taxonomy based on nuclear restriction fragment length polymorphisms (RFLPs). III. Genome relationships in *Brassica* and related genera and the origin of *B. oleracea* and *B. rapa* (syn. *campestris*), *Theor. Appl. Genet.*, 79, 497, 1990.
13. **Palmer, J. D., Shield, C.R., Cohen, D.B., and Orton, T. J.**, Chloroplast DNA evolution and the origin of amphidiploid *Brassica* species, *Theor. Appl. Genet.*, 65, 181, 1983.
14. **Prakash, S. and Hinata, K.**, Taxonomy, cytogenetics and origin of crop *Brassicas*, a review, *Opera Bot.*, 55, 1, 1980.
15. **Vaughan, J. G.**, A multidisciplinary study of the taxonomy and origin of *Brassica* crops, *BioScience*, 27, 35, 1977.
16. **Quiros, C. F., Ochoa, O., Kianian, S. F., and Douches, D.**, Analysis of the *Brassica oleracea* genome by generation of *B. campestris-oleracea* chromosome addition lines: characterization by isozymes and rDNA genes, *Theor. Appl. Genet.*, 74, 758, 1987.
17. **U. N.**, Genome analysis of *Brassica* with special reference to the experimental formation of *B. napus* and its peculiar mode of fertilization, *Jpn. J. Bot.*, 7, 389, 1935.
18. **Demeke, T., Adams, R. P., and Chibbar, R.**, Potential taxonomic use of random amplified polymorphic DNA (RAPD): a case study in *Brassica*, *Theor. Appl. Genet.*, 84, 990, 1992.
19. **Halward, T., Stalker, T., LaRue, E., and Kochert, G.**, Use of single primer DNA amplifications in genetic studies of peanut (*Arachis hypogaea* L.), *Plant Mol. Biol.*, 18, 315, 1992.
20. **Kazan, K., Manners, J. M., and Cameron, D. F.**, Genetic variation in agronomically important species of *Stylosanthes* determined using random amplified polymorphic DNA markers, *Theor. Appl. Genet.*, 85, 882, 1993.
21. **Adams, R. P., Demeke, T., and Abulfatih, H. A.**, RAPD DNA fingerprints and terpenoids: clues to past migrations of *Juniperus* in Arabia and east Africa, *Theor. Appl. Genet.*, 87, 22, 1993.
22. **Kerfoot, O.**, Origin and speciation of the *Cupressaceae* in Sub-Sahara Africa, *Boissiera*, 24, 145, 1975.
23. **Van Heusden, A. W. and Bachmann, K.**, Genotype relationships in *Microseris elegans* (*Asteraceae*, *Lactuceae*) revealed by DNA amplification from arbitrary primers (RAPDs), *Plant Syst. Evol.*, 179, 221, 1992.
24. **Brauner, S., Crawford, D. J., and Stuesy, T. F.**, Ribosomal DNA and RAPD variation in the rare plant family *Lactoridaceae*. *Am. J. Bot.*, 79, 1436, 1992.
25. **Demeke, T. and Adams, R. P.**, The use of RAPDs to determine germplasm collection strategies in the African species, *Phytolacca dodecandra*, in *Conservation of Plant Genes II: Utilization of Ancient and*

Modern DNA, Adams, R. P., Miller, J. S., Golenberg, E. M., and Adams, J. E., Eds., Missouri Botanical Garden Press, St. Louis, 1994, chap. 11.

26. **Adams, R. P., Neisess, K. R., Parkhurst, R. M., Makhubu, L. P., and Wolde Yohannes, L.,** *Phytolacca dodecandra* (*Phytolaccaceae*) in Africa: geographical variation in morphology, *Taxon,* 38, 17, 1989.
27. **Adams, R.P., Parkhurst, R. M., Wolde Yohannes, L., and Makhubu, L. P.,** *Phytolacca dodecandra* (*Phytolaccaceae*) in Africa: geographical variation in leaf chemistry, *Biochem. Syst. Ecol.,* 18, 429, 1990.
28. **Parkhurst, R. M., Thomas, D. W., Adams, R. P., Makhubu, L. P., Mthupha, B. M., Wolde Yohannes, L., Mamo E., Heath, G. E., Stobaeus, J. K., and Jones, W. O.,** Triterpene aglycones from various *Phytolacca dodecandra* populations, *Phytochemistry,* 29, 1171, 1990.
29. **Guthrie, P. A. I., Magill, C. W., Frederiksen, R. A., and Odovody, G. N.,** Random amplified polymorphic DNA markers: A system for identifying and differentiating isolates of *Colletotrichum graminicola*, *Phytopathology,* 82, 832, 1992.
30. **Stiles, J. I., Lemme, C., Sondur, S., Morshidi, M. B., and Manshardt R.,** Using randomly amplified polymorphic DNA for evaluating genetic relationships among papaya cultivars, *Theor. Appl. Genet.,* 85, 697, 1993.
31. **Tinker, N. A., Fortin, M. G., and Mather, D. E.,** Random amplified polymorphic DNA and pedigree relationships in spring barley, *Theor. Appl. Genet.* 85, 976, 1993.
32. **Gale, M. D., Chao, S., and Sharp, P. J.,** RFLP mapping in wheat — progress and problems, in *Gene Manipulation in Plant Improvement,* Vol. 2, Gustafson, J. P., Ed., Plenum Press, New York, 353, 1990.
33. **Weining, S. and Langridge, P.,** Identification and mapping of polymorphisms in cereals based on the polymerase chain reaction, *Theor. Appl. Genet.,* 82, 209, 1991.
34. **Vierling, R. A. and Nguyen, H. T.,** Use of RAPD markers to determine the genetic diversity of diploid, wheat genotypes, *Theor. Appl. Genet.,* 84, 835, 1992.
35. **Fisher, S. G. and Lerman, L. S.,** DNA fragments differing by single base-pair substitutions are separated in denaturing gradient gels: correspondence with melting theory, *Proc. Natl. Acad. Sci. U.S.A.,* 80, 1579, 1983.
36. **Riedel, G. E., Swanberg, S. L., Kuranda, K. D., Marquette, K., LaPan, P., Bledsoe, P., Kennedy, A., and Lin B. Y.,** Denaturing gradient gel-electrophoresis identifies genomic DNA polymorphism with high frequency in maize, *Theor. Appl. Genet.,* 80, 1, 1990.
37. **Dweikat, I., Mackenzie, S., Levy, M., and Ohm, H.,** Pedigree assessment using RAPD-DGGE in cereal crop species, *Theor. Appl. Genet.,* 85, 497, 1993.
38. **Devos, K. M. and Gale, M. D.,** The use of random amplified polymorphic DNA markers in wheat, *Theor. Appl. Genet.,* 84, 567, 1992.
39. **Paran, I. and Michelmore, R. W.,** Development of reliable PCR-based markers linked to downy mildew resistance genes in lettuce, *Theor. Appl. Genet.,* 85, 985, 1993.
40. **Fukuoka, S., Hosaka, K., and Kamijima, O.,** Use of random amplified polymorphic DNAs (RAPDs) for identification of rice accessions, *Jpn. J. Genet.,* 67, 243, 1992.
41. **Glaszmann, J. C.,** Geographic pattern of variation among Asian native rice cultivars (*Oryza sativa* L.) based on fifteen isozyme loci, *Genome,* 30, 782, 1988.
42. **Wang, Z. Y. and Tanksley, S. D.,** Restriction fragment length polymorphism in *Oryza sativa* L., *Genome,* 32, 1113, 1989.
43. **Kochert, G. D., Halward, T. M., Branch, W. D., and Simpson, C. E.,** RFLP variability in peanut (*Arachis hypogaea* L.) cultivars and wild species, *Theor. Appl. Genet.,* 81, 565, 1991.
44. **Halward, T. M., Stalker, H. T., LaRue, E. A., and Kochert, G. D.,** Genetic variation detectable with molecular markers among unadapted germplasm resources of cultivated peanut and related wild species, *Genome,* 34, 1013, 1991.
45. **Koller, B., Lehmann, A., McDermott, J. M., and Gessler, C.,** Identification of apple cultivars using RAPD markers, *Theor. Appl. Genet.,* 85, 901, 1993.
46. **Kaemmer, D., Afza, R., Weising, K., Kahl, G., and Novak, F. J.,** Oligonucleotide and amplification fingerprinting of wild species and cultivars of banana (*Musa* spp.), *Bio/Technology,* 10, 1030, 1992.
47. **Yang, X. and Quiros, C.,** Identification and classification of celery cultivars with RAPD markers, *Theor. Appl. Genet.,* 86, 205, 1993.
48. **Tulsieram, L. K., Glaubitz, J. C., Kiss, G., and Carlson, J. E.,** Single tree genetic linkage mapping in conifers using haploid DNA from megagametophytes, *Bio/Technology,* 10, 686, 1992.
49. **Isabel, N., Tremblay, L., Michaud, M., Tremblay, F. M., and Bousquet, J.,** RAPDs as an aid to evaluate the genetic integrity of somatic embryogenesis-derived populations of *Picea mariana* (Mill.) B. S. P., *Theor. Appl. Genet.,* 86, 81, 1993.
50. **Lewis, P. O. and Snow, A. A.,** Deterministic paternity exclusion using RAPD markers, *Mol. Ecol.,* 1, 155, 1992.
51. **Crawford, D. J., Brauner, S., Cosner, M. B., and Steussy, T. F.,** Use of RAPD markers to document the origin of the intergeneric hybrid *Margyracaena skottsbergii* (*Rosaceae*) on the Juan Fernandez Islands, *Am. J. Bot.,* 80, 89, 1993.

52. **Arnold, M. L., Buckner, C. M., and Robinson, J. J.,** Pollen-mediated introgression and hybrid speciation in Louisiana irises, *Proc. Natl. Acad. Sci. U.S.A.,* 88, 1398, 1991.
53. **Waugh, R., Baird, E., and Powell, W.,** The use of RAPD markers for the detection of gene introgression in potato, *Plant Cell Rep.,* 11, 466, 1992.
54. **Carlson, J. E., Tulsieram, L. K., Glaubitz, J. C., Luk, V. W. K., Kauffeldt, C., and Rutledge, R.,** Segregation of random amplified DNA markers in F_1 progeny of conifers, *Theor. Appl. Genet.,* 83, 194, 1991.
55. **Mosseler, A., Egger, K. N., and Hughes, G. A.,** Low levels of genetic diversity in red pine confirmed by random amplified polymorphic DNA markers, *Can. J. For. Res.,* 22, 1332, 1992.
56. **Lanham, P. G., Fennell, S., Moss, J. P., and Powell, W.,** Detection of polymorphic loci in *Arachis* germplasm using random amplified polymorphic DNAs, *Genome,* 35, 885, 1992.
57. **Echt, C. S., Erdahl, L. A., and McCoy, T. J.,** Genetic segregation of random amplified polymorphic DNA in diploid cultivated alfalfa, *Genome,* 35, 84, 1992.
58. **Martin, G. B., Williams, J. G. K., and Tanksley, S. D.,** Rapid identification of markers linked to a *Pseudomonas* resistance gene in tomato by using random primers and near isogenic lines, *Proc. Natl. Acad. Sci. U.S.A.,* 88, 2336, 1991.
59. **Paran, I., Kesseli, R., and Michelmore, R. W.,** Identification of restriction fragment length polymorphism and random amplified polymorphic DNA markers linked to downy mildew resistance genes in lettuce, using near-isogenic lines, *Genome,* 34, 1021, 1991.
60. **Ronald, P. C., Albano, B., Tabien, R., Abenes, L., Wu, K. S., McCouch, S., and Tanksley, S. D.,** Genetic and physical analysis of the rice bacterial blight disease resistance locus, Xa21, *Mol. Gen. Genet.,* 236, 113, 1992.
61. **Rogers, S. O. and Bendich, A. J.,** Extraction of DNA from milligram amounts of fresh, herbarium and mummified plant tissues, *Plant. Mol. Biol.,* 5, 69, 1985.
62. **Brown, T. A., Allaby, R. G., Brown, K. A., O'Donoghue, K., and Sallares, R.,** DNA in wheat seeds from European archaeological sites, in *Conservation of Plant Genes II: Intellectual Property Rights and the Utilization of Ancient and Modern DNA,* Adams, R. P., Miller, J. S., Golenberg, E. M., and Adams, J. E., Eds., Missouri Botanical Garden Press, St. Louis, 1994, chap. 20.
63. **Pääbo, S., Irwin, D. M., and Wilson, A. C.,** Enzymatic amplification from modified DNA templates, *J. Biol. Chem.,* 265, 4718, 1990.
64. **Rollo, F., Asci, W., and Sassaroli, S.,** Assessing the genetic variation in pre-Columbian maize at the molecular level, in *Conservation of Plant Genes II: Intellectual Property Rights and the Utilization of Ancient and Modern DNA,* Adams, R. P., Miller, J. S., Golenberg, E. M., and Adams, J. E., Eds., Missouri Botanical Garden Press, St. Louis, 1994, chap. 19.
65. **Golenberg, E. M., Giannassi, D. E., Clegg, M. T., Smiley, C.J., Durbin, M., Henderson, D., and Zurawski, G.,** Chloroplast DNA sequence from a Miocene *Magnolia* species, *Nature,* 344, 656, 1990.
66. **Pääbo, S. and Wilson A. C.,** Miocene DNA sequences- a dream come true, *Curr. Biol.,* 1(1), 45, 1991.
67. **Soltis, P. S., Soltis, D. E., and Smiley, C. J.,** An rbcL sequence from a Miocene *Taxodium* (bald cypress), *Proc. Natl. Acad. Sci. U.S.A.,* 89, 449, 1992.
68. **Pääbo, S.,** Molecular cloning of ancient Egyptian mummy DNA, *Nature,* 314, 644, 1985.
69. **Pääbo, S.,** Ancient DNA: Extraction, characterization, molecular cloning and enzymatic amplification, *Proc. Natl. Acad. Sci. U.S.A.,* 86, 1939, 1989.
70. **Pääbo, S., Gifford, J. A., and Wilson, A. C.,** Mitochondrial DNA sequences from a 7000-year old brain, *Nucleic Acids Res.,* 16(20), 9775, 1988.
71. **Gower, J. C.,** A general coefficient of similarity and some of its properties, *Biometrics,* 27, 857, 1971.
72. **Adams, R. P.,** Statistical character weighting and similarity stability, *Brittonia,* 27, 305, 1975.
73. **Gower, J. C.,** Some distance properties of latent root and vector methods used in multivariate analysis, *Biometrika,* 53, 315, 1966.
74. **Jaccard, P.,** Nouvelles recherches sur la distribution florale, *Bull. Soc. Vaud. Sci. Nat.,* 44, 223, 1908.
75. **Nei, M. and Li, W-H.,** Mathematical model for studying genetic variation in terms of restriction endonucleases, *Proc. Natl. Acad. Sci. U.S.A.,* 76, 5269, 1979.
76. **Jackson, D. A., Somers, K. M., and Harvey, H.H.,** Similarity coefficients: measures of co-occurrence and association or simply measures of occurrence? *Am. Nat.,* 133, 436, 1989.
77. **Rogers, J. S.,** Measures of similarity and genetic distance, studies in genetics, *Univ. Tex. Publ.,* 7213, 145, 1972.
78. **Sneath, P. H. A. and Sokal, R. R.,** *Numerical Taxonomy,* W. H. Freeman, San Francisco, 1973.
79. **Felsenstein, J.,** *PHYLIP* (Phylogeny Inference Package) — *Version 3.0 Manual,* University of Washington Press, Seattle, 1987.

Chapter 22

OPTIMIZATION OF DNA-EXTRACTION AND PCR PROCEDURES FOR RANDOM AMPLIFIED POLYMORPHIC DNA (RAPD) ANALYSIS IN PLANTS

Kangfu Yu and K. Peter Pauls

TABLE OF CONTENTS

I. Introduction 193

II. Material and Methods 194
   A. Genomic DNA Extraction 194
   B. Optimization of the PCR Program 195

III. Results and Discussion 195
   A. Genomic DNA Extraction 195
   B. Optimization of the PCR Program 197
   C. Optimization of Amplification Conditions 198

IV. Acknowledgments 200

References 200

I. INTRODUCTION

In random amplified polymorphic DNA (RAPD) analysis, fragments of DNA are synthesized in a PCR reaction mixture containing genomic DNA and an arbitrarily chosen primer that has binding sites on the complementary strands of the genomic DNA within approximately 3 kb. This method of analysis has been used for genome mapping, gene tagging, and relatedness studies.[1-3] For most of the applications large numbers of samples need to be analyzed; therefore, it is important to minimize the time required for each analysis. The efficiency of the RAPD procedure is determined by the techniques used for DNA extraction and PCR amplification.

Many of the genomic DNA-extraction methods described in the literature are based on the time-consuming cesium chloride density-gradient technique or the cetyltrimethylammonium bromide (CTAB) extraction procedure, which uses hazardous chemicals such as phenol and chloroform.[4-7] Since most PCR products have a moderate size, it is not essential for the template DNA used for PCR to be of high molecular weight. Therefore, a number of DNA-extraction procedures that are simple and rapid[8,9] can be employed for RAPD analysis. The PCR program commonly used for RAPD analysis has a 1 min template denaturing step at 94°C, a 1-min primer annealing step at 36°C, and a 2-min primer extension step at 72°C and is repeated 45 times.[1] This program can take as long as 5 h to complete, depending on the thermal cycler used.[10] The length of the program can significantly limit the usefulness of RAPD analysis when large numbers of samples need to be analyzed. We have found that the PCR program can be shortened without changing the RAPD pattern.[10]

Our objectives in writing this chapter were to outline a quick DNA-extraction protocol and a short PCR program for RAPD analysis as well as to discuss the amplification reaction conditions that need be optimized at the beginning of a study in order to efficiently use the RAPD procedure for genome mapping, gene tagging, and population studies of plants. The procedures will be illustrated by discussing results we obtained from alfalfa (*Medicago sativa*).

II. MATERIALS AND METHODS

A. GENOMIC DNA EXTRACTION

Extraction procedures for plant genomic DNA include treatments that disrupt plant cell walls and cell membranes to release the cellular constituents into an extraction buffer that contains compounds to protect the DNA from the activity of endogenous nucleases.[11] The method presented here is a modification of a rapid procedure for plant genomic DNA extraction described by Edwards et al.[8] The procedure does not require the use of expensive and hazardous chemicals and uses only milligram quantities of leaf tissue or a single seed. Approximately 4 to 8 h are needed to process 50 to 100 samples, and each sample contains enough DNA for at least 100 PCRs.

An alfalfa leaflet was placed in 400 µl of extraction buffer (made by mixing 10 ml of 1 M Tris-HCl, pH 7.4, 2.5 ml of 5 M NaCl, 2.5 ml of 0.5 M EDTA, pH 8.0, 2.5 ml of 10% SDS, and 32.5 ml of double-distilled H_2O) in a 1.5-ml sterile Eppendorf tube. The leaf tissue was homogenized in the buffer using a pestle that fit the Eppendorf tube tightly and was attached to a motor revolving at approximately 500 rpm (a new pestle was used for each sample or it was washed thoroughly with distilled water between samples to avoid contamination). The homogenate can be left at room temperature for several hours until all the samples have been prepared. After centrifugation of the extracts in the microcentrifuge for 1 min, 300 µl of the supernatant were transferred to a fresh Eppendorf tube. The DNA was precipitated at room temperature for 5 min by adding (1) 1 vol of isopropanol, or (2) 2 vol of 100% ethanol, or (3) 1 vol of 20% PEG (polyethylene glycol) containing 6 M NaCl. After centrifugation for 5 min, the pellet was air dried and then allowed to dissolve in 100 µl of distilled water (from 10 min to overnight at 4°C), without agitation. The sample was subsequently centrifuged for 1 min, and the supernatant, containing the DNA, was collected. One microliter of the solution, which contains about 20 ng of DNA, was sufficient for a 25-µl PCR containing 10 mM Tris-HCl pH 8.8, 50 mM KCl, 1.5 mM $MgCl_2$, 0.1% Triton X-100, 0.1 mM each of dATP, dCTP, dGTP, and dTTP (Pharmacia), 0.36 µM random primer (Biotechnology Laboratory, University of British Columbia), and 1.5 U of *Taq* DNA polymerase (Promega).

When seeds were used to extract DNA, the seeds were soaked in water overnight before use. A hydrated seed was ground in a sterile 1.5-ml Eppendorf tube, and 350 µl of extraction buffer were added. The samples were vortexed for 30 s. Thereafter, extraction and PCR procedures were carried out in the same way as described above for leaves.

B. OPTIMIZATION OF THE PCR PROGRAM

The PCR program normally used for RAPD analysis is a typical PCR program except that the stringency during primer annealing is much lower than usual because of the primer used in the RAPD procedure is relatively short and has an arbitrary sequence. For random 10-mers, primer annealing at 36°C was found to work well for many organisms, including plants.[1] Therefore, optimization of the PCR program for a RAPD analysis entails a search for the best combination of step lengths and cycle numbers, with a view to minimizing the amplification time without losing any information. In this study the effects of different denaturing times (5, 30, and 60 s), annealing times (5, 30, and 60 s), and extension times (5, 30, and 60 s) on the RAPD patterns from several primers were determined[10] using an M.J. Research, Inc., PTC-100 Programmable Thermal Controller and block control to regulate the reaction tempera-

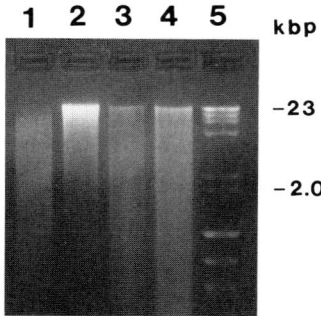

Figure 1. Effect of the presence or absence of buffer during the homogenization step on the recovery of genomic DNA from A70-34 alfalfa. Leaf samples were homogenized immediately after sampling with (lane 4) or without buffer (lane 3). Leaf samples were homogenized 3 h after being placed into a tube with (lane 2) or without (lane 1) buffer; 5 µl of the DNA solution were electrophoresed through a 0.8% agarose gel and stained with ethidium bromide. Molecular weight markers are in lane 5.

tures. In addition, the effects of different numbers of cycles (35, 55, and 75) in the PCR program on the quantity of DNA-amplification products were tested.

III. RESULTS AND DISCUSSION

A. GENOMIC DNA EXTRACTION

The genomic DNA obtained from alfalfa with the present method was generally of high molecular weight. With the standard DNA-isolation procedure,[11] a day is usually required to process approximately 10 samples, and more than 1 gram of leaf tissue is needed for each sample. By contrast, 50 to 100 samples can easily be prepared in one working day with the modified rapid procedure, and only 20 mg of leaf tissue or a single seed are required per sample. Another advantage of the present method is that it does not use expensive and hazardous chemicals, such as phenol and chloroform.

An important modification to the procedure described by Edwards et al.[8] was the addition of the extraction buffer to the sample before homogenization. Figure 1 indicates that more high-molecular-weight DNA was obtained from an alfalfa leaf sample ground in the presence (lane 4) than in the absence (lane 3) of the buffer. In fact, incubation of alfalfa leaflets in the buffer for 3 h prior to homogenization enhanced the amount of DNA recovered (Figure 1, lane 2). Conversely, a 3-h dry incubation period followed by dry homogenization decreased the amount of high molecular weight DNA observed in the sample (Figure 1, lane 3 vs. lane 1). The buffer probably protected the DNA from degradation by endogenous nucleases;[11] it also prevented desiccation of the tissue and may have helped to lyse the cells and release the DNA into solution.

A second modification to the Edwards et al.[8] procedure that improved the purity of the DNA preparation was the inclusion of a centrifugation step after dissolving the isopropanol-precipitated DNA in water. Figure 2 shows that centrifugation removed undissolved material that remained trapped in the well of a lane loaded with a noncentrifuged sample. Our studies indicated that the debris that remains in preparations obtained by the unmodified Edwards et al.[8] procedure results in variable PCR patterns (data not presented).

The three methods tested for precipitating DNA (i.e., isopropanol, ethanol, or polyethylene glycol) were found to be equivalent. Furthermore, RAPD patterns obtained from DNA precipitates that were or were not washed with 70% ethanol were comparable (Figure 3).

When the quick-extraction procedure was used, the same PCR-amplification patterns were observed repeatedly with different DNA preparations from the same plant (Figure 3). Further-

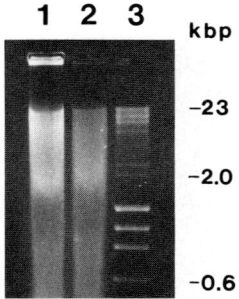

Figure 2. Effect of centrifugation on the purity of the DNA sample used for PCR analysis. DNA from an alfalfa leaf sample precipitated with isopropanol was dissolved in water and electrophoresed (lane 1) or centrifuged before electrophoresis (lane 2); 15 µl of each of the samples were electrophoresed through a 0.8% agarose gel and stained with ethidium bromide. Molecular weight markers are in lane 3.

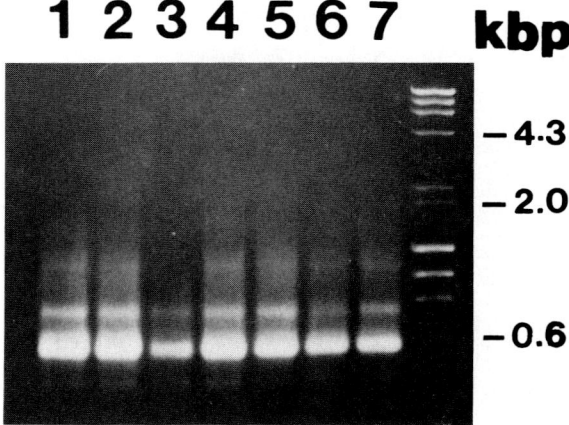

Figure 3. Random amplification of A70-34 alfalfa genomic DNA prepared in various ways. Lane 7 contains the PCR products of A70-34 DNA isolated with the procedure of Rogers and Bendich[11] and primed with 5'CCTGGGTTCC3'. Lanes 1, 2, and 3 contain the PCR products of A70-34 DNA isolated with the present method and precipitated with 1 vol of 20% PEG containing 6 M of NaCl, or 1 vol of isopropanol, or 2 vol of 100% ethanol, respectively. Lanes 4, 5, and 6 contain the PCR products of A70-34 DNA prepared in the same way as lanes 1, 2, and 3, respectively, except that a final wash with 70% ethanol is included.

more, the PCR patterns were the same for DNA prepared by the present method or by a conventional method[11] employing phenol/chloroform extraction (Figure 3).

Most of the genomic DNA extraction procedures that have been described for plants utilize leaf tissue. For some applications it would be advantageous to prepare DNA samples directly from seeds rather than waiting for a seedling to develop leaves. The present procedure was found to be applicable for isolating DNA from single alfalfa seeds (Figure 4a). The polymorphisms observed in the RAPD patterns obtained from two alfalfa seeds (Figure 4b) indicate that PCR analyses could be used to assess the variability within alfalfa seed lots.

The simplicity and speed of the present DNA extraction procedure makes it particularly suited for PCR analysis of large numbers of individuals. The modifications described in the present method enhance its applicability to plant breeding programs since samples can be rapidly collected in vials containing buffer from large numbers of individuals. This is particularly important when leaf tissue from plants growing in the field are required for analysis. As indicated above, the time in the buffer between sampling and homogenization would enhance recovery of

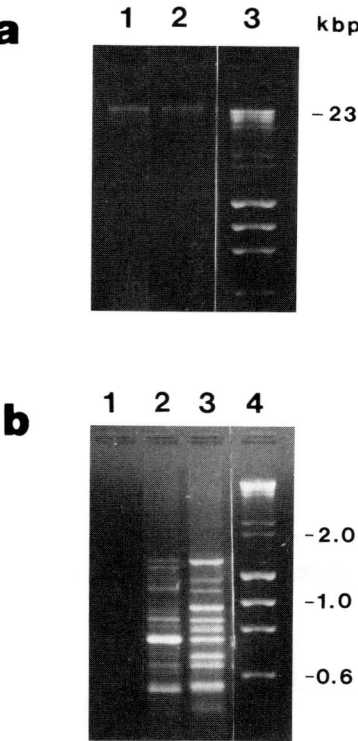

Figure 4. Genomic DNA (a) and RAPD patterns (b) obtained from two seeds of alfalfa (cv. Peace) primed with 5'TGACCCCTCG3'. The RAPD patterns in lanes b2 and b3 correspond to genomic DNA preparations shown in lanes a1 and a2, respectively. Lane b1 is a control, in which genomic DNA was replaced by H_2O.

DNA, whereas a delay between sampling and homogenization without buffer reduces the amount of DNA that can be obtained from the samples. The procedure described here was also found to yield DNA suitable for PCR analysis from rapeseed[16] white beans,[17] and barley[18] although two isopropanol DNA-precipitation steps are recommended for the latter.

B. OPTIMIZATION OF THE PCR PROGRAM

A PCR program with a denaturing time of 5 s gave better PCR product yields than programs with 30- or 60-s denaturation times (Figure 5). This result can probably be attributed to the fact that *Taq* DNA polymerase has a limited lifespan at high temperatures.[12] Therefore, it is better to use a denaturing time that is as short as possible.

Figure 6 shows that there is an interaction between the time required for primer annealing and the GC content of the primer. For all the primers used in this study (which had GC contents from 50 to 80%), 30 s of annealing time appeared to be sufficient to obtain a complete RAPD pattern (Figure 6). However, for the primers containing 50 or 60% GC, the amounts of PCR products were reduced considerably when a 5-s annealing time was used. In contrast, the amounts of PCR products from primers with 70 or 80% GC were no different with 5 or 60 s of annealing time (Figure 6). Figure 7 shows that there is a correspondence between the extension time and the maximum size of fragment that is amplified in the PCR reaction. Between 5 and 60 s, this relationship is linear (Figure 7b), and 5 s are sufficient for amplification of a 0.6-kb fragment while 60 s are required for a 3-kb fragment.

No large differences in band intensity were found among PCR products obtained from programs run for 35, 55, or 75 cycles (Figure 8). This result may be attributable to *Taq* DNA

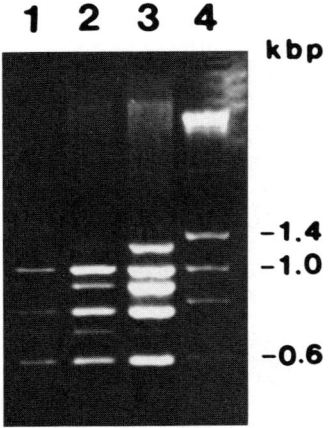

Figure 5. Effects of denaturing time on RAPD patterns obtained from alfalfa genomic DNA and primer 5'AGCAGCGTGG3'. The PCR program consisted of 35 cycles of denaturing at 94°C for 5 s (lane 3), 30 s (lane 2), or 60 s (lane 1), annealing at 36°C for 30 s and extension at 72°C for 60 s.

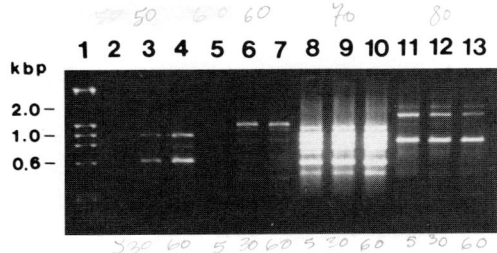

Figure 6. Effects of primer/template annealing time and primer GC content on RAPD patterns obtained from alfalfa genomic DNA. The PCR program consisted of 35 cycles of denaturing at 94°C for 5 s, annealing at 36°C for 5 s (lanes 2, 5, 8, and 11), 30 s (lanes 3, 6, 9, and 12), or 60 s (lanes 4, 7, 10, and 13) and extension at 72°C for 60 s. The patterns in lanes 2 to 4, 5 to 7, 8 to 10, and 11 to 13 were obtained with primers 5'TTAGCGGTCT3' (50% GC), 5'CAAGGGAGGT3' (60% GC), 5'AGCAGCGTGG3' (70% GC), and 5'CTGGCGGCTG3' (80% GC), respectively.

polymerase inactivation over time or be indicative that some other components in the reaction mixture become limiting at high cycle numbers.

The optimized 35-cycle program that was used to produce RAPD patterns with alfalfa genomic DNA and random 10-mers has a 5-s denaturing step at 94°C, a 30-s annealing step at 36°C, and a 60-s extension step at 72°C. The 35-cycle program requires only 2.5 h to complete, and it results in sharper banding patterns for most of the primers than the original program. The time for each step may need to be optimized for differently sized random primers and/or DNA templates from different species, but the present results suggest that there is considerable opportunity for increasing the efficiency of the PCR program used for RAPD analyses.

C. OPTIMIZATION OF AMPLIFICATION CONDITIONS

Although RAPD patterns can be easily obtained, several parameters of the amplification reaction need to be optimized in order to produce reproducible results. Generally, the concentration of template, magnesium, primer, enzyme, and dNTPs need be optimized.[13,14] The enzyme source is also critical. For alfalfa, 1 ng/μl of template DNA, 1.5 mM Mg^{2+}, 0.36 μm primer, 1.5 to 2.0 U/25 μl reaction of Promega *Taq* DNA polymerase, and 0.1 mM each of dNTPs was found to be optimal. The use of more than 1 ng/μl of template were found to be necessary for generating reproducible fingerprints in bacteria.[14] For conifers, a reaction

Optimization of DNA-Extraction and PCR Procedures

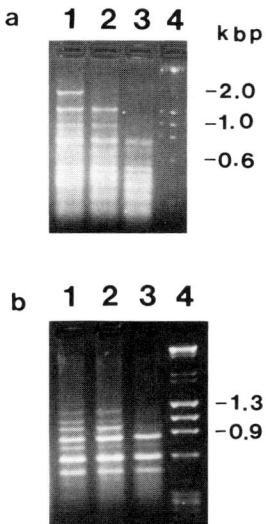

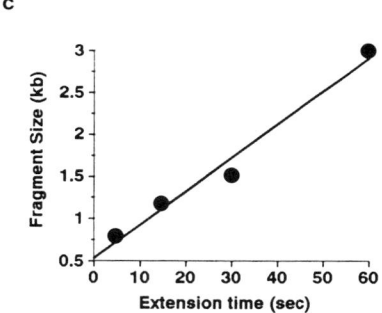

Figure 7. Effects of primer extension time on RAPD patterns obtained from alfalfa genomic DNA. The PCR program consisted of 35 cycles of denaturing at 94°C for 5 s, annealing at 36°C for 30 s, and extension at 72°C for 5 s (lanes 3a and 3b), 30 s (lanes 2a and 2b), or 60 s (lanes 1a and 1b). The patterns in (a) and (b) were obtained with primers 5'GCTTGTGAAC3' and 5'AGCAGCGTGG3', respectively. (c) Relationship between primer extension time and maximum size of PCR product determined from patterns obtained from three primers.

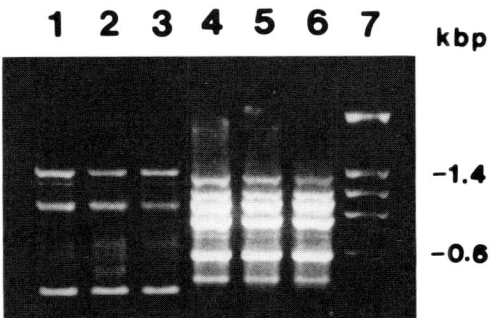

Figure 8. Effects of the number of amplification cycles on yield of PCR products. The RAPD patterns were obtained from alfalfa genomic DNA with a PCR program that consisted of 35 cycles (lanes 3 and 6), 55 cycles (lanes 2 and 5), or 75 cycles (lanes 1 and 4) of denaturing at 94°C for 5 s, annealing at 36°C for 30 s, and extension at 72°C for 60 s. The patterns in lanes 1 to 3 and 4 to 6 were obtained with primers 5'CAAGGGAGGT3' and 5'AGCAGCGTGG3', respectively.

mixture containing 8 ng/μl DNA, 1.9 mM Mg^{2+}, 0.2 μm random 10-mer, 1 U/100 μl reaction of Cetus Amp iTaq enzyme, and 0.2 mM each of dNTPs was reported to be optimal for obtaining repeatable RAPD patterns.[15] The examples cited above indicate that at the beginning of a study with the RAPD procedure, it is essential to test a variety of reaction conditions to determine those that are best suited for the particular plant species under study.

ACKNOWLEDGMENTS

Assistance with preparation of the manuscript by Jennifer Kingswell is gratefully acknowledged. The work performed in the authors' laboratory was supported financially by the Ontario Ministry for Food and Agriculture.

REFERENCES

1. **Williams, J. G., Kubelik, A. R., Livak, K. J., Rafalski, J. A., and Tingey, S. V.**, DNA polymorphisms amplified by arbitrary primers are useful as genetic markers, *Nucleic Acids Res.*, 18, 6531, 1990.
2. **Martin, G. B., Williams, J. G. K., and Tanksley, S. D.**, Rapid identification of markers linked to a *Pseudomonas* resistance gene in tomato by using random primers and near-isogenic lines, *Proc. Natl. Acad. Sci. U.S.A.*, 88, 2336, 1991.
3. **Goodwin, P. H. and Annis, S. L.**, Rapid identification of genetic variation and pathotype of *Leptosphaeria maculans* by random amplified polymorphic DNA assay, *Appl. Environ. Microbiol.*, 57, 2482, 1991.
4. **Bendich, A. J., Anderson, R. S., and Ward, B. L.**, Plant DNA: long, pure and simple, in *Genome Organization and Expression*, Leaver, C. J., Ed., Plenum Press, New York, 1988, 31.
5. **Murray, H. G. and Thompson, W. F.**, Rapid isolation of higher molecular weight DNA, *Nucleic Acids Res.*, 8, 4321, 1980.
6. **Taylor, B. and Powell, A.**, Isolation of plant DNA and RNA, *Focus*, 4, 4, 1982.
7. **Ausubel, F. M., Brent, R., Kingston, R. E., Moore, D. D., Seidman, J. G., Smith, J. A., and Struhl, K.**, in *Current Protocols in Molecular Biology*, John Wiley & Sons, New York, 1990, chap. 15.
8. **Edwards, K., Johnstone, C., and Thompson, C.**, A simple and rapid method for the preparation of plant genomic DNA for PCR analysis, *Nucleic Acids Res.*, 19, 1349, 1991.
9. **Higuchi, R.**, Simple and rapid preparation of samples for PCR, in *PCR Technology: Principles and Applications for DNA Amplification*, Erlich, H. A., Ed., Stockton Press, New York, 1989, 31.
10. **Yu, K. and Pauls, K. P.**, Optimization of the PCR program for RAPD analysis, *Nucleic Acids Res.*, 20, 2606, 1992.
11. **Rogers, S. O. and Bendich, A. J.**, Extraction of DNA from plant tissues, in *Plant Molecular Biology Manual*, Gelvin, S. B. and Schilperoort, R. A., Eds., Kluwar Academic, Dordrecht, the Netherlands, 1988, A6:1–10.
12. **Gelfand, D. H.**, in *PCR Technology: Principles and Applications for DNA Amplification*, Erlich, H. A., Ed., Stockton Press, New York, 1989, 17–22.
13. **Caetano-Anollés, G., Bassam, B. J., and Gresshoff, P. M.**, in DNA amplification fingerprinting using very short arbitrary oligonucleotide primers, *Biotechnology*, 9, 553, 1991.
14. **Caetano-Anollés, G., Bassam, B. J., and Gresshoff, P. M.**, DNA amplification fingerprinting: a strategy for genome analysis, *Plant Mol. Biol. Rep.*, 9, 294, 1991.
15. **Carlson, J. E., Tulsieram, L. K., Glaubitz, J. C., Luk, V. W. K., Kauffeldt, C., and Rutledge, R.**, Segregation of random amplified DNA markers in F1 progeny of conifers, *Theor. Appl. Genet.*, 83, 194–200, 1992.
16. **Deynze and Pauls**, unpublished.
17. **Fuchs et al.**, unpublished.
18. **Xu and Kasha**, unpublished.

Chapter 23

THE USE OF RAPD ANALYSIS TO TAG GENES AND DETERMINE RELATEDNESS IN HETEROGENEOUS PLANT POPULATIONS USING TETRAPLOID ALFALFA AS AN EXAMPLE

Kangfu Yu and K. Peter Pauls

TABLE OF CONTENTS

I. Introduction 201

II. RAPD Analysis to Identify Molecular Markers 202
   A. Alfalfa Genetics 202
   B. Inheritance of RAPD Markers in Tetraploid Alfalfa 202
   C. Detection of Linkage Among RAPD Markers in Tetraploid Alfalfa ... 202
   D. Estimation of Recombination Fraction between RAPD Markers 204
   E. Identification of a RAPD Marker Linked to Somatic Embryogenesis ... 206
      1. Detection of Linkages 207
      2. Estimation of Recombinant Fraction 207

III. The Use of RAPD Markers to Determine Similarities and
Genetic Distances 209
   A. Determination of Genetic Distances Among Individuals 209
   B. Determination of Genetic Distances Among Populations 209

Acknowledgments 213

References 213

I. INTRODUCTION

Significant advances in the applicability of the PCR procedure were made possible by the demonstration that a variety of DNA fragments are usually produced in reactions containing genomic DNA and a single, short DNA primer sequence. Procedures based on this general principle have been given a number of names, including random amplified polymorphic DNA (RAPD) analysis;[1] arbitrarily primed polymerase chain reaction;[2] and DNA amplification fingerprinting (DAF).[3] The first report of the RAPD procedure, by Williams et al.,[1] also showed how it could be used to identify molecular markers linked to genes of specific interest and discussed the possibility of using this technique to identify genotypes or plant varieties. In this chapter we will illustrate the applicability of the RAPD procedure to analysis of heterogeneous plant populations by describing the identification of linkages among RAPD markers and between markers and traits in tetraploid alfalfa as well as describing the usefulness of RAPD markers for determining relatedness among populations of this species.

II. RAPD ANALYSIS TO IDENTIFY MOLECULAR MARKERS

The mode of inheritance of many plant traits has been determined but for most traits the locations and molecular characteristics of the genes that control them are not known. The identification of DNA markers that are closely linked to these genes represents an important first step toward the molecular analysis of plant traits. Molecular markers are particularly useful for tracking traits in a population that are determined by (1) recessive genes, (2) a gene complex, or (3) genes whose expression is highly sensitive to environmental conditions. DNA-based molecular markers are ideal for gene tagging because they are dominant and they are not affected by the conditions under which the plants are grown or the developmental stage of the tissue from which the DNA is extracted.

The identification of molecular tags for genes is usually a three-step process that involves (1) characterization of the modes of inheritance of the trait and the molecular marker, (2) demonstration that the marker and trait are linked through a segregation analysis, and (3) estimation of the recombinant fraction to determine the closeness of the linkage. These steps will be illustrated in the following sections, which describe the identification of linkages between two RAPD markers and the identification of a RAPD marker linked to somatic embryogenesis in alfalfa.

A. ALFALFA GENETICS

Genetic analysis in alfalfa has been hampered because it is an autotetraploid and it exhibits severe inbreeding depression. Furthermore, very few molecular markers have been identified for linkage studies in this species.[4] However, a recent study with diploid alfalfa[5] and our own study with tetraploid alfalfa[6] showed that RAPD markers segregate in a Mendelian fashion at both ploidy levels and can be utilized for linkage analyses as well as to identify gene tags linked to phenotypic traits.

The RAPD markers used in our study were present in one parent but absent in the other (Figure 1). Segregation of the markers and traits was studied in an F_1 population. Because of the dominant nature of RAPD markers,[1] the genotype of the parent without the marker was assumed to be homozygous recessive. Only those markers that were in a simplex or duplex condition in the parental line were expected to segregate in the F_1 and the ratio of F_1s with the marker to those without the marker was expected to be 1:1 for the former condition (Table 1) and 5:1 for the latter condition if random chromosome segregation is assumed for tetraploid alfalfa. This assumption is supported by previous cytological evidence[7] and by the fact that only one example of a double reduction was observed in this study for the 32 RAPD markers and 100 F_1 plants that were screened.[6]

B. INHERITANCE OF RAPD MARKERS IN TETRAPLOID ALFALFA

Table 2 gives the observed segregation frequencies in the F_1 population for 32 RAPD markers. Of the 32 markers that were scored, 29 segregated as Mendelian traits, and of these, 21 were simplex and 8 were duplex. Three markers did not fit 1:1 or 5:1 segregation ratios. The lack of fit may be attributable to contamination of the genomic DNA with chloroplast or mitochondrial DNA, some degree of double reduction or preferential pairing, gametophytic selection or linked deleterious mutations. Of the RAPD markers previously tested in diploid alfalfa, 24% also did not segregate as expected.[5]

C. DETECTION OF LINKAGE AMONG RAPD MARKERS IN TETRAPLOID ALFALFA

The identification of molecular tags for plant traits is based on demonstrations of cosegregation between a trait and a molecular marker. The linkage analysis procedure for

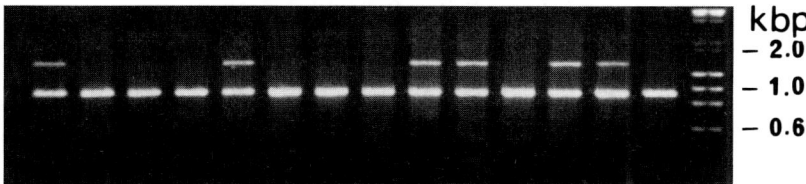

Figure 1. Example of segregation of a RAPD marker in an F_1 population of alfalfa. Lanes 1 to 12 were obtained from F_1s, lane 13 was obtained from the male parent, and lane 14 was obtained from the female parent. Lane 15 contains molecular weight markers. The sequence of the primer that was used was 5′CCGTCATTGG3′. Each 25-μl reaction contained 1 μl of the DNA extract (approximately 20 ng), 10 mM Tris-HCl, pH 8.8, 50 mM KCl, 1.5 mM MgCl$_2$, 0.1% Triton X-100, 0.1 mM each of dATP, dCTP, dGTP, and DTTP (Pharmacia), 0.36 μM random primer (Biotechnology Lab, University of British Columbia) and 1 U of *Taq* polymerase (Promega). Amplification was performed in an MJ Research Programmable Thermal Controller (PTC-100), programmed for 45 cycles of 1 min at 94°C, 1 min at 36°C, and 2 min at 72°C. The RAPD fragments were separated by electrophoresis using a 1.4% agarose gel and visualized with ethidium bromide.

TABLE 1.
Expected Segregation Frequencies in the F_1 for RAPD Markers Present in One Parent and Absent in the Other

Zygotic genotypes	Genotypes of gametes	Frequencies of gametes	F_1 phenotypes[b]	Phenotypic frequencies
Mmmm (simplex)	MM	0	M+	1/2
	Mm	1/2		
	mm	1/2	M−	1/2
MMmm (duplex)	MM	1/6	M+	5/6
	Mm	4/6		
	mm	1/6	M−	1/6
MMMm (triplex)	MM	1/2	M+	1
	Mm	1/2		
	mm	0	M−	0
MMMM (tetraplex)	MM	1	M−	1

[a] The F_1 phenotypes assume that the gametes from the parent without the marker are recessive (mm); M+ and M− indicate F_1 with and without marker, respectively.

tetraploid alfalfa can be understood by considering the occurrence of cosegregation among RAPD markers. Since only simplex and duplex markers derived from the same parent are useful for a linkage study in an F_1 population, only seven combinations of marker pairs are expected in the F_1. Table 3 gives the possible genotypes and ratios among genotypes containing pairs of markers that are either linked or unlinked in an F_1 population of tetraploid alfalfa (assuming chromosome segregation).

A linkage between a pair of markers (M1 and M2) is signified by a χ^2 value that is larger than expected for a group of individuals distributed randomly among the four genotype classes, namely; M1M2, M1m2, m1M2, and m1m2. In the present study two simplex markers derived from A70 were found to be linked. In particular, the expected ratios for genotypes defined by the presence or absence of two simplex RAPD markers segregating independently is 1:1:1:1:1 for the genotypes M1M2:M1m2:m1M2:m1m2 (Table 3), whereas a 28:11:21:27

TABLE 2.
Occurrence of RAPD Markers in PCR Patterns Obtained with Genomic DNA from F_1 Progeny; X^2 Values for Expected Segregation Ratios Are Given

RAPD allele[a]	F_1s with marker	F_1s without marker	X^2 Cs 1:1	X^2 Cs 5:1
A108-2	49	45	0.10	63.68[b]
A108-3	57	38	3.41	35.58[b]
A110-4	51	43	0.52	55.15[b]
A112-5	77	22	29.45[b]	1.82
A112-6	80	19	36.36[b]	0.29
A112-7	47	52	0.16	89.09[b]
A117-8	54	46	0.49	59.86[b]
A117-9	77	19	33.84[b]	0.47
A119-10	58	40	2.95	39.43[b]
A135-12	52	46	0.26	62.50[b]
A145-13	45	52	0.37	92.67[b]
A147-14	74	24	24.50[b]	3.77
A148-15	56	43	1.45	49.16[b]
A149-16	71	28	17.82[b]	8.80[b]
A171-19	54	44	0.83	54.22[b]
A171-20	65	33	9.81[b]	19.20[b]
A172-21	79	18	37.11[b]	0.13
A172-22	83	15	45.81[b]	0.05
A181-23	47	51	0.09	85.77[b]
A183-24	62	37	5.82[b]	20.09[b]
A188-27	50	39	1.12	45.31[b]
A192-30	56	43	1.45	49.16[b]
A198-35	45	54	0.65	99.56[b]
Ar108-36	47	47	0.00	72.82[b]
Ar117-37	48	52	0.09	87.36[b]
Ar117-38	48	52	0.09	87.36[b]
Ar119-39	85	14	49.49[b]	0.29
Ar119-40	46	53	0.36	94.25[b]
Ar120-41	45	7	26.33[b]	0.19
Ar147-42	51	47	0.09	66.86[b]
Ar175-43	16	19	0.11	33.01[b]
Ar178-44	23	30	0.68	58.02[b]

[a] Bands from A70-34 (female parent) and Arrow36 (male parent) are indicated by "A" and "Ar" in front of the markers, respectively.
[b] Significant at the 5% probability level.

ratio was observed among these classes for individuals segregating for the simplex markers A191-23 and A188-27.

D. ESTIMATION OF RECOMBINATION FRACTION BETWEEN RAPD MARKERS

After two markers or a marker and a trait have been shown to be linked, it is of interest to determine the distance that separates them on a chromosome. The distance can be defined in terms of a recombination frequency since the farther they are apart the larger is the chance that they will be separated by recombination during meiosis. The recombination frequency (p) is simply the percent of recombinants in a segregating population. For two linked simplex RAPD

TABLE 3.
Gamete Types with Respect to Pairs of Markers from One Parent and Their Expected Frequencies in an F_1 Population When the Two Markers Are Linked or Independent

Marker pair conditions in one parent	Types of gametes from the parent	Gamete ratios[a] General	M1 and M2 linked	M1 and M2 not linked
Simplex coupling M1M2.(m1m2)$_3$	M1M2 = e	$1-p$	1	1
	M1m2 = f	p	0	1
	m1M2 = g	p	0	1
	m1m2 = h	$1-p$	1	1
Simplex repulsion M1m2.m1M2.(m1m2)$_2$	M1M2 = e	$1+p$	1	1
	M1m2 = f	$2-p$	2	1
	m1M2 = g	$2-p$	2	1
	m1m2 = h	$1+p$	1	1
Asymmetrical coupling M1M2.M1m2.(m1m2)$_2$	M1M2 = e	$3-p$	6	5
	M1m2 = f	$2+p$	4	5
	m1M2 = g	p	0	1
	m1m2 = h	$1-p$	2	1
Asymmetrical repulsion (M1m2)$_2$.m1M2.m1m2	M1M2 = e	$2+p$	4	5
	M1m2 = f	$3-p$	6	5
	m1M2 = g	$1-p$	2	1
	m1m2 = h	p	0	1
Duplex coupling (M1M2)$_2$(m1m2)$_2$	M1M2 = e	$5-2p+p^2$	35	25
	M1m2 = f	$2p-p^2$	0	5
	m1M2 = g	$2p-p^2$	0	5
	m1m2 = h	$1-2p+p^2$	1	1
Duplex repulsion (M1m2)$_2$.(m1M2)$_2$	M1M2 = e	$4+p^2$	24	25
	M1m2 = f	$1-p^2$	6	5
	m1M2 = g	$1-p^2$	6	5
	m1m2 = h	p^2	0	1
Duplex coupling/repulsion M1m2.M1M2.m1M2.m1m2m1	M1M2 = e	$8+p-p^2$	24	25
	M1m2 = f	$2-p+p^2$	6	5
	m1Ma = g	$2-p+p^2$	6	5
	m1m2 = h	$p-p^2$	0	1

[a] The gamete ratios are equal to the phenotypic ratios observed in each population because one parent is double recessive for all the useful markers.

markers derived from one parent and segregating in an F_1 population, the recombinant genotypes can be easily recognized because they are the genotypes that have only one marker, i.e., M1m2 and m1M2. Therefore, the recombination frequency in the simplex coupling case is defined by

$$p = \frac{M1m2 + m1M2}{M1M2 + M1m2 + m1M2 + m1m2} \times 100 = \frac{f+g}{e+f+g+h} \times 100 \quad (1)$$

For the marker pairs A191-23 and A188-27, e, f, g, and h were equal to 28, 11, 21, and 27, respectively; therefore, $p = [(11 + 21)/28 + 11 + 21 + 27] \times 100 = 36.8$.

In some cases the recombinant fraction is not as easy to identify as in the simplex coupling case because the recombinants cannot be distinguished from nonrecombinants when the markers are tightly linked. For example, a cosegregation analysis of the F_1 alfalfa population for markers listed in Table 2 revealed that there were five linked marker pairs in simplex repulsion conditions. The expected ratios among the genotype classes are 1:1:1:1 or 1:2:2:1

for M1M2:M1m2:m1M2:m1m2, respectively, if M1 and M2 are independent or linked. The recombination fraction for these linked marker pairs cannot be determined directly since there is no way to separate recombinant M1m2, m1M2 from nonrecombinant. In this case, a maximum likelihood equation, which compares the obtained distribution of individuals among classes to the expected distribution assuming linkage, is required. The equation applicable to a specific case can be derived from the general likelihood function:[8]

$$\Theta = n!/(a1!\,a2!\ldots aj!) \times (k1)^{a1} \times (k2)^{a2} \times \ldots \times (kj)^{aj} \qquad (2)$$

where $n = a1 + a2 \ldots aj$ (as are the observed numbers for each class) and ks are the expected ratios for each class. For example, in the simplex repulsion condition in the F_1 population, the probability (Θ) of obtaining the recombinants e and h and nonrecombinants f and g in a group of size n is

$$\Theta = n!/(e!\,f!\,g!\,h!) \times (1+p)^e \times (2-p)^f \times (2-p)^g \times (1+p)^h \qquad (3)$$

where e, f, g, and h are the observed numbers for each of the M1M2, M1m2, m1M2, and m1m2 phenotypes, respectively, and $(1+p)$, $(2-p)$, $(2-p)$, and $(1+p)$ are the expected ratios (Table 3). A log transformation gives

$$L = \log[n!/(e!\,f!\,g!\,h!)] + e\log(1+p) + f\log(2-p) + g\log(2-p) + h\log(1+p) \qquad (4)$$

Differentiating the equation and equating it to 0 results in

$$dL/dP = e/(1+p) - f/(1-p) - g/(1-p) + h/(1+p) = 0 \qquad (5)$$

A rearrangement of this equation gives

$$p = (2e - f - g + 2h)/(e + f + g + h) \qquad (6)$$

where $0 \leq p \leq 0.5$. For instance, the p value for marker pair A1171-19 and A192-30 can be determined by substituting e, f, g, and h in the equation with the observed numbers of each phenotype, which were 18, 36, 25, and 19, respectively. Therefore,

$$p = (2e - f - g + 2h)/(e + f + g + h)$$
$$= (2 \times 18 - 36 - 25 + 2 \times 19)/(18 + 36 + 25 + 19)$$
$$= 0.133 \qquad (7)$$

The standard error for p can be determined as described by Mather.[8]

E. IDENTIFICATION OF A RAPD MARKER LINKED TO SOMATIC EMBRYOGENESIS

The ability of alfalfa tissue cultures to regenerate plants is genetically controlled.[9] Wan et al.[10] obtained results that led them to propose that somatic embryogenesis in tetraploid alfalfa is a qualitative trait under the control of two dominant genes with complementary effects. Based on the assumptions of this genetic model, an embryogenic individual has the genotype A— B—.

Somatic embryogenesis is a potentially useful trait in alfalfa because it could be used to obtain large quantities of parental material for the production of limited synthetic or hybrid seed.[11] Vegetative propagation through somatic embryogenesis would circumvent the self-incompatibility and inbreeding depression that have hindered traditional alfalfa improvement. A prerequisite for this approach is a method for recognizing embryogenic genotypes in alfalfa breeding populations. A simple tissue culture test that utilizes a single cotyledon is an effective method for screening a population for embryogenic plants,[12] but the utilization of a molecular marker that is tightly linked to the trait would simplify and accelerate the identification process.[13] Furthermore, molecular markers for each of the two genes would enable plant breeders to track their segregation, individually, in populations without resorting to test crosses.

The female parent used in the present study (A70) was an embryogenic genotype and the male parent was a nonembryogenic genotype (Arrow36). Of the 83 F_1 plants that were screened for their ability to form somatic embryos, 34 were observed to be embryogenic. The ratio of embryogenic to nonembryogenic genotypes in this population best fits a two-locus model, assuming that the genotypes of the female parent (A70-34) and the male parent (Arrow36) were duplex/simplex (AAaaBbbb) and double recessive (aaaabbbb), respectively.

The use of RAPD markers to tag genes controlling complex traits is illustrated in the following study aimed at identifying a RAPD marker linked to somatic embryogenesis in alfalfa. Because the linkage analysis was performed in an F_1 population derived from two heterozygous plants, only those markers that were derived from the embryogenic parent and were in a simplex or a duplex condition were useful for detecting linkages between the RAPD markers and the somatic embryogenesis trait.

1. Detection of Linkages

Table 4 gives the possible phenotypes, including embryogenic with band (E$^+$), embryogenic without band (E$^-$), nonembryogenic with band (N$^+$) and nonembryogenic without band (N$^-$). The table also gives their expected frequencies in the F_1 population derived from a cross between A70-34 (AAaaBbbb) and Arrow36 (aaaabbbb) alfalfa when the RAPD markers are independent or linked to either of the two loci controlling the embryogenesis trait. It is evident that if there is any association between a RAPD marker and the embryogenic trait, the ratio of E$^+$:N$^+$:E$^-$:N$^-$ F_1s should be significantly different from 5:7:5:7. For the 19 RAPD markers that were analyzed, only marker A171-19 appeared to be linked to the somatic embryogenesis trait (Figure 2). The observed distribution of individuals among the four phenotypic classes was 24:17:10:32 for E$^+$:N$^+$:E$^-$:N$^-$, respectively, which was significantly different (at the 5% level) from the expected distribution of 16.8:24.2:17.2:24.8 for the 83 individuals tested.

From Table 4 it is evident that for a RAPD marker linked to the A locus, the expected ratio between F_1s that are embryogenic with a marker (E$^+$) to F_1s that are nonembryogenic with the marker (N$^+$) is 1:1, whereas a 5:1 ratio is expected for the same phenotypes if the marker is linked to the B locus. The observed ratio of 24:17 (E$^+$:N$^+$) was not significantly different from 1:1 but was significantly different from 5:1, indicating that marker A171-19 is linked to the A locus.

2. Estimation of Recombination Fraction

To estimate the linkage distance (p) between the embryogenesis gene A and marker A171-19, the number of individuals in four categories with respect to the gene (A) and the marker (M), namely, dominant-dominant (A—M—), dominant-recessive (A—mmmm), recessive-dominant (aaaaM—) and recessive-recessive (aaaammmm) must be determined. This is not possible to do directly because the somatic embryogenesis trait in alfalfa is controlled by two dominant and complementary genes (A and B). That is, the expression of gene A cannot be observed when gene B is absent. Consequently, the group of nonembryogenic F_1s contained a mixture of individuals with AAaabbbb, Aaaabbbb, aaaaBbbb, and aaaabbbb genotypes (Table 4) which could not be partitioned. However, by screening more primers in the fashion

TABLE 4.
Expected Genotypic and Phenotypic Frequencies in an F_1 Population Derived from a Cross Between A70-34 (AAaaBbbb) and Arrow36 (aaaabbbb) Alfalfa When the Marker (M) Is Independent or Linked to the A or B Gene Controlling the Somatic Embryogenesis Trait

Phenotypes and genotypes in F_1	Phenotypic frequencies		
	M not linked ($p = 0.5$)	M linked to A ($p = 0$)	M linked to B ($p = 0$)
Embryogenic with band (E⁺)[a]			
AAaaBbbbMmmm	1/24	2/24	2/24
AaaaBbbbMmmm	4/24	4/24	8/24
(Class total)	(5/24)	(6/24)	(10/24)
Embryogenic without band (E⁻)[a]			
AAaaBbbbmmmm	1/24	0	0
AaaaBbbbmmmm	4/24	4/24	0
(Class total)	(5/24)	(4/24)	(0)
Nonembryogenic with band (N⁺)[a]			
AAaabbbbMmmm	1/24	2/24	0
AaaabbbbMmmm	4/24	4/24	0
(Subclass total)	(5/24)	(6/24)	(0)
aaaaBbbbMmmm	1/24	0	0
aaaabbbbMmmm	1/24	0	2/24
(Subclass total)	(2/24)	(0)	(2/24)
(Class total)	(7/24)	(6/24)	(2/24)
Nonembryogenic without band (N⁻)[a]			
AAaabbbbmmmm	1/24	0	2/24
Aaaabbbbmmmm	4/24	4/24	8/24
(Subclass total)	(5/24)	(4)	(10)
aaaaBbbbmmmm	1/24	2/24	0
aaaabbbbmmmm	1/24	2/24	2/24
(Subclass total)	(2/24)	(4)	(2)
(Class total)	(7/24)	(8/24)	(12/24)

[a] "E⁺, N⁺, E⁻, and N⁻ are embryogenic with band, nonembryogenic with band, embryogenic without band, and nonembryogenic without band, respectively.

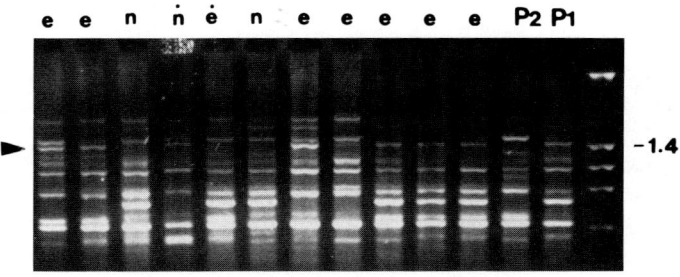

Figure 2. Cosegregation between the somatic embryogenesis trait and marker A171-19 in an F_1 population derived from a cross between embryogenic (A70-34) and nonembryogenic (Arrow36) parents; n = nonembryogenic, e = embryogenic. PCR reaction conditions as described in Figure 1. Lanes labeled with ė represent a recombinant. Lanes labeled with ṅ may or may not represent a recombinant (see text).

described above, it is highly likely that a marker linked to the B gene for somatic embryogenesis can also be identified. After both loci are tagged with tightly linked RAPD markers, it will be possible to identify the genotypic classes and calculate the recombination fractions. When this has been achieved, a molecular screen for embryogenic plants in a plant breeding program will be possible.

In general, the results obtained with alfalfa illustrate the utility of RAPD markers for linkage studies in plants with complex modes of inheritance.

III. THE USE OF RAPD MARKERS TO DETERMINE SIMILARITIES AND GENETIC DISTANCES

Information about germplasm diversity and relationships among elite breeding materials is fundamental in plant breeding.[14] This is especially true for a species like alfalfa (*Medicago sativa* ssp.) which is an outcrosser and suffers severe inbreeding depression. Ideally, methods for elucidating genetic relationships in alfalfa should be based on comparisons of plants for monogenic traits the expression of which is not affected by plant development or growth environment. Biochemical methods, such as isozyme analysis, have been used for this purpose but they are limited by the small number of isozyme markers that are available, a general lack of polymorphism for these traits in elite breeding materials, and variability in the electrophoretic patterns that is due to plant development.[15-17] DNA-based methods of identity analysis, like restriction fragment length polymorphism (RFLP) determinations,[18] are generally unaffected by plant development, but their detection is expensive, time consuming, and technically complex. Recently, RAPD markers were used to determine genetic relationships among peanuts (*Arachis hypogaea* L.)[19] and beans.[20] The major advantages of the RAPD analysis approach are that (1) prior DNA sequence information is not required, (2) the laboratory manipulations are simple to perform, (3) large numbers of samples can be processed quickly, and (4) no radioactive reagents are utilized in the assay.

A. DETERMINATION OF SIMILARITY AMONG INDIVIDUALS

Molecular markers provide a virtually limitless set of descriptors with which to compare individual plants. The degree of similarity can be quantified by determining the fraction of shared markers in the total number of markers that have been compared. A comparison of two individuals with respect to the presence or absence of a particular RAPD marker leads to four possible outcomes, or contingencies: present in genotype 1 and present in genotype 2 (1,1); present in genotype 1 and absent in genotype 2 (1,0); absent in genotype 1 and present in genotype 2 (0,1); or absent in genotype 1 and absent in genotype 2 (0,0). Skroch et al.[20] suggested that all four contingencies provide nearly equivalent information and described a simple computational method for determining genetic distance using RAPD data. According to their method, each comparison is given a value of 1 for a difference and 0 for a similarity, and the genetic distance is equal to the numerical mean of the set of observations (Figure 3). Calculating genetic distances in this manner allows their associated variances and standard errors to also be calculated and tests of the significance of differences among genetic distances among pairs of genotypes to be performed.

B. DETERMINATION OF SIMILARITY AMONG POPULATIONS

Many applications of similarity testing require comparisons to be made among populations of plants. This is the case, for example, when comparisons among heterogeneous varieties or natural populations of plants are being made. In these cases a single plant cannot represent the complete spectrum of genetic diversity that exists in the population, and the RAPD patterns of a number of plants from the population need to be determined. One way to sample the

a)

RAPD marker numbers	Genotype A	Genotype B
1	—	
2	—	
3		—
4	—	
5		—
6	—	—

b)

RAPD marker number	Genotype A contingency condition	Genotype B contingency condition	|A-B|
1	1	1	0
2	1	0	1
3	0	1	1
4	1	0	1
5	0	1	1
6	1	1	0
Σ|A-B|			4

c) Genetic Distance (A,B)

$$= \frac{\sum_{i=n}^{i=1} |A_i - B_i|}{n}$$

$$= \frac{4}{6}$$

$$= 0.67$$

Figure 3. Method for determining genetic distances from a comparison of RAPD markers. (a) Hypothetical RAPD patterns obtained from two genotypes. (b) Contingency conditions for genotypes A and B based on the patterns shown in (a). (c) Genetic distance calculation.

genetic variability in a population is to bulk genomic DNA samples obtained from several individuals in the population and to use this mixture as the template DNA in the PCR. Bulked DNA samples from segregating populations have been used as templates to rapidly identify RAPD markers linked to disease-resistance genes in lettuce.[21] Comparisons among RAPD patterns obtained from bulked samples containing DNA from 7, 10, or 15 individuals of the alfalfa breeding population V0 or P3 indicated that, in most cases, the RAPD patterns obtained from the bulks of seven individuals could represent the patterns seen in the entire population (Figure 4).

When using bulked DNA samples to represent a population the genetic distance (D) between two populations can be calculated from RAPD patterns of bulked DNA samples using[22]

$$D = -\ln(F) \qquad (8)$$

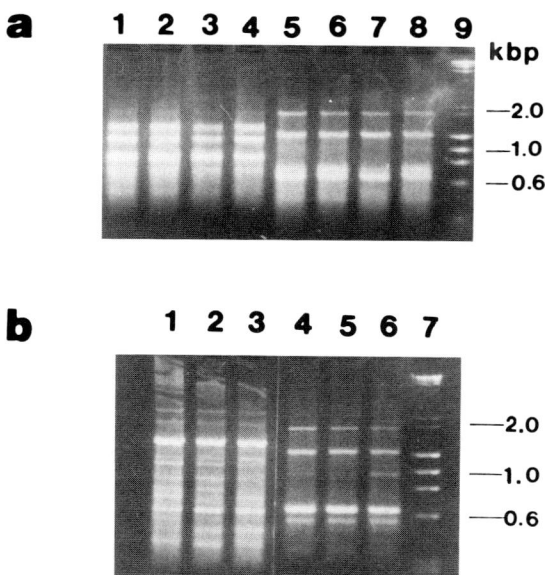

Figure 4. RAPD patterns obtained from bulked genomic DNA samples containing DNA from different numbers of plants from one alfalfa population: lanes 1a, 5a, 1b, and 4b from 15 plants; lanes 2a, 6a, 2b and 5b from ten plants; lanes 3a and 7a from seven plants; and lanes 4a, 8a, 3b, and 6b from another random sample of seven plants. The patterns in lanes 1a to 4a, lanes 5a to 8a, lanes 1b to 3b, and lanes 4b to 6b, respectively, were obtained using primers with the following sequences: 5'GTGCGTCGCT3', 5'GCTTGTGAAC3', 5'GATCTCAGCG3', 5'GTGCGTCCTC3'. PCR conditions were as described in Figure 1.

where F is an estimate of similarity, which is based on the fraction of shared RAPD markers between populations; it can be calculated with the formula[23]

$$F = 2X_{1,2}/X_1 + X_2 \qquad (9)$$

where $X_{1,2}$ is the number of amplified DNA fragments with the same molecular weight found in both populations, X_1 is the total number of fragments found in one population, and X_2 is the total number found in the other.

To test the usefulness of the RAPD patterns from bulked DNA samples for determining genetic relatedness, the F values were calculated for comparisons among three cultivars (Dupuits, Peace, and Anik) and a breeding population (V0). The relationships among these populations are known, and Figure 5 shows that the genetic distance values obtained from the F values [$D = -\ln F$] reflect the fact that Anik (which is 100% *Medicago falcata*) is a greater distance from Dupuits (which is 100% Flemish *M. sativa*) than Peace or V0 (which have both *M. falcata* and *M. sativa* in their background).

The test of the procedure was extended to comparisons among breeding populations in *M. sativa*. Figure 6 shows that the percentage of shared bands in bulked RAPD profiles for related populations were significantly higher than for unrelated populations. In particular, the comparisons V0/V3 and P0/P3 had approximately 88% shared bands compared to 75% for the V0/P0, V3/P0, and V0/P3 comparisons. Thus, the results of these comparisons indicate that the RAPD patterns from bulked DNA samples can be analyzed to estimate relatedness among alfalfa populations. Previous estimates of relatedness have been based on comparisons of RFLP or RAPD patterns obtained from individual plant, animal or fungal samples.[19,23-25] The intra-cultivar variability seen in alfalfa makes a comparison based on analysis of individuals an arduous task. Bulking samples from many individuals in a population allows the workload

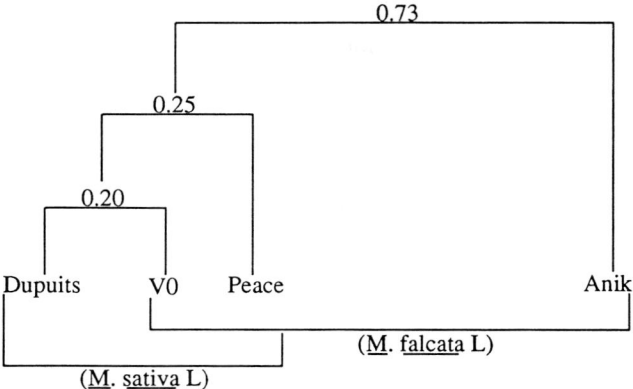

Figure 5. Phylogeny for three cultivars (Dupuits, Peace, and Anik) and one breeding population (V0) based on genetic distances calculated from RAPD patterns obtained from bulked samples containing DNA from seven plants per cultivar or population and ten different PCR primers. The known sources of the materials are indicated on the bottom.

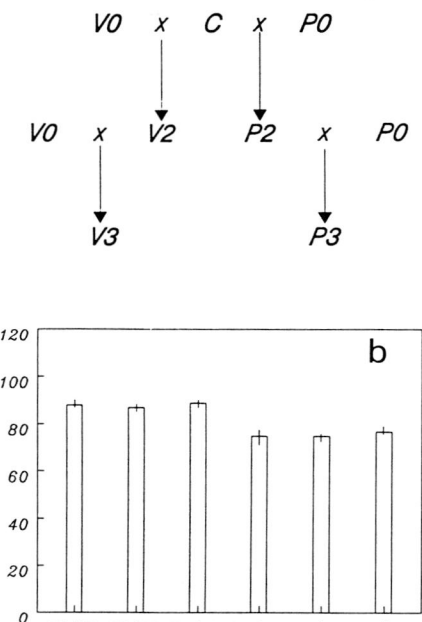

Figure 6. Relationships among four breeding populations of alfalfa (V0, V3, P0, P3). (a) The common parent in the crosses P0 × C and V0 × C is A70-34 alfalfa, an embryogenic genotype selected from Rangelander (D.C.W. Brown, Agriculture Canada). V3 and P3 were screened for their ability to produce somatic embryos.[27] (b) Comparisons of percent shared bands (%SB) in RAPD patterns between pairs of alfalfa breeding populations (V0, V3, P0, P3). Genomic DNAs from seven plants per population were bulked, and separate PCR reactions were primed with 20 different 10-mers. The mean of %SB between two populations and their standard errors are shown. The values for comparisons V0/V3, P0/P3, and V3/P3 were significantly different from the V0/P3, V3/P0, and V0/P0 comparisons; according to paired t-tests at the 5% level.

to be decreased substantially. The bulking procedure should also be suited to comparisons among wild species represented by several accessions or heterogeneous natural populations.

Questions of relatedness arise in plant breeding when parents for crosses are being selected. A potential application of the analysis described above in alfalfa breeding (or breeding of any other species that suffers inbreeding depression) would be to determine the relatedness among potential parents so that combinations that maximize genetic distance and heterosis can be made. This suggestion is supported by previous findings that RFLP-based genetic distances between maize inbreds are good predictors of heterosis in their F_1 hybrids.[26]

ACKNOWLEDGMENTS

Assistance with the preparation of the manuscript by Jennifer Kingswell is gratefully acknowledged. The work performed in the authors' laboratory was supported financially by the Ontario Ministry for Food and Agriculture.

REFERENCES

1. **Williams, J. G. K, Kubelik, A. R., Livak, K. J., Rafalski, J. A., and Tingey, S. V.**, DNA polymorphisms amplified by arbitrary primers are useful as genetic markers, *Nucleic Acids Res.,* 18, 6531, 1990.
2. **Welsh, J., Honeycutt, R. J., McClelland, M., and Sobral, B. W. S.**, Parentage determination in maize hybrids using the arbitrarily primed polymerase chain reaction (AP-PCR), *TAG,* 82, 473, 1991.
3. **Caetano-Anolles, G. C., Bassam, B. J., and Gresshoff, P. M.**, DNA amplification fingerprinting using very short arbitrary oligonucleotide primers, *Biotechnology,* 9, 553, 1991.
4. **Quiros, C. F. and Bauchan, G. R.**, The genus *Medicago* and the origin of the *Medicago sativa* complex, in *Alfalfa and Alfalfa Improvement,* Hanson, A. A., Barnes, D. K., and Hill, R. R., Jr., Eds., American Association of Agronomy, 1988, 93.
5. **Echt, C. S., Erdahl, L. A., and McCoy, T. J.**, Genetic segregation of random amplified polymorphic DNA in diploid cultivated alfalfa, *Genome,* 35, 84, 1991.
6. **Yu, K. F.**, The Application of Random Amplified Polymorphic DNA (RAPD) Markers to Determine Relatedness and Genetic Markers in Alfalfa, Ph.D. thesis, University of Guelph, Guelph, Ontario, Canada, 1992.
7. **McCoy, T. J. and Bingham, E. T.**, Cytology and cytogenetics of alfalfa, in *Alfalfa and Alfalfa Improvement,* Hanson, A. A., Barnes, D. K., and Hill, R. R., Jr., Eds., American Association of Agronomy, 1988, 737.
8. **Mather, W. B.**, *Principles of Quantitative Genetics,* Burgess Publishing, Minnesota, 1965.
9. **Saunders, J. W. and Bingham E. T.**, Production of alfalfa plants from callus tissue, *Crop Sci.,* 12, 804, 1972.
10. **Wan, Y., Sorenson, E. L., and Liang, G. H.**, Genetic control of *in vitro* regeneration in alfalfa (*Medicago sativa* L.), *Euphytica,* 39, 3, 1988.
11. **Senaratna, T., McKersie, B. D., and Bowley, S. R.**, Artificial seeds of alfalfa (*Medicago sativa* L.). Induction of desiccation tolerance in somatic embryos, *In Vitro Cell Dev. Biol.,* 26, 85, 1990.
12. **Kielly, G. A. and Bowley, S. R.**, Genetic control of somatic embryogenesis in alfalfa, *Genome,* 35, 474, 1992.
13. **Martin, G. B., Williams, J. G. K., and Tanksley, S. D.**, Rapid identification of markers linked to a *Pseudomonas* resistance gene in tomato by using random primers and near-isogenic lines, *Proc. Natl. Acad. Sci. U.S.A.,* 88, 2336 1991.
14. **Hallauer, A. R. and Miranda, J. B.**, Eds., *Quantitative Genetics in Maize Breeding,* 2nd ed., Iowa State University Press, Ames, 1988.
15. **Tanksley, S. D.**, Molecular markers in plant breeding, *Plant Mol. Biol. Rep.,* 1, 3, 1983.
16. **Tanksley, S. D. and Orton, T. J.**, *Isozymes in Plant Genetics and Breeding,* Parts A and B, Elsevier, Amsterdam, 1983.
17. **Tanksley, S. D., Young, N. D., Paterson, A. H., and Bonierbale, M. W.**, RFLP mapping in plant breeding: new tools for an old science, *BioTechnology,* 7, 257, 1989.
18. **Apuya, N. R., Frazier, B. L., Keim, P., Roth, E. J., and Lark, K. G.**, Restriction fragment length polymorphisms as genetic markers in soybean *Glycine max* (L.) merrill, *Theor. Appl. Genet.,* 75, 889, 1988.
19. **Halward, T., Stalker, T., LaRue, E., and Kochert, G.**, Use of single-primer DNA amplifications in genetic studies of peanut (*Arachis hypogaea* L.), *Plant Mol. Biol.,* 18, 315, 1992.
20. **Skroch, P., Tivang, J., and Nienhaus, J.**, Analysis of genetic relationships using RAPD marker data, in J. Plant Breed. Symp. Applications of RAPD Technology to Plant Breeding, Minneapolis, 1992, 26.

21. **Michelmore, R. W., Paran, I., and Kesseli, R. V.,** Identification of markers linked to disease-resistance genes by bulked segregant analysis: a rapid method to detect markers in specific genomic regions by using segregating populations, *Proc. Natl. Acad. Sci. U.S.A.,* 88, 9828, 1991.
22. **Swafford, D. L. and Olson, G. J.,** Phylogeny reconstruction, in *Molecular Systematics,* Hillis, D. M. and Moritz, C., Eds., Sinauer Associates, Sunderland, MA, 1990, chap. 11.
23. **Packer, C., Gilbert, D. A., Pusey, A. E., and O'Brien, S. J.,** A molecular genetic analysis of kinship and cooperation in African lions, *Nature,* 351, 562, 1991.
24. **Goodwin, P. H. and Annis, S. L.,** Rapid identification of genetic variation and pathotype of *Leptosphaeria maculans* by random amplified polymorphic DNA assay, *Appl. Environ. Microbiol.,* 57, 2482, 1991.
25. **Nei, M.,** *Molecular Evolutionary Genetics,* Columbia University Press, New York, 1987.
26. **Smith, O. S., Smith, J. S. C., Bowen, S. L., Tenborg, R. A., and Wall, S. J.,** Similarities among a group of elite maize inbreds as measured by pedigree, F_1 grain yield, grain yield, heterosis, and RFLPs, *Theor. Appl. Genet.,* 80, 833, 1990.
27. **Bowley, S.,** personal communication.

Chapter 24

DNA RECOMBINATION IN THE COURSE OF PCR

Andrey B. Vartapetian

TABLE OF CONTENTS

I. Introduction 215

II. Materials and Methods 215
    A. Generation of a Set of Truncated Primers 215
    B. Performance of PCR 216
    C. Cloning of PCR Products 216

III. Results and Discussion 216

IV. Conclusion 218

Acknowledgments 218

References 218

I. INTRODUCTION

One of the advantages of PCR which makes it the method of choice for DNA amplification is the ability to introduce the desired alterations in the amplified sequence. In the most common case, these are nucleotide additions and substitutions introduced in the primers. In the more complicated versions, which utilize two- or even three-stage PCR, creation of deletions, insertions, and formation of chimeric DNA molecules is feasible.[1]

Here I describe a spontaneous rearrangement of the amplified sequence in the course of PCR resulting in duplication of the terminal portion of the amplified DNA fragment due to the presence of two imperfect direct repeats in the sequence under study.

II. MATERIALS AND METHODS

A. GENERATION OF A SET OF TRUNCATED PRIMERS

The rationale for this step is as follows. The nucleotide sequence of the primers for amplification of rat prothymosin α cDNA was derived from human prothymosin α cDNA, which is very similar, yet not identical, to the corresponding region of rat sequence.[2] To better fit the rat prothymosin α cDNA sequence, the upstream 20-nucleotide-long primer was statistically truncated from its 3' end by snake venom phosphodiesterase treatment to generate a mixture of oligonucleotides 10 to 20 bases in length having uniform 5' ends.

To this end, 10 μg of the primer dATGTCAGACGCAGCCGTAGA were dissolved in 150 μl of 1×SVPD buffer (20 mM Tris-acetate, pH 8.7, 20 mM MgCl$_2$) and incubated with 2 μg of snake venom phosphodiesterase (Sigma) at 37°C for 10 min. The reaction was stopped by addition of EDTA to 25 mM with subsequent phenol-chloroform extraction. Sodium acetate (pH 5.2) and 5 vol of ethanol were added to ensure precipitation of the rather short oligonucleotides.

B. PERFORMANCE OF PCR

The reaction mixture for PCR contained, in a final volume of 100 μl containing 1xPCR buffer[3] (50 mM KCl, 20 mM Tris-HCl, pH 8.4, 2.5 mM MgCl$_2$, 0.1 mg/ml BSA), 1 ng pRSpro-187 plasmid bearing a rat prothymosin α cDNA, 0.2 mM of each dNTP, 0.2 μg of the 20-nucleotide-long downstream primer dTAGTCATCCTCGTCGGTCTT, 1 μg of the truncated upstream primer, and 2 U of *Taq* polymerase (Fermentas). Forty cycles of PCR were performed on the Perkin-Elmer Cetus DNA Thermal Cycler with the following temperature profile: denaturation at 94°C for 1 min; annealing temperature was increased from the initial 48°C by 1°C per cycle until it reached 55°C, for 2 min; extension at 72°C for 2 min, except for the final cycle, where extension proceeded for 5 min.

C. CLONING OF PCR PRODUCTS

After completion of PCR, an aliquot of the reaction mixture was withdrawn and analyzed by a 1.5% agarose gel electrophoresis. A band of 420 bp was excised, DNA eluted, and treated with polynucleotide kinase and ATP as described by Maniatis et al.[4]

Cloning of PCR products sometimes presents a problem. Here, the blunt-ended 420-bp fragment was cloned into *Sma*I-site of pUC19 vector by an approach that has been very successful for cloning of PCR products. After completion of the ligation step, DNA ligase was inactivated by heating at 65°C for 10 min, and DNA was precipitated by addition of sodium acetate (pH 5.2) and 2.5 vol of ethanol. The pellet was dissolved in 20 μl of *Sma*I buffer, 0.5 U of *Sma*I restriction endonuclease were added, and the reaction mixture was incubated at 37°C for 1 h followed by heat inactivation of the enzyme at 65°C for 15 min. An aliquot of the digest was used directly for transformation of the JM109 strain of *Escherichia coli*. Redigestion with *Sma*I endonuclease linearized self-ligated vector DNA but left vector molecules containing inserts in the circular form, which is far more efficient in cell transformation. This approach is a convenient alternative to the widely used procedure of dephosphorylation of vector; it avoids introduction of restriction endonuclease recognition sites in primers and permits the cloning of even trace amounts of PCR products. This approach, of course, cannot be employed when the PCR product itself contains a recognition site for *Sma*I, yet taking in account the rather small length of the amplified DNA, SmaI sites should occur infrequently in them.

The clones with the insert were picked up, and DNA sequence of PCR product was established by dideoxy sequencing.

III. RESULTS AND DISCUSSION

Amplification of the DNA segment encompassing the whole rat prothymosin α-coding region resulted in formation, alongside with large quantities of a DNA fragment of estimated size (340 bp), a minor amount of a DNA fragment of greater length (420 bp). Sequence analysis of this 420-bp product revealed that it possesses an 80-bp duplication of the N terminal prothymosin α-coding sequence. Such a duplication could have arisen through the process depicted in Figure 1, where parts of the direct repeat of the upstream primer sequence (A) are also present in the target molecule. This would result in annealing of the very 3′ terminus of minus strand (A′) to the repetitive sequence (designated *a* in Figure 1; see structure 3). Its subsequent elongation should lead to the formation of a DNA strand with a duplication of the region positioned between the annealing site of the upstream primer and the repetitive *a* sequence. The DNA structure thus formed (structure 4 in Figure 1) can be further amplified by PCR, yielding a 420-bp double-stranded DNA molecule (structure 5).

It should be mentioned that formation of structure 3 (Figure 1) is likely to occur only at an early stage of PCR (when a low annealing temperature is used) since it should be rather unstable due to potential mismatches between *a* and the upstream primer sequences.

DNA Recombination in the Course of PCR

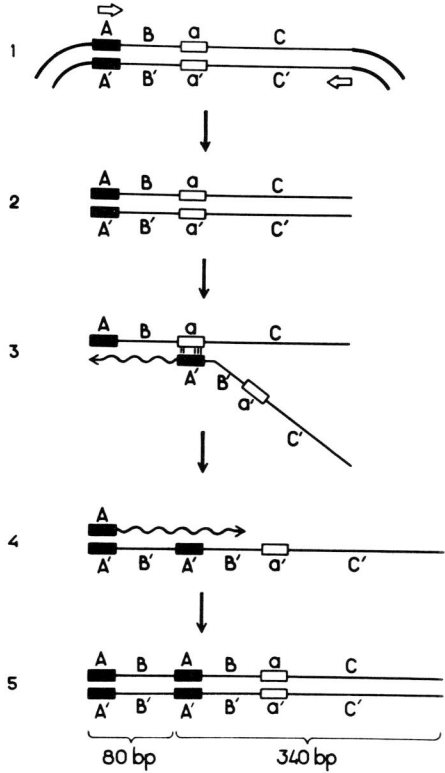

Figure 1. A scheme of DNA recombination in the course of PCR leading to duplication of the terminal portion of the amplified DNA fragment. Primers are shown by empty arrows. a denotes an imperfect direct repeat of A.

Closer inspection of the prothymosin α-coding sequence revealed an imperfect direct repeat of a portion of the upstream primer sequence at the predicted position, 80 bp apart (see Figure 2a). An important difference between the predicted and actual structures of the amplified DNA molecules was observed at the site of junction of the duplicated domains: instead of the tetranucleotide TCTC, a dinucleotide AA was present (see Figure 2b). It can be hypothesized that the two A residues were added to the 3′ terminus of the newly synthesized minus strand by *Taq* polymerase in the non-template manner at stage 2 (Figure 1). How these A nucleotides could substitute for TCTC at the elongation step (Figure 1, structure 3) is unclear. Whether this is due to a-A′ duplex defects, is not known.

Since the regions of human and rat prothymosin α cDNA corresponding to the upstream primer differ in a number of positions, the length of the truncated upstream primer utilized at different steps of PCR could be derived from the sequence of the PCR product. It turned out that an upstream primer of no more than 14 nucleotides in length was recruited at stage 1 (Figure 1) and subseqently inserted into the beginning of the second duplicated domain. In contrast, an 18- to 20-nucleotide primer was utilized at the later stage (stage 4, Figure 1) and is located at the terminus of the 420-bp fragment. This is in accord with the temperature profile of PCR employed (gradual increase of the annealing temperature during the first eight cycles). It should be mentioned, however, that the calculated T_m for the 14-mer primer is below the initial annealing temperature used. It appears that *Taq* polymerase is capable of trapping and extending even such unstable duplexes. This is also true for the postulated duplex a-A′ formed at stage 3 (Figures 1 and 2b), which should be very easy to melt at the annealing temperature employed. In the latter case, the possibility that random interactions of the preceding part of

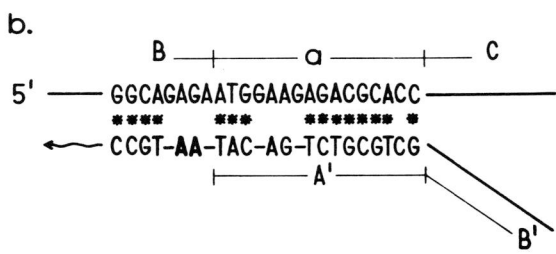

Figure 2. (a) Sequence similarity of *A* (an upstream primer) and *a* (an imperfect direct repeat); (b) The observed difference between the expected and actual sequences of the recombinant DNA (a detailed description of structure 3, Figure 1). The two *A* residues inserted instead of TCTC and subsequently amplified by PCR are shown in bold type.

the invading molecule with the 3' end of the positive strand provide some additional stability to this duplex cannot be excluded.

IV. CONCLUSION

DNA recombination in the course of PCR described here deserves some comments. On the one hand, it may be regarded as an artifact of PCR. It is more interesting, however, to exploit this property of PCR for constructing DNA molecules with duplications of interest. The only requirement is that they should contain an imperfect direct repeat of the terminal sequence, even of limited similarity. Besides, DNA recombination in the reaction tube reflects that in the living cell and is therefore of interest in itself.

ACKNOWLEDGMENTS

I thank Drs. B. Horecker and M. Clinton for providing pRSpro-187 plasmid.

REFERENCES

1. **Higuchi, R.,** Using PCR to engineer DNA, in *PCR Technology: Principles and Applications for DNA Amplification,* Erlich, H., Ed., Stockton Press, New York, 1989, chap. 6.
2. **Frangou-Lazaridis, M., Clinton, M., Goodall, G. J., and Horecker, B. L.,** Prothymosin α and parathymosin: amino acid sequences deduced from the cloned rat spleen cDNAs, *Arch. Biochem. Biophys.,* 263, 305, 1988.
3. **Kawasaki, E. S. and Wang, A. M.,** Detection of gene expression, in *PCR Technology: Principles and Applications for DNA Amplification,* Erlich, H., Ed., Stockton Press, New York, 1989, chap. 8.
4. **Maniatis, T., Fritsch, E. F., and Sambrook, J.,** *Molecular Cloning: A Laboratory Manual,* Cold Spring Harbor Laboratory Press, Cold Spring Harbor, NY, 1982, 124.

Chapter 25

THERMOSTABLE DNA POLYMERASES FOR *IN VITRO* DNA AMPLIFICATIONS

Asim K. Bej and Meena H. Mahbubani

TABLE OF CONTENTS

I. Introduction 221

II. DNA Polymerase from *Thermus aquaticus* (Taq™) 221
    A. Isolation 221
    B. Biochemical Characteristics 221
        1. Nuclease Activity 223
        2. Template 223
        3. Inhibitors 223
        4. Thermostability 223
        5. Reaction Parameters 223
    C. Applications 224

III. DNA Polymerase from *Thermus aquaticus* (Stoffel Fragment™ of Taq™) 224
    A. Isolation 224
    B. Biochemical Characteristics 224
        1. Nuclease Activity 224
        2. Thermostability 224
        3. Reaction Parameters 224
    C. Applications 224

IV. DNA Polymerase from *Thermus thermophilus* (Tth™) 225
    A. Isolation 225
    B. Biochemical Characteristics 225
        1. Nuclease Activity 225
        2. Template 225
        3. Thermostability 225
        4. Reaction Parameters 225
    C. Applications 226

V. DNA Polymerase from *Thermococcus litoralis* (Tli/Vent™ DNA Pol) 226
    A. Isolation 226
    B. Biochemical Characteristics 226
        1. Nuclease Activity 226
        2. Thermostability 227
        3. Reaction Parameters 227
    C. Applications 227

VI. DNA Polymerase from *Pyrococcus furiosus* (pfu™) 227
    A. Isolation 227
    B. Biochemical Characteristics 228
        1. Nuclease Activity 228

   2. Thermostability 228
   3. Reaction Parameters 228
  C. Applications 228

VII. DNA Polymerase from *Bacillus* spp. (*Bst*™) 228
  A. Isolation 228
  B. Biochemical Characteristics 229
   1. Nuclease Activity 229
   2. Inhibitors 229
   3. Thermostability 229
   4. Reaction Parameters 229
  C. Applications 229

VIII. DNA Polymerase from *Sulfolobus acidocaldarius* (*Sac*) 230
  A. Isolation 230
  B. Biochemical Characteristics 230
   1. Nuclease Activity 230
   2. Inhibitors 230
   3. Reaction Parameters 230
   4. Thermostability 230
  C. Applications 230

IX. DNA Polymerase from *Thermoplasma acidophilium* (*Tac*™ DNA Pol) 231
  A. Isolation 231
  B. Biochemical Characteristics 231
   1. Nuclease Activity 231
   2. Template 231
   3. Inhibitors 231
   4. Thermostability 231
   5. Reaction Parameters 231
  C. Applications 232

X. DNA Polymerase from *Thermus flavus* (*Tfl/Tub*™) 232
  A. Isolation 232
  B. Biochemical Characteristics 232
   1. Nuclease Activity 232
   2. Inhibitors 232
   3. Template 232
   4. Reaction Parameters 232
   5. Thermostability 232
  C. Applications 232

XI. DNA Polymerase from *Thermus ruber* (*Tru*) 233
  A. Isolation 233
  B. Biochemical Characteristics 233
   1. Nuclease Activity 233
   2. Inhibitors 233
   3. Template 233
   4. Reaction Parameters 233
   5. Thermostability 233
  C. Applications 234

XII. DNA Polymerase from *Thermotoga* sp. (*Tsp*) 234
    A. Isolation 234
    B. Biochemical Characteristics 234
        1. Nuclease Activity 234
        2. Inhibitors 234
        3. Reaction Parameters 234
        4. Thermostability 234
    C. Applications 235

XIII. DNA Polymerase from *Methanobacterium thermoautotrophicum* (*Mth*) 235
    A. Isolation 235
    B. Biochemical Characteristics 235
        1. Nuclease Activity 235
        2. Template 235
        3. Inhibitors 235
        4. Thermostability 235
        5. Reaction Parameters 235
    C. Applications 236

References 236

I. INTRODUCTION

The thermostable DNA polymerase is the key ingredient of the polymerase chain reaction (PCR) *in vitro* DNA amplification method. Early experiments used the thermolabile Klenow fragment, which was required to be added every cycle.[1] The introduction of the thermostable DNA polymerase from *Thermus aquaticus* (*Taq*™ DNA pol)[2-4] allowed the automation of the process. Since 1988, a host of thermostable DNA polymerases, each with characteristic properties, have been reportedly used for PCR. These include the enzyme from *Thermus thermophilus* (*Tth*™ DNA pol),[5,6] *Thermococcus litoralis* (*Tli*/Vent™ DNA pol),[7-9] *Pyrococcus furiosus* (*pfu*™ DNA pol),[10-14] *Bacillus stearothermophilus* (*Bst*™ DNA pol),[15,16] *Sulfolobus acidocaldarius* (*Sac* DNA pol),[17-19] *Thermoplasma acidophilum* (*Tac* DNA pol),[20] *Thermus flavus* (*Tfl*/Tub™ DNA pol),[21-23] *Thermus ruber* (*Tru* DNA pol),[24] *Thermotoga* sp. (*Tsp* DNA pol),[25] and *Methanobacterium thermoautotrophicum* (*Mth* DNA pol).[26] Some of the important properties and applications of these thermostable DNA polymerases are summarized in Table 1.

II. DNA POLYMERASE FROM *THERMUS AQUATICUS* (*TAQ*™)

A. ISOLATION
This is the most widely used PCR DNA polymerase. The *Taq*™ DNA pol I was isolated from *T. aquaticus* strain YT1,[3] a thermophilic eubacterium capable of growth at 70 to 75°C, which was isolated from a hot spring at Yellowstone National Park and first described in 1969.[27]

B. BIOCHEMICAL CHARACTERISTICS
The 62 to 68 kDa *Taq*™ DNA pol I was originally purified by a six-step column chromatography method.[28,3] Recently, the *Taq*™ DNA pol from the same source has been purified with a molecular weight of 94 kDa.[29] The reason for the two different molecular weights of these DNA pols isolated from the same bacterial source is not clear at this time. However, it

TABLE 1.
Comparisons of the Properties and Applications of Various Thermostable DNA Polymerases

Enzyme (DNA pol)	Source	Molecular weight (kDa)	Optimum range of reaction conditions				Exonuclease activity		General applications	Ref.
			$MgCl_2$ (mM)	KCl (mM)	pH	Temperature (°C)	$5' \rightarrow 3'$	$3' \rightarrow 5'$		
Taq™	*Thermus aquaticus*	94	2–4	50–55	7.8–9.4	70–80	Yes	No	PCR, cloning, direct sequencing, mutagenesis	2–4
Stoffel Fragment™ [a]	*Thermus aquaticus*	61.3	2–10	10	8.3	70–80	No	No	DNA sequencing, PCR of template with secondary structures, amplification of large quantity of PCR product, higher thermostability allows greater number of amplification cycles	29
Tth™ [a]	*Thermus thermophilus*	110–120	1.5–2.5	100	8–9.3	50–60	Yes	No	PCR from mRNA template with high sensitivity (Mn^{2+} is required for reverse transcription), cloning genes from cellular and viral RNA, PCR detection from blood samples	5,6
Tli/Vent™ [b]	*Thermococcus litoralis*	92–97	2–8	0–50	8–9	72–80	No	Yes	PCR DNA synthesis with high fidelity, ligation of blunt-end DNA fragments, DNA sequencing, gene cloning, and site-directed mutagenesis	7–9
pfu™ [c]	*Pyrococcus furiosus*	92	1.5–8	10	8–9	70–80	Yes	Yes	PCR DNA synthesis with high fidelity, cloning, sequencing, and site-directed mutagenesis	10–14
Bst™ [d]	*Bacillus stearothermophilus*	95	10–30	100–200	8–9	60–65	No	No	PCR and general sequencing with high sensitivity	15,16
Sac	*Sulfolobus acidocaldarius*	100	2–8	ND	7–8	70–80	No	Yes	DNA sequencing, site-directed mutagenesis, gene fusions by PCR	17–19
Tac	*Thermoplasma acidophilum*	88	2–4	ND [c]	8–9	65	No	Yes	PCR for A-T rich templates in shorter cycling time	20
Tfl/Tub™ [e]	*Thermus flavus*	66	10–15	5–10	7–8	70 [a] 50 [a]	No	No	PCR amplification of large-template DNA (>10 kb)	21–23
Tru	*Thermus ruber*	70	2–3	15	7–12	50	No	No	Potential for PCR amplification of contaminated samples, DNA sequencing	24
Tsp	*Thermotoga sp.*	85	10	10	7.5–8	80	No	No	Potential for PCR amplification	25
Mth	*Methanobacterium thermoautotrophicum*	72	10–20	100–300	7–9	65	Yes	Yes	Potential for DNA sequencing, gene cloning, site-directed mutagenesis, PCR amplification of A-T rich DNA	26

[a] Taq™, Tth™, and Stoffel Fragment™ DNA pols are registered trademarks of Perkin Elmer Corporation, 761 Main Ave., Norwalk, CT 06859.
[b] Vent™ DNA pol is a registered trademark of New England Biolabs, Inc., 32 Tozer Rd. Beverly, MA 01915-5599.
[c] pfu™ DNA pol is a registered trademark of Stratagene Cloning Systems, Inc., 11099 N. Torrey Pines Rd, La Jolla, CA 92037.
[d] Bst™ DNA pol is a registered trademark of BioRad Laboratories, 3300 Regatta Blvd., Richmond, CA 94804.
[e] Tfl/Tub™ DNA pol is a registered trademark of Promega Corporation, 2800 Woods Hollow Road, Madison, WI 53711-5399.
[f] ND = not determined.
[g] DNA template.
[h] RNA template.

is predicted that the Taq™ pol I may be the product of a distinct gene (or genes) to provide a 94-kDa fragment or that the two DNA polymerases are partially purified proteolytic degradation fragments of the same translation product.[29]

1. Nuclease Activity

The native Taq™ pol I does not have any 3' → 5' exonuclease activity. It creates both single base substitution errors and transitions, transversions, frameshifts, and/or deletion mutations during in vitro DNA synthesis.[30] However, the native Taq™ pol I enzyme possesses a polymerization-dependent 5' → 3' exonuclease activity.[29]

2. Template

The Taq™ DNA pol I enzyme has been shown to be active when activated double- or single-stranded DNA, poly(dA)-poly(dT), poly(dA)-oligo(dT)$_{10}$, and poly(rA)-oligo(dT)$_{10}$ are used as templates, and to be inactive on the native and denatured DNAs as well as on the native molecules of RNA and poly(rC)-oligo(dG)$_{12-180}$.[3,29]

3. Inhibitors

Presence of a high (2 M) concentration of urea inhibits polymerase activity to 18%, 20% dimethylsulfoxide to 89%, 20% dimethylformamide to 83%, 20% formamide to 51%, and 0.1% SDS to 99.9%.[31]

4. Thermostability

The Taq™ DNA pol has an apparent T_{opt} of 75 to 80°C with a K_{cat} approaching 150 nucleotides/s/enzyme molecule.[31] It has an extension rate of >60 nt/s at 70°C, 24 nt/s at 55°C, 1.5 nt/s at 37°C, and ~0.25 nt/s at 22°C for a GC-rich 30-mer primer on M13. Little DNA synthesis is seen at temperatures above 90°C in vitro, which may be limited by the stability of the primer-template duplex. However, the enzyme is very stable at high temperatures. It retains 50% of its activity after 130, 40, and 9 min at 92.5, 95, and 97.5°C, respectively.[29]

5. Reaction Parameters

Taq™ pol activity is sensitive to the concentration of Mg^{2+}. In a standard 10-min assay, using minimally activated salmon sperm DNA, 2.0 mM $MgCl_2$ maximally stimulates Taq™ pol activity at 0.7 to 0.8 mM total dNTP. The polymerase activity is inhibited at higher Mg^{2+} concentrations, with 40 to 50% inhibition at 10 mM $MgCl_2$. The Mg^{2+} concentration required for maximum activation of the enzyme depends on the dNTP concentration, since dNTPs bind Mg^{2+}. Also, at the optimal Mg^{2+} concentration, the synthesis rate of Taq™ polymerase decreases by as much as 20 to 30% as the total dNTP concentration is increased to 4 to 6 mM.[29] Although Mg^{2+} is the preferred cofactor, substitution of Mg^{2+} with 0.6 mM $CoCl_2$ and 0.6 mM $MnCl_2$ showed 80% of the maximal activity stimulated by the optimum concentration of Mg^{2+}.[29]

Low and balanced concentrations of dNTPs give increased specificity and fidelity as well as satisfactory yield of PCR products. Such concentrations also facilitate labeling of PCR products with radioactive or biotinylated precursors. Very low dNTP concentrations may adversely affect the processivity of Taq™ DNA pol. The optimum concentration of each of the dNTPs is 200 µM.[32] Modest concentrations of a monovalent cation such as KCl stimulate the synthesis rate of Taq™ pol by 50 to 60% with an apparent optimum at 50-55 mM. The activation of the Taq™ DNA pol activity has been found to be template dependent.[29] The presence of >200 mM KCl inhibits the enzyme activity.[29]

The optimum polymerization temperature for the in vitro DNA-amplification method is 70 to 72°C. The optimum pH range for the Taq™ DNA pol I is 7.8 to 9.4.[29]

The addition of a low quantity of gelatin, bovine serum albumin, NaCl or non-ionic detergent such as NP40 or Triton X-100 has been shown to stimulate the activity of the *Taq*™ DNA pol.[31]

C. APPLICATIONS

Taq™ pol is the most widely used thermostable DNA polymerase for *in vitro* DNA-amplification method (PCR). It has also been used for the direct sequencing of bacteriophage T4 DNA.[29] The method uses ^{32}P-end-labeled primers to produce extension products that allow the analysis of at least 200 nucleotides on a single sequencing gel. The *Taq*™ DNA pol has been used for a wide variety of biochemical processes such as gene cloning; mutagenesis; identification of various microbial pathogens in clinical, environmental, and food samples; detection of human genetic diseases; forensics; identification and characterization of ancient human DNA; and in plant biotechnology.

III. DNA POLYMERASE FROM *THERMUS AQUATICUS* (STOFFEL FRAGMENT™ OF *TAQ*™)

A. ISOLATION

The Stoffel Fragment™ is a deletion derivative of the full-length 832-amino-acid *Taq*™ pol I, and it encodes a 544-amino-acid translation product. The molecular weight of the gene is 61.3 kDa.[29]

B. BIOCHEMICAL CHARACTERISTICS

Divalent metal ion titration data using $MgCl_2$, $MnCl_2$, and $CoCl_2$ indicate that the Stoffel Fragment™ has a preference for $MgCl_2$. Unlike the full-length form, Stoffel Fragment™ enzyme activity was maximally stimulated over a broad concentration range of $MgCl_2$ (2 to 10 mM), with a maximum of 4 mM. Similarly, its activity was stimulated over a broad concentration range of $MnCl_2$, with a maximum of 0.92 mM (60% of the maximal activity achieved with $MgCl_2$). Cobalt chloride stimulated Stoffel Fragment™ activity maximally at 0.88 mM (27% of the maximal activity achieved with $MgCl_2$).[29]

Potassium chloride titration showed that with activated DNA template, maximum activity was achieved with 0 mM KCl, although polymerase activity remained above 90% of maximum up to 45 mM KCl and remained over 50% at 100 mM KCl. With M-13 primer-template, activity declined rapidly at >10 mM KCl, to 50% at 40 mM KCl and to 0 at 100 mM KCl.[29]

1. Nuclease Activity

The Stoffel Fragment™ is completely devoid of exonuclease activity.[29]

2. Thermostability

The Stoffel Fragment™ has an activity half-life of 21 min at 97.5°C.[29]

3. Reaction Parameters

The optimum reaction temperature is 80°C, pH is 8.3, $MgCl_2$ concentration is 4 mM and KCl concentration is 10 mM.[29]

C. APPLICATIONS

Due to the lack of 5' → 3' exonuclease activity, the Stoffel Fragment™ may be preferred for performing dideoxy nucleotide sequencing. It may also be useful in PCR with templates that have a stable secondary structure and in reactions where a large PCR product is desired.

IV. DNA POLYMERASE FROM *THERMUS THERMOPHILUS* (*TTH*™)

A. ISOLATION

The *Tth*™ thermostable DNA polymerase was isolated from a thermophilic eubacterium *Thermus thermophilus* HB-8.[5,6]

B. BIOCHEMICAL CHARACTERISTICS

Three isoenzymes, A, B, and C, of the thermostable DNA polymerase *Tth*™ have been purified from *T. thermophilus* with the relative molecular masses in the range of 110 to 120 kDa.[6]

1. Nuclease Activity

All three isoenzyme forms of *Tth*™ polymerase showed no exonuclease activity at low salt concentration.[6] Carballeira et al.[5] isolated a 67-kDa protein from *T. thermophilus* HB 8 strain without any 5' → 3' exonuclease activity. However, Myers and Gelfand[33] showed that the *Tth*™ polymerase has optimally active 5' → 3' exonuclease activity at high salt concentrations. At low salt concentrations, the *Tth*™ polymerase does not manifest any exonuclease activity, which is in agreement with the study conducted by Carballeira et al.[5] No proofreading exonuclease activity (3' → 5') has been detected for *Tth*™ polymerase.

2. Template

The template-primer specificity of all three isoenzyme forms of *Tth*™ has been tested in the presence of activated DNA. All three *Tth*™ polymerases (A, B, and C) showed 100% specificity at 50, 63, and 63°C, respectively, in the presence of Mg^{2+}. However, when Mn^{2+} was used, the specificity of *Tth*™ pol A, B, and C was 100% at 50°C, 60% at 63°C, and 50% at 63°C, respectively, in the presence of activated DNA.[6]

3. Thermostability

The stability of these three forms of *Tth*™ polymerases at 90°C varies, and it was found that the B form is the most stable of the three and the A form is the least stable. A study by Rüttiman et al.[6] showed that incubation of enzymes A, B, and C of the *Tth*™ at 90°C for 0 to 60 min reduces the polymerase activity to >5, ~35, and ~45%, respectively.

Recently, the modified *Tth*™ gene was cloned and expressed in *Escherichia coli* to produce large quantities of this thermostable DNA polymerase[34,29] The cloned *Tth*™ gene in the *E. coli* host is called r*Tth*™, and it produces a 94-kDa protein with a half-life of approximately 20 min at 95°C.

4. Reaction Parameters

The *Tth*™ DNA polymerase can use Mg^{2+} or Mn^{2+} as co-factor. In the presence of Mg^{2+}, the highest incorporation was obtained at 50°C with enzyme A and at 60°C with enzymes B and C. If Mg^{2+} is replaced with Mn^{2+}, the incorporation temperature remains unchanged for all three enzyme forms. However, enzyme A has been found to be stimulated approximately twofold, whereas the activity decreases to one-half for enzymes B and C.[6]

The *Tth*™ or r*Tth*™ polymerases have been shown to synthesize cDNA from mRNA template in the presence of Mn^{2+} at an elevated temperature followed by PCR amplification of the cDNA in the presence of Mg^{2+} after deactivation of the Mn^{2+} with a chelating agent such as EGTA. Although, the use of Mn^{2+} has increased the reverse transcriptase activity, the fidelity of this enzyme for the synthesis of cDNA may be decreased as compared to the conventional method.[33] However, the use of EGTA to chelate the Mn^{2+}, elevated reaction temperatures for cDNA synthesis, low dNTP and $MgCl_2$ concentrations compensate for the lower fidelity for Tth^{TM} DNA polymerase.[33] The reaction conditions for PCR amplification following reverse transcription-cDNA synthesis or for direct PCR amplification of target DNA are the same as those for Taq^{TM} DNA polymerase.

C. APPLICATIONS

The Tth^{TM} polymerase extends 25 nucleotides per second which is much slower than Taq^{TM} DNA polymerase, which extends 60 nucleotides per second.[5] An extension of up to 1.3 kb was achieved by using Tth^{TM} DNA polymerase at 70, 75, and 80°C, with no apparent difference in the yield of the amplified product.[5] When compared with Taq^{TM} DNA polymerase, PCR amplification coupled with reverse trancriptase-cDNA synthesis using Tth^{TM} DNA polymerase showed 200-fold greater sensitivity of detection (100 copies) than did conventional methods of amplification of an RNA template (10^4 copies).[33] The higher sensitivity of detection is predicted to be due to higher reverse transcriptase activity instead of increase in PCR amplification efficiency.[33] Although the amino acid sequence comparisons showed that there is significant sequence homology between the Taq^{TM} DNA polymerase and Tth^{TM} polymerase, they differ in their salt sensitivity and ability to utilize RNA templates.[33] For cloning and PCR amplification of mRNA, the Tth^{TM} polymerase is extremely useful. The presence of RNA secondary structures interferes with the synthesis of complete cDNA from mRNA. The reverse transcriptase activity of the Tth^{TM} polymerase allows synthesis of cDNA at higher temperature, which eliminates the secondary structures allowing the synthesis of the complete message from the mRNA. In addition to the reverse transcriptase activity, the presence of DNA polymerase activity makes it extremely convenient to amplify the message by PCR method in the same reaction simply by inactivating the co-factor Mn^{2+} and adding the Mg^{2+}. The Tth^{TM} polymerase has the advantage of being usable for cloning, detecting, or characterizing the biological messages and gene expression in both prokaryotes and eukaryotes. Also, it has been reported that the Tth^{TM} DNA polymerase is more resistant to blood components than are other thermostable DNA polymerases.[34,35] Therefore, the Tth^{TM} polymerase is useful in detection, in study of gene expression by amplifying mRNA following cDNA synthesis, and in cloning genes from cellular and viral RNA. Also, since this polymerase is resistant to blood components, the use of this polymerase can be helpful in detection of pathogens in blood samples.

V. DNA POLYMERASE FROM *THERMOCOCCUS LITORALIS* (TLI/VENT™ DNA POL)

A. ISOLATION

The Tli/Vent™ DNA polymerase was isolated from an extremely thermophilic marine archaebacterium *Thermococcus litoralis*,[7] from a submarine thermal vent, where the temperature is approximately 98°C. The molecular weight of this DNA polymerase is 92 to 97 kDa.

B. BIOCHEMICAL CHARACTERISTICS

Three different forms of Vent™ DNA polymerases were studied for their biochemical characteristics: (1) a native form called Vent™$_N$, which was isolated directly from *T. litoralis*; (2) a recombinant form of Vent™ DNA polymerase, which is the same as the native Vent™$_N$ DNA polymerase; and (3) a genetically engineered Vent™ DNA polymerase, named Vent™$_{RExo}$.

1. Nuclease Activity

Two of the three forms of the Vent™ DNA polymerase, native Vent™N and Vent™R, have the $3' \rightarrow 5'$ proofreading exonuclease activity. However, the genetically engineered form, Vent™$_{RExo}$, has been shown to lack the $3' \rightarrow 5'$ exonuclease activity. The production of a relatively high percentage of medium blue plaques in a transfection experiment, using the products generated by the Vent™$_N$ and the Vent™$_R$ DNA polymerases, indicated 94 to 100% excision of terminal mispaired nucleotides at the 3'-OH ends of the primers.[8]

2. Thermostability

The *Tli*/Vent™ DNA polymerase has a very high level of thermostability, with a half-life of 2 h at 100°C.[8]

3. Reaction Parameters

The mutational frequency for the Vent™$_N$ and the Vent™$_R$ increased approximately twofold when the concentration of the dNTP was increased from 200 μM to 4 mM in the reaction.[8] This indicated that the 3' → 5' proofreading exonuclease activity was inhibited by the increased quantity of dNTP. Interestingly, it has also been reported that increase in dNTP concentration did not affect the mutation rate in Vent™$_{RExo}$.[8] In another study, it was shown that the fidelity of the Vent™ DNA polymerase is unaffected by varying the dNTP concentration between 10 and 200 μM.[36] When ≤10 μM dNTP was used, a smear of the amplified DNA was evident in a denaturing gradient gel analysis, possibly due to the incomplete extension of the template, resulting in truncated ends. The mutation frequencies were at least fourfold higher in the reactions with 10 mM Mg^{2+} and total 16 mM dNTP concentration than in reactions carried out with 1 mM Mg^{2+} concentration. Mutation frequency was minimum with low Mg^{2+} concentration (1 mM) with the dNTP concentration between 1 and 16 mM. This suggests that the fidelity of the Vent™$_N$ DNA polymerase depends on the relative concentrations of both Mg^{2+} and dNTPs in the reaction mixture.[8] The optimum Mg^{2+} concentration for this polymerase ranges from 2 to 8 mM.[8] Although the presence of monovalent cations such as KCl or NaCl is not required for its polymerization activity, the addition of 10 to 50 mM KCl or NaCl in the reaction mixture has been found to slightly increase the quantity of the DNA amplification products.[8] The pH range for this polymerase is 8 to 9.[8]

C. APPLICATIONS

The mutational frequency for this polymerase is reported to be 30×10^{-6}, which is 5 to 10 times lower than the other reported thermostable polymerases lacking proofreading activity.[8] The mutational frequency of the Vent™$_N$ and the Vent™$_R$ was reported to be between 11×10^{-4} and 34×10^{-4}, which is 2 to 11 times lower than that of the Vent™$_{RExo}$.[8] High thermostability with the 3' → 5' proofreading exonuclease activity facilitates a highly accurate DNA synthesis *in vitro*.

In a study conducted by Lohff and Cease,[37] it was shown that the cloning of *Taq*™-amplified DNA treated with T4 kinase produced 49% white colonies, and the amplified DNA using Vent™ DNA polymerase without any further modifications yielded 53% white colonies against a background of 10% blue-green colonies. This suggests that a thermostable DNA polymerase with 3' → 5' exonuclease activity, such as Vent™ DNA polymerase, yields PCR products with ends that are both blunt and flush with the ends of the primer sequence, which are suitable for efficient and accurate ligation into a blunt-cut vector, for the non-directional cloning of PCR products without any further end modification.[9,37]

VI. DNA POLYMERASE FROM *PYROCOCCUS FURIOSUS* (*PFU*™)

A. ISOLATION

Pyrococcus furiosus ("rushing fireball") is an anaerobic, hyperthermophilic marine archaebacterium that was isolated from geothermally heated marine sediments in Vulcano, Italy, by Fiala and Stetter.[11] The optimum growth temperature of this microorganism is 100°C.[10]

B. BIOCHEMICAL CHARACTERISTICS

The pfu™ DNA polymerase is a 92-kDa monomeric enzyme that was purified by multiple steps of column chromotagraphy.[12]

1. Nuclease Activity

The pfu™ DNA polymerase exhibits both 5' → 3' DNA strand displacement and 3' → 5' proofreading exonuclease activity, which ensures the higher degree of amplification fidelity during the replication process. Unlike other thermostable DNA pols such as Vent™ DNA pol, the 3' → 5' exonuclease activity of the pfu™ DNA pol peaks sharply at its standard polymerization temperature (75 to 80°C), minimizing undesirable primer-degradation activity at low temperatures.[13,14]

2. Thermostability

After incubation at 95°C for 60 min, 95% of the polymerase activity is retained.[13,14] The pfu™ DNA pol exhibits maximum polymerase activity at 75°C, and at lower temperature such as 30°C, negligible activity was detected. Therefore, during in vitro amplification undesirable low-temperature polymerization reactions are less likely to occur with pfu™ DNA polymerase.[13,14]

3. Reaction Parameters

Presence of Mg^{2+} is essential for the polymerase activity. The $MgCl_2$ range for its activity is 1.5 to 8 mM. However, the optimum $MgCl_2$ concentration for in vitro DNA-amplification methods has been determined to be 2 mM. The optimum in vitro DNA-amplification reaction conditions for pfu™ DNA pol requires the presence of 10 mM KCl, 2 mM $MgCl_2$, 6 mM $(NH_4)_2SO_4$. Also, the addition of a non-ionic detergent such as 0.1% Triton X-100 increases the stability of the enzyme. The optimum polymerization temperature is near 75°C, and the pH range is 8 to 9.[13,14]

C. APPLICATIONS

During PCR DNA amplification, the pfu™ DNA polymerase exhibits 11 to 12-fold greater replication fidelity than does Taq™ DNA polymerase.[12] The average error rate for pfu™ DNA polymerase was determined to be 1.6×10^{-6}, and for Taq™ DNA polymerase, 2.0×10^{-5}.[12] Therefore, after 20 c of PCR DNA amplification of a 1-kb DNA fragment using Taq™ DNA polymerase, it is expected that 40% of the amplification products will have both silent and phenotypically detectable mutations, whereas under similar conditions, the use of pfu™ DNA polymerase will generate only 3.2% mutations.

The application of the pfu™ DNA polymerase is significant in research areas where cloning, sequencing, site-directed mutagenesis, and characterizations of genes are performed by using the PCR DNA-amplification method. Efficient proofreading activity of pfu™ DNA polymerase provides a high level of replication fidelity, providing a gene sequence without any mutations. Also, a higher level of thermostability provides longer enzyme stability during the in vitro DNA-amplification method.

VII. DNA POLYMERASES FROM BACILLUS SPP. (BST™)

A. ISOLATION

Several Bacillus DNA pols have been isolated from a number of thermophilic Bacillus spp. such as B. stearothermophilus,[15,38,39] B. caldotenax,[39] B. caldovelox,[39] and B. licheniformis.[38]

B. BIOCHEMICAL CHARACTERISTICS

Following extraction by the conventional five-step procedure using several column chromatographs, further purification of the DNA polymerase from *B. caldotenax*, *B. caldovelox*, and *B. stearothermophilus* (*Bst*™) yielded a 95-kDa protein.

1. Nuclease Activities

No exonuclease activity was found to be associated with the purified enzymes.[15]

2. Inhibitors

The DNA polymerases were inhibited by a single dideoxyribonucleoside triphosphate (ddNTP) in the presence of the four dNTPs.[15] By using ratios of ddNTP to dNTP of 1:1 and 5:1, the lowest inhibition was observed with DNA polymerase from *B. caldotenax* of 38 ± 5.1% and 52 ± 9.2%. The enzymes from *B. stearothermophilus* (49 ± 2.8% and 59 ± 4.7%) and *T. thermophilus* (51 ± 2.1% and 61 ± 4.7%) were about equally sensitive. Each of the latter enzymes was used for two-step DNA sequencing by the method. For this purpose, ratios between dNTPs and ddNTPs of 1:7 (G), 1:40 (A), 1:40 (T), and 1:7 (C) were utilized.[16]

3. Thermostability

The thermostability of each enzyme was tested by preincubating for 10 min at the appropriate temperature in the standard reaction mix omitting dNTPs and DNA. The reactions were then started by adding the missing components and the extent of incorporation measured. The enzyme activities were not affected if the preincubation temperatures did not exceed the optimal temperature by more than 10°C. The activities were reduced significantly at higher temperatures, and the enzymes were inactivated at 80°C (*B. stearothermophilus*) and 85°C (*B. caldotenax* and *B. caldovelox*). *Bacillus* DNA polymerases differ in their optimal reaction temperatures between 65 and 70°C and are inactivated at temperatures between 70 and 75°C.[15,16]

4. Reaction Parameters

The optimal temperature for *B. stearothermophilus* DNA polymerase was 60 to 65°C, whereas the enzymes from *B. caldotenax* and *B. caldovelox* showed the highest incorporation rate at 65 to 70°C. The optimal pH for all enzymes was determined to be 8.8. The enzymes require Mg^{2+} (10 to 30 mM) for optimal activity, although 0.4 mM Mn^{2+} could substitute for magnesium. The use of potassium (KCl) as a monovalent cation resulted in a relatively broad plateau between 100 and 200 mM for the *Bacillus* DNA polymerases.[15,16]

C. APPLICATIONS

The use of *Bst*™ DNA polymerase permits rapid sequence analysis from nanogram amounts of template.[40] The large fragment of *Bst*™ polymerase is heat stable and permits sequencing reactions to be carried out at a high temperature, melting secondary structure.[41] This results in uniform band intensities, including the first dC where there are runs of dCs, which in most cases is much weaker than the rest of the dCs. *Bst*™ polymerase can be handled for a few days at room temperature, 14 days at 25°C or 7 days at 37°C without loss of sequencing activity. The enzyme can be used in the standard Sanger one-step protocol or in a two-step protocol and can be used successfully in double-stranded[16] and reverse-sequencing.[42] *Bst*™ polymerase has also been used in automated fluorescent sequencing.[43]

VIII. DNA POLYMERASE FROM *SULFOLOBUS ACIDOCALDARIUS* (*SAC*)

A. ISOLATION

A thermostable DNA polymerase has been isolated from the thermoacidophilic archaebacterium *Sulfolobus acidocaldarius*.[17]

B. BIOCHEMICAL CHARACTERISTICS

Sodium dodecyl sulfate gel electrophoresis of the DNA polymerase revealed a polypeptide of 100 kDa. The enzyme is a monomer.

1. Nuclease Activities

An endonuclease activity was detected up to the blue-sepharose chromatography step, but was completely removed following sucrose-gradient centrifugation. No 5' → 3' exonuclease activity was detected.[18] However, 3' → 5' exonuclease activity has been detected on double- or single-stranded substrates at either 37, 60, or 70°C.

2. Inhibitors

The DNA polymerase activity is resistant to aphidicolin but inhibited by ddTTP; 17 and 33% inhibition were observed with ddTTP/dTTP ratios of 1:1 and 5:1, respectively. Preincubation of the DNA polymerase for 15 min at 70°C with 2 and 5 mM N-ethylmaleimide inhibited the activity by 35 and 66%, respectively, whereas no inhibition was observed when the preincubation was performed at 0°C.[18]

3. Reaction Parameters

The *Sac* DNA polymerase can utilize a broad range of $MgCl_2$ concentrations (2 to 8 mM). Also, this thermostable DNA polymerase is not stringently dependent upon the concentration of dNTP in the reaction, as found in the case of *Taq*™ DNA polymerase from *T. aquaticus*. For example, the *Taq*™ DNA polymerase is inhibited when 0.3 mM dNTP is used in the presence of 2 mM $MgCl_2$ whereas the DNA polymerase from *Sac* produces significant amounts of amplified DNAs under the same parameters.[19]

4. Thermostability

The nucleotide incorporation by *Sac* DNA polymerase was found to increase as the temperature was increased up to 70°C and rapidly decreased after 10 min at higher temperatures.[17] The initial velocity of incorporation during the first 8 to 10 min was the same as at 70 and at 80°C, during which point the activity of the enzyme decreased and eventually stopped at temperatures higher than 70°C. The *in vitro* thermostability and residual DNA polymerase activity of the *Sac* DNA polymerase depend upon the protein concentration in the reaction mixture. Preincubation, in the absence of dNTP, of the *Sac* DNA polymerase in 400 µg/ml of bovine serum albumin, for 40 min at both 70 and 80°C, has been shown to have retained >80% residual polymerase activity, whereas the activity dropped to 20% at 80°C when preincubation was performed in the presence of activated calf thymus DNA.[17] The activity remained the same (>80%) at 70°C under the same experimental parameters. Therefore, the optimum temperature at which the *Sac* DNA polymerase is stable and has the highest incorporation rate is between 70 and 80°C.[17] The *Sac* DNA polymerase was heat inactivated and 50% residual activity was obtained when the enzyme was preincubated at 87°C for 15 min.[17]

C. APPLICATIONS

The sensitivity of this enzyme to ddNTP allows use of *Sac* DNA polymerase for DNA sequencing by the Sanger dideoxy chain termination method at 70°C; this could avoid problems arising from hairpin formation at 37°C when mesophilic DNA polymerases such as the Klenow fragment of the *E. coli* DNA polymerase I, and T7 DNA polymerase are used. Also, other properties such as optimum themostability, low sensitivity to variation in $MgCl_2$ concentrations without being affected by the concentration of dNTPs in the reaction, high rate of polymerization, equal extension of all hybridized primers, and $3' \rightarrow 5'$ exonuclease activity make this enzyme a useful tool for *in vitro* DNA amplification, site-directed mutagenesis, gene fusions, and other biotechnological procedures at higher temperatures.

IX. DNA POLYMERASE FROM *THERMOPLASMA ACIDOPHILUM* (*TAC* DNA POL)

A. ISOLATION

A thermophilic DNA polymerase has been purified from the thermoacidophilic archaebacterium *Thermoplasma acidophilum*.[20]

B. BIOCHEMICAL PROPERTIES

The DNA polymerase is monomeric and is composed of a 88-kDa polypeptide.

1. Nuclease Activities

No endonuclease activity has been detected in the purified fractions of the polymerase. A $3' \rightarrow 5'$ exonuclease activity was detected when the enzyme was assayed on [$3'$-^{3}H]poly(dA) in the absence of dNTPs: 0.7-U DNA polymerase activity solubilized 80% of the input radioactivity in 30 min at 65°C; this exonuclease was also active on double-stranded DNA: 20% of linearized SV40 [^{3}H]DNA was rendered acid-soluble/0.9-U DNA polymerase in 30 min at 65°C. This $3' \rightarrow 5'$ exonuclease activity cosedimented with the DNA polymerase activity and the 88-kDa polypeptide. No $5' \rightarrow 3'$ exonuclease activity was observed on [$5'$-^{32}P]poly(dA) or on SV40[^{3}H]DNA.[20]

2. Template

At 40 or 65°C, the DNA polymerase was able to use neither poly(rA)·(dT)12 as template-primer in the presence of dTTP (reverse transcriptase assay) nor M13 single-stranded DNA as template.[20]

3. Inhibitors

The DNA polymerase was resistant to high aphidicolin concentrations (up to 100 µg/ml) and only slightly sensitive to ddTTP: 25% inhibition was observed with a ddTTP/dTTP ratio of 50. The enzyme was inhibited by *N*-ethylmaleimide only when it was preincubated with the drug at 65°C.[20]

4. Thermostability

The DNA polymerase and exonuclease activities were optimal at 65°C.[20] At 37°C, the polymerase was completely inactive and the exonuclease retained 66% of its optimal activity. The enzyme is thermostable up to 65°C. At higher temperatures, the polymerase activity decreased more rapidly than the exonuclease activity: 50% residual activity was obtained after a 30-min preincubation at 75 and 84°C, for DNA polymerase and exonuclease activities, respectively; the polymerase activity was completely inactivated after a 30-min preincubation at 85°C, whereas residual exonuclease activity was still detected after a 30-min preincubation at 95°C.[20]

5. Reaction Parameters

Optimal activity was obtained in weak monovalent salt concentration, at pH 8 in the presence of 4 m*M* $MgCl_2$.[20] The addition of glycerol or ethyleneglycol to the reaction mixture produced a stimulation of 30%, whereas no stimulation was obtained with polyethylene glycol.[20]

C. APPLICATIONS

The relatively lower optimum polymerization temperature (65°C) of the *Tac* DNA pol as compared to other widely used thermostable DNA polymerases can be useful in *in vitro* DNA amplification of the templates that do not require high temperatures for denaturation (such as in AT-rich templates). Thus the cycling time can be reduced, and the amplification can be performed within a relatively short period of time.

X. DNA POLYMERASE FROM *THERMUS FLAVUS* (*TFL/TUB*™)

A. ISOLATION

The *Tfl/Tub*™ DNA pol was isolated from a thermophilic bacterial species, *Thermus flavus*.[21]

B. BIOCHEMICAL CHARACTERISTICS

The *Tfl/Tub*™ DNA pol was purified by a five-step procedure using several column chromatographs. The molecular weight of this polymerase was determined to be 66 kDa.[21]

1. Nuclease Activities

Neither $3' \rightarrow 5'$ nor $5' \rightarrow 3'$ exonuclease activities were detected.[21,22]

2. Inhibitors

Unlike *Tru* DNA pol, *Tfl/Tub*™ DNA pol was inhibited by up to 50% of its original specific activity in the presence of 21% ethanol, 15 µM/ml actinomycin D, 30 µM/ml acriflavine, 11 mM ara-CTP, 85 mM spermidine, or 5 mM spermine.[21]

3. Template

The maximal activity of the *Tfl/Tub*™ pol requires the presence of the template, dNTPs, and monovalent and divalent cations in the reaction mixture. The enzyme is highly active when activated DNA, poly(dA)-poly(dT), poly(dA)-oligo(dT)$_{10}$, and poly(rA)-oligo(dT)$_{10}$ are used as templates, moderately active with single- and double-stranded DNAs, and inactive when poly(rC)-oligo(dG)$_{12-18}$ and native RNA molecules are used.[21]

4. Reaction Parameters

The *Tfl/Tub*™ DNA pol utilizes 10 to 15 mM of divalent cation, MgCl$_2$, and 0.5 mM MnCl$_2$.[21] The optimum pH of this DNA pol for its maximum activity is 7.5. The optimum polymerization temperature for the *Tfl/Tub*™ DNA pol is 70°C for DNA templates. However, for the poly(rA)-poly(dT)$_{10}$ RNA template the optimum polymerization temperature was determined to be 50°C.[21] The presence of a monovalent cation such as KCl increases the activity of the *Tfl/Tub*™ DNA pol activity. The optimum KCl concentration is 5 to 10 mM. Increase in the KCl concentrations to 250 mM inhibits 50% of the original polymerase activity of the *Tfl/Tub*™ DNA pol.[21]

5. Thermostability

The optimum temperature for the *Tfl/Tub*™ pol on the DNA template has been described as 70°C and that on RNA template as 50°C.[21]

C. APPLICATIONS

The *Tfl/Tub*™ DNA polymerase has been used to amplify large target DNA fragments up to 15.6 kb with a two-step PCR-amplification method with a denaturation temperature of 94°C for 1 min and an appropriate primer annealing and extension temperature (>60°C) between 4 and 12 min.[22] The *Tfl/Tub*™ DNA pol is useful for the amplification of large DNA templates with an average size of >10 kb. Also, the *Tfl/Tub*™ DNA pol has been used for direct nucleotide sequence analysis.[23]

XI. DNA POLYMERASE FROM *THERMUS RUBER* (*TRU*)

A. ISOLATION

The *Tru* DNA pol was isolated from an extremely thermophilic bacterial species *Thermus ruber* No. 21.[24]

B. BIOCHEMICAL CHARACTERISTICS

The *Tru* DNA pol was isolated by a six-stage purification method, which included initial concentration of the enzyme by ammonium sulfate followed by several column chromatography methods such as DEAE cellulose, hydroxyapatite, heparin-sepharose, and single-stranded DNA cellulose.[24] The molecular weight of the *Tru* DNA pol was determined to be 70 kDa.

1. Nuclease Activity

In the homogenous preparation of the *Tru* DNA pol, no apparent $3' \rightarrow 5'$ or $5' \rightarrow 3'$ exonuclease activity was detected.[24]

2. Inhibitors

On the activated DNA template, the *Tru* DNA pol is inhibited up to 75 to 80% by the presence of high concentrations of KCl (150 mM), NaCl (125 mM), NaF (35 mM), polyamines (e.g., spermidine, spermine, putrescine, cadaverine, etc.), and intercalating agents (i.e., ethidium bromide).[24]

3. Template

Among the DNA templates, the poly(dA)-poly(dT) has been found to be preferred over activated native and denatured DNAs. Also, the *Tru* DNA pol can utilize poly(rA)-oligo(dT)$_{10}$ RNA template, but remains inactive on poly(rC)-oligo(dG)$_{12-18}$ and natural RNA templates.[24]

4. Reaction Parameters

The pH range of the *Tru* DNA pol activity is 7 to 12. The optimum pH for DNA templates has been described ad 9.0.[24] On a poly(rA)-oligo(dT)$_{10}$ template, the pH optimum has been described as 7.5. In the absence of a divalent cation the *Tru* DNA pol does not exhibit any specific activity. The optimum MgCl$_2$ and MnCl$_2$ concentrations are 2.5 and 0.15 mM, respectively. In the presence of MgCl$_2$, the enzyme is 50% more active than in the presence of MnCl$_2$.[24] Other divalent cations such as Ca^{2+} or Zn^{2+} do not catalyze the activity of the *Tru* DNA pol enzyme.[24] Addition of monovalent cations such as Na$^+$ (NaCl) or K$^+$ (KCl) to an optimum concentration of 15 mM increases the activity of the *Tru* DNA pol. Manifestation of maximum *Tru* DNA pol activity requires the presence of 0.5 mM of each of the dNTPs with an activated template DNA.

5. Thermostability

The *Tru* DNA pol can retain 90% of its initial activity after 2 h of incubation at 70°C on activated DNA template. Although, the optimum enzyme activity was observed at 70°C, the activity can be preserved for 1 to 2 h between 50 and 85°C. The optimum temperature for poly(rA)-oligo(dT)$_{10}$ RNA template is 50°C.[24]

C. APPLICATIONS

Although the *Tru* DNA pol has not been extensively used for the *in vitro* DNA-amplification method, the use of optimum concentrations of Mn^{2+} in the reaction mixture has the potential to resolve DNA sequence analysis with more clarity and to be used in cDNA synthesis from cellular messages followed by PCR DNA amplification of the target cDNA. Also, the lack of inhibition of enzyme activity by ethanol, non-intercalating agents, and sulfhydryl reagents allows its use in DNA amplification of samples that may be contaminated with such materials.

XII. DNA POLYMERASE FROM *THERMOTOGA* SP. (*TSP*)

A. ISOLATION

The *Tsp* pol was isolated from an anaerobic bacterial species *Thermotoga* sp. strain FjSS3-B.1 from an intertidal hot spring on Savusavu beach in Fiji.[25]

B. BIOCHEMICAL CHARACTERISTICS

The *Tsp* DNA polymerase was purified by a five-step procedure using polyethylenimine to precipitate the nucleic acids, followed by DEAE sepharose, CM-sephrarose, hydroxyapatite column chromatography, and SDS-polyacrylamide gel electrophoresis methods.[25] The molecular weight of the *Tsp* pol was determined to be 85 kDa.

1. Nuclease Activities

No $3' \rightarrow 5'$ or $5' \rightarrow 3'$ exonuclease activity nor any endonuclease activity has been detected.[25]

2. Inhibitors

High ionic strength in the reaction mixture inhibits the enzyme activity. At a concentration of ~80 mM NaCl in the polymerization reaction, 50% of the enzyme activity is inhibited. Only 10% of the enzyme activity has been detected when 200 mM NaCl is used in the reaction.[25]

3. Reaction Parameters

The optimum pH range for its polymerase activity is between 7.5 and 8. The optimum $MgCl_2$ concentration for polymerization is 10 mM, whereas optimum $MnCl_2$ concentration is 15 mM.[25] When $CaCl_2$ or $MnCl_2$ was substituted for $MgCl_2$, less than 10% of the control activity was observed. No activity was evidenced when divalent cations were replaced by monovalent cations such as Na^+, K^+, Rb^+, and Cs^+ at a concentration of 10 mM.[25] At least 20 mM Tris-Cl (pH 7.5) is required for the polymerase activity of the *Tsp* pol. Also, addition of β-mercaptoethanol, gelatin, and Triton X-100 in the reaction mixture increases the polymerization activity by stabilizing the enzyme for a longer period of time.[25]

4. Thermostability

The *Tsp* DNA pol was assayed within the temperature range of 30 to 80°C, and the maximum activity observed in a 30-min assay was at 80°C.[25] The incorporation of [^{32}P]dCTP by the *Tsp* DNA pol was nine times higher at 80 than at 30°C. When the *Tsp* DNA pol was preincubated in the absence of substrate DNA at 50, 70, and 95°C, the half-lives of the enzyme were 60, 45, and 3 min, respectively. Addition of various cosolvents such as Triton X-100 (0.01%), Tween-80 (0.001%), or mannitol (5%) has been shown to increase the half-life of the enzyme to 80 min at 50°C, and 7 min at 95°C.[25] Furthermore addition of a combination of these cosolvents in the reaction mixture has been shown to increase the thermostability of this enzyme by up to 85% at 50°C.

C. APPLICATIONS

The *Tsp* pol has the potential for use in *in vitro* DNA amplification because of its high optimum incorporation temperature (80°C).

XIII. DNA POLYMERASE FROM *METHANOBACTERIUM THERMOAUTOTROPHICUM* (MTH)

A. ISOLATION

The *Mth* DNA pol was isolated from a thermophilic methanogenic archaebacterium *Methanobacterium thermoautotrophicum*.[26]

B. BIOCHEMICAL CHARACTERISTICS

The *Mth* DNA pol was purified by a six-step purification method consisting of purification by several column chromatographs followed by a glycerol gradient centrifugation method.[26] The molecular weight of the enzyme was determined to be 72 kDa, and it remained as a single polypeptide.

1. Nuclease Activity

The *Mth* DNA pol possesses both 3' → 5' and 5' → 3' exonuclease activity in the presence of activated target DNA.[26] The 3' → 5' exonuclease activity can be stimulated up to 170% by addition of a monovalent cation such as 100 mM KCl. Also, at 57°C the activity of the 3' → 5' exonuclease was 1.6 times higher than at 37°C. Activity of this exonuclease was 1.5 times higher on single-stranded DNA template than on native double-stranded DNA. The 5' → 3' exonuclease was inhibited up to 90% of its specific activity by the presence of 100 mM KCl. This exonuclease activity was found to be increased to five times at 56°C as compared to incubation at 37°C. The 5' → 3' exonuclease did not show any obvious preference for the form (single- vs. double-stranded) of the substrate.

2. Template

Activated double- or single-stranded DNA is the template for the *Mth* DNA pol. No activity was apparent in the absence of primer, that is with poly(dA) and poly(dT) as templates and with the addition of corresponding priming ribonucleotides, or with the poly-ribonucleotide templates, poly(rA)-oligo(dT).[26]

3. Inhibitors

The activity of the *Mth* DNA pol was completely (100%) inhibited by the presence of 20 μg/ml aphidicolin, 89% inhibited by the presence of 25 μM ddNTPs, 74% inhibited by the presence of a sulfhydryl blocking agent, and 32% inhibited by the presence of 1 mM arabinosyl-CTP.[26]

4. Thermostability

The optimum temperature at which maximum stability of the *Mth* DNA pol with specific activity of 160% was observed was 65°C.[26] However, at 56°C the enzyme is stable and retains its activity up to 100%. At a higher temperature such as 80°C, the *Mth* DNA pol is unstable, and its activity drops to only 25% within 10 min.[26] After pretreatment of the enzyme without the substrate for 10 min at 100°C, the *Mth* DNA pol loses its polymerization activity completely.[26]

5. Reaction Parameters

The *Mth* DNA pol has a broad pH range with an optimum pH of 8.0; approximately 60% of the activity has been observed between pH 6.0 and 9.5.[26] The activity of the *Mth* DNA pol is dependent upon the presence of divalent cations such as Mg^{2+} with the maximum activity determined between 10 and 20 mM $MgCl_2$ in the reaction. Presence of Mn^{2+} at the optimum quantity of 0.5 mM showed moderate activity of the *Mth* DNA pol.[26] The polymerase activity was stimulated by the presence of 100 mM KCl[26] to 165%. Up to 140% of the activity was retained in the presence of 200 mM KCl, 90% in the presence of 300 mM KCl, and 20% when 400 mM KCl was used in the reaction mixture.[26] The optimum polymerization temperature was 65°C.

C. APPLICATIONS

Although the *Mth* DNA pol was isolated from a methanogenic bacterial species, the enzyme was found not to be sensitive to oxidizing atmosphere,[26] which makes this enzyme a good candidate for storage and use in the *in vitro* DNA amplification method. The presence of both 3' → 5' and 5' → 3' exonuclease activities is useful for DNA sequence analysis, gene cloning, and site directed mutagenesis methods that utilize the *in vitro* DNA-amplification method. However, the low thermostability (65°C) is a disadvantage for this enzyme for multiple cycles in the *in vitro* DNA amplification method.

REFERENCES

1. **Saiki, R. K., Scharf, S., Faloona, F., Mullis, K. B., Horn, G. T., Erlich, H. A., and Arnheim, N.,** Enzymatic amplification of β-globin genomic sequences and restriction site analysis for diagnosis of sickle cell anemia, *Science,* 230, 1350, 1985.
2. **Bej, A. K., Mahbubani, M. H., and Atlas, R. M.,** Amplification of nucleic acids by polymerase chain reaction (PCR) and other methods and their applications, *Crit. Rev. Biochem. Mol. Biol.,* 26(3/4), 301, 1991.
3. **Kaledin, A. S., Slyusarenko, A. G., and Gorodetskii, S. I.,** Isolation and properties of DNA polymerase from extremely thermophilic bacterium *Thermus aquaticus* YT1, *Biokhimiya,* 45, 644, 1980.
4. **Saiki, R. K., Gelfand, D. H., Stoffel, S., Scharf, S., Higuchi, R., Horn, G. T., Mullis, K. B., and Erhich, H. A.,** Primer-directed enzymatic amplification of DNA with a thermostable DNA polymerase, *Science,* 239, 487, 1988.
5. **Carballeira, N., Nazabal, M., Brito, J., and Garcia, O.,** Purification of a thermostable DNA piolymerase from *Thermus thermophilus* HB8, useful in the polymerase chain reaction, *BioTechniques,* 9, 276, 1990.
6. **Rüttiman, C., Cotasras, M., Zaldivar, J., and Vicuna, R.,** DNA polymerases from the extremely thermophilic bacterium *Thermus thermophilus* HB-8, *Eur. J. Biochem.,* 149, 41, 1985.
7. **Neuner, A., Jannasch, H. W., Belkin, S., and Stetter, K. O.,** *Thermococcus litoralis* sp. nov.: A new species of extremely thermophilic marine archaebacteria, *Arch. Microbiol.,* 153, 205, 1990.
8. **Matilla, P., Korpela, J., Tenkanen, T., and Pitkanen, K.,** Fidelity of DNA synthesis by the *Thermococcus litoralis* DNA-polymerase-an extremely heat stable enzyme with proofreading activity, *Nucleic Acids Res.,* 19, 4967, 1991.
9. **Skerra, A.,** Phosphorothioate primers improve the amplification of DNA sequences by DNA polymerases with proofreading activity, *Nucleic Acids Res.,* 20, 3551, 1992.
10. **Adams, M. W. W.,** The metabolism of hydrogen by extremely thermophilic, sulfue-dependent bacteria, *FEMS Microbiol. Rev.,* 75, 219, 1990.
11. **Fiala, G., and Stetter, K. O.,** *Pyrococcus furiossus* sp. nov. represents a novel genus of marine heterotrophic archaebacteria growing optimally at 100°C, *Arch. Microbiol.,* 145, 56, 1986.
12. **Lundberg, K. S., Shoemaker, D. D., Adams, M. W. W., Short, J. M., Sorge, J. A., and Mathur, E. J.,** High fidelity amplification using a thermostable DNA polumerase isolated from *Pyrococcus furiossus, Gene,* 108, 1, 1991.
13. **Mathur, E., Shoemaker, D., Scott, B., Rombouts, J., Bergseid, M., and Nielson, K.,** Pfu DNA polymerase update, *Strategies,* 5, 11, 1992a.
14. **Mathur, E. J.,** Applications of thermostable DNA polymerases in molecular biology, in *Biocatalysis at Extreme Temperatures, Enzyme Systems Near and Above 100°C*, Adams, M. W. W. and Kelly R. M., Eds., American Chemical Society, Washington, DC, 189, 1992b.
15. **Kaboev, O. K., Luchkina, L. K., Akhmedov, A. T., and Bekker, M. L.,** Purification and properties of deoxyribonucleic acid polymerase from *Bacillus stearothermophilus, J. Bacteriol.,* 145, 21, 1981.
16. **McClary, J., Ye, S. Y., Hong, G. F., and Whitney, F.,** DNA sequence, *J. DNA Sequenc. Map.,* 1, 173, 1991.
17. **Elie, C., Salhi, S., Rossignol, J. M., Forterre, P., and de Recondo, A. M.,** A DNA polymerase from a thermoacidophilic archaebacterium: evolutionary and technological interests, *Biochim. Biophys. Acta,* 951, 261, 1988.

18. **Klimczak, L. J., Grummt. F., and Burger, K. J.,** Purification and characterization of DNA polymerase from the archaebacterium *Sulfolobus acidocaldarius*, *Nucleic Acids Res.*, 13, 5269, 1985.
19. **Salhi, S., Elie, C., Forterre, P., de Recondo, A. M., and Rossignol, J. M.,** DNA polymerase from *Sulfolobus acidocaldarius*: replication at high temperature of long stretches of single-stranded DNA, *J. Mol. Biol.*, 209, 635, 1989.
20. **Hamal, A., Forterre, P., and Elie, C.,** Purification and characterization of a DNA polymerase from the archaebacterium *Thermoplasma acidophilum*, *Euro. J. Biochem.*, 190, 517, 1990.
21. **Kaledin, A. S., Slyusarenko, A. G., and Gorodetskii, S. I.,** Isolation and properties of DNA polymerase from the extremely thermophilic bacterium *Thermus flavus*, *Biokhimiya*, 46, 1576, 1981.
22. **Kainz, P., Schmiedlechner, A., and Strack, H. B.,** In vitro amplification of DNA fragments >10 Kb, *Anal. Biochem.*, 202, 46, 1992.
23. **Rao, V. B. and Saunders, N. B.,** A rapid polymerase-chain-reaction-directed sequencing strategy using a thermostable DNA polymerase from *Thermus flavus*, *Gene*, 113, 17, 1992.
24. **Kaledin, A. S., Slyusarenko, A. G., and Gorodetskii, S. I.,** Isolation and properties of DNA polymerase from the extremely thermophilic bacterium *Thermus ruber*, *Biokhimiya*, 47, 1785, 1982.
25. **Simpson, H. D., Coolobear, T., Vermue, M., and Daniel, R, M.,** Purification and some properties of a thermostable DNA polymerase from a *Thermotoga* species, *Biochem. Cell. Biol.*, 68, 1292, 1990.
26. **Klimczak, L. J., Grummt, F., and Burger, K. J.,** Purification and characterization of DNA polymerase from the archaebacterium *Methanobacterium thermoautotrophicum*, *Biochemistry*, 25, 4850, 1986.
27. **Brock, T. D. and Freeze, H.,** *Thermus aquaticus* gen. n., and sp. n., a non-sporulating extreme thermophile, *J. Bacteriol.*, 98, 289, 1969.
28. **Chien, A., Edgar, D. B., and Trela, J. M.,** Deoxyribonucleic acid polymerase from the extreme thermophile *Thermus aquaticus*, *J. Bacteriol.*, 127, 1550, 1976.
29. **Lawyer, F. C., Stoffel, S., Saiki, R. K., Chang, S. Y., Landre, P. A., Abramson, R. D., and Gelfand, D. H.,** High-level expression, purification, and enzymatic characterization of full-length *Thermus aquaticus* DNA polymerase and a truncated form deficient in 5′ to 3′ exonuclease activity, *PCR Methods Appl.*, 2, 275, 1993.
30. **Tindall, K. R. and Kunkel, T. A.,** Fidelity of DNA synthesis by the *Thermus aquaticus* DNA polymerase, *Biochemistry*, 27, 6008, 1988.
31. **Gelfand, D. H.,** *Taq* DNA polymerse, in *PCR Technology: Principles and Applications for DNA Amplification*, Erlich, H. A., Ed., Stockton Press, NY, 1989, 17.
32. **Lawyer, F. C., Stoffel, S., Saiki, R. K., Myambo, K., Drummond, R., and Gelfand, D. H.,** Isolation, characterization, and expression in Escherichia coli of the DNA polymerase gene from *Thermus aquaticus*, *J. Biol. Chem.*, 264, 6247, 1989.
33. **Myers, T. W. and Gelfand, D. H.,** Reverse transcription and DNA amplification by a *Thermus thermophilus* DNA polymerase, *Biochemistry*, 30, 7661, 1991.
34. **Erlich, H. A., Gelfand, D. H., and Sninsky, J. J.,** Recent advances in the polymerase chain reaction, *Science*, 252, 1643, 1991.
35. **Bej, A. K. and Mahbubani, M. H.,** Applications of the polymerase chain reaction in environmental microbiology, *PCR Methods Appl.*, 1, 151, 1992.
36. **Cariello, N. F., Swenberg, J. A., and Skopek, T. R.,** Fidelity of *Thermococcus litoralis* DNA polymerase (Vent™) in PCR determined by denaturing gradient gel electrophoresis, 19, 4193, 1991.
37. **Lohff, C. J. and Cease, K. B.,** PCR using a thermostable polymerase with 3′ to 5′ exonuclease activity generates blunt products suitable for direct cloning, *Nucleic Acids Res.*, 20, 144, 1991.
38. **Stenesh, J. and Roe, B. A.,** DNA polymerase from mesophilic and thermophilic bacteria. I. Purification and properties of DNA polymerase from *Bacillus licheniformis* and *Bacillus stearothermophilus*, *Biochim. Biophys. Acta*, 272, 156, 1972.
39. **Sellmann, E., Schroder, K. L., Knoblich, I. M., and Westermann, P.,** Purification and characterization of DNA polymerases from *Bacillus* species, *J. Bacteriol.*, 174, 4350, 1992.
40. **Mead, D. A., McClary, J. A., Luckey, J. A., Kostichka, A. J., Witney, F. R., and Smith, L. M.,** Bst DNA polymerase permits rapid sequence analysis from nanogram amounts of template, *Biotechniques*, 11, 76, 1991.
41. **Lu, Y., Ye, S., and Hing, G.,** Large fragment of DNA polymerase I from *Bacillus stearothermophilus* (Bst polymerase) is stable at ambient temperature, *Biotechniques*, 11, 464, 1991.
42. **Hong, G. F.,** A method for DNA sequencing single-stranded cloned DNA in both directions, *Biosci. Rep.*, 1, 243, 1981.
43. **Mardis, E. R. and Roe, B. A.,** Automated methods for single-stranded DNA isolation and dideoxynucleotide DNA sequencing reactions on a robotic workstation, *BioTechniques*, 7, 840, 1989.

include false-positives caused by cross-hybridization (attributable to the presence of repetitive sequences or several copies of highly homologous sequences in a region). In some regions containing homologous repeats, contig assembly has been achieved only with the aid of PCR-based probes.[21] In contrast PCR-based mapping is specific, sensitive, and convenient, and it has the potential for automation.

Using YACs, several targeted regions, including one third of the X chromosome (done mainly by hybridization),[12] and the Y chromosome and long arm of human chromosome 21 (done by PCR-based methods), have been mapped.[22,23]

II. STS-CONTENT MAPPING STRATEGIES

In practice, long-range physical mapping of genomes with the help of STSs as landmarks includes the cloning of fragments of genomic DNA, screening the clones for the presence of common STSs, and then reconstructing a physical map from the cloned DNAs based on the relative position of STSs.[6,24] The basic outline of STS-content mapping with YACs is schematized in Figure 1.

One formulation of STS-content mapping involves (1) development of STSs, (2) PCR-based screening to detect the YAC clones containing particular STSs in a genomic library, (3) assembling YACs using the presence or absence of additional STSs (see below), and finally, (4) shifting from the use of random STSs to the use of end-fragment STSs of terminal YACs in a contig and thus closing the gaps. Figure 2 represents different steps involved and their interrelation.

Random STSs can be derived independent of YACs by a method assuring wide distribution in the region of interest.[25,26] Such random STSs allow coverage of large regions. These STSs need not be localized to specific regions at the start. Rather localization can be done subsequently, using growing contigs and new STSs supplemented with those from genes and linkage mapping probes of defined, ordered locations. Such an approach was adapted in the mapping of human chromosome Y and the long arm of chromosome 21.[22,23]

A second basic approach can be taken in STS-content mapping. It involves screening with some defined STSs, and deriving others from YAC inserts or from the ends of those inserts. Thus, coverage of targeted regions up to several megabases or more can be reached, for example, using a feedback loop in which STSs from YAC insert ends are screened only when they strike out into still-unmapped regions.[26]

Targeted STS-content mapping can be more efficiently employed if a collection of YACs specific for the chromosome or chromosomal region of interest are used; and of course mapping is more efficient with larger YACs.[13,23]

III. STS DEVELOPMENT

A. SOURCE OF DNA AND GENERATION OF STSS

Two essential components for STSs are generation and sequencing of DNA segments from the region of interest, and development of reproducible PCR assays. DNA segments of about 200 bp containing unique sequences are usually sufficient for the generation of STSs. The DNA segments can be derived from various source. These include flow-sorted chromosomes, hybrid cells containing region of interest, microdissection clones, known sequences specific for genes, linkage probes, single copy probes (assigned to a locus and partially sequenced), and insert segments from YACs.[25,26]

Sequences derived from these probes are employed in the next step of STS development, primer pair design. Several methods use computer-assisted analyses to detect repeated sequences using FASTA,[27] which are then excluded from further processing. Additional software, such as "*OSP*",[28] is used to assist in primer selection. Such programs select possible

18. **Klimczak, L. J., Grummt. F., and Burger, K. J.,** Purification and characterization of DNA polymerase from the archaebacterium *Sulfolobus acidocaldarius, Nucleic Acids Res.,* 13, 5269, 1985.
19. **Salhi, S., Elie, C., Forterre, P., de Recondo, A. M., and Rossignol, J. M.,** DNA polymerase from *Sulfolobus acidocaldarius*: replication at high temperature of long stretches of single-stranded DNA, *J. Mol. Biol.,* 209, 635, 1989.
20. **Hamal, A., Forterre, P., and Elie, C.,** Purification and characterization of a DNA polymerase from the archaebacterium *Thermoplasma acidophilum, Euro. J. Biochem.,* 190, 517, 1990.
21. **Kaledin, A. S., Slyusarenko, A. G., and Gorodetskii, S. I.,** Isolation and properties of DNA polymerase from the extremely thermophilic bacterium *Thermus flavus, Biokhimiya,* 46, 1576, 1981.
22. **Kainz, P., Schmiedlechner, A., and Strack, H. B.,** *In vitro* amplification of DNA fragments >10 Kb, *Anal. Biochem.,* 202, 46, 1992.
23. **Rao, V. B. and Saunders, N. B.,** A rapid polymerase-chain-reaction-directed sequencing strategy using a thermostable DNA polymerase from *Thermus flavus, Gene,* 113, 17, 1992.
24. **Kaledin, A. S., Slyusarenko, A. G., and Gorodetskii, S. I.,** Isolation and properties of DNA polymerase from the extremely thermophilic bacterium *Thermus ruber, Biokhimiya,* 47, 1785, 1982.
25. **Simpson, H. D., Coolobear, T., Vermue, M., and Daniel, R, M.,** Purification and some properties of a thermostable DNA polymerase from a *Thermotoga* species, *Biochem. Cell. Biol.,* 68, 1292, 1990.
26. **Klimczak, L. J., Grummt, F., and Burger, K. J.,** Purification and characterization of DNA polymerase from the archaebacterium *Methanobacterium thermoautotrophicum, Biochemistry,* 25, 4850, 1986.
27. **Brock, T. D. and Freeze, H.,** *Thermus aquaticus* gen. n., and sp. n., a non-sporulating extreme thermophile, *J. Bacteriol.,* 98, 289, 1969.
28. **Chien, A., Edgar, D. B., and Trela, J. M.,** Deoxyribonucleic acid polymerase from the extreme thermophile *Thermus aquaticus, J. Bacteriol.,* 127, 1550, 1976.
29. **Lawyer, F. C., Stoffel, S., Saiki, R. K., Chang, S. Y., Landre, P. A., Abramson, R. D., and Gelfand, D. H.,** High-level expression, purification, and enzymatic characterization of full-length *Thermus aquaticus* DNA polymerase and a truncated form deficient in 5′ to 3′ exonuclease activity, *PCR Methods Appl.,* 2, 275, 1993.
30. **Tindall, K. R. and Kunkel, T. A.,** Fidelity of DNA synthesis by the *Thermus aquaticus* DNA polymerase, *Biochemistry,* 27, 6008, 1988.
31. **Gelfand, D. H.,** *Taq* DNA polymerse, in *PCR Technology: Principles and Applications for DNA Amplification,* Erlich, H. A., Ed., Stockton Press, NY, 1989, 17.
32. **Lawyer, F. C., Stoffel, S., Saiki, R. K., Myambo, K., Drummond, R., and Gelfand, D. H.,** Isolation, characterization, and expression in Escherichia coli of the DNA polymerase gene from *Thermus aquaticus, J. Biol. Chem.,* 264, 6247, 1989.
33. **Myers, T. W. and Gelfand, D. H.,** Reverse transcription and DNA amplification by a *Thermus thermophilus* DNA polymerase, *Biochemistry,* 30, 7661, 1991.
34. **Erlich, H. A., Gelfand, D. H., and Sninsky, J. J.,** Recent advances in the polymerase chain reaction, *Science,* 252, 1643, 1991.
35. **Bej, A. K. and Mahbubani, M. H.,** Applications of the polymerase chain reaction in environmental microbiology, *PCR Methods Appl.,* 1, 151, 1992.
36. **Cariello, N. F., Swenberg, J. A., and Skopek, T. R.,** Fidelity of *Thermococcus litoralis* DNA polymerase (Vent™) in PCR determined by denaturing gradient gel electrophoresis, 19, 4193, 1991.
37. **Lohff, C. J. and Cease, K. B.,** PCR using a thermostable polymerase with 3′ to 5′ exonuclease activity generates blunt products suitable for direct cloning, *Nucleic Acids Res.,* 20, 144, 1991.
38. **Stenesh, J. and Roe, B. A.,** DNA polymerase from mesophilic and thermophilic bacteria. I. Purification and properties of DNA polymerase from *Bacillus licheniformis* and *Bacillus stearothermophilus, Biochim. Biophys. Acta,* 272, 156, 1972.
39. **Sellmann, E., Schroder, K. L., Knoblich, I. M., and Westermann, P.,** Purification and characterization of DNA polymerases from *Bacillus* species, *J. Bacteriol.,* 174, 4350, 1992.
40. **Mead, D. A., McClary, J. A., Luckey, J. A., Kostichka, A. J., Witney, F. R., and Smith, L. M.,** Bst DNA polymerase permits rapid sequence analysis from nanogram amounts of template, *Biotechniques,* 11, 76, 1991.
41. **Lu, Y., Ye, S., and Hing, G.,** Large fragment of DNA polymerase I from *Bacillus stearothermophilus* (Bst polymerase) is stable at ambient temperature, *Biotechniques,* 11, 464, 1991.
42. **Hong, G. F.,** A method for DNA sequencing single-stranded cloned DNA in both directions, *Biosci. Rep.,* 1, 243, 1981.
43. **Mardis, E. R. and Roe, B. A.,** Automated methods for single-stranded DNA isolation and dideoxynucleotide DNA sequencing reactions on a robotic workstation, *BioTechniques,* 7, 840, 1989.

Chapter 26

PCR IN SEQUENCE-TAGGED SITE (STS) CONTENT GENOME MAPPING

Anand K. Srivastava

TABLE OF CONTENTS

I. Introduction 239

II. STS-Content Mapping Strategies 240

III. STS Development 240
    A. Source of DNA and Generation of STSs 240
    B. PCR-based Isolation of Unique YAC Insert Sequences 242
        1. Isolation of Terminal Sequences from YACs 242
    C. Amplification-Driven Sequencing of Double-Stranded PCR Products 243

IV. PCR-Based Screening and Assembly of Map 244

V. Conclusions 245

Acknowledgments 245

References 245

I. INTRODUCTION

Genome mapping has been propelled by several recent technological advances. First, yeast artificial chromosomes (YACs)[1] have provided isolated large DNA segments of up to megabase size, with genes conserved in their normal genomic context.[2-8] This has bridged the gap between conventional cloning, at the kilobase level, and linkage mapping, at the centimorgan/megabase level. Second, the polymerase chain reaction (PCR) provides exponential accumulation of discrete fragments bounded by two specific oligonucleotide primers.[9,10] This led to the formulation of the notion of sequence-tagged sites (STSs, defined as "short single copy DNA sequences that can be detected by using PCR") as landmarks for the genome.[11] Taken together, the use of STSs and YACs has provided a new route to the mapping of genomes as disparate as human,[12,13] *C. elegans*,[14] and *Drosophila*.[15]

YACs, by allowing long-range coverage of the genomic DNA in cloned form, have facilitated the recovery of several genes implicated in genetic diseases.[16-18] To constitute a continuous physical map of larger regions, YAC clones are overlapped with others in a growing "contig". Overlaps are determined by the common content of DNA segments among YACs.[7] The "common DNA segments" could be single-copy hybridization probes or STSs; but mapping with STSs has the advantage that it both assembles and formats the map in a way that is easily communicated and reproduced at many sites.[6,11]

DNA-DNA hybridization and the polymerase chain reaction have both been utilized in the map assembly. Hybridization-based methods, which have been applied extensively in chromosome walking, use various probes (including end fragments of YAC inserts).[7] They are also applied in the verification of the integrity of a contig by "fingerprinting" using repetitive sequences like Alu, L1, or LF1.[19,20] However, hybridization has several limitations. These include false-positives caused by cross-hybridization (attributable to the presence of repetitive sequences or several copies of highly homologous sequences in a region). In some regions containing homologous repeats, contig assembly has been achieved only with the aid of PCR-based probes.[21] In contrast PCR-based mapping is specific, sensitive, and convenient, and it has the potential for automation.

Using YACs, several targeted regions, including one third of the X chromosome (done mainly by hybridization),[12] and the Y chromosome and long arm of human chromosome 21 (done by PCR-based methods), have been mapped.[22,23]

II. STS-CONTENT MAPPING STRATEGIES

In practice, long-range physical mapping of genomes with the help of STSs as landmarks includes the cloning of fragments of genomic DNA, screening the clones for the presence of common STSs, and then reconstructing a physical map from the cloned DNAs based on the relative position of STSs.[6,24] The basic outline of STS-content mapping with YACs is schematized in Figure 1.

One formulation of STS-content mapping involves (1) development of STSs, (2) PCR-based screening to detect the YAC clones containing particular STSs in a genomic library, (3) assembling YACs using the presence or absence of additional STSs (see below), and finally, (4) shifting from the use of random STSs to the use of end-fragment STSs of terminal YACs in a contig and thus closing the gaps. Figure 2 represents different steps involved and their interrelation.

Random STSs can be derived independent of YACs by a method assuring wide distribution in the region of interest.[25,26] Such random STSs allow coverage of large regions. These STSs need not be localized to specific regions at the start. Rather localization can be done subsequently, using growing contigs and new STSs supplemented with those from genes and linkage mapping probes of defined, ordered locations. Such an approach was adapted in the mapping of human chromosome Y and the long arm of chromosome 21.[22,23]

A second basic approach can be taken in STS-content mapping. It involves screening with some defined STSs, and deriving others from YAC inserts or from the ends of those inserts. Thus, coverage of targeted regions up to several megabases or more can be reached, for example, using a feedback loop in which STSs from YAC insert ends are screened only when they strike out into still-unmapped regions.[26]

Targeted STS-content mapping can be more efficiently employed if a collection of YACs specific for the chromosome or chromosomal region of interest are used; and of course mapping is more efficient with larger YACs.[13,23]

III. STS DEVELOPMENT

A. SOURCE OF DNA AND GENERATION OF STSS

Two essential components for STSs are generation and sequencing of DNA segments from the region of interest, and development of reproducible PCR assays. DNA segments of about 200 bp containing unique sequences are usually sufficient for the generation of STSs. The DNA segments can be derived from various source. These include flow-sorted chromosomes, hybrid cells containing region of interest, microdissection clones, known sequences specific for genes, linkage probes, single copy probes (assigned to a locus and partially sequenced), and insert segments from YACs.[25,26]

Sequences derived from these probes are employed in the next step of STS development, primer pair design. Several methods use computer-assisted analyses to detect repeated sequences using FASTA,[27] which are then excluded from further processing. Additional software, such as "*OSP*",[28] is used to assist in primer selection. Such programs select possible

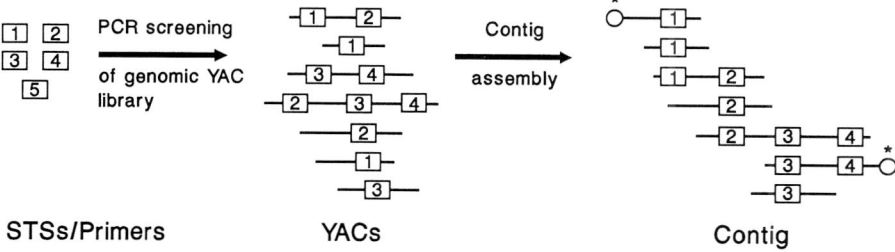

Figure 1. General strategy for constructing STS-content maps. (*Circled ends are potential candidates for the next cycle of STS development and chromosome walking.)

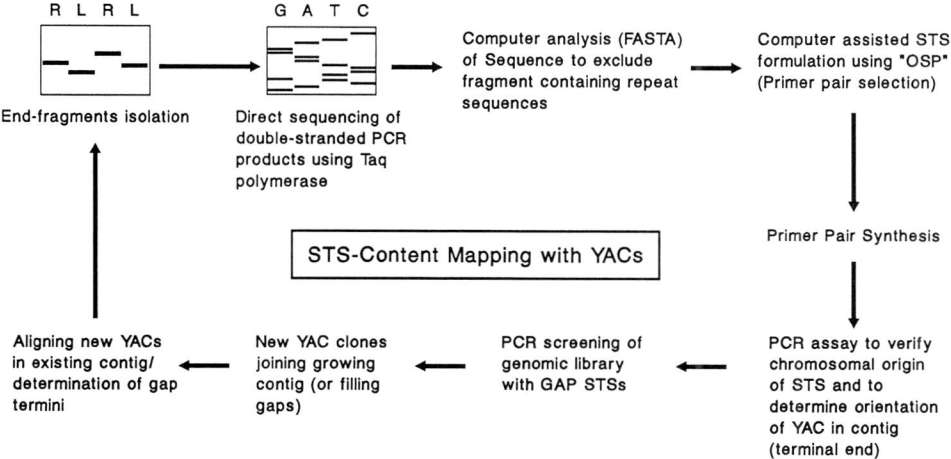

Figure 2. A flow diagram for PCR-based STS-content mapping with yeast artificial chromosomes. R and L indicate right and left end fragments of a YAC clone, respectively.

primer pairs optimized for parameters that include a number of physical properties of DNA (including G/C content and melting temperature of primers and products, etc.) that affect the efficiency of PCR. Oligodeoxynucleotide primers are then synthesized and tested before they are used in screening.

Optimal buffer conditions for PCR are determined for primer pairs. In an approach designed to favor uniform conditions to permit automation of large numbers of PCR reactions, three alternative buffers vary in their KCl concentration: TNK100 (100 mM), TNK50 (50 mM), or TNK 25 (25 mM).[29] All buffers contain 1.5 mM MgCl$_2$, 5 mM NH$_4$Cl, and 10 mM Tris-HCl, pH 8.6; the varying monovalent cation levels provide a range of stringency that gives adequate conditions for most primers. Standard temperature-cycling conditions in a Perkin Elmer thermocycler (94°C for 30 s, 55°C for 45 s, and 72°C for 45 s with 30 c) are normally adequate. (An additional denaturation for 2.5 min at 94°C prior to the first cycle has given more uniform results.)

Each primer pair is tested with several templates before YAC library screening. They include yeast genomic DNA, YAC DNA pools, human genomic DNA, and DNA from hybrid cells containing individual chromosomes or regions of interest, for example, an X chromosome (if mapping chromosome X) or chromosome 21 (if mapping chromosome 21). Inclusion of a deletion hybrid panel in subsequent PCR tests can localize particular STSs to defined intervals or can confirm their origin if localization is already known.

Under optimal PCR conditions, each STS must amplify the product of appropriate size (as determined by sequence of origin) from total human DNA and from a pool of anonymous YACs supplemented with human DNA. It should also give a positive signal from DNA of a somatic cell hybrid containing only the putative chromosome of origin. Critical negative controls show that it does not amplify a corresponding product from yeast DNA or a pool of random YACs.

B. PCR-BASED ISOLATION OF UNIQUE YAC INSERT SEQUENCES

From existing YACs, unique probes can be isolated in several ways. Recently, PCR-based methods have been widely used to generate such unique probes or sequences and develop STSs from YAC vector-insert junctions, or from within YAC inserts. Any unique sequence/probe from a YAC can be used for step-wise walking. However, end fragments have several advantages as sources of STSs. For example, their defined position along the YAC of origin, orienting YACs in a contig, can facilitate the accurate formatting of the map. They detect chimeric YACs (containing insert DNA from two different regions of the genome), and a contig developed with end STSs from longer YACs or terminal YAC tends to cover the maximum possible distance with each screening. Furthermore, when STSs used for screening are selected from YACs that have not been detected by previous STSs (see above), one ensures the finding of new YACs (or gaps) in screening.

Human-specific STSs have been developed using inter-Alu PCR.[30,31] Inter-Alu products can be amplified with the use of two Alu primers (orienting outward in an Alu repeat). Alu primers have been efficiently used to amplify products both from human-rodent hybrid cell lines[30] and from YAC inserts.[21] Such STSs have proven useful both to assemble contigs and to verify their integrity.

YAC DNA can also be amplified using Alu-vector PCR,[30,32] inverse PCR,[33,34] or ligation-mediated PCR.[35,36] Of course, end clones can also be derived by conventional subcloning,[7] but this method is time consuming and involves intensive manual operations. It is thus not the method of choice. PCR products developed using any of these methods are sequenced (see below) to develop STSs.

1. Isolation of Terminal Sequences from YACs

Several protocols have been developed to achieve this task. However, ligation-mediated methods for YAC end recovery have shown better success. An approach developed by Riley et al.[36] has successfully done end-fragment isolation using a "vectorette" (an oligonucleotide duplex containing a region of non-complementarity) as linker. A second method, adapted from the primer-ligation method of Mueller and Wold[35] has also been successfully used for the recovery of YAC ends by Kere et al.[26] It also consistently recovers "true ends". In principle, the YAC clone is digested with a range of restriction enzymes, a common linker is ligated to the DNA fragments, and terminal sequences are amplified using a vector-specific primer and a linker-specific primer.

More specifically, this method (essentially as described in Kere et al.,[26]) involves digestion of total yeast DNA (ca. 100 to 200 ng) containing a YAC (in a slice of an agarose plug or in agarose beads or as DNA in solution) with restriction enzymes RsaI or AluI (PvuII or ScaI can also be used) in a 17.6-µl reaction volume for 1 to 2 h. After cooling of the reaction mixture to room temperature, 2.4 µl of mix containing linker (1 pmol), ligation buffer (10X), and T4 DNA ligase (1 to 2 U) are added. Incubation is continued at room temperature for another 1 to 2 h. The ligation mix is diluted with 80 µl of water and heated to 90°C for 5 to 10 min to destroy residual enzyme activity. Then 2.0 µl of the ligation mixture are amplified by PCR in a total volume of 10 µl with primer specific for right and left arm of YAC vector (10 pmol, each in separate tubes) and linker primer (1 pmol). All primer sequences are described in Kere et al.[26] Amplification is performed using conditions 94°C for 30 s, 65°C for 45 s, and 72°C for 1.5 min for 35 c in TNK50 buffer (see above) containing 0.25 U of AmpliTaq DNA polymerase (Perkin Elmer Cetus) and 125 µM dNTPs. Each PCR mix is diluted to 250 µl with water, and 1.0 µl of mix are reamplified using 10 pmol each of linker primer and primer for corresponding arms, internal to the one used in first PCR in a total volume of 10 µl under identical condition (see above). Amplified products are purified on a low-melting-point agarose gel. Each reaction produces a single specific band from each YAC; occasionally multiple bands are seen, but they normally result from incomplete digestion by restriction enzymes or from several YACs in the same strain.

This method relies on the presence of restriction sites in close proximity to a YAC end. In some instances where only small products are seen (due to multiple nearby restriction sites), the use of different enzyme usually permits the recovery of a reasonably sized end fragment.

The capture PCR technique has a similar basis.[37] It utilizes ligation-mediated recovery of DNA in combination with physical enrichment of extension products, using biotin-streptavidin interaction to aid in subsequent amplification, as reported by Rosenthal and Jones.[38] Here, multiple extension reactions are performed using biotinylated primers, permitting the subsequent isolation of the amplified product on a streptavidin-coated support.

Inverse PCR involves digestion of a source DNA of interest and circularization of cleavage products (under conditions that favor the formation of monomeric circles) prior to amplification using primers that are reversed in direction in respect to their normal orientation for PCR.[33,34] For YACs, purified YAC DNAs are digested with EcoRV and HincII to ensure a convenient length of vector sequence at the insert junction.[39] This technique relies on circularization and is dependent on the presence of two restriction sites within the size range of PCR amplification. It has several attractive features (for example, it can be used to amplify regions of unknown sequence flanking any characterized segment of the genome), but it is more cumbersome and less certain than other end-recovery procedures.

Another method for YAC end-fragment isolation utilizes Alu-vector PCR, based on the fact that Alu repeat sequences are present frequently in human genome.[40] Presence of Alu sequences within a short distance of the end in right orientation is prerequisite for the success of this method. Specificity of the reaction is assured with the use of one primer specific for the vector arm. Reamplification of product with nested primer from a vector arm increases the recovery of "true ends". The technique is simple, but works on only 30 to 80% of YACs in various regions.

C. AMPLIFICATION-DRIVEN DIRECT SEQUENCING OF DOUBLE-STRANDED PCR PRODUCTS

An easy and rapid sequencing method is necessary to generate large numbers of STSs. The polymerase chain reaction has provided an effective way to bypass subcloning steps in the preparation of templates and sequencing. Amplification-driven sequencing methods (temperature-cycle sequencing with Taq polymerase) have been extensively used for a variety of tasks, including the direct sequencing of double-stranded PCR products.

Amplification-based sequencing permits the rapid sequencing of large numbers of templates with minimal manipulation. The amount of template required is very low (nanogram quantities). Other significant features are (1) either a primer used for template preparation or an internal primer; (2) efficacy with either double- or single-stranded templates, thus eliminating the need for asymmetric reamplification of double-stranded PCR products prior to sequencing; (3) adequacy of DNA present in agarose gel bands, with no further DNA purification required; and (4) utility of either ^{32}P or ^{35}S as label. This method has also been successfully adapted for fluorescence-based sequencing, eliminating the use of radioactive materials.

Two variations of sequencing reactions have been used to label sequencing products. In one approach, the sequencing primer is kinase labeled at its 5'-end with [γ-^{32}P]dATP. Alternatively, an unlabeled primer can be used with either [α-^{32}P]dATP or [α-^{35}P]dCTP in the sequencing reactions. Detailed methodology has been described, for example, in Reference 41. In brief, a sequencing mix is prepared in a total volume of 32 μl containing 8 μl 5X TNK50 buffer (see above), 4.5 μl 25 μM dNTPs solution, 1.0 U of AmpliTaq, 0.5 pmol of end-labeled sequencing primer. Then 8.0 μl of mix are added to each of four tubes (labeled G, A, T, and C) containing 1.0 μl of template DNA (50 to 200 ng) and 1.0 μl of 0.25 mM ddGTP, 2.5 mM ddATP, 3.75 mM ddTTP, or 2.5 mM ddCTP, respectively. The sequencing reaction is performed for 20 to 30 c at 94°C for 1 min, 55°C for 2 min, and 72°C for 2 min. The reaction is terminated with 4 μl of stop solution (95% formamide, 0.1% xylene cyanol, and 0.1% bromophenol blue) at the end of cycle, and the content is heated at 90°C for 2 min before loading 3 to 4 μl of each reaction mix on standard denaturing sequencing gel. It should be noted that if the template DNA is too dilute, the sequencing reaction volume can be adjusted to accommodate template DNA in more than 1.0 μl per reaction. If, instead, large quantities of template DNA are used, the number of cycles in amplification reaction can be reduced to 10 to 15.

IV. PCR-BASED SCREENING AND ASSEMBLY OF MAP

For many purposes, including mapping and hunting for individual genes, it may be necessary to recover YACs that contain particular sequences from large collections of clones ("libraries").

As discussed above, of the two major alternatives for screening, PCR-based screening has some advantages over hybridization-based assays. The choice of method is based on available materials and technology as well as personal preference. Hybridization-based assays are less expensive, rapid, and permit re-use of individual filters bearing lysed colonies of yeast (though such filters show great variability in reusability compared to filters made with bacterial lysates or purified DNA).

In order to minimize work and numbers of PCR assays for screenings, libraries of YAC clones are generally organized in an ordered fashion in pools or arrays. PCR-based screening can be performed directly with yeast cells or extracted DNA. Initial formulations of PCR-based screening involved a combination of PCR and hybridization steps;[42] newer strategies are completely PCR based.[22,43] DNAs from YAC clones are pooled in groups to yield a screening tree. Positive signals in any of these pools are followed up by screening subpools of decreasing complexity, with the number of clones tested decreased in each subpool by a factor of 2 to 4.[22,43]

Alternative formulations use combinatorial steps, in which individual clones are included in a number of pools (such as rows and columns of a matrix array). The address of a clone is inferred from the ensemble of positive pools.[44]

Pure screening through "tree" assemblies has the advantage that every positive is scored against negatives at every level, and the method tends to minimize the number of PCR assays.[22,43] Combinatorial schemes use fewer "levels" of screening, and are therefore faster and easier to automate; but this ease is achieved at the cost of more assays and some difficulty discriminating the identity of multiple positives with nearby addresses.[44]

From a YAC library, using appropriate PCR assays, YACs containing specific STSs can be isolated by a variety of such schemes. Following their isolation, YACs are also sized by pulsed-field gel electrophoresis (PFGE).[45] The information about STS content and size of a collection of clones can align and orient them in an overlapping manner, i.e., a map. As indicated above, the process can be greatly facilitated using end-clone STSs. Isolation of both end fragments from several YACs (usually a terminal one in a contig) and subsequent PCR assays, as described for testing STSs, of other YACs determines the orientation of YACs and identifies any chimeric ends.

Resolution and depth in a contig can be further improved by screening YAC panels with more STSs. All this information also makes it easier to deduce the relative order and spacing of the STSs and the YAC ends.

To verify YAC matches, inter-Alu probes or Alu-Alu fingerprinting using PCR is used widely. Since Alu sequences occur scattered in the human genome,[40] the pattern of Alu-Alu fragments is characteristic of the segments of chromosomal DNA under study, providing a kind of "fingerprint" of individual YACs or sets of YACs.[13,19]

V. CONCLUSIONS

It should be noted that most of the mapping data can be obtained with PCR as the analytical tool (though of course sizing and restriction fragment analysis for further refinement of a map requires the use of PFGE). Besides the utility of PCR in physical mapping of the genome, it has been extensively utilized in the other aspects of genome-related work, for example, in functional analysis. Analogous to STSs, systematic sequencing of cDNA clones has been used to generate expressed sequence tags (ESTs).[46] These ESTs can be localized on specific chromosomes by fluorescent *in situ* hybridization (FISH) or by PCR tests using hybrid cell lines harboring individual chromosome or more accurately by the use of deletion hybrid panel.[47] ESTs can be used directly for PCR-based screening of YAC libraries, or can be placed on existing YAC contigs (if the approximate localization of ESTs is known).[48] In an approach very similar to PCR-based VAC screening[42,44] primer pairs for ESTs, transcribed sequences can be used in a hierarchical, tree-based scheme for the completely PCR-based screening of cDNA libraries to recover full length cDNAs. In another application, STSs developed from conserved regions of a known gene may permit one to isolate syntenic regions of other genomes.[49]

The practical advances of PCR and YACs have joined with the theoretical formulation of STSs to revolutionize the mapping of DNA across long distances. Combined with the use of radiation hybrids, somatic cell hybrids, patient deletion panels, cytogenetic placement, and other methods to place STSs in intervals, current protocols are carrying the weight of the genome initiative, and further advances, such as increased automation, improvement of the quality of YACs, and extension of the product size achievable with PCR,[50] will surely make genome mapping increasingly straightforward.

ACKNOWLEDGMENTS

I thank David Schlessinger for his valuable suggestions, critical evaluation and support. I also thank Fatima Abidi and my colleagues for their comments. This work was supported by grants from NSF and NIH.

REFERENCES

1. **Burke, D. T., Carle, G. F., and Olson, M. V.,** Cloning of large segments of exogenous DNA into yeast by means of artificial chromosome vectors, *Science,* 236, 806, 1987.
2. **Albertsen, H. M., Abderrahim, H., Cann, H. M., Dausset, J., Le Paslier, D., and Cohen, D.,** Construction and characterization of a yeast artificial chromosome library containing seven haploid human genome equivalents, *Proc. Natl. Acad. Sci. U.S.A.,* 87, 4256, 1990.
3. **Anand, R., Villasante, A., and Tyler-Smith, C.,** Construction of yeast artificial chromosome libraries with large inserts using fractionation by pulsed-field gel electrophoresis, *Nucleic Acids Res.,* 17, 3425, 1989.
4. **Anand, R., Ogilvie, D. J., Butler, R., Riley, J. H., Finniear, R. S., Powell, S. J., Smith, J. C., and Markham, A. F.,** A yeast artificial chromosome contig encompassing the cystic fibrosis locus, *Genomics,* 9, 124, 1991.
5. **D'Urso, M., Zucchi, I., Ciccoricola, A., Palmieri, G., Abidi, F. and Schlessinger, D.,** A yeast artificial chromosome expresses enzymatically active glucose 6-phosphate dehdrogenase, *Genomics,* 7, 531, 1990.

6. **Green, E. D. and Olson, M. V.,** Chromosomal region of the cystic fibrosis gene in yeast artificial chromosomes: a model for human genome mapping, *Science,* 250, 94, 1990.
7. **Little, R. D., Pilia, G., Johnson, S., Zucchi, I., D'Urso, M., and Schlessinger, D.,** Yeast artificial chromosomes spanning 8 Mb and 10-15 centiMorgans of human cytogenetic band Xq26, *Proc. Natl. Acad. Sci. U.S.A.,* 89, 177, 1992.
8. **Schlessinger D.,** Yeast artificial chromosomes: tools for mapping and analysis of complex genomes, *Trends Genet.,* 6, 248, 1990.
9. **Saiki, R. K., Scharf, S. J., Faloona, F., Mullis, K. B., Horn, G. T., Erlich, H. A., and Arnheim, N.,** Enzymatic amplification of β-globin genomic sequences and restriction site analysis for diagnosis of sickle cell anemia, *Science,* 230, 1350, 1985.
10. **Saiki, R. K., Gelfand, D. H., Stoffel, S., Scharf, S. J., Higuchi, R., Horn, G. T., Mullis, K. B., and Erlich, H. A.,** Primer-directed enzymatic amplification of DNA with a thermostable DNA polymerase, *Science,* 239, 487, 1988.
11. **Olson, M. V, Hood, L., Cantor, C., and Botstein, D.,** A common language for physical mapping of the human genome, *Science,* 245, 1434, 1990.
12. **Schlessinger, D., Little, R. D., Freije, D., Abidi, F., Zucchi, I., Porta, G., Pilia, G., Nagaraja, R., Johnson, S. K., Yoon, J. Y., Srivastava, A., Kere, J. Palmieri, G., Ciccodicola, A., Montanaro, V., Romano, G., Casamassimi, A., and D'Urso, M.,** Yeast artificial-based genome mapping: some lessons from Xq24-28, *Genomics,* 11, 783, 1991.
13. **Bellanne-Chantelot, C., Lacroix, B., Ougen, P., Billault, A., Beaufilis, S., Bertrand, S., Georges, I., Glibert, F., Gros, I., Lucotte, G., Susini, L., Codani, J-J., Gesnouin, P., Pook, S., Vaysseix, G., Lu-Kuo, J., Ried, T., Ward, D., Chumakov, I., Le Paslier, D., Barillot, E. and Cohen, D.,** Mapping the human genome by fingerprinting yeast artificial chromosomes, *Cell,* 70, 1059, 1992.
14. **Coulson, A., Waterston, R., Kiff, J., Sulston, J., and Kohara, Y.,** Genome linking with yeast artificial chromosomes, *Nature,* 335, 184, 1988.
15. **Garza, D., Ajioka, J. W., Burke, D. T., and Hartl, D. L.,** Mapping the Drosophila genome with yeast artificial chromosomes, *Science,* 246, 641, 1989.
16. **Hirst, M. C., Rack, K., Nakahori, Y., Roche, A., Bell, M. V., Flynn, G., Christadoulou, Z., MacKinnon, R. N., Francis, M., Littler, A. J., Anand, R., Poustka, A.-M., Lehrach, H., Schlessinger, D., D'Urso, M., Buckle, V. J., and Davies, K. E.,** A YAC contig across the fragile X site defines the region of fragility, *Nucleic Acids Res.,* 19, 3283, 1991.
17. **Attree, O., Olivos, I. M., Okabe, I., Bailey, L. C., Nelson, D. L., Lewis, R. A., McInnes, R. R., and Nussbaum, R. L.,** The Lowe's oculocerebrorenal syndrome gene encodes a protein highly homologous to inositol polyphosphate-5-phosphatase, *Nature,* 358, 239, 1992.
18. **Franco, B., Guioli, S., Pragliola, A., Incerti, B., Bardoni, B., Tonlorenzi, R., Carrozzo, R., Maestrini, E., Pierett, M., Taillon-Miller, P., Brown, C. J., Willard, H. W., Lawrence, C., Persico, M. G., Camerino, G., and Ballabio, A.,** A gene deleted in Kallmann's Syndrome shares homology with neural cell adhesion and axonal path-finding molecules, *Science,* 353, 529, 1991.
19. **Porta, G., Zucchi, I., Hillier, L., Green, P., Nowotny, V., D'Urso, M., and Schlessinger, D.,** Alu and L1 sequence distribution in Xq24-q28, and their comparative utility in YAC contig assembly and verification, *Genomics,* 16, 417, 1993.
20. **Zucchi, I. and Schlessinger, D.,** Distribution of moderately repetitive sequence PTR5 and LF1 in Xq24-28 human DNA, and their use in assembling YAC contigs, *Genomics,* 12, 264, 1992.
21. **Freije, D. and Schlessinger, D.,** A 1.6 Mb contig of yeast artificial chromosomes around the human factor VIII gene reveals a locus homologous to DXYS64 as well as several duplicated sequences, *Am. J. Hum. Genet.,* 51, 62, 1992.
22. **Foote, S., Vollarth, D., Hilton, A., and Page, D.,** The human Y chromosome: overlapping DNA clones spanning the euchromatic region, *Science,* 258, 60, 1992.
23. **Chumakov, I., Rigault, P., Guillou, S., Ougen, P., Billaut, A., Guasconi, G., Gervy, P., LeGall, I., Soularue, P., Grinas, L., Bougueleret, L., Bellanné-Chantelot, C., Lacroix, B., Barillot, E., Gesnouin, P., Pook, S., Vaysseix, G., Frelat, G., Schmitz, A., Sambucy, J.-L., Bosch, A., Estivill, X., Weissenbach, J., Vignal, A., Riethman, H., Cox, D., Patterson, D., Gardiner, K., Hattori, M., Sakaki, Y., Ichikawa, H., Ohki, M., Le Paslier, D., Heilig, R., Antonarakis, S., and Cohen, D.,** Continuum of overlapping clones spanning the entire human chromosome 21q, *Nature,* 359, 380, 1992.
24. **Green, E. D. and Green, P.,** Sequence-tagged site (STS) content mapping of human chromosomes: theoretical considerations and early experiences, *PCR Methods Appl.,* 1, 77, 1991.
25. **Green, E. D., Mohr, R. M., Idol, J. R., Jones, M., Buckingham, J. M., Deaven, L. L., Moyzis, R. K., and Olson, M. V.,** Systematic generation of sequence-tagged sites for physical mapping of human chromosomes — application to the mapping of human chromosome 7 using yeast artificial chromosomes, *Genomics,* 11, 548, 1991.

26. **Kere, J., Nagaraja, R., Mumm, S., Ciccodicola, A., D'Urso, M., and Schlessinger, D.,** Mapping human chromosomes by walking with sequenced-tagged sites from end fragments of Yeast Artificial Chromosomes, *Genomics,* 14, 241, 1992.
27. **Pearson, W. R. and Lipman, D. J.,** Improved tools for biological sequence comparison, *Proc. Natl. Acad. Sci. U.S.A.,* 85, 2444, 1988.
28. **Hillier, L. and Green, P.,** OSP: a computer program for choosing PCR and DNA sequencing primers. *PCR Methods Applic.,* 1, 124, 1991.
29. **Blanchard, M. M., Miller, P. T., Nowotny, P., and Nowotny, V.,** PCR-buffer optimization with uniform temperature regimen to facilitate automation, *PCR Methods Appl.,* 2, 234, 1993.
30. **Nelson, D. L., Ledbetter, S. A., Corbo, L., Victoria, M. F., Ramirez-Solis, R., Webster, T. D., Ledbetter, D. H., and Caskey, T. C.,** Alu polymerase chain reaction: a method for rapid isolation of human-specific sequences from complex sources, *Proc. Natl. Acad. Sci. U.S.A.,* 86, 6686, 1989.
31. **Cole, C. G., Goodfellow, P. N., Bobrow, M., and Bentley, D. R.,** Generation of novel sequence tagged sites (STSs) from discrete chromosomal regions using Alu-PCR, *Genomics,* 10, 816, 1991.
32. **Breukel, C., Wijnen, J., Tops, C., Klift, H. Dauwerse, H., and Khan, P. M.,** Vector-Alu PCR: a rapid step in mapping cosmid and YACs, *Nucleic Acids Res.,* 18, 309, 1990.
33. **Ochman, H., Gerber, A. S., and Hartl, D. L.,** Genetic application of an inverse polymerase chain reaction, *Genetics,* 120, 621, 1988.
34. **Triglia, T., Peterson, M. G., and Kemp, D. J.,** A procedure for *in vitro* amplification of DNA segments that lie outside the boundaries of known sequences, *Nucleic Acid Res.,* 16, 8186, 1988.
35. **Mueller, P. R. and Wold, B.,** *In vivo* footprinting of a muscle specific enhancer by ligation mediated PCR, *Science,* 246, 780, 1989.
36. **Riley, J., Butler, R., Ogilvie, D., Finniear, R., Jenner, D., Powell, S., Anand, R., Smith, J. C., and Markham, A. F.,** A novel, rapid method for the isolation of terminal sequences from yeast artificial chromosome (YAC) clones, *Nucleic Acids Res.,* 18, 2887, 1990.
37. **Lagerstrom, M., Parik, J., Malmgren, H., Stewart, J., Pettersson, U., and Landegren, U.,** Capture PCR: efficient amplification of DNA fragments adjacent to a known sequence in human and YAC DNA, *PCR Methods Appl.,* 1, 111, 1991.
38. **Rosenthal, A. and Jones, D. S. C.,** Genomic walking and sequencing by oligo-cassette mediated polymerase chain reaction, *Nucleic Acids Res.,* 18, 3095, 1990.
39. **Ochman, H., Medhora, M., Garza, D., and Hartl, D.,** Amplification of flanking sequences by inverse PCR, in *PCR Protocols: A Guide to Methods and Applications,* Innis, M. A., Gelfand, D. H., Sninsky, J. J., and White, T. J., Eds., Academic Press, San Diego, CA, 1990, 219.
40. **Jelinek, W. R. and Schmidt, C. W.,** Repetitive sequences in eukaryotic DNA and their expression, *Ann. Rev. Biochem.,* 51, 813, 1982.
41. **Srivastava, A. K., Montanaro, V., and Kere, J.,** Simplified template preparation and improved direct sequencing using Taq polymerase, *PCR Methods Appl.,* 1, 255, 1992.
42. **Green, E. D. and Olson, M. V.,** Systematic screening of yeast artificial-chromosome libraries by use of the polymerase chain reaction, *Proc. Natl. Acad. Sci. U.S.A.,* 87, 1213, 1990.
43. **Brownstein, B.H. and co-workers,** Personal communication, 1992.
44. **Nowotny, V. and co-workers,** Personal communication, 1992.
45. **Carle, G. F. and Olson, M. V.,** An electrophoretic karyotype for yeast. *Proc. Natl. Acad. Sci. U.S.A.,* 82, 3756, 1985.
46. **Adams, M. D., Kelley, J. M., Gocayne, J. D., Dubnick, M., Polymeropolous, M. H., Xiao, H., Merril, C. R., Wu, A., Olde, B., Moreno, R. F., Kerlavage, A. R., McCombie, W. R., and Venter, J. C.,** Complementary DNA sequencing: expressed sequence tags and human genome project, *Science,* 252, 1651, 1991.
47. **Polymeropoulos, M. H., Xiao, H., Glodek, A., Gorski, M., Adams, M. D., Moreno, R. F., Fitzgerald, M. G., Venter, J. C., and Merril, C. R.,** Chromosomal assignment of 46 Brain cDNAs, *Genomics,* 12, 492, 1992.
48. **Mazzarella, R. and Srivastava, A. K.,** Expressed sequence tags (ESTs) in the transcriptional mapping of the x chromosome, submitted, 1993.
49. **Mazzarella, R., Motanaro, V., Kere, J., Reinbold, R., Ciccodicola, A., D'Urso, M., and Schlessinger, D.,** Conserved sequence-tagged sites: a phylogenetic approach to genome mapping, *Proc. Natl. Acad. Sci. U.S.A.,* 89, 3681, 1992.
50. **Ohler, L. D. and Rose, E. A.,** Optimization of long-distance PCR using a transposon-based model system, *PCR Methods Appl.,* 2, 51, 1992.

Chapter 27

FALSE-POSITIVES AND CONTAMINATION IN PCR

E. P. H. Yap, Y.-M. O. Lo, K. A. Fleming, and J. O'D. McGee

TABLE OF CONTENTS

I. Introduction 249

II. General Precautions 251
    A. Laboratory 251
    B. Personnel 252
    C. Equipment 252
    D. Reagents 252

III. Specific Prevention 252
    A. Pre-PCR Sterilization 252
        1. Enzymatic Digestion 253
            a. Endonuclease Digestion 253
            b. DNase I Digestion 253
            c. Exonuclease Digestion 253
        2. Ultraviolet Irradiation 253
        3. Ionizing Radiation 253
        4. Psoralens 254
    B. Post-PCR Sterilization 254
        1. Isopsoralens 254
        2. Uridine Incorporation and Glycosylation 254

IV. Contamination Detection and Result Verification 254
    A. Negative Controls 254
    B. Sequence Polymorphism 255
        1. Sequencing 255
        2. Heteroduplex Analysis 255
        3. Single-Strand Conformation Polymorphism 256
    C. Repetition 256

V. Management of Contamination 256

Acknowledgments 257

References 257

I. INTRODUCTION

The susceptibility of the polymerase chain reaction to false-positive results due to accidental contamination of the sample or reagents is a limitation that currently restricts its use in routine diagnostic analytical applications. False-negative results, on the other hand, are less common in PCR due to the extremely high sensitivity of this reaction. Where present, they are

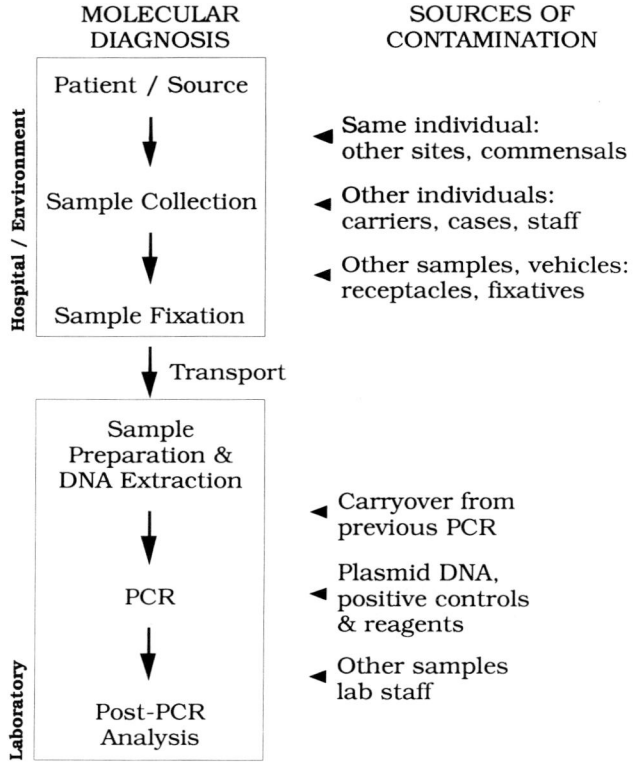

Figure 1. Contamination in diagnostic PCR.

usually due to inadequate sample, unremoved inhibitors in the sample, or inappropriate choice of target sequences.

False-positive results are usually due to amplification of specific target from sources other than the original sample. The sources and types of contamination (Figure 1) are (in decreasing order of importance) as follows:

- carry-over products from previous PCR, by far the most significant, due to degree of amplification (10^6- to 10^{12}-fold)[1]
- plasmid, phage, or cosmid vectors containing cloned target DNA, especially where used as probes for hybridization experiments[2]
- positive controls (cloned DNA, highly infective tissue with high copy number)
- cross-contamination between samples
- contamination from other sources during collection of samples (for instance, human papillomavirus DNA can be detected in the air and surgical gloves during cervical biopsy)
- contamination of *Taq* polymerase with bacterial DNA, an important consideration when amplifying conserved ribosomal RNA sequences[3]

It should be noted that while most instances of contamination occur in the laboratory, for which general and specific precautions are necessary and available (see below), contamination may also occur during sample collection, processing, and transport, during which it may not be possible to implement stringent sterile procedures.

Recent methodological advances in PCR to improve sensitivity (e.g., nested PCR) and specificity (e.g., Hot-Start PCR[4]) involve more manipulative steps, reduce the threshold target copy number for a positive result, and/or increase the concentration and amount of DNA products. These, per se, pose added risk of carry-over. Added care should be taken in such instances, and the potential advantages of these adaptations should be balanced against the risks. In many applications, for instance, PCR specificity is better confirmed by a sensitive hybridization assay rather than a further round of nested amplification, which increases risk of contamination. Alternative approaches may also reduce the occurrence of false-positives (automated Hot-Start;[4] *in situ* PCR[5]). However, the problem of contamination exists to some extent with other forms of exponential amplification methods (ligase chain reaction (LCR), Q-β replicase-mediated amplification, self-sustained sequence replication (3SR), nucleic acid sequence-based amplification (NASBA™), strand-displacement amplification (SDA), and repair chain reaction).

Negative controls, repetition of test, and sequence analysis are required to verify the result, especially for diagnostic purposes. Once a general contamination problem is detected, it is often more efficient to discard existing reagent stocks than to perform exhaustive troubleshooting. The practical aspects of these procedures are discussed below.

Examples of reasons for false-positive results other than exogenous contamination include the following:

- presence of amplifiable DNA during reverse-transcription PCR for RNA target
- amplification of similar sequences that bind primers, such as pseudogenes, and homologues
- non-specific amplification
- presence of specific target not of biological or medical significance (e.g., commensal non-pathogenic bacteria, dead non-viable pathogens)

Most of these problems may be obviated by judicious primer design, optimization of PCR conditions, nested PCR, post-PCR sequence analysis (e.g., hybridization), and proper interpretation in the clinical/biological context. These factors are specific to individual assays and are not discussed further in this chapter.

II. GENERAL PRECAUTIONS

The principle underlying contamination prevention is the physical separation of pre- and post-PCR reactions, including reagents, equipment, and personnel (and their effects). Transmission may range from the obvious (e.g., pipettes, water for dilution, gloves) to the subtle (e.g., microcentrifuges, iceboxes, pens used for labeling, microtome blades). Meticulous laboratory precautions and specific preventive procedures are therefore necessary. Often, the degree of laboratory discipline and precautions needed equals those required for sterile procedures common in the operating theater or microbiological laboratory.[6] The importance of such general measures cannot be over-emphasized.

A. LABORATORY

- Ideally, pre- and post-PCR procedures are carried out in different labs, so as to minimize inadvertent sharing of resources. If this is not feasible, separate parts of the same lab should be designated for pre-PCR (preparing and diluting stocks/PCR mixes, processing DNA samples, storing reagents and equipment) and for post-PCR (opening PCR tubes, storage of products, gel electrophoresis, and all subsequent steps).
- Containment units and laminar flow hoods for all handling procedures (with air flowing toward operator) may be effective.
- Good planning should minimize handling of liquids and pipetting.

- Pre-PCR procedures should be scheduled before post-PCR procedures in daily or weekly routines.

B. PERSONNEL

- Some research groups have different staff to handle pre- and post-PCR materials
- Laboratory coats, notebooks, and stationery are some of the personal items that may escape notice.
- Gloves should be changed frequently, as they cost less than reagents and salaries.
- Each person using PCR should have his/her own set of reagents and disposable items, including such things as mineral oil overlay, polypropylene tubes. Ground rules should be established for the use of common resources.

C. EQUIPMENT

- Microcentrifuges, water baths, autopipettes, iceboxes, tube racks, and labeling pens should be dedicated for PCR use.
- Smaller items may be UV-irradiated at regular intervals to cross-link contaminating DNA by exposure to a UV source for at least 2 h (see below). UV transilluminators for viewing gels must not be used as they are a definite contamination risk.
- Disposable items such as polypropylene tubes and pipette tips should be used from bags sealed at point of manufacture and need not be autoclaved.
- Positive displacement pipettes or barrier pipette tips help eliminate aerosol carry-over.

D. REAGENTS

- Reagents should be made in large batches from molecular biology grade chemicals kept separate from general use.
- Reagents should be diluted and stored in small aliquots that are serially numbered for identification.
- When assembling reactions, reagents should be returned to refrigerated storage before DNA samples are pipetted.
- The use of strongly positive controls (e.g., insufficiently diluted plasmid DNA) should be avoided.

III. SPECIFIC PREVENTION

Several methods have been developed to ensure that no DNA other than from the intended sample is present in PCR reagents and reactions. These may either remove contaminating DNA in the components of the PCR before the reaction (pre-PCR sterilization) or inactivate the PCR products after the reaction so that they are unable to serve subsequently as effective template (post-PCR sterilization), thus specifically preventing carry-over. Conventional microbiological sterilization methods such as filtration and autoclaving are generally not effective for removal of DNA.

A. PRE-PCR STERILIZATION

Pre-PCR methods to remove DNA are performed on the components of PCR, before addition of sample DNA. This assumes that sample DNA is not contaminated during processing (e.g., cell lysis, proteolysis, protein removal). Some of these methods affect the activity of *Taq* polymerase and/or oligonucleotide primers, and may have a deleterious effect on the efficiency of PCR.

1. Enzymatic Digestion
a. Endonuclease Digestion

Restriction enzymes that cleave within the target sequence for PCR may be used to restrict any contaminating sequence prior to addition of the target.[7,8] After incubation and before PCR, the enzyme is destroyed by thermal denaturation (e.g., 94°C for 10 min); thus thermostable restriction enzymes such as *Taq*I should not be used for this purpose. In a model system, restriction with *Msp*I (10 U for 1 h) reduced contamination by a factor of 5 to 10 without impairing the efficiency of the PCR.[7]

b. DNase I Digestion

This approach is similar in principle to restriction enzyme digestion except that DNase I is used. Prior treatment with 0.5 U of DNase I for 30 min was shown to reduce contamination by a factor of 1000 without impairing the efficiency of the PCR.[7]

c. Exonuclease Digestion

Certain exonucleases, e.g., exonuclease III and T7 exonuclease, when added to fully assembled PCR reactions containing genomic target DNA, are able to render carry-over PCR product molecules non-amplifiable but spare identical target sequences in genomic DNA.[9,10] In a model system a 30-min incubation with exonuclease III was able to degrade 5×10^5 copies of carry-over amplicons.[9] Several mechanisms have been postulated for the selectivity against PCR products: (1) The relatively long chain length of genomic DNA resists degradation by exonucleases better than the comparatively short PCR products.[9] (2) Carry-over DNA and genomic target are geometrically different; primer-binding sites are located at opposite ends of the molecule in PCR products but are likely to be on the same side of the geometric center for genomic DNA (which is randomly sheared during DNA extraction). Since T7 exonuclease digestion of double-stranded DNA results in two single-stranded half molecules, primer-binding sites will be on different strands in digested carry-over DNA but on the same strand in digested genomic target. Thus genomic target, but not PCR carry-over, remains amplifiable.[10]

Exonuclease treatment for the prevention of PCR carry-over thus possesses the advantage that the completed reaction tubes do not have to be re-opened for addition of the target and/or *Taq* polymerase. Furthermore, exonuclease treatment has the added advantage of being able to destroy even non-uridine-containing PCR products and thus would be useful in an already-contaminated environment.

2. Ultraviolet Irradiation

Exposure of PCR reagents to short-wave UV radiation (254 to 300 nm) for 5 to 20 min results in inactivation of contaminating DNA.[1] This is thought to occur through the dimerization of pyrimidine residues, resulting in inter-strand cross-links or premature termination of *Taq* polymerase extension. While deoxynucleotide triphosphates are chemically resistant to (despite absorbing) UV, *Taq* polymerase activity is reduced by UV irradiation, and the sensitivity of oligonucleotide primers to UV modification varies.[11] However, the efficacy of UV for making contaminating template unamplifiable appears to be dependent on the length and sequence of the DNA, shorter fragments generally being less susceptible than longer ones to UV inactivation.[12] While it has been recommended that PCR bench surfaces and equipment be exposed to UV light when not in use,[13] and UV hood-containment bench tops are available commercially, UV inactivation is less effective in inactivating dried DNA compared with DNA in aqueous solution.[12]

3. Ionizing Radiation

Gamma radiation has also been used to eliminate trace amounts of DNA contamination in PCR reagents, presumably by its interaction with free radicals generated by ionization of water.[14] Doses of 150 to 200 krad are apparently optimal while higher doses reduce amplification efficiency, but conditions should be individually established.

4. Psoralen

Psoralens (linear furocoumarins) intercalate between double-stranded nucleic acids, and form inter-strand cross-links on photo-activation. They can be used to cross-link contaminating DNA before PCR and prevent subsequent denaturation and template activity. Incubation of PCR components with 0.46 mM 8-methoxypsoralen for at least 0.5 h followed by UV irradiation at 365 nm for 1 h has been reported to remove more than 99% of contaminating DNA, but it is not known if the sensitivity of the PCR is affected.[15]

B. POST-PCR STERILIZATION

Post-PCR methods modify the PCR products so that they are unable to participate as templates in subsequent PCR. A prerequisite is that they do not significantly affect post-PCR analysis, such as electrophoretic mobility and hybridization. They specifically prevent product carry-over but not other sources of contamination.

1. Isopsoralens

Isopsoralens (angular furocoumarins), unlike psoralens, do not cross-link DNA strands but only form single-stranded monoadducts with pyrimidines in a photochemical reaction. Such modified bases terminate DNA polymerase extension and prevent template activity without significantly affecting hybridization. Isopsoralens can be added to the PCR before cycling and apparently do not significantly affect amplification efficiency. After completion of the reaction, long-wave UV photoactivation will sufficiently inactivate the PCR products so that carry-over risk is reduced considerably.[16]

2. Uridine Incorporation and Glycosylation

A combination of pre- and post-PCR sterilization involves incorporating dUTP in all PCR products by substituting dUTP for TTP. Before commencing all PCR, fully pre-assembled reactions are treated with uracil DNA glycosylase (UDG), which cleaves the uracil base from the phosphodiester backbone of uracil-containing DNA but has no effect on naturally occurring DNA template containing thymidine bases.[17] The resulting apyrimidinic sites block replication by DNA polymerases and are labile to acid/base hydrolysis. Because UDG does not react with dUTP and is inactivated by heat denaturation prior to the actual PCR, contamination of PCR with dUTP-containing contaminants may be effectively controlled. While dUTP-containing PCR products can be used for a variety of applications including sequencing and cloning (with the use of appropriate endonucleases and bacterial hosts), PCR amplification efficiency may be reduced compared with conventional TTP use.[18]

IV. CONTAMINATION DETECTION AND RESULT VERIFICATION

Systematic contamination of all samples by a common reagent/step can be detected with negative controls. However, sporadic false-positives (e.g., cross-contamination of samples during transport) are harder to detect, and sequence analysis for variation, performing the PCR again, or using other confirmatory tests may be required.

A. NEGATIVE CONTROLS

The use of negative controls is the single most important step in detecting/excluding false-positive PCR results. Blank (water) controls without sample DNA can be used during assembly of the PCR, and detect contaminated reagents. Similarly, blank controls that have been processed with the other samples during DNA extraction/purification serve to eliminate contamination of samples. Ideally, controls should be taken through the entire process from specimen collection, using water or samples from a healthy volunteer. While reaction conditions for PCR assays would have been optimized before routine implementation, the use of genomic DNA samples known not to contain template can be used as a quality control for specificity of the primer (e.g., mix-up or mis-labeling) and/or reaction conditions (e.g., substringent annealing temperature). This may take the form of DNA from a cell line with a gene deletion or without the infectious agent.

Multiple negative controls should be used in order to detect sporadic contamination or a low level of contamination that is around the threshold of detection.

B. SEQUENCE POLYMORPHISM

The precautions and procedures described above are useful for minimizing contamination risk in the laboratory. Exogenous contamination of the reaction is usually detectable in the negative controls. However, false-positive results may also arise due to inadvertent contamination of samples during collection, biopsy fixation, and transport before arrival at the laboratory. While it may be possible to use negative controls for some parts of this process (by processing water blanks together with samples), verification of results by other means is usually necessary.

The natural variation (polymorphism) of target DNA being amplified can be exploited for contamination exclusion by showing that the sequences of amplicons differ from each other, and hence are not due to amplification from a common contaminating source. Such sequence polymorphisms occur within and around genes, including hypervariable microsatellite oligonucleotide repeats, oligo-allelic single base substitutions, restriction site polymorphisms, and minisatellite repeats. Polymorphic loci in constitutional genes being studied may be selected, so that significant deviation from expected frequencies may point to systematic contamination. Sequence polymorphisms also occur frequently in non-coding regions such as the mitochondrial D-loop, and in a large variety of microbiological pathogens and viruses. Several methods have been developed to demonstrate sequence differences in PCR products.

1. Sequencing

Nucleic acid sequence determination of PCR products amplified from loci that exhibit genetic variation may be used to demonstrate the uniqueness of the original templates. This may be performed either indirectly by cloning amplified DNA and sequencing one or more inserts, or directly by using purified PCR DNA as a template for the sequencing reaction.

Dideoxynucleotide chain termination protocols and kits for direct sequencing of PCR products are available from various suppliers, as is the linear amplification modification of this method using thermostable polymerases and thermal cycling. The latter method is particularly suitable for the relatively short double-stranded products of PCR.

In a very sensitive PCR assay for hepatitis B virus (HBV) DNA in human serum, capable of detecting as few as three viral genomes, a variable region of the precore/core reading frame was amplified.[19] The presence of viral DNA in PCR-positive cases, particularly those that were negative by hybridization methods, was confirmed by sequencing. Sequence differences between 2 and 15% were detected in this 198-bp fragment. One of the samples shared exactly the same sequence as the recombinant clone used in that laboratory, strongly suggesting exogenous contamination.

2. Heteroduplex Analysis

Screening methods for detecting small sequence differences may be useful since it is usually only necessary to show that each sample is unique. Double-stranded heteroduplexes form between closely homologous but not identical complementary strands. Heteroduplexes

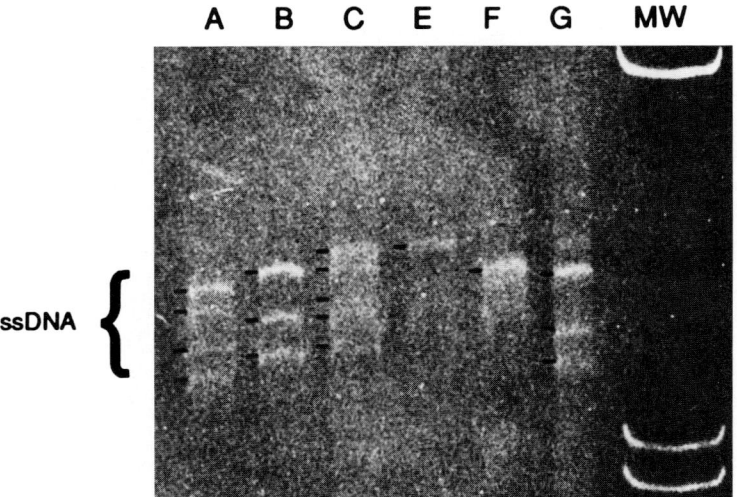

Figure 2. Exclusion of contamination in virological PCR by SSCP. The presence of hepatitis B virus in routinely biopsied paraffin-embedded hepatocellular carcinoma samples from South Africa, an HBV-endemic area, was detected by a highly sensitive nested PCR assay.[21] Positive samples, yielding 285-bp products amplified from the pre-core/core region, were analyzed by single-strand conformation polymorphism (SSCP) analysis to show that the product (and hence target) sequence was different for each case. The single-stranded DNA (ssDNA) bands (indicated by black lines) after 4 h of non-denaturing polyacrylamide gel electrophoresis show sequence-dependent mobility differences, which are unique in each of the six samples (A to F).

with mismatches of a few base pairs may be detected by electrophoresis on non-denaturing polyacrylamide gels. The retarded mobility of these bands compared to homoduplexes may be shown non-isotopically by ethidium bromide staining. This method has been used to exclude contamination in PCR for HBV, by denaturing and reannealing paired mixtures of PCR samples.[20]

3. Single-Strand Conformation Polymorphism

Another rapid and inexpensive method for demonstrating sequence differences, single-strand conformation polymorphism (SSCP) analysis (see Chapter 20), has been used to exclude exogenous contamination.[21] SSCP is based on the sequence-dependent secondary folding of single-stranded DNA. Sequence variation, even at the single nucleotide level, results in different conformations of varying mobilities on non-denaturing polyacrylamide gel electrophoresis. Non-isotopic SSCP analysis of PCR-amplified HBV DNA showed unique banding profiles in a study of hepatocellular carcinomas (Figure 2).

C. REPETITION

False-positives (and -negatives) may be detected by repeated analysis of samples taken on a separate occasion, and processed and amplified separately where possible. Results may also be verified with confidence if PCR for several fragments of the same template are performed, or if validated by other methods such as probe hybridization (e.g., Southern blots, *in situ* hybridization) or protein detection (e.g., immunohistochemistry), though these may not be of comparable sensitivity. Clearly these reduce the advantages of PCR by increasing the duration and cost of the assay.

V. MANAGEMENT OF CONTAMINATION

Almost every laboratory that performs large numbers of PCR assays eventually experiences problems with contamination. Trouble-shooting is extremely time consuming and perhaps unnecessary. A complete renewal of all reagents and disposables, examination of current lab practices, and a break from PCR are usually more effective and efficient. Obviously, the quality-assurance measures and contamination-management procedures employed depend on the nature of the research/service performed, number of personnel involved and tolerance (error rate) permissible, and a host of other factors. In a medium-sized research laboratory, we use those guidelines:

- Discard aliquots of deoxynucleotides, water, and buffer in current use after an experiment with a false-positive control.
- If the contamination is still present for the second time, UV-irradiate small pieces of equipment (racks, pipettes), wipe down work areas, microcentrifuges, and thermocyclers with alkali (0.5 M NaOH) or sodium hypochlorite (liquid bleach).[22] Sodium hydroxide may damage some metallic surfaces. Start with new aliquots of primers and a new stock of enzyme.
- If contamination persists with no obvious source, new primers to other parts of the gene studied may have to be synthesized. Alternatively, cease work on that amplicon for a few months.

ACKNOWLEDGMENTS

EPHY received support as a Rhodes Scholar, Wellcome Trust Advanced Training Fellow (grant 035401/Z/92/Z/14T) and Junior Research Fellow at Wolfson College, Oxford. Y-MDL acknowledges the support of the Wellcome Trust and Foulkes Foundation. We also acknowledge the support of the Cancer Research Campaign, U.K.

REFERENCES

1. **Sarkar, G. and Sommer, S.,** Shedding light on PCR contamination, *Nature,* 343, 27, 1990.
2. **Lo, Y.-M. D., Mehal, W. Z., and Fleming, K. A.,** False-positive results and the polymerase chain reaction, *Lancet,* ii, 679, 1988.
3. **Bottger, E. C.,** False positive reactions in PCR, in *PCR Topics: Usage of Polymerase Chain Reaction in Genetic and Infectious Diseases,* Rolfs, A., Schumacher, H. C., and Marx, P., Eds., Springer-Verlag, Berlin, 1991, 66.
4. **Chou, Q., Russel, M., Birch, D. E., Raymond, J., and Bloch, W.,** Prevention of pre-PCR mis-priming and primer dimerization improves low-copy-number amplifications, *Nucleic Acids Res.,* 20, 1717, 1992.
5. **Haase, A. T., Retzel, E. F., and Staskus, K. A.,** Amplification and detection of lentiviral DNA inside cells, *Proc. Natl. Acad. Sci. U.S.A.,* 87, 4971, 1990.
6. **Kwok, S.,** Procedures ot minimize PCR-product carry-over, in *PCR Protocols,* Innis, M. A., Gelfand, D. H., Sninsky, J. J., and White, T. J., Eds., Academic Press, San Diego, CA, 1990, 142.
7. **Furrer, B., Candrian, U., Wieland, P., and Luthy, J.,** Improving PCR efficiency, *Nature,* 346, 324, 1990.
8. **Lo, Y.-M. D., Patel, P., Wainscoat, J. S., Sampietro, M., Gillmer, M. D. G., and Fleming, K. A.,** Prenatal sex determination by DNA amplification from maternal peripheral blood, *Lancet,* ii, 1353, 1989.
9. **Zhu, Y. S., Isaacs, S. T., Cimino, G. D., and Hearst, J. E.,** The use of exonuclease III for polymerase chain reaction sterilization, *Nucleic Acids Res.,* 19, 2511, 1991.
10. **Muralidhar, B. and Steinman, C. R.,** Geometric differences allow differential enzymatic inactivation of PCR product and genomic targets, *Gene,* 117, 107, 1992.
11. **Ou, C. Y., Moore, J. L., and Schochetman, G.,** Use of UV irradiation to reduce false positivity in polymerase chain reaction, *BioTechniques,* 10, 442, 1991.
12. **Sarkar, G. and Sommer, S.,** More light on PCR contamination, *Nature,* 347, 340, 1990.
13. **Shibata, D.,** Polymerase chain reaction and the molecular genetic analysis of tissue biopsies, in *Diagnostic Molecular Pathology: A Practical Approach,* Vol. 2, Herrington, C. S. and McGee, J. O'D., Eds., IRL Press, Oxford, 1992, 85.
14. **Deragon, J.-M., Sinnett, D., Mitchell, G., Potier, M., and Labuda, D.,** Use of gamma irradiation to eliminate DNA contamination for PCR, *Nucleic Acids Res.,* 18, 6149, 1990.
15. **Jinno, Y., Yoshiura, K., and Niikawa, N.,** Use of psoralen as extinguisher of contaminated DNA in PCR, *Nucleic Acids Res.,* 18, 6739, 1990.

16. **Cimino, G. D., Metchette, K. C., Tessman, J. W., Hearst, J. E., and Isaacs, S. T.,** Post-PCR sterilization: a method to control carryover contamination for the polymerase chain reaction, *Nucleic Acids Res.,* 19, 99, 1991.
17. **Longo, M. C., Berninger, M. S., and Hartley, J. L.,** Use of uracil DNA glycosylase to control carry-over contamination in polymerase chain reactions, *Gene,* 93, 125, 1990.
18. **Pang, J., Modlin, J., and Yolken, R.,** Use of modified nucleotides and uracil-DNA glycosylase (UNG) for the control of contamination in the PCR-based amplification of RNA, *Mol. Cell. Probes,* 6, 251, 1992.
19. **Kaneko, S., Miller, R. H., Feinstone, S. M., Unoura, M., Kobayashi, K., Hattori, N., and Purcell, R. H.,** Detection of serum hepatitis B virus DNA in patients with chronic hepatitis using the polymerase chain reaction assay, *Proc. Natl. Acad. Sci. U.S.A.,* 86, 312, 1989.
20. **Lo, Y.-M. D., Lo, E. S.-F., Patel, P., Tse, C. H., and Fleming, K. A.,** Heteroduplex formation as a means to exclude contamination in virus detection using PCR, *Nucleic Acids Res.,* 19, 6653, 1991.
21. **Yap, E. P. H., Lo, Y.-M. D., Cooper, K., Fleming, K. A., and McGee, J. O'D.,** Exclusion of false-positive PCR viral diagnosis by single-strand conformation polymorphism, *Lancet,* 340, 726, 1992.
22. **Prince, A. M. and Andrus, L.,** PCR: how to kill unwanted DNA, *BioTechniques,* 12, 358, 1992.

Chapter 28

THE APPLICATION OF PCR-BASED TECHNOLOGIES TO FORENSIC ANALYSIS*

Lawrence A. Presley and Bruce Budowle

TABLE OF CONTENTS

I. Introduction 260
    A. PCR: A Unique Sample-Preparation Technique 260
    B. Methodological Variation 261

II. Validation 261

III. Extraction Methods 262

IV. Quantitation 262

V. The Enzyme Used in Amplification Reactions 262

VI. Typing 263
    A. Dot Blot Analysis 263
    B. Gel Electrophoresis 264
    C. Sequencing 265

VII. Minisatellite Variant Repeat (MVR)-PCR Analysis 266

VIII. Limitations 267

IX. Contamination 267

X. Forensic Analysis 268
    A. Typing DNA after Other Forensic Processing 270

XI. Quality Control/Quality Assurance in Forensic DNA Typing 270
    A. Future Considerations for QA/QC in Forensic DNA Typing 273

XII. Conclusion 273

References 273

---

* This is a publication of the Laboratory Division of the Federal Bureau of Investigation. Names of commercial manufacturers are provided for identification only and inclusion does not imply endorsement by the Federal Bureau of Investigation.

I. INTRODUCTION

The first widely used DNA technology in forensic casework was restriction fragment length polymorphism (RFLP) analysis of variable number tandem repeat (VNTR) loci. Although RFLP analysis has been shown to yield reliable results in forensic examinations, the technique has several limitations that may be commonly encountered in routine casework. RFLP analysis generally (1) requires a sufficient quantity of high-molecular-weight DNA, usually at least 50 ng, as suggested by Budowle and Baechtel, 1990;[1] (2) needs isotopically labeled probes to obtain a high level of sensitivity of detection (this requirement of radioactive materials can impede the transfer of RFLP technology to some application-oriented laboratories); (3) is a labor-intensive as well as time-consuming process, which can take six to eight weeks to obtain results on four VNTR loci, and (4) cannot resolve unequivocally the alleles of many VNTR loci.

To circumvent some of the limitations of RFLP analysis in forensic testing, typing methodologies based on polymerase chain reaction (PCR) analysis have been investigated. PCR is simply a unique sample-preparation method. Although there are many approaches for PCR analysis, it is basically an *in vitro* enzymatic synthesis of millions of copies of a target DNA sequence, as reported by Saiki et al., 1985.[2] PCR amplifies specific DNA sequences by repetitively denaturing template DNA, hybridizing specific DNA primers (designed from DNA sequences flanking the region to be amplified) to the template DNA, and extending the primers. In this way, multiple copies of the target region can be generated. PCR provides a method of obtaining relatively large amounts of specific DNA sequences from relatively small (picogram or nanogram) quantities of DNA and does not require isotopic DNA detection methods. Additionally, degraded DNA samples (for example, from forensic samples, museum samples, biopsy samples, Guthrie cards, and so on) can be amplified by PCR and subsequently typed in many cases.

The advantages of a PCR-based technology are, when compared with the RFLP approach, augmented sensitivity and specificity, ability to detect small alleles in highly degraded samples, decreased assay time and labor, nonisotopic labeling, and less consumption of evidentiary material. Also, PCR analytical methods take only a few days compared with weeks for RFLP analysis, and PCR systems are generally amenable to automation. Additionally, small amounts of DNA template and several distinct loci can be amplified in one PCR reaction mixture simultaneously, generally referred to as multiplexing, as described by Chamberlain et al.[3] These qualities combine to make PCR an extremely useful tool for analyzing small or degraded biological materials found in criminal investigations. For the purposes of applying PCR-based technology to characterization of forensic evidence, defined polymorphic loci and relatively facile and robust analytical techniques are required; population studies on various genetic markers are needed; and validation studies, particularly environmental insult studies, should be performed. While PCR is newer than other genetic marker tests, the literature is becoming replete with DNA markers that are suitable for PCR analysis. In addition, significant data already exist to support the utility of PCR for forensic analyses. This chapter provides a discussion of a portion of those data and suggests that PCR-based techniques do and will continue to provide valid and reliable approaches for characterizing biological evidence associated with criminal investigations.

A. PCR: A UNIQUE SAMPLE PREPARATION TECHNIQUE

PCR analysis is based on the annealing and extension of two oligonucleotide primers that flank a specific target DNA segment. Denaturization is achieved by heating the sample to approximately 95°C using a thermal cycler. After denaturation of the double-stranded DNA molecules, each primer hybridizes to one of the separated strands. This primer annealing is generally accomplished by lowering the thermal cycler temperature to between 37 and 72°C.

The specific annealing temperature is empirically determined for each primer set/target region combination. The next temperature in the cycle is generally 72°C, the temperature at which *Thermus aquaticus* (*Taq*) polymerase, a commonly used thermostable DNA polymerase, can then extend the primers. These three steps (denaturation, primer binding, and primer elongation) represent a single PCR cycle. When the newly synthesized strand extends through the region complementary to the other primer, it can serve as a primer binding site and template for a subsequent PCR cycle. Upon repeated cycles of denaturation, primer annealing, and primer extension, an exponential accumulation of a discrete DNA fragment is generated. By repeating the cycle typically 25 to 30 times, millions of copies of target sequence can be obtained.

Generally for PCR to work, information of the DNA sequence (at least the flanking region) is necessary for primer synthesis. Primers are single-stranded DNA oligonucleotides 20 to 30 bp long, and they can be obtained commercially or synthesized in house.

PCR analysis, then, is easy to accomplish in principle. One needs only a DNA template sample, primers, a mixture of four dNTPs, buffer, and a thermostable DNA polymerase. All ingredients can then be placed in a reaction tube and inserted into a programmable thermal cycler, which permits the temperature to vary, and routinely, PCR can be carried out in a matter of hours.

B. METHODOLOGICAL VARIATION

For some genetic loci that have been subjected to PCR, artifactual products can be generated. One cause for this phenomenon is the nonspecific annealing of primers at ambient temperatures, which can occur in the amplification mixture prior to the first PCR cycle. If this phenomenon presents itself as an issue, hot-start PCR offers a viable approach for circumventing nonspecific annealing and DNA synthesis prior to PCR. Hot-start PCR analysis, as suggested by Sparkman, 1992,[4] is a method in which a critical component of the amplification reaction mixture is kept separate from the reaction mixture by creating a wax barrier. Melted wax beads are added to the amplification mix and then allowed to solidify. The separated critical ingredient is then placed on the solid layer of wax covering the reaction mix. Once the melting temperature for the wax is reached, it will liquify, and the ingredients necessary for PCR to occur will mix together.

II. VALIDATION

Currently, there are continued efforts to validate the forensic applications of PCR-based methods of amplified fragment length polymorphisms (AMP-FLPs) such as D1S80, and sequence polymorphisms such as HLA DQα, Amplitype Polymarker (PM), and mitochondrial (mt)DNA. While some studies are still under way, there exist substantial data that have demonstrated numerous loci and systems to be valid and reliable for the characterization of biological evidence. Various sources of DNA have been typed successfully and reported using PCR-based systems, and these include hair by Higuchi et al.;[5] sperm by Sajantila et al.;[6] bloodstains by Jinks et al.,[7] Williams et al.,[8] and Sajantila et al.;[6] urine by Gasparini et al.;[9] bone by Hochmeister et al.,[10] Hagelberg et al.,[11] and Sullivan et al.;[12] saliva by Hochmeister, et al.;[13] and formalin-fixed paraffin-embedded tissues by Impriam et al.[14] Notably, Comey et al.,[15] Comey and Budowle,[16] and Presley et al.,[17] have demonstrated the validity, reliability, and robustness of the PCR DQα typing system for a wide range of forensic specimens. Hochmeister et al.,[10] demonstrated the reliability of AMP-FLP typing of DNA extracted from compact bone of decomposed human remains. Additionally, Sajantila et al.,[18] showed that DNA extracted from muscle tissue, bone marrow, and blood of severely burned bodies can be typed for the HLA-DQα and D1S80 loci. These studies, as well as those of Hagelberg et al.,[11] support the reliability of AMP-FLP and HLA-DQα typing by demonstrating concordance of

Mendelian inheritance of the loci in the individuals using family members. Thymann et al.,[19] showed the applicability of PCR of the D1S80 locus for paternity testing, and also demonstrated the high degree of inter-laboratory reproducibility of D1S80 phenotyping. The utility of PCR-based systems in forensics is demonstrated further by a study described by Hochmeister et al.,[13] in which DNA was extracted from saliva on cigarette butts. Only limited genetic information can be obtained from saliva typed by conventional serological means, and in general, saliva samples yield insufficient quantities of DNA for RFLP analysis. DNA isolated from 100 cigarettes smoked by ten different individuals (ten cigarettes per individual) and three cigarettes recovered from two crime scenes were analyzed for the D1S80 and HLA-DQα loci. The DNA from 99 out of 100 samples, as well as the casework cigarette butts, could be amplified and were consistent with control samples. Sajantila et al.,[6] described the utility of AMP-FLP analysis of the D1S80 locus in 36 forensic cases, consisting of 18 rapes, 14 homicides, and four other violent crimes. They found 88.2% of the semen samples and 72.1% of the bloodstains typeable. In no case was there evidence of false-positive or false-negative results. Blake et al.,[20] analyzed DNA by PCR-based HLA DQα typing for more than 250 criminal cases and more than 2000 forensic samples, which included blood and semen stains, hairs, bone fragments, and tissue samples.

III. EXTRACTION METHODS

There have been several methods used for the extraction and subsequent amplification of DNA for PCR. These methods include but are not limited to Chelex, Organic (phenol/chloroform/isoamyl alcohol), and Centricon extractions.[21-23] Bovine serum albumin (BSA) has also been suggested for the enhancement of PCR-amplification products of forensic samples that may contain PCR inhibitors that were not removed by the extraction process.[24,17,13] All these cited methods have been shown to be effective in yielding DNA that can be amplified and typed.

IV. QUANTITATION

Slot blot assays for quantifying DNA can be used to augment the success of obtaining PCR-amplified product by ensuring that the proper amounts of target DNA are available for PCR. Several slot blot approaches have been used to semiquantitatively assess the amounts of DNA extracted from forensic samples. These approaches are both radioactive, as described by Waye et al.,[25] and nonradioactive, as described by Walsh et al.,[26] and Klevan et al.,[27] detection methods, which employ a primate-specific alphoid probe.

V. THE ENZYME USED IN AMPLIFICATION REACTIONS

Initial PCR amplifications were reported by Saiki et al.,[2] using the DNA polymerase I Klenow fragment isolated from *Escherichia coli*. However, because of its thermal lability, Klenow had to be added after each PCR cycle. Until a thermostable DNA polymerase was available, PCR was not generally a practical method for sample preparation. The use of thermostable polymerases eliminated the need to add fresh enzyme after the denaturation step and, consequently, allowed the development of thermal cycling devices, which enabled automation of the PCR process. Consequently, a number of thermostable DNA polymerases have been reported, including *Taq* polymerase from *Thermus aquaticus* by Saiki et al.,[28] which is commonly used in PCR applications. *Taq* polymerase can synthesize DNA at a higher temperature than Klenow, and these higher temperatures for primer annealing and extension have significantly increased the specificity of the PCR. *Taq* has $5' \rightarrow 3'$ exonuclease activity but lacks a $3' \rightarrow 5'$ exonuclease activity. The genetically engineered Stoffel fragment of *Taq* also lacks the 5' → 3' exonuclease activity and appears to be more thermostable than *Taq* isolated from bacteria.

VI. TYPING

Once the sample has been amplified by PCR, there are three general approaches for analyzing the product, each dictated by the genetic marker and its particular type of polymorphism. The analytical methods that have been and are being developed for the characterization of forensic materials are (1) dot blot assays using allele-specific oligonucleotide (ASO) probes to detect sequence polymorphisms; (2) electrophoretic separations in agarose and polyacrylamide gels to resolve size polymorphisms; and (3) direct sequencing of the PCR product.

A. DOT BLOT ANALYSIS

Allele specific oligonucleotide (ASO) probes can be used to distinguish single nucleotide differences in PCR products under hybridization stringency and wash conditions, in which only completely complementary probe sequences will bind stably to target sequences immobilized on nylon membranes, as reported by Saiki et al.[29,30] Basically, the amplified DNA was spotted onto a nylon membrane, and ASO probes were allowed to hybridize with the target DNA. A panel of oligonucleotide probes could then be used to type multiple target sequences or alleles. The assay is interpreted on the presence or absence of visible dots: however, this dot blot approach can become cumbersome as the number of alleles increases.

Historically, the first PCR-related approach used for forensic purposes was detection of sequence polymorphisms of the HLA-DQα locus by use of ASO hybridization probes in a dot blot format. A separate nylon strip was required for each allele to be detected for the DNA marker system. To make the typing of samples more manageable, a reverse dot blot format was developed by Saiki et al.[31] In this approach, the ASO probes that detect each allele of a particular locus are fixed to a nylon membrane strip, and the amplified alleles of the sample are hybridized to the immobilized probes. Detection can occur via a biotin tag on the 5' end of one of the primer sequences, which after hybridization can bind with a strepavidin-horseradish peroxidase assay complex. The cumbersome nature of a test with a number of probes is greatly reduced because the amplified material is hybridized against a number of allelic probes in a single hybridization assay.

HLA-DQα, as suggested by Saiki et al.,[31] Erlich et al.,[32] and Walsh et al.,[33] is one of the most characterized and reliable PCR-based systems using the reverse dot blot format for the analysis of forensic specimens. The HLA-DQ molecule is a heterodimer composed of one alpha chain (encoded by the HLA-DQα locus) and one beta chain. It is expressed in B lymphocytes, marcrophages, thymic epithelial cells, and activated T cells. The HLA-DQ protein serves as an integral membrane protein for binding as well as presenting antigen peptide fragments to the T cell receptor of CD4+ T lymphocytes. The polymorphisms that determine the alleles of this class II HLA gene are located in a 242-bp region (or 239-bp length for alleles 2 and 4) in the second exon of the HLA-DQα gene. Eight alleles have been identified; they are designated 1.1, 1.2, 1.3, 2, 3, 4.1, 4.2, and 4.3. A kit is commercially available (Roche Molecular Systems, Alameda, CA) for typing six of the alleles. Four of the probes are designed to detect alleles 1, 2, 3, and 4 (which detects 4.1, 4.2, and/or 4.3). The 1 allele can be subtyped further as a 1.1, 1.2, or a 1.3 allele by using a second set of ASO probes present on the same typing strip. Thus, one nylon strip permits phenotyping of the HLA-DQα locus for an individual. The reverse dot blot format for HLA-DQα typing uses biotin-labeled primers in the PCR. During hybridization, a streptavidin-horseradish peroxide complex is added so that duplex formation between the amplified DNA and immobilized ASO probe can be detected by subsequent oxidation of tetramethylbenzidine which results in a blue dot. Thus, there is no need for an isotopic label. This approach also has been used for typing mitochondrial (mt) DNA variants by Stoneking et al.[34]

B. GEL ELECTROPHORESIS

Currently, polymorphic loci that have alleles resulting from VNTRs with repeat units 2 to 80 bp long are the most informative genetic markers for attempting to individualize biological material. The size of the amplified fragment (or allele) is dictated by the number and size of the repeat sequences contained within it. With appropriate VNTR loci and high-resolution electrophoretic systems, amplification by PCR of specific VNTR sequences could prove useful for identity-testing purposes. In forensic analysis, several VNTR loci have been and are being evaluated. These include D1S80 by Budowle et al.,[35] and Kasai et al.;[36] D17S5 [or D17S30] by Horn et al.;[37] D19S20 by Odelberg et al.;[38] the 3' hypervariable region of the apolipoprotein B gene by Boerwinkle et al.,[39] and Ludwig et al.;[40] Col2A1 by Wu et al.;[41] short tandem repeat loci, which include HUMTH01 by Edwards et al.;[42] SE33 and TC11 by Weigand et al.;[43] myelin basic protein by Polymeropolous et al.;[44] and VWA by Kimpton et al.;[45] as well as sex-typing markers.[46-48]

The general approach for the analysis of PCR-amplified VNTR polymorphisms or AMP-FLPs has been separation of the amplified products in agarose gels with subsequent ethidium bromide staining or separation in polyacrylamide sequencing gels with isotopic detection as reported by Allen et al.,[49] and Budowle et al.[35] The detection system has been accomplished either by incorporation of radioisotopes into the amplified product or by hybridization to a locus-specific probe. Agarose gel electrophoresis does not provide the resolution necessary to separate some VNTR alleles, particularly short tandem repeat (STR) loci, into discrete entities. Although sequencing gels provide high resolution of DNA fragments, they can be cumbersome to cast. Moreover, while STR loci are of a size amenable for sequencing gel format analysis, AMP-FLPs with larger repeats may not be easily typed with this electrophoretic approach, and consequently, a variety of gel formats may be necessary. Furthermore, ethidium bromide and radioisotopes are hazardous materials, which require special handling and disposal.

Allen et al.,[49] Budowle and Allen,[50] and Budowle et al.[35] have described a simple high-resolution, discontinuous buffer, horizontal polyacrylamide gel system that renders AMP-FLP analysis relatively easy and inexpensive. Ultrathin-layer polyacrylamide gels can be cast by the flap technique, according to Allen.[51] No specialized gel-casting equipment is required, and there is a high success rate for producing usable gels. Alternatively, rehydratable polyacrylamide gels can be used, as reported by Allen and Lack,[52] and Allen et al.[49] These are essentially dried, empty gels that can be conveniently rehydrated with any separation buffer. Both types of gels are bound to Mylar films to facilitate handling. A discontinuous buffer system permits manipulation of the resolution potential by changing the ionic strength and viscosity of the resolving gel buffer, by adjusting the pH of the buffers, by altering the trailing ion, and/or by adjusting acrylamide concentration. Therefore, a wide variety of AMP-FLP loci can be accommodated with one general analytical system. Alternatively, vertical polyacrylamide gel systems can be employed with results comparable to those obtained with horizontal gels, as reported by Sajantila et al.[6] Visualization of the separated AMP-FLPs is achieved by silver staining, which is an inexpensive and nonmutagenic means of detection. Finally, a permanent record is obtained by subsequent drying of the horizontal gel, which is backed on Mylar.

Since the size of fragments of VNTR loci amenable to PCR are small (generally less than 2 kb), high-resolution electrophoretic systems enable alleles to be resolved more effectively. Thus, AMP-FLP analysis will reduce the level of measurement imprecision encountered with routine RFLP analysis. Furthermore, with the AMP-FLP approach, alleles can be designated specifically without estimating base-pair size. An unknown sample can be compared with an allelic ladder consisting of a composite of the common alleles of a particular VNTR, and the alleles can be named generally based on repeat sequence number or with an arbitrary nomenclature.

An alternative approach to high-resolution, discontinuous polyacrylamide gel electrophoresis and silver staining is automated analysis of fluorescently tagged AMP-FLPs, as reported by Robertson et al.[53] To detect AMP-FLPs, the primers for PCR are tagged with fluorescein derivatives. The amplified product is loaded on a polyacrylamide sequencing gel. Electrophoresis is carried out in an automated DNA sequencer (for example, Applied Biosystems 373A DNA Sequencer, Foster City, CA, which is a real-time fluorescent detection instrument). The labeled AMP-FLPs are detected by argon laser excitation as they migrate, during electrophoresis, past a designated window. The emissions, between 540 and 610 nm, depending on the dye, are filtered through 10-nm band pass filters and collected by a photomultiplier tube. The signal is digitized and analyzed by a computer, which assigns fragment size and, if desired, can quantify fluorescence. As few as 50 to 400 amol of DNA fragments per band can be detected, according to Robertson et al.[53]

This automated approach provides some advantages when compared to the more routine manual approach. An internal standard, for example, labeled with a red dye, can be placed in every gel lane. This standard can serve as an internal electrophoretic lane control to evaluate potential lane-to-lane differences. This obviates the need to subject a known and evidentiary sample to a coelectrophoresis experiment to confirm whether or not their AMP-FLP types, for example, labeled with a blue dye, are operationally the same. It could also eliminate the need to create specific allelic ladders for several VNTR loci. Additionally, multiple samples, each labeled with a different colored dye, can be typed for the same locus within a single electrophoresis lane. Also, multiplexing of several VNTR loci amplified together in one sample application can be achieved.

C. SEQUENCING

Direct sequencing could provide the most detailed information of a DNA segment obtained by PCR amplification. PCR creates sufficient template for sequencing and thus eliminates the need for cloning.

For sequencing, it is desirable to use a single-stranded PCR product. Strands can be made single-stranded by asymmetric PCR, as described by Gyllensten and Erlich,[54] in which one of the primers is limited during the PCR, or by biotinylating one of the primers such that one strand of the amplified product can be isolated by a streptavidin bead, as described by Hultman et al.,[55] or by magnetic or polystyrene bead. Cycle sequencing with a thermostable polymerase, in which repeated extension and denaturation cycles result in a linear accumulation of PCR products, can increase the sensitivity of the sequencing reaction, as reported by Lee,[56] and Ruano and Kidd.[57]

Generally, chain-termination sequencing is carried out with either labeled primers (generally radioactive or fluorescent), labeled dNTPS, or labeled dideoxynucleotide terminators, as suggested by Sanger et al.[58] Sequencing was not considered a practical alternative method for revealing genetic variation for forensic purposes until instruments that provided automated analysis of fluorescently labeled sequence reaction products were made available. Various Sanger approaches, with primers labeled either by the approach described by Smith et al.,[59] or terminator dideoxynucleotide analogues described by Prober et al.,[60] which are tagged with fluorescent labels, offer viable approaches for forensic analysis. The sequence reaction products can be separated in polyacrylamide gels and detected in real time with an automated DNA sequencer. This approach permits automation of a cumbersome portion of the assay, that is, sequence interpretation. It eliminates the need for radioactivity, and because different fluorescent dyes are used in each chain-termination reaction, it also allows an entire analysis to be run in one lane instead of four lanes of a gel, as is generally done with conventional Sanger sequencing.

Nuclear genes, with the possible exception of those on the Y chromosome, may not be the most practical candidates for a sequencing approach in forensic analyses because the presence of two copies of each gene can make interpretation more complex. This could be confounded further when the evidence contains mixed biological fluids from several individuals, which are encountered in evidence associated with violent crimes. It was first suggested by Merril,[61] that mitochondrial DNA (mtDNA) could serve as a useful genetic marker(s) for the characterization of certain evidentiary material. MtDNA is approximately 16.5 kb long, generally maternally inherited, has no recombination, and is homogeneous and hemizygous. The mitochondrial genome is an extrachromosomal, circular piece of DNA, which has been sequenced completely and has had all genes mapped. It consists of 13 subunits for respiratory chain enzymes, two ribosomal RNAs, and 22 transfer RNAs. These features make mtDNA highly desirable for certain forensic analyses: (1) mtDNA can occur in more than 5000 copies per cell, which offers a greater potential DNA target source than does nuclear genomic DNA; (2) it is maternally inherited and, therefore, monoclonal; (3) it generally appears to be homogeneous within different tissues of an individual, which facilitates comparisons of DNA derived from different body tissues such as blood and semen; and (4) portions of the noncoding D-loop region are highly polymorphic. PCR has been used to obtain mtDNA sequences especially from the polymorphic D loop region from a variety of individuals. Sequencing of amplified mtDNA has been successfully accomplished for such specimens as 7000-year-old brain tissue by Paabo et al.,[62] various types of old tissue by Paabo et al.,[63] and 4000-year-old mummified tissue by Paabo.[24,64] In addition to biological tissues, an important potential application of mtDNA sequencing in forensics is the analysis of hair. Since individual hairs contain very small quantities of DNA, mtDNA sequence analysis may be the only viable technique for analysis. Hugichi et al.[5] have sequenced amplified mtDNA from a single hair.

Sullivan et al.[65] described a two-round PCR procedure for preparing single-stranded mtDNA in which during the first round a hypervariable region of the D-loop was amplified by standard PCR. In the second round unequal concentrations of nested primers were used during PCR. The DNA then was sequenced via the Sanger method using a fluorescently labeled universal sequencing primer. Automated sequence analysis enabled a 403-bp region of the D loop to be sequenced in a single electrophoresis lane. Using this approach, it has been possible to type DNA successfully from hair, bone fragments, and necrotic skin.

VII. MINISATELLITE VARIANT REPEAT (MVR)-PCR ANALYSIS

Minisatellite alleles vary not only in repeat copy number, but also in the interspersion pattern of variant repeat units along the alleles. These repeat units differ in some instances by single base substitutions, which can be detected by digesting with a restriction enzyme, which has been carefully chosen to identify a restriction site within the repeat variants along the alleles. A DNA variant repeat sequence can be amplified by PCR followed by partial digestion of the alleles with a restriction enzyme. Subsequent electrophoresis of the digested product results in a pattern array in a binary code format for the allele. This binary code information demonstrates a high degree of allelic variation and consequently a high degree of discrimination.

Jeffreys et al.[66] eliminated the need for restriction digestion of the amplified product and yet could still detect the interspersion pattern of variant repeat units, and this technique was termed *MVR-PCR*. Parallel PCRs are performed with different primer sets; each set consists of one primer that is homologous to the flanking region of the VNTR and the other that is specific for a type of repeat unit. By running the products of the two PCRs in adjacent lanes of a gel, a binary pattern will be produced. A band in one lane, but not in the other lane at the same position, suggests that both alleles carry the same repeat unit variant. A band in both lanes at the same gel position suggests that the sample is heterozygous for that repeat unit. The use of MVR-PCR has been suggested as a technique applicable to forensic analysis by Jeffreys et al.[66] MVR-PCR analysis offers a relatively sensitive (ca. 10 ng of human DNA) technique, which can be applied to highly degraded samples, even to the point that truncated diploid code information could be derived for severely degraded samples. The technique has also been suggested for use with mixed samples and for paternity testing in forensic cases.

Although the technique may provide a high degree of discrimination, it has not yet been widely adopted by the forensic community. MVR-PCR may produce ambiguities arising from mixtures that may not be readily resolved, and databasing may be cumbersome.

VIII. LIMITATIONS

It is important to evaluate typing systems to understand the limitations of an analytical technique so that when applied in a forensic context, proper interpretations can be made. A number of general issues need to be addressed to determine the validity of PCR-based assays for forensic analyses. These include allele dropout, laboratory contamination, and the impact of environmental insults on DNA typing, as described by Budowle and Allen,[67] Adams et al.,[68] and Comey and Budowle.[16]

It has been possible to amplify DNA up to 10 kb in size, but generally for efficiency of yield, PCR-amplified fragments of interest tend to be less than 2 kb in size. Erlich et al.[69] expressed concern regarding the preferential amplification of the smaller allelic PCR product of an AMP-FLP heterozygous profile. Additionally, Horn et al.[37] observed for the D17S30 locus that larger alleles could be amplified to a significantly less extent than smaller alleles. This suggests that there is a potential of dropout of the larger allele of a heterozygous profile. However, Budawie et al.[96] demonstrated that using the Perkin Elmer Gene Amp 9600 thermal cycler for PCR and reducing the quantity of template DNA in the PCR can reduce the effect of preferential amplification. Walsh et al.[70] has also observed similar preferential amplification effects.

For typing purposes, it is important to minimize the possibility of detecting only one of the two alleles of a heterozygous individual. When DNA isolated from blood was amplified in a TC-1 Perkin Elmer DNA Thermal Cycler (Norwalk, CT) and typed for HLA-DQα using reverse dot blot strips, sometimes there was failure to detect one of the alleles of a heterozygous individual, as reported by Comey and Budowle.[16] Dropout was observed with the 1.1, 1.2, and 1.3 alleles and was attributed to insufficient denaturation during PCR. Since the 1 alleles are more G/C rich than alleles 2, 3, and 4, as reported by Gyllensten and Erlich,[71] the duplex DNA of the 1 alleles could be more difficult to denature. If the wells of a thermal cycler do not reach the prescribed denaturation temperature during PCR, a HLA-DQα heterozygote (carrying the 1 allele) may appear as a homozygote. Fortunately, avoidance of placing samples in the front two rows of this TC-1 thermal cycler or using and pretesting later Perkin Elmer thermal cycler models (the 480 and 9600 which are more efficient at heat exchange) has obviated the problem of allele dropout due to inefficient denaturing temperatures for HLA-DQα typing.

Alternatively, Comey et al.[72] reported that the use of the DNA denaturant formamide facilitates the amplification of the more G/C-rich 1 allele of the HLA-DQα locus even when the desired denaturing temperature is not achieved. The presence of 5% formamide in the PCR permits effective denaturation of all HLA-DQα alleles, even at 90°C. All samples typed correctly and *Taq* polymerase did not appear to be inhibited.

IX. CONTAMINATION

The high sensitivity of PCR means that small amounts of DNA accidently introduced into the PCR will be amplified. The potential main sources of DNA contamination are from PCR products from previous amplification reactions, and from sample to sample in a specific case analysis. By keeping separate the laboratory spaces for DNA extraction and PCR preparation from the laboratory space used for PCR amplifications and typings, by using face masks and gloves during PCR setup, by analyzing known and questioned samples separately, and by using aerosol-resistant tip (ART) pipette tips, the effects of contamination can be minimized. The routine use of positive, negative, and reagent blank controls are also important for the detection of potential contamination.

If contamination persists and it is determined to be the result of PCR product carry-over, then uracil-N-glycosylase (UNG) can be used, as described by Longo et al.[73] In this situation, dUTP is substituted for dTTP during PCR. Any dU-containing PCR product that is inadvertently carried over to new reactions can be cleaved by using UNG prior to a subsequent PCR.

Comey and Budowle[16] showed that bloodstains contaminated with chemicals such as gasoline, motor oil, chemical base, bleach, soap, and salt, and with the microorganisms *Escherichia coli*, *Bacillus subtilis*, and *Candida albicans*, all typed correctly for the HLA DQα locus using PCR. They also demonstrated that routine handling, routine crime scene processing, or coughing on samples that were then PCR HLA DQα typed did not reveal the presence of any contaminating DNA source.

X. FORENSIC ANALYSIS

One of the first cases in the U.S. involving DNA typing, *Pennsylvania vs. Pestinikas*, used PCR analysis, as reported by Blake et al.[20] and personal communication with Ernest Preate, Jr., and Ed Blake. In this case, the alleged victim was reported to have been starved to death by the defendants. The defendants, who were funeral directors, entered into a "contract of care" agreement with the victim to ensure that the victim was fed and cared for properly. The prosecution's case focused on the premise that the victim was starved to death and that the defendants violated their contract. The defendants were charged with third-degree murder and were convicted. Due to very specific, technical "contract of care" issues, an appellate court overturned the conviction, and the state subsequently appealed.

The DNA testimony revolved around tissue samples taken from the victim. The victim's body was exhumed in order to perform autopsies and examine stomach/intestine samples to determine if food was present. The first autopsy revealed no fecal material was present, suggesting that the victim had been starved to death. A second exhumation was conducted to allow another group of experts to examine the stomach/intestinal materials. It appeared that the second group of body parts did not look like the first group of body parts that had been examined. The pathologist who performed the first autopsy indicated that during the second autopsy, the physical condition of the body and samples did not resemble his work. For example, the stitching of the chest did not look like he had done it. It was alleged that the funeral directors tried to place foreign body parts with the victim's body in order to cover up possible evidence of starvation. Tissue samples from both sets of body parts had been preserved in formaldehyde.

Samples of the apparent first and second body parts were sent to Forensic Science Associates (FSA) in Richmond, CA, for testing. The first tests, ABO blood grouping serological tests, yielded inconclusive blood type results for the intestine samples. Formaldehyde may have adversely affected these initial test results. The intestine sample revealed degraded DNA about 100 bp in length, which was too short for the conventional DQα primers, which would require 242-bp lengths of DNA. Primers surrounding approximately 140 bp of the "nominal" DQα typing region were used to amplify the samples. The DNA DQα tests indicated that the body parts had the same DQα "nominal" type 1, 1. The defense counsel argued that this was exculpatory evidence and consistent with the fact that all the body parts were from the victim and not substituted by the defendants. The prosecutor attempted to minimize the effect of this result.

The DNA evidence was admitted without any admissibility hearing. The prosecutor indicated that the judge was aware of the exculpatory nature of the DNA test and was predisposed to admit it for that reason.

It appeared that the jury did not consider the DNA evidence as meaningful or as relevant as the eyewitness testimony in this case, and consequently, it convicted the defendants.

In another criminal investigation, which involved the armed robbery and murder case of an FBI Agent, a stick, a plastic bag, and a cap were found near the scene of the crime. It was suspected that the cap belonged to the suspect. Latent print, serological, and chemical testing of the cap could not associate it with the suspect. After all these forensic tests had been completed, the cap was submitted for HLA DQα analysis. Although the DNA extracted from the cap was highly degraded, typing by HLA DQα was possible.

The DNA DQα results were as follows:

Specimens	DNA DQα type
Victim's blood	1.2, 3
Cap from scene	1.2, 4

The cap found at the crime scene did not have the same type as the victim.

As the investigation continued, known blood samples from potential suspects were submitted for DNA analysis. The DNA DQα types listed below were detected for those specimens:

Specimens	DNA DQα type
Suspect 1	2, 4
Suspect 2	3, 4
Suspect 3	4, 4
Suspect 4	4, 4

Based on these DNA DQα results, suspects 1 through 4 were excluded as potential contributors of the DNA detected on the cap. Their exclusions early on in the investigation allowed the investigation to move on to other possible suspects.

Eventually, a blood sample from a fifth suspect was submitted for analysis, and the following DNA DQα result was obtained:

Specimen	DNA DQα type
Suspect 5	1.2, 4

Based on these HLA DQα results, the fifth suspect was included as a potential contributor of the DNA detected on the cap. The DQα results ultimately corroborated the confession of the suspect that he had been wearing the cap found at the crime scene.

In another homicide case, a Drug Enforcement Administration Special Agent was shot and killed while sitting in a car. The bullet that killed the Agent was not recovered immediately because it had gone through the victim's body. A systematic search of the crime scene resulted in the recovery of a bullet consistent with being shot by the probable murder weapon. A very small amount of blood was recovered from this bullet. Previously, the FBI Laboratory would have been able to determine only if the blood was human, and it would have been highly unlikely that any other information could be obtained. The minute bloodstain was typed successfully using the HLA DQα technique, and was found to match the HLA DQα type of the victim. When a weapon is recovered, this HLA DQα typing information will become important in associating the bullet and weapon with the death of the victim.

The PCR technology used in these cases provided a new avenue for obtaining information important to solving the case. Without the use of PCR technology, little or no information would have been obtained from the biological evidence in these cases.

A novel PCR-based technique referred to as random amplified polymorphic DNA (RAPD) analysis uses a PCR primer(s) with an arbitrary sequence under conditions of reduced stringency. This approach yields a number of DNA fragments that can be detected by gel electrophoresis, as reported by Welsh and McClelland[74] and Williams et al.[75] RAPD analysis can produce a reproducible pattern under defined conditions and has been used in plant studies, as reported by Deragon and Landry.[76] RAPD was used in a murder case in Arizona, when RAPD DNA profiles developed from paloverde seed pods found in the back of the suspect's truck were used to place it back at the crime scene. These DNA plant profiles linked the seed pods to a plant located in the desert where the victim's body had been abandoned. The PCR-generated RAPD DNA profiles were admitted as evidence in the trial as reported by Yoon.[77]

Pennsylvania vs. Pestinikas was the beginning of the emerging PCR technology and its application to forensic casework. By 1992, PCR analysis was being used routinely in forensic laboratories on an international level. A survey of a small sampling of laboratories throughout the world using PCR technology in forensic casework reveals the widespread application of PCR technology (Table 1).

A. TYPING DNA AFTER OTHER FORENSIC PROCESSING

Some types of evidence require special handling due to the nature of the evidence. For example, saliva stains on envelopes, letters, and stamps, commonly encountered in mail bomb, extortion, and kidnapping cases, can provide useful investigative information. Questioned document examinations, latent fingerprint examinations, and DNA analysis are all possible procedures to be used on these types of evidence. However, the order of these examinations and whether or not these tests adversely affect the ability to DNA type the samples should be determined.

One of the initial examinations that may be done in some criminal cases involving letters and envelopes is using an electrostatic detection apparatus (ESDA) method to visualize indented writing, as described by Foster and Morantz,[78] and by Noblett and James.[79] The ESDA method allows the document examiner to make indented impressions visible using electrostatic detection. The technique has been shown to have no subsequent effect on latent fingerprint examinations as reported by Noblett and James.[79]

Typically, after ESDA processing, envelopes and related types of evidence are sent to a latent fingerprint section for processing. Envelopes, stamps, and letters, as well as other types of evidence, are routinely examined using three separate processes: (1) DFO (1,8-Diazafluoren-9-one), (2) ninhydrin, and (3) physical developer.[97] DFO (1,8-Diazafluoren-9-one) is a fluorescent chemical process used to visualize latent fingerprints.[80] Ninhydrin processing, which is designed to react with amino acids present in latent fingerprints, utilizes a variety of chemical solvents including acetone, methanol, ethanol, ethyl ether, ethylene glycol, petroleum ether, naphtha, and Freon 113.[81-85] Physical developer is a chemical process used to develop latent fingerprints on porous surfaces such as paper. This process is generally performed after DFO and ninhydrin processing and is done in lieu of the conventional silver nitrate method.[86]

Presley et al.[17] have shown that DNA from evidence such as envelopes and stamps can successfully withstand certain evidence processing, such as latent fingerprint examinations with DFO and ninhydrin and/or questioned document processing using the ESDA method, and still yield a sufficient quality and quantity of DNA for PCR HLA DQα typing. However, physical developer fingerprint processing did adversely affect HLA DQα typing, and no HLA DQα type was obtained for samples processed using physical developer.

XI. QUALITY CONTROL/QUALITY ASSURANCE IN FORENSIC DNA TYPING

The use of RFLP and PCR analysis to help solve violent crimes has revolutionized the impact of forensic analyses of biological specimens in the criminal justice system. The RFLP

TABLE 1.
PCR Data from Survey of Forensic Laboratories

Laboratory[b]	Total number of cases[a]		
	1991	1992	1993[c]
FBI	—	65	48
UZ-I	4	40	—
SPS	161	301	—
IML-UC	5	20	—
ISFM-S	—	1	—
NISI[d]	10[e]	100[e]	—
DDEJ	—	started	—
GL	—	7	17
ILM-UM	150	200	—
DFM	—	64	—
FSA	100[a]	100[e]	—
SFSL	50	79	—
Totals	480	997	65

Note: Loci used include DQα, D1S80, APO-B, D17S5, SE 33, TC11, YNZ22, THO 11, MBP, COL 2A, HUMTH 01, D2S6, D17S30, and VWA.

a   Total for all laboratories, January 1991 through March 1993 equals 1522 cases.
b   Data from the following forensic laboratories: Federal Bureau of Investigation (FBI) Laboratory, Washington, DC; University of Zurich-Irchel (UZ-I), Zurich, Switzerland; Servizio Polizia Scientifica (SPS), Rome, Italy; Instituto di Medicina Legale-Universitat Cattolica del S. Cuore (IML-UC), Rome, Italy; Institute of Scientific and Forensic Medicine (ISFM-S), Singapore; National Institute of Scientific Investigation (NISI), Seoul, Korea; Direction des Expertises Judiciares (DDEJ), Montreal, Canada; Gerechtelyk Laboratorium (GL), Netherlands; Institute of Legal Medicine-University of Munster (ILM-UM), Munster, Germany; Department of Forensic Medicine (DFM), Bern, Switzerland; Forensic Science Associates (FSA), Richmond, CA; State Forensic Science Laboratory (SFSL), Adelaide, Australia.
c   1993 data for January through March only.
d   Cases estimated based on five samples per case and 50 samples in 1991 and 586 samples in 1992.
e   Approximate value.

and PCR technologies provide highly discriminating data that can be used to include or exclude potential suspects in criminal investigations. Because of the high discriminating power afforded by these techniques, much attention by the forensic community has been directed to the proper use of these techniques. Quality control (QC) and quality assurance (QA) measures represent important aspects of the forensic application of RFLP and PCR methodologies.

QC and QA have been defined as two different and specific functions.[87-90] The QC measures employed by a laboratory are designed to insure that the quality of the product, the DNA type in the forensic application, meets and satisfies specified criteria. The function of a QA program is to provide the criteria that can be used to establish with confidence that the QC functions are being performed adequately.

PCR analysis of DNA was the first use of DNA typing in the U.S. in 1986, as previously described by Blake et al.[20] Subsequently, the PCR-based typing of the HLA-DQα gene region and other genetic loci for forensic casework has become a widespread routine application. Between 1986 and 1992, several QC issues for PCR analysis were addressed in the literature by von Beroldingen et al.[91] and Higuchi and Kwok[92] and guidelines for PCR-based technologies were proposed by the Technical Working Group on DNA Analysis Methods (TWGDAM) in 1991,[93] and in 1992 by the International Society for Forensic Haemogenetics.[94]

Validation studies of PCR-based technologies for forensic casework, as have been previously discussed, brought about many QC considerations, and casework practice necessitated additional QC measures. For example, it has become useful to use a slot blot method to determine the approximate quantity of DNA in a forensic sample. Too little or too much DNA in the amplification process may result in little or no amplified product or artifactual products. Routine casework also required documentation of sample processing and analysis based on a detailed protocol.

In 1992, the FBI Laboratory implemented PCR HLA DQα typing for casework analysis and proposed the following outline of QC measures based on the literature, validation studies, and simulated and routine casework by Comey et al.[15] and Presley et al.[95]

I. Physical separation of locations for pre- and post-amplification procedures; DNA extraction and PCR set-up separated in time.
II. Extraction of DNA from known and questioned samples at different times.
III. Extraction blank(s) to test extraction reagents for DNA contamination.
IV. Slot blot hybridization with a probe for D17Z1 or other appropriate locus for human DNA quantitation. This can be documented by film or photographic record.

    A. Quantitation standards run on each blot (K562 cell line or other appropriate samples of known concentrations).
    B. Quality control of membrane lot for slot blot hybridization.

V. Amplification/typing controls.

    A. Amplification blank to test for DNA contamination of PCR reagents.
    B. Positive control with each thermal cycler run (DQα type 1.1, 3, or other appropriate control) to test for success of amplification, including information with regard to the detection of allele dropout and potentially different allele intensities.
    C. Duplicate amplification/typing of DNA extracts from questioned samples where feasible.
    D. Post-amplification test gel to look for the presence of the appropriate-sized amplified DNA product (pBR322 DNA digested with HAE III can be used as molecular weight size marker). This can be documented by a photograph.
    E. Quality control of AmpliType kit lot (amplify positive control, amplification blank, extraction blank, and DNA extract of known type). A run time segment printout record will be generated for each AmpliType kit lot tested. (This procedure is suggested for kits)
    F. Examination of dot blots for presence of sensitivity dot (i.e., C dot) for interpretation.
    G. Results for samples and controls can be photographically documented.

VI. Equipment controls.

    A. Thermal cycler (Refer to User's Manual, DNA Thermal Cycler 480, 9600, or other models, Perkin Elmer).
        1. Using Perkin Elmer Temperature Verification System, run periodic temperature-calibration verification test.
        2. Using Perkin Elmer Temperature Verification System, run periodic temperature-uniformity test, service yearly or as dictated by results of periodic tests.
        3. Within-instrument consistency using replicate samples of the same DNA quantity will be tested periodically.
        4. Between-instrument consistency using range of DNA concentration controls will be tested periodically.

5. Using thermal cycler diagnostic files for Perkin Elmer model 480 Thermal Cycler, the following checks will be performed periodically: chiller test; heater test; sensor test; overshoot test; and undershoot test.
6. The sample block will be cleaned periodically with cotton swabs that have been dipped in an approximately 50% v/v solution of isopropanol.
7. To verify cycle time reproducibility, a standard two-temperature incubation cycle will be performed periodically and record maintained.

B. Hot shaker water baths (check temperature routinely prior to use with NIST-traceable thermometer).

VII. Cleaning procedures for amplified DNA work area can be included in the protocol.

A. FUTURE CONSIDERATIONS FOR QA/QC IN FORENSIC DNA TYPING

The emergence of PCR-based multiplexing systems used for short tandem repeat and amplitype PM (Polymarker) systems and of PCR-based sequencing systems used for mtDNA may require additional QA/QC criteria before routine casework is performed. Because these PCR-based multiplexing and sequencing approaches are based wholly or in part on current PCR-based methodologies, the transition to multiplexing and sequencing strategies should require very little that is novel in the criteria for QA/QC.

XII. CONCLUSION

Technologies still relatively untapped by the forensic community, and new and developing biotechnologies, such as riboprinting, RAPD analysis, automated systems for sequencing mitochondrial or genomic DNA, multiplexing systems, and methodologies for amplifying single or few copies of DNA, will advance the development of new forensic applications. Appropriate QA/QC measures to assure their reliability and validity will be incorporated into the applications. The forensic community has wholly embraced the concepts and implementation of quality control and assurance measures in order to foster and consistently produce quality results. These continuing efforts will ensure the sound application and use of new and emerging DNA technologies in forensic analysis.

REFERENCES

1. **Budowle, B. and Baechtel, F. S.,** Modifications to improve the effectiveness of restriction fragment length polymorphism typing, *Appl. Theor. Electrophoresis,* 1, 181, 1990.
2. **Saiki, R. K., Scharf, S., Faloona, F., Mullis, K. B. Horn, G. T., Erlich, H. A., and Arnheim, N.,** Enzymatic amplification of beta-globin genomic sequences and restriction site analysis for diagnosis of sickle cell anemia, *Science,* 230, 1350, 1985.
3. **Chamberlain, J. S., Gibbs, R. A., Ranier, J. E., Nguyen, N., and Caskey, C. T.,** Deletion screening of Duchenne muscular dystrophy locus via multiplex DNA amplification, *Nucleic Acids Res.,* 16, 11141, 1988.
4. **Sparkman, D. R.,** Paraffin wax as a vapor barrier for the PCR, *PCR Methods Appl.,* 2(2), 180, 1992.
5. **Hugichi, R., von Beroldingen, C. H., Sensabaugh, G. F., and Erlich, H. A.,** DNA typing from single hairs, *Nature,* 322, 543, 1988.
6. **Sajantila, A., Budowle, B., Strom, M., Johnsson, V., Lukka, M., Peltonen, L., and Ehnholm, C.,** PCR Amplification of alleles at the D1S80 locus, comparison of a Finnish and North American Caucasian population sample, and forensic casework evaluation, *Am. J. Hum. Gen.,* 50, 816, 1992.
7. **Jinks, D. C., Minter, M., Tarver, D. A., Vanderford, M., Hejtmancik, J. F., and McCabe, E. R. B.,** Molecular genetic diagnostics of sickle-cell disease using dried blood specimens on blotters used for newborn screening, *Hum. Genet.,* 81, 363, 1989.

8. Williams, C. Weber, L., Williamson, R., and Hjelm, M., Guthrie spots for DNA-based carrier testing in cystic fibrosis, *Lancet,* ii, 693, 1988.
9. Gasparini, P., Savoia, A. Pignatti, P. F., Dallapiccola, B., and Novelli, G., Amplification of DNA from epithelial cells in urine, *New Engl. J. Med.,* 320, 809, 1989.
10. Hochmeister, M. N., Budowle, B., Borer, U. V., Eggmann, U., Comey, C. T., and Dirnhofer, R., Typing of DNA extracted from compact bone from human remains, *J. Forensic Sci.,* 36(6), 1649, 1991.
11. Hagelberg, E., Sykes, B., and Hedge, R., Ancient bone DNA amplified, *Nature,* 342, 485, 1989.
12. Sullivan, K. M., Hopgood, R,. and Gill, P., Identification of human remains by amplification and automated sequencing of mitochondrial DNA, *Int. J. Legal Med.,* 105, 83, 1992.
13. Hochmeister, M. N., Budowle, B., Jung, J., Borer, U. V., Comey, C. T., and Dirnhofer, R., PCR-based typing of DNA extracted from cigarette butts, *Int. J. Legal Med.,* 104, 229, 1991.
14. Impriam, C. C., Saiki, R. K., Erlich, H. A., and Teplitz, R. L., Analysis of DNA extracted from formalin-fixed, parrafin-embedded tissues by enzymatic amplification and hybridization with sequence-specific oligonucleotides, *Biochem. Biophys. Res. Commun.,* 41, 710, 1987.
15. Comey, C. T., Budowle, B., Adams, D. E., Baumstark, A. L., Lindsey, J. A., and Presley, L. A., PCR amplification and typing of the HLA DQα gene in forensic samples, *J. Forensic Sci.,* 38(2), 239, 1993.
16. Comey, C. T. and Budowle, B., Validation studies on the analysis of the HLA-DQα locus using the polymerase chain reaction, *J. Forensic Sci.,* 36(6), 1633, 1991.
17. Presley, L. A., Baumstark, A. L., and Dixon, A., The effects of specific latent fingerprint and questioned document examinations on the amplification and typing of the HLA DQα gene region in forensic casework, *J. Forensic Sci.,* 38(5), 1028, 1993.
18. Sanantila, A., Strom, M., Budowle, B., Karhunen, P. J., and Peltonen, L., The polymerase chain reaction and post-mortem forensic identity testing, application of amplified D1S80 and HLA-DQα loci to the identification of fire victims, *Forensic Sci. Int.,* 51, 22, 1991.
19. Thymann, M., Nellemann, L. J., Masumba, G., Irgens-Moller, L., and Morling, N., Analysis of the locus D1S80 by amplified fragment length polymorphism technique (AMP-FLP). Frequency distribution in Danes. Intra- and inter-laboratory reproducibility of the technique, *Forensic Sci. Int.,* 60, 47, 1993.
20. Blake, E., Mihalovich, J., Higuchi, R., Walsh, P. S., and Erlich, H., Polymerase chain reaction (PCR) amplification and human leukocyte antigen (HLA)-DQα oligonucleotide typing on biological evidence samples, casework experience. *J. Forensic Sci.,* 37(3), 700, 1992.
21. Walsh, P. S., Metzger, D. A., and Higuchi, R., Chelex 100 as a medium for simple extraction of DNA for PCR-based typing from forensic material, *Biotechniques,* 10, 506, 1991.
22. Singer-Sam, J., Tanguay, R. L., and Riggs, A., Use of chelex to improve the PCR signal from a small number of cells *Amplifications,* 3, 11, 1989.
23. Jung, J. M., Comey, C. T., Baer, D. B., and Budowle, B., Extraction strategy for obtaining DNA from bloodstains for PCR amplification and typing of the HLA-DQα gene, *Int. J. Legal Med.,* 104, 145, 1991.
24. Paabo, S., Ancient DNA, Extraction, characterization, molecular cloning, and enzymatic amplification *Proc. Natl. Acad. Sci. U.S.A.,* 86, 1939, 1989.
25. Waye, J. S., Presley, L. A., Budowle, B., Shutler, G. G., and Fourney, R. M., A simple and sensitive method for quantifying human genomic DNA in forensic specimen extracts, *Biotechniques,* 7, 852, 1989.
26. Walsh, P. S., Varlaro, J., and Reynolds, R., A rapid chemiluminescent method for quantitation of human DNA, *Nucleic Acids Res.,* 20(19), 5061, 1992.
27. Klevan, L., Horton, L., Carlson, D. P., and Eisenberg, A. J., Chemiluminescent detection of DNA probes in forensic analysis, 2nd Int. Symp. Forensic Aspects of DNA Analysis, FBI Academy, Forensic Science and Research Center, Quantico, VA, March 29 to April 2, 1993.
28. Saiki, R. K., Gefland, D. H., Stoffel, S., Scharf, S., Higuchi, R., et al., Primer directed enzymatic amplification of DNA with thermostable DNA polymerase, *Science,* 239, 487, 1988.
29. Saiki, R. K., Bugawan, T. L., Horn, G. T., Mullis, G. K., and Erlich, H. A., Analysis of enzymatically amplified beta-globin and HLA-DQ alpha with allele specific probes, *Nature,* 324, 163, 1986.
30. Saiki R. K., Chang, C. A., Levenson, C. H., Warren, T. C., Boehm, C. D., et al., Diagnosis of sickle-cell anemia and beta-thalassemia with enzymatically amplified DNA and non-radioactive allele-specific oligonucleotide probes, *New Engl. J. Med.,* 319, 537, 1988.
31. Saiki, R. K., Walsh, P. S., Levenson, C. H., and Erlich, H. A., Genetic analysis of amplified DNA with immobilized sequence-specific oligonucleotide probes, *Proc. Natl. Acad. Sci. U.S.A.,* 86, 6230, 1989.
32. Erlich, H. A., Higuchi, R., Lictenwalter, K., Reynolds, R., and Sensabaugh, G., Reliability of the HLA-PCR-based oligonucleotide typing system, *J. Forensic Sci.,* 36(5), 1017, 1990.
33. Walsh, S. P., Fildes, N., Louie, A. S., and Higuchi, R., Report of the blind trial of the cetus amplitype HLA DQ alpha forensic deoxyribonucleic acid (DNA) amplification and typing kit, *J. Forensic Sci.,* 36(5), 1551, 1991.
34. Stoneking, M., Hedgecock, D., Higuchi, R., Vigilant, L., and Erlich, H. A., Population variation of human mtDNA control region sequences detected by enzymatic amplification and sequence-specific oligonucleotide probes, *Am. J. Hum. Genet.,* 48, 370, 1991.

35. **Budowle, B., Charkraborty, R., Guisti, A. W., Eisenberg, A. J., and Allen, R. C.**, Analysis of the VNTR locus D1S80 by PCR followed by high resolution PAGE, *Am. J. Hum. Genet.*, 48, 943, 1991.
36. **Kasai, K., Nakamura, Y., and White, R.**, Amplification of a variable number of tandem repeat (VNTR) locus (pMCT118) by the polymerase chain reaction (PCR) and its application to forensic science, *J. Forensic Sci.*, 35, 1196, 1990.
37. **Horn, G. T., Richards, B., and Klinger, K. W.**, Amplification of a highly polymorphic VNTR segment by the polymerase chain reaction, *Nucleic Acids Res.*, 17, 2140, 1989.
38. **Odelberg, S. J., Plaetke, R., Eldridge, J. R., Ballard, P., et al.**, Characterization of eight VNTR loci by agarose gel electrophoresis, *Genomics*, 5, 915, 1989.
39. **Boerwinkle, E., Xiong, W., Fourest, E., and Chan, L.**, Rapid typing of tandemly repeated hypervariable loci by the polymerase chain reaction, application to the apolipoprotein B 3 prime hypervariable region *Proc. Natl. Acad. Sci. U.S.A.*, 86, 212, 1989.
40. **Ludwig, E. H., Friedl, W., and McCarthy, B. J.**, High-resolution analysis of a hypervariable region in the human apolipoprotein B gene, *Am. J. Hum. Genet.*, 48, 458, 1989.
41. **Wu, S., Senio, S., and Bell, G. I.**, Human Collagen, Type II, Alpha 1 (Col2A1) gene, VNTR polymorphism detected by gene amplification, *Nucleic Acids Res.*, 18, 3102, 1990.
42. **Edwards, A., Civitello, A., Hammon, H. A., and Caskey, C. T.**, DNA typing and genetic mapping with trimeric and tetrameric tandem repeats, *Am. J. Hum. Genet.*, 49, 746, 1991.
43. **Weigand, P., Budowle, B,. Rand, S., and Brinkmann, B.**, Forensic validation of STR systems SE33 and TC11, *Int. J. Legal Med.*, 105, 315, 1993.
44. **Polymeropoulos, M. H., Xiao, H., and Merril, C. R.**, Tetranucleotide repeat polymorphism at the human myelin basic protein gene (MBP) *Hum. Mol. Genet.*, 1, 658, 1992.
45. **Kimpton, C. P., Walton, A., and Gill, P.**, A further tetranucleotide repeat polymorphism in the VWA *Hum. Mol. Genet.*, 1, 28, 1992.
46. **Akane, A., Shiono, H., Matsubara, K., Nakahori, Y., Seki, S., et al.**, Sex identification of forensic specimens by polymerase chain reaction (PCR), two alternative methods, *Forensic Sci. Int.*, 49, 81, 1991.
47. **Nakahori, Y., Hamono, K., Iwayam, M., and Nakagoma, Y.**, Sex identification by polymerase chain reaction using X-Y homologous primer *Am. J. Hum. Genet.*, 39, 472, 1991.
48. **Pascal, O., Aubert, D., Gilbert, E., and Moisan, J. P.**, Sexing of forensic samples using PCR, *Int. J. Legal Med.*, 104, 205, 1991.
49. **Allen, R. C., Graves, G., and Budowle, B.**, Polymerase chain reaction amplification products separated on rehydratable polyacrylamide gels and stained with silver, *BioTechniques*, 7, 736, 1989.
50. **Budowle, B. and Allen, R. C.**, Discontinuous polyacrylamide gel electrophoresis of DNA fragments, in *Protocols in Human Molecular Genetics-Methods in Molecular Biology*, Vol. 9, Mathew, C. G., Ed., Human Press, Clifton, NJ, 1991.
51. **Allen, R. C.**, Ed., Rapid isoelectric focusing and detection of nanogram amounts of proteins from body tissues and fluids, *Electrophoresis*, 1, 32, 1980.
52. **Allen, R. C. and Lack, M.**, Standardization in isoelectric focusing on ultrathin-layer rehydratable polyacrylamide gels, in *New Directions in Electrophoretic Methods*, Jorgenson, J. W. and Phillips, M., Eds., American Chemical Society, Washington, DC, 1987.
53. **Robertson, J., Schafer, T., Kronick, M., and Budowle, B.**, Automated analysis of fluorescent amplified fragment length polymorphism for DNA typing, in *Advances in Forensic Haemogenetics*, Vol. 4, Springer-Verlag, Berlin, New York, 1991.
54. **Gyllensten, U. B. and Erlich, H. A.**, Generation of single-stranded DNA by the polymerase chain reaction and its application to direct sequencing of the HLA-DQ alpha locus *Proc. Natl. Acad. Sci. U.S.A.*, 86, 9986, 1988.
55. **Hultman, et al.**, *BioTechniques*, 10, 84, 1991.
56. **Lee, J. S.**, Alternative dideoxy sequencing of double-stranded DNA by cyclic reactions using Tag polymerase, *DNA Cell Biol.*, 10, 67, 1991.
57. **Ruano, G. and Kidd, K. K.**, Coupled amplification and sequencing of genomic DNA, *Proc. Natl. Acad. Sci. U.S.A.*, 88, 2815, 1991.
58. **Sanger, F., Nicklen, S., and Coulson, A. R.**, DNA Sequencing with chain-terminating inhibitors, *Proc. Natl. Acad. Sci. U.S.A.*, 74, 5463, 1977.
59. **Smith, L. M., Sanders, J. Z., Kaiser, R. J., Hughes, P., Dodd, C., et al.**, Fluorescent detection in automated DNA sequence analysis, *Nature*, 321, 674, 1986.
60. **Prober, J. M., Trainor, G. L., Dam, R. J., Hobbs, F. W., Robertson, C. E., et al.**, A system for rapid DNA sequencing with fluorescent chain-terminating dideoxynucleotides, *Science*, 238, 336, 1987.
61. **Merril, C. R.**, Genetics, forensics and electrophoresis, in Proc. Int. Symp. Forensic Applications of Electrophoresis, U.S. Government Printing Office, Washington, DC, 1984.
62. **Paabo, S. Gifford, J. A., and Wilson, A. C.**, Mitochondrial DNA sequences for a 7000 year old brain, *Nucleic Acids Res.*, 16, 9775, 1988.

63. **Paabo, S., Higuchi, R., and Wilson, A. C.,** Ancient DNA and the polymerase chain reaction, *J. Biol. Chem.,* 264, 9709, 1989.
64. **Paabo, S.,** Molecular cloning of ancient Egyptian mummy DNA, *Nature,* 314, 644, 1985.
65. **Sullivan, K. M., Hopgood, R., Lang, B., and Gill, P.,** Automated amplification and sequencing of human mitochondrial DNA, *Electrophoresis,* 12, 17, 1991.
66. **Jeffreys, A. J., MacLeod, A., Tamaki, K., Neil, D. L., and Monckton, D. G.,** Minisatellite repeat coding as a digital approach to DNA typing, *Nature,* 354, 204, 1991.
67. **Budowle, B. and Allen, R. C.,** Electrophoresis reliability. I. The contaminant issue, *J. Forensic Sci.,* 32, 1537, 1987.
68. **Adams, D. E., Presley, L. A., Baumstark, A. L., Hensley, K. W., Hill, A. L., et al.,** DNA analysis by restriction fragment length polymorphisms of body fluid stains subjected to contamination and environmental insults, *J. Forensic Sci.,* 36, 1284, 1991.
69. **Erlich, H. A., Gefland, D., and Sninsky, J. J.,** Recent advances in the polymerase chain reaction, *Science,* 252, 1643, 1991.
70. **Walsh, P. S., Erlich, H. A., and Higuchi, R.,** Preferential PCR amplification of alleles, mechanisms and solutions *PCR Meth. Appl.,* 1(4), 241, 1992.
71. **Gyllensten, U. and Erlich, H. A.,** Ancient roots for polymorphism at the HLA-DQ alpha locus in primates, *Proc. Natl. Acad. Sci. U.S.A.,* 86, 9986, 1989.
72. **Comey, C. T., Jung, J. M., and Budowle, B.,** Use of formamide to improve amplification of HLA-DQ alpha sequences, *BioTechniques,* 10, 60, 1991.
73. **Longo, M., Beringer, M. C., and Hartley, J. L.,** Use of uracil DNA glycosylase to control carry-over contamination in polymerase chain reactions, *Gene,* 93, 125, 1990.
74. **Welsh, J. and McClelland, M.,** Fingerprinting genomes using PCR with arbitrary primers, *Nucleic Acids Res.,* 18, 6531, 1990.
75. **Williams, J. G. K., Kubelik, K. J., Livak, J. A., Rafalski, J. A., and Tingey, S. V.,** DNA polymorphisms amplified by arbitrary primers are useful as genetic markers, *Nucleic Acids Res.,* 18, 6532, 1990.
76. **Deragon, J-M. and Landry, B. S.,** RAPD and other PCR-based analyses of plant genomes using DNA extracted from small leaf disks, *PCR Methods Appl.,* 1(3), 175, 1992.
77. **Yoon, C. K.,** *Science,* 260, 894, 1993.
78. **Foster, D. J. and Morantz, D. J.,** An electrostatic imaging technique for the detection of indented impressions in documents, *Forensic Sci. Int.,* 13, 51, 1979.
79. **Noblett, M. G. and James, E. L.,** Optimum conditions for examination of documents using an electrostatic detection apparatus (ESDA) device to visualize indented writings, *J. Forensic Sci.,* 28, 697, 1983.
80. **Pounds, C. A., Grigg, R., and Mongkolaussavaratana, T.,** The use of 1,8-diazafluoren-9-one (DFO) for the fluorescent detection of latent fingerprints on paper: a preliminary evaluation, *J. Forensic Sci.,* 35, 169, 1990.
81. **Mooney, D. G., Currin, T. J., and Matheny, D.,** Naphtha-ninhydrin method *Identification News,* 27, 6, 1977.
82. **Mooney, D. G.,** Additional notes on the use of ninhydrin, *Identification News,* 23, 9, 1973.
83. **Crown, D. A.,** The Development of latent fingerprints with ninhydrin, *J. Crim. Law Criminol.,* 60, 258, 1969.
84. **Linde, H. G.,** Latent fingerprints by a superior ninhydrin method, *J. Forensic Sci.,* 20, 581, 1975.
85. **Morris, J. R. and Goode, G. C.,** NFN-An improved ninhydrin reagent for the detection of latent fingerprints, *Police Res. Bull.,* 24, 45, 1974.
86. **Phillips, C. E., Cole, D. O., and Jones, G. W.,** Physical developer: a practical and productive latent print developer, *J. Forensic Identification,* 40(3), 135, 1990.
87. **Anon.,** Quality Systems Terminology, American National Standard ANSI/ASQC A3-1978, American Society for Quality Control, Milwaukee, 1978.
88. **Kilshaw, D.,** Quality assurance. I. Philosophy and basic principles, *Med. Lab. Sci.,* 43, 377, 1986.
89. **Kilshaw, D.,** Quality assurance. I. Internal quality control, *Med. Lab. Sci.,* 44, 73, 1987.
90. **Kilshaw, D.,** Quality assurance. I. External quality assessment, *Med. Lab. Sci.,* 44, 178, 1987.
91. **von Beroldingen, C. H., et al.,** Application of PCR to the analysis of biological evidence, in *PCR Technology, Principles and Applications for DNA Amplifications,* Erlich, H. A., Ed., Stockton Press, New York, 209–223, 1989.
92. **Higuchi, R. and Kwok, S.,** Avoiding false positives with PCR, *Nature,* 339, 237, 1989.
93. **Technical Working Group on DNA Analysis Methods (TWGDAM),** Guidelines for a quality assurance program for DNA analysis, *Crime Lab. Dig.,* 18(2), 44, 1991.
94. **Anon.,** Recommendations of the DNA Commission of the International Society for Forensic Haemogenetics Relating to the Use of PCR-Based Polymorphisms, *Forensic Sci. Int.,* 55 (Editorial), 1, 1992.
95. **Presley, L. A., et al.,** The implementation of polymerase chain reaction (PCR) HLA DQα typing by the FBI laboratory, Proc. 3rd Int. Symp. Human Identification 1992, Promega Corporation, Madison, WI, 245, 1992.
96. **Budowle, B. et al.,** unpublished data.
97. FBI Laboratory, Latent Fingerprint Section, personal communication.

Chapter 29

APPLICATION OF PCR-AMPLIFIED DNA MARKERS IN IDENTIFICATION OF INDIVIDUALS

Antti Sajantila and Leena Peltonen

TABLE OF CONTENTS

I. Introduction 277

II. Analysis of Multiallelic Mini- and Microsatellite Loci 278
    A. Construction of Allelic Standards for AMP-FLP Analysis 278
    B. Separation of Amplified Mini- and Microsatellite Alleles 279
    C. Detection of PCR Products from Multiallelic Loci 279

III. Analysis of Biallelic DNA Markers 282
    A. Principle of the Method 282
    B. Evaluation of Solid-Phase Minisequencing Method for
       Forensic Medicine 282

IV. Application of PCR to Forensic Casework and Paternity Testing 283
    A. General Considerations 283
    B. Population Genetic Issues 284
    C. Forensic Case-Work 284
    D. Identification of Human Remains 285

V. Conclusion 285

References 285

I. INTRODUCTION

The application of DNA typing to the identification of individuals provides objective evidence in the analysis of biological material obtained in criminal cases as well as in disputed paternity cases. Since the advent of the polymerase chain reaction (PCR),[1,2] amplification of DNA from the evidence samples has been applied in genetic characterization of individuals.

Several features of PCR make it particularly suitable for forensic analysis. The amplification of DNA fragments by PCR does not necessarily require purified DNA,[3,4] and degraded DNA from various biological specimens, even old, decomposed samples can be typed.[5-8] Amplification of degraded DNA is possible because the degraded DNA may contain intact short fragments of the specific region, and also because the DNA polymerase can use short, degraded DNA templates as primers in the amplification process.[9] The sensitivity of PCR is unique when compared to other methods used to identify DNA sequences in biological samples, which enables the analysis of minute amounts of DNA. Higuchi et al.[10] have shown that it is possible to amplify DNA fragments even from a single hair. Also, DNA fragments from single human sperm and diploid cells[11] have been amplified using PCR. The PCR technique is also time saving and economical when compared to the other techniques commonly used in forensic medicine (i.e., analysis of restriction fragment length polymorphisms by Southern blotting). Because of these advantages, it is obvious that several forensic and paternity-testing laboratories have added PCR-based methods to their repertories.[12-16]

Generally, PCR can be considered as a technique for sample preparation, since it enables amplification of the specific DNA region from subanalytical quantities present in crude biological material to a level at which conventional DNA methods can be directly applied to analyze the sample. The choice of detection method of the amplified product is determined by the type of polymorphism to be analyzed. Currently, the PCR-based DNA methods applied in forensic science include analysis of minisatellite[17] or microsatellite[18,19] loci, allele-specific oligonucleotide hybridization using the reverse dot blot method for the detection of sequence polymorphism at the HLA-DQα locus,[20] and nucleotide sequence determination of mitochondrial DNA (mtDNA).[21] Also analysis of biallelic DNA markers by oligonucleotide ligation assay[22,23] or solid-phase minisequencing[24] can be effectively used in individual identification. The latest development of PCR-based methods to identity testing is minisatellite variant repeat (MVR) mapping.[25] This novel technique takes into account both length and sequence polymorphism at a particular locus. Thus, an extremely high level of individuality can be achieved by analyzing only one single locus.

This chapter briefly describes some aspects in the genetic characterization of individuals in the context of forensic medicine. A particular emphasis is drawn to the analysis of multiallelic mini/microsatellites by amplified fragment length polymorphism (AMP-FLP) technique[26] and to the analysis of biallelic DNA markers by solid-phase minisequencing technique.[27]

II. ANALYSIS OF MULTIALLELIC MINI- AND MICROSATELLITE LOCI

To be useful for identification of individuals, the markers should fulfill certain criteria (Table 1). Mini- and microsatellite loci are particularly suitable for forensic purposes, since they are highly polymorphic and they exhibit a high degree of heterozygosity. These polymorphisms arise from differences in the copy number of tandemly organized stretches of specific DNA sequences.[17,28] In addition to the evaluation of the population genetic parameters of mini/microsatellites, the technique for analysis of these informative markers has to be simple, rapid, and reproducible. A widely used strategy in PCR-based analysis of forensic samples is to detect mini/microsatellite loci by the AMP-FLP technique. In the AMP-FLP analysis, these loci are amplified by the PCR, after which the alleles are separated in a high-resolution polyacrylamide gel electrophoresis (PAGE) and visualized by silver staining.[29] Examples of suitable mini/microsatellite loci for forensic and paternity analysis are given in Table 2.

The evaluation of the AMP-FLP method for forensic purposes includes the construction of allelic standards for multiallelic loci, the development of an effective separation of different sized mini/microsatellite alleles, and the application of suitable detection methods for the PCR products.

A. CONSTRUCTION OF ALLELIC STANDARDS FOR AMP-FLP ANALYSIS

A prerequisite for the identification of the alleles at mini-or microsatellite loci and for standardization of the AMP-FLP analysis is a locus-specific allele standard composed of a suitable set of known alleles. An allelic ladder can be constructed from a mixture of genomic DNA with different genotypes for each locus or by cloning individual alleles at the mini/microsatellite loci. A necessity for the first alternative is a continuous source of genomic DNA from individuals with suitable genotypes, and the latter one is too laborious for a routine forensic laboratory. In our laboratory, we have developed a practical and simple reamplification procedure for producing allelic ladders for AMP-FLP analysis[30] (Figure 1). This inexpensive

TABLE 1.
Characteristics of a Useful Forensic Genetic Marker

Pattern of inheritance well established
High degree of polymorphism
High degree of heterozygosity
Independent inheritance from the other markers used
Mutation rate very low
Reliable allele detection
Population data of allele, phenotype, and/or genotype frequency established
A simple, rapid, and reproducible analytical method available
Little material needed for the analysis

TABLE 2.
Examples of Mini/Microsatellite DNA Markers That Can Be Utilized in Identity Testing

	No. of alleles	Ref.
Minisatellites		
D1S80	>20	49
apoB	>20	50
D17S30	>13	40
col2A1	>10	51
Microsatellites		
vWA	>7	52
HUMTH01	>5	53

procedure can be easily applied for any locus showing length polymorphism, and the allelic standards can serve as an unlimited supply for several laboratories. The use of common ladders in different forensic and paternity testing laboratories is essential for comparison of results and for sharing population data. Internationally shared allele markers would be valuable for standardization of forensic analyses as well as in other genetic studies.

B. SEPARATION OF AMPLIFIED MINI- AND MICROSATTELLITE ALLELES

After amplification, the mini/microsatellite alleles are separated according to their size in electrophoresis. As the size of these alleles is usually less than 1 kb, many laboratories have chosen high-resolution PAGE for this purpose. When using proper, locus-specific allelic standards, different laboratories obtain comparable results, independently, whether they have employed vertical[31] or horizontal[26] PAGE formats or used denaturing or non-denaturing gels. An example of AMP-FLP analysis of mini- and microsatellite loci is given in Figure 2.

C. DETECTION OF PCR PRODUCTS FROM MULTIALLELIC LOCI

The amplified alleles at mini/microsatellite loci can be detected after the electrophoretic size separation using autoradiography,[18,19] staining with ethidium bromide[32] or silver nitrate.[29] The alleles can also be detected during the electrophoresis with the aid of a fluorescence detector.[33] Currently several laboratories utilize silver staining in forensic analysis.[13-16] As a novel and sensitive alternative to silver staining, fluorescent detection of PCR products using an automatic DNA sequencer has been evaluated.[33] This approach needs the introduction of a fluorescence label to the PCR product by either fluorescently labeled PCR primer or

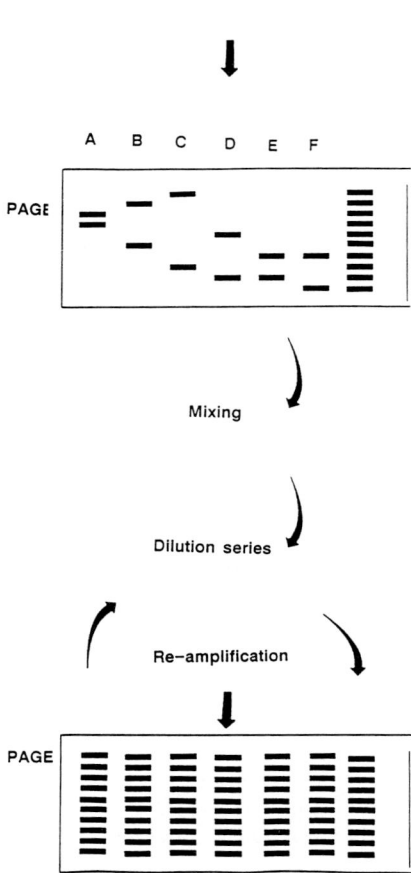

Figure 1. First, to obtain DNA fragments of equal intensity in the final allelic ladder, the amplification products of the initial PCR are run in separate lanes in PAGE. An amount of each PCR product estimated to result in equal intensity of the different alleles is then mixed according to visual evaluation, and a dilution series from $1:10^5$ to $1:10^{10}$ is made from this DNA mixture. Each dilution is reamplified at the same PCR conditions as the individual samples. For each of the loci a "window" of suitable dilution, usually ranging from $1:10^6$ to $1:10^9$, can thus be found. For details, see Reference 30.

fluorescently labeled nucleotides. In addition to the locus-specific allelic ladders described above, internal molecular weight standards also run in the same lane as the sample can be used.[33,34] Practically any molecular weight standard with alleles suitably sized outside the range of the allele sizes of the amplified sample can be used. The use of internal standards facilitates automatic sizing of alleles that differ only 2 to 4 bp in length.[34]

In addition to automatic sizing of alleles, the potential for quantitative analysis of PCR products in fluorescent-based techniques can also be utilized in forensic and population genetic studies. We have developed a novel quantitative method allowing rapid determination of allele frequencies at mini- or microsatellite loci.[35] This method involves amplification of the polymorphic loci from pooled DNA samples (hundreds or thousands of individuals) using fluorescently labeled primer in the PCR. After electrophoresis, the relative amount of each PCR product is quantitatively determined with the aid of the fluorescent detector and a computer program designed for this purpose. The relative amount of an amplified DNA fragment obtained from the pooled sample represents the frequency of the individual allele in

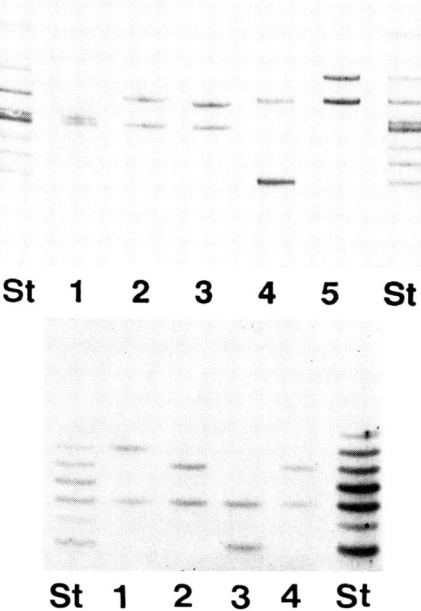
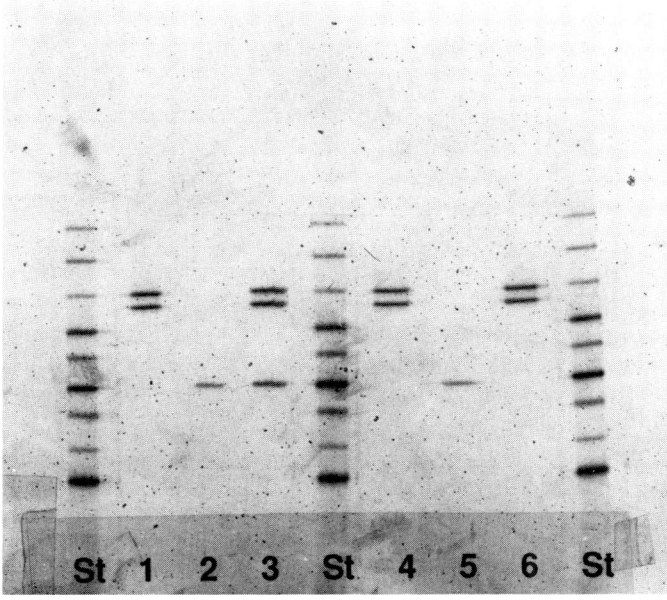

Figure 2. Examples of AMP-FLP analysis using a vertical 16-cm discontinuous PAGE. For technical details, see Reference 31. (A) An example of AMP-FLP analysis at the minisatellite col2A1 locus. St = allelic standard composed of col2A1 alleles seen in the Finnish population, 1 to 5 DNA samples. The alleles consist of 31 to 33 bp repeats. (B) An example of AMP-FLP analysis of the microsatellite at the vWA locus. St = allelic standard composed of vWA alleles seen in the Finnish population, 1 to 4 DNA samples. The alleles consist of tetranucleotide repeats. (C) An example of AMP-FLP analysis of the minisatellite at the D1S80 locus from a sexual assault. St = allelic standard composed of every second D1S80 allele seen in the Finnish population, 1 to 6 DNA samples. Sample 3 is mixed sample (vaginal swab) showing both male fragments (samples 1, 4, and 6) and female fragments (samples 2 and 5). The alleles consist of 16 bp repeats.

the population. The distribution of the alleles obtained from pooled samples was shown to directly correspond to that of the alleles obtained from individual samples analyzed separately.[35] From a practical point of view, it would be rational to initially prescreen potential DNA markers, after which the allele frequencies for the most informative ones would also be analyzed from individual samples.

III. ANALYSIS OF BIALLELIC DNA MARKERS

In addition to the multiallelic DNA loci, single nucleotide variations resulting in biallelic sequence polymorphism can also be utilized in identification of individuals. Solid-phase minisequencing is a novel method previously described for the detection of single nucleotide variations and small deletions or insertions in PCR amplified DNA fragments.[24] Solid phase minisequencing has been frequently applied to the detection of point mutations causing human genetic diseases,[37,38] but can be equally well applied to the identification of individuals.[27]

A. PRINCIPLE OF THE METHOD

In the solid-phase minisequencing method the DNA fragment spanning the polymorphic site is amplified by PCR using one biotinylated and one non-biotinylated primer. The biotinylated strand of the PCR product is then captured in a streptavidin coated microtiter plate well, as the non-biotinylated DNA strand is removed by alkaline denaturation. Detection of the variable nucleotide at the polymorphic site is performed in two separate minisequencing reactions. In the minisequencing reaction a detection step primer, which anneals immediately adjacent to the variable site, is elongated by a DNA polymerase with one single ^{3}H-labeled dNTP complementary to the nucleotide at the polymorphic site. The incorporated label is measured in a scintillation counter, and the result of the solid-phase minisequencing analysis is given as the ratio between the labels incorporated in the two reactions. The principle of the method is presented in Figure 3.

B. EVALUATION OF SOLID-PHASE MINISEQUENCING METHOD FOR FORENSIC MEDICINE

The frequency distribution of alleles and genotypes at each of the biallelic DNA markers used in the solid-phase minisequencing format can be determined by analyzing both individual DNA samples and by quantitative analysis of pooled DNA samples. As in the case of mini/microsatellite loci, this quantitative approach has been shown to facilitate a rapid and accurate estimation of the allele frequencies of biallelic markers in the population.[27]

Since biallelic markers are much less informative than multiallelic polymorphisms, it is necessary to analyze several loci from each sample to achieve discrimination between individuals as good as that obtained with multiallelic DNA markers. Thus, design of primers suitable for multiplex PCR is crucial. With a panel of 12 markers (three sets of multiplex PCRs for four loci per reaction) the discrimination power in forensic casework and the exclusion power in paternity testing is similar to that obtained by analyzing three minisatellite loci (D1S80, D17S30, apoB) by the AMP-FLP technique.[27]

In identity testing the main advantages of the solid-phase minisequencing method are that the test can easily be automated and that the results are obtained as numeric values which unequivocally define the genotypes of the samples. The microtiter plate format of the test is particularly efficient for the detection of multiple single nucleotide variations from a large number of samples, and this setup can be easily combined with automated pipettors and washers used routinely for immunochemical analyses in clinical laboratories. Additionally, a problem of preferential amplification of smaller alleles is avoided. Two disadvantages of analyzing biallelic markers, compared to analyses of multiallelic length polymorphisms, are that less information is obtained, requiring simultaneous analyses of numerous biallelic loci,

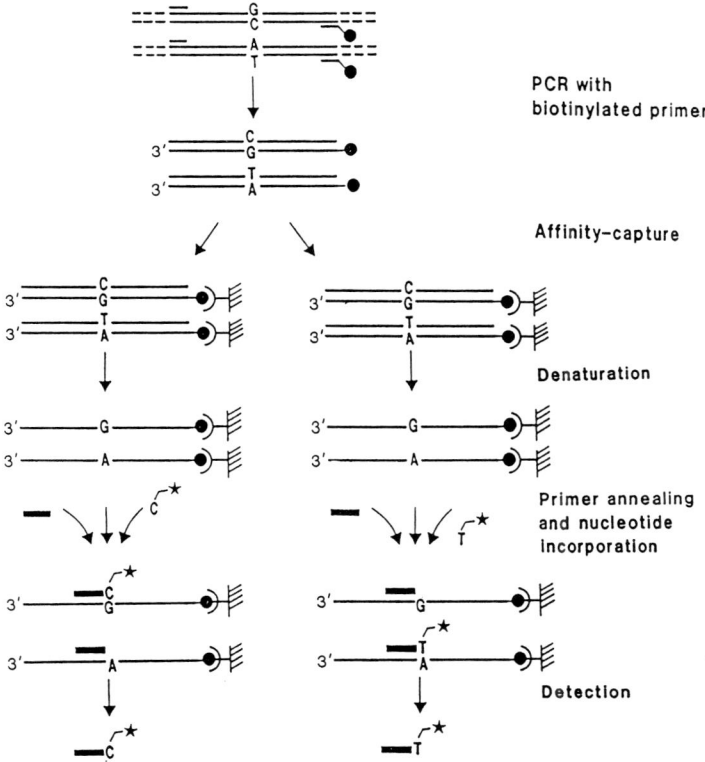

Figure 3. Principle of the solid-phase minisequencing technique. For details, see text and Reference 27.

and that sensitivity to detection of the contaminating DNA often present in forensic casework is lower. However, the quantitative nature of the solid-phase minisequencing can also be utilized to solve this latter problem.[27]

IV. APPLICATION OF PCR TO FORENSIC CASEWORK AND PATERNITY TESTING

A. GENERAL CONSIDERATIONS

Since PCR is extremely sensitive, the PCR-based analyses are prone to cross-contamination between samples and especially to contamination by PCR products from the laboratory. A minimal amount of previously amplified DNA may exist as a contaminant in the laboratory air and contain more copies of the DNA fragment to be amplified than the actual evidence sample. In routine forensic and paternity-testing laboratories, this problem is emphasized, since same DNA regions are repeatedly amplified from a large number of samples. One way to avoid contamination is to perform the three stages of the analysis (sample treatment, PCR process, and analysis of PCR products) in separate rooms. Precautions in sample handling, using separate equipment in each of the steps, are crucial. In addition, potential contaminations of samples should be routinely monitored using negative controls and controls of known DNA type.

To avoid mis-priming or formation of primer dimers in the crucial first cycle of the PCR, adoption of a "hot-start" procedure[38] in the routine laboratory protocol is highly recommended. The hot start has also been demonstrated to improve the sensitivity of the PCR. In practice, the hot start is generally performed by withholding one reagent from the reaction mixture at temperatures below the proper primer annealing temperature. A feasible way to amplify DNA using hot start is to heat the reaction mixture containing the DNA template and all reagents except DNA polymerase to 95°C for 5 min, after which the DNA polymerase is added at 80°C.[40]

When amplifying minisatellite alleles of different lengths, the amplification may proceed less efficiently for longer DNA fragments. More efficient amplification of one of the alleles can also occur if the template DNA is extensively degraded or when the initial copy number of target DNA is very small. Preferential amplification of one of the alleles in a heterozygous sample is seen as uneven intensities of the alleles or even as a complete loss of the larger allele. This phenomenon of an "allelic drop-out" may result in incorrect interpretation of homozygosity. The problem of preferential amplification has been encountered especially when amplifying the D17S30 locus.[40,41] However, preferential amplification can be avoided by adjusting the template to DNA polymerase ratio.[41,54] On the other hand, when analyzing biallelic DNA markers, with alleles of equal length, the problem of preferential amplification of either of the alleles does not exist.

B. POPULATION GENETIC ISSUES

Analysis of a forensic sample has three potential interpretations of comparisons between known and unknown sample: (1) inclusion, i.e., the sample in question potentially originates from the same biological source as the reference sample, (2) exclusion, i.e., the questioned sample and the reference sample cannot originate from same biological source, and (3) inconclusive, when data obtained are insufficient to interpret the samples. When the DNA profile obtained from a forensic specimen matches that obtained from a suspect, the weight of the evidence is based on the frequency of the genotypes (or alleles) in the relevant population. The genotype frequency data in the appropriate population are essential for calculating the probability that the same DNA profile could be derived from a random, unrelated person.

One issue in population genetics connected to genetic evidence is the assumption of Hardy-Weinberg (H-W) equilibrium.[42,43] A H-W population is one that meets the criteria of (1) no selection, (2) no immigration, (3) no emigration, (4) no mutation, and (5) large population. Essentially this means that if a locus meets H-W expectations, the alleles associate randomly. It is important to address the concept of H-W expectation when analyzing the highly polymorphic loci, since the assumption of H-W equilibrium permits estimation of the frequency of occurrence of a particular genotype that has not been observed in the population previously.

C. FORENSIC CASE-WORK

In cases of sexual assault the evidence sample often contains a mixture of DNA from the victim and the suspect(s). A differential extraction procedure[44] can been applied to separate the DNA from female epithelial cells from that of sperm cells. Experience from analysis of the minisatellite loci by the AMP-FLP technique have shown the usefulness of the differential extraction procedure for PCR based analysis of vaginal swabs.[39]

Occasionally the sperm fraction of the samples also contains female DNA in addition to the sperm cell DNA. This results in a DNA profile of more than two bands on the silver-stained gels. However, this does not confound the interpretation of the results, but rather aids in the process, since the female bands in the male DNA fraction serve as internal controls (Figure 2C).

DNA extracted from blood from various substrates consisting of clothing, wood, paper, human skin, rubber, leather, and steel has been shown to amplify reliably using different techniques to detect the amplified DNA samples.[39,45] Also DNA from cigarette butts can be a potential source of evidence.[46] In samples where amplification of the original DNA sample is unsuccessful, nested primers can be applied to improve the efficiency and specificity of the amplification process.[32]

D. IDENTIFICATION OF HUMAN REMAINS

Due to the effects of temperature, moisture, and autolytic process within the body, the identification of human remains by conventional forensic means is not always possible. Several reports of DNA amplification from human cadavers have shown the utility of PCR in identifying unknown human remains.[6,7] A potential source of DNA from human remains is compact bones,[8] but soft tissue samples can also be used if available.[47] In the identification of human remains, reference blood samples from relatives are essential for interpretation of the results. Ideally, the blood samples are obtained from the putative parents. However, in cases where samples from the parents are unavailable, blood from other relatives can be analyzed.[48]

V. CONCLUSION

Despite some technical problems encountered due mainly to the high sensitivity of DNA amplification, PCR-based techniques will remain the best alternative for DNA typing in different forensic cases in the future. Virtually any defined short polymorphic DNA sequence can potentially be utilized in the analysis. Low cost, technical feasibility, the fact that the quantity and quality of the analyzed DNA are not a limiting factor, and the possibility for discrete determination of alleles are the main advantages of PCR.

In the future, PCR-based pretreatment of the samples, combined with the technical development toward automation of the DNA analyses and the interpretation of the results, will help forensic laboratories in performing effective and highly reliable identity testing.

REFERENCES

1. **Mullis, K. B. and Faloona, F. A.,** Specific synthesis of DNA *in vitro* via a polymerase-catalyzed chain reaction, *Methods Enzymol.,* 155, 335, 1987.
2. **Saiki, R. K., Scharf, S., Faloona, F., Mullis, K. B., Horn, G. T., Erlich, H. A., and Arnheim, N.,** Enzymatic amplification of beta-globin genomic sequences and restriction analysis for diagnosis of sickle cell anemia, *Science,* 230, 1350, 1985.
3. **Higuchi, R.,** Simple and rapid preparation of samples for PCR, in *PCR Technology: Principles and Applications for DNA Amplification,* Erlich, H. A., Ed., Stockton Press, New York, 1989, 31.
4. **Walsh, P. S., Metzger, D. A., and Higuchi, R.,** Chelex-100 as a medium for simple extraction of DNA for PCR-based typing from forensic material, *BioTechniques,* 10, 506, 1991.
5. **Bugawan, T. L., Saiki, R. K., Levenson, C. H., Watson, R. M., and Erlich, H. A.,** The use of non-radioactive oligonucleotide probes to analyze enzymatically amplified DNA for prenatal diagnosis and forensic HLA typing, *BioTechnology,* 6, 943, 1988.
6. **Hagelberg, E., Sykes, B., and Hedges, R.,** Ancient bone DNA amplified, *Nature,* 342, 485, 1990.
7. **Haglund, W. D., Reay, D. T., and Tepper, S. L.,** Identification of decomposed human remains by deoxyribonucleic acid (DNA) profiling, *J. Forensic Sci.,* 35, 724, 1990.
8. **Hochmeister, M. N., Budowle, B., Jung, J., Borer, U. V., Comey, C. T., and Dirnhofer, R.,** Typing of DNA extracted from compact bone tissue from human remains, *J. Forensic Sci.,* 36, 1649, 1991.
9. **Pääbo, S., Irwin, D. M., and Wilson, A. C.,** DNA damage promotes jumping between templates during enzymatic amplification, *J. Biol. Chem.,* 265, 4718, 1990.
10. **Higuchi, R., von Beroldigen, C. H., Sensabaugh, G. F., and Erlich, H. A.,** DNA typing from single hairs, *Nature,* 332, 543, 1988.
11. **Li, H., Gyllensten, U. B., Cui, X., Saiki, R. K., Erlich, H. A., and Arnheim, N.,** Amplification and analysis of DNA sequences in single human sperm and diploid cells, *Nature,* 335, 414, 1988.
12. **Chimera, J.,** Interlaboratory assessment of the MCT118 AMPFLP locus in paternity testing, in Proc. 3rd Int. Symp. Human Identification, Scottsdale, AZ, 1993, 199.
13. **Kloosterman, A. D.,** PCR amplification of the D1S80 and the HLA DQα loci in the Dutch population in comparison to other Caucasian populations, in Proc. 3rd Int. Symp. Human Identification, Scottsdale, AZ, 1993, 329.
14. **Parkin, B.,** The development of forensic DNA profiling in the Metropolitan Police Forensic Science Laboratory, in Proc. 3rd Int. Symp. Human Identification, Scottsdale, AZ, 1993, 345.

15. **Brinkmann, B.,** The use of STR's in stain analysis, in *Proc. 3rd Int. Symp. Human Identification,* Scottsdale, AZ, 1993, 357.
16. **Sajantila, A.,** PCR amplification applied to forensic medicine, in *Proc. 3rd Int. Symp. Human Identification,* Scottsdale, AZ, 1993, 375.
17. **Jeffreys, A. J., Wilson, V., and Thein, S. L.,** Hypervariable "minisatellite" regions in human DNA, *Nature,* 314, 67, 1985.
18. **Litt, M. and Luty, J. A.,** A hypervariable microsatellite revealed by *in vitro* amplification of a dinucleotide repeat within the cardiac muscle actin gene, *Am. J. Hum. Genet.,* 44, 397, 1989.
19. **Weber, J. L. and May, P. E.,** Abundant class of human DNA polymorphisms which can be typed using the polymerase chain reaction, *Am. J. Hum. Genet.,* 44, 388, 1989.
20. **Saiki, R. K., Walsh, P. S., Lewenson, C. H., and Erlich, H. A.,** Genetic analysis of amplified DNA with immobilized sequence-specific oligonucleotide probes, *Proc. Natl. Acad. Sci. U.S.A.,* 86, 6230, 1989.
21. **Sullivan, K. M., Hopgood, R., Lang, B., and Gill, P.,** Automated amplification and sequencing of human mitochondrial DNA, *Electrophoresis,* 12, 17, 1991.
22. **Landegren, U., Kaiser, R., Sanders, J., and Hood, L.,** A ligase-mediated gene detection technique, *Science,* 241, 376, 1988.
23. **Nickerson, D.A., Kaiser, R., Lappin, S., Stewart, J., Hood, L., and Landegren, U.,** Automated DNA diagnostics using ELISA-based oligonucleotide ligation assay, *Proc. Natl. Acad. Sci. U.S.A.,* 87, 8923, 1990.
24. **Syvänen, A.-C., Aalto-Setälä, K., Harju, L., Kontula, K., Söderlund, H.,** A primer-guided nucleotide incorporation assay in the genotyping of apolipoprotein E, *Genomics,* 8, 684, 1990.
25. **Jeffreys, A. J., MacLeod, A., Tamaki, K., Neil, D. L., and Monckton, D. G.,** Minisatellite repeat coding as a digital approach to DNA typing, *Nature,* 354, 204, 1991.
26. **Budowle, B., Chakraborty, R., Giusti, A. M., Eisenberg, A. J., and Allen, R.,** Analysis of the VNTR locus D1S80 by the PCR followed by high resolution PAGE, *Am. J. Hum. Genet.,* 48, 137, 1991.
27. **Syvänen, A.-C., Sajantila, A., and Lukka, M.,** Identification of individuals by analysis of biallelic DNA markers, using PCR and solid-phase minisequencing, *Am. J. Hum. Genet.,* 52, 46, 1993.
28. **Nakamura, Y., Leppert, M., O'Connell, P., Wolff, R., Holm, T., Culver, M., Martin, C., Fujimoto, E., Hoff, M., Kumlin, E., and White, R.,** Variable number of tandem repeat (VNTR) markers for human gene mapping, *Science,* 235, 1616, 1987.
29. **Allen, R.C., Graves, G., and Budowle, B.,** Polymerase chain reaction amplification products separated on rehydratable polyacrylamide gels stained with silver, *BioTechniques,* 7, 736, 1989.
30. **Sajantila, A., Puomilahti, S., Johnsson, V., and Ehnholm, C.,** Amplification of reproducible allele markers for amplified fragment length polymorphism (Amp-FLP) analysis, *BioTechniques,* 12, 16, 1992.
31. **Sajantila, A. and Lukka, M.,** Improved separation of PCR amplified VNTR alleles by a vertical polyacrylamide gel electrophoresis, *Int. J. Legal Med.,* 105, 355, 1993.
32. **Vuorio, A., Sajantila, A., Hämäläinen, T., Syvänen, A.-C., Ehnholm, C., and Peltonen, L.,** Amplification of the hypervariable region close to the apolipoprotein B gene: application to forensic problems, *Biochem. Biophys. Res. Commun.,* 170, 616, 1990.
33. **Robertson, J., Ziegle, J., Kronick, M., Madden, D., and Budowle, B.,** Genetic typing using automated electrophoresis and fluorescence detection, in *DNA Fingerprinting: Approaches and Applications,* Burke, T., Dolf, G., Jeffreys, A. J., and Wolff, R., Eds., Birkhäusen, Basel, 1991, 391.
34. **Mayrand, P. E., Corcoran, K. P., Ziegle, J. S., Robertson, J. M., Hoff, L. B., and Kronick, M. N.,** The use of fluorescence detection and internal lane standards to size PCR products automatically, *Appl. Theor. Electrophoresis,* 3, 1, 1992.
35. **Pacek, P., Sajantila, A., and Syvänen, A.-C.,** Determination of allele frequencies at loci with length polymorphism by quantitative analysis of DNA amplified from pooled samples, *PCR Methods Appl.,* 2, 313, 1993.
36. **Syvänen, A.-C., Söderlund, H., Laaksonen, E., Bengtström, M., Turunen, M., and Palotie, A.,** N-ras gene mutations in acute myeloid leukemia: accurate detection by solid-phase minisequencing, *Int. J. Cancer,* 50, 713, 1992.
37. **Jalanko, A., Ranki, M., and Söderlund, H.,** Detection of four cystic fibrosis mutations with solid-phase minisequencing, *J. Cell. Biochem.,* Suppl. 16F, 12, 1992.
38. **Chou, Q., Russell, M., Birch, D. E., Raymond, J., and Bloch, W.,** Prevention of pre-PCR mis-priming and primer dimerization improves low-copy-number amplifications, *Nucleic Acids Res.,* 20, 1717, 1992.
39. **Sajantila, A., Budowle, B., Ström, M., Johnsson, V., Lukka, M., Peltonen, L., and Ehnholm, C.,** PCR amplification of alleles at the D1S80 locus: comparison of a Finnish and a North American caucasian population sample, and forensic casework evaluation, *Am. J. Hum. Genet.,* 50, 816, 1992.
40. **Horn, G. T., Richards, B., and Klinger, K. W.,** Amplification of a highly polymorphic VNTR segment by polymerase chain reaction, *Nucleic Acids Res.,* 17, 2140, 1989.
41. **Walsh, P. S., Erlich, H. A., and Higuchi, R.,** Preferential PCR amplification of alleles: mechanisms and solutions, *PCR Methods Appl.,* 2, 241, 1992.

42. **Hardy, G. H.,** Mendelian proportions in mixed populations, *Science,* 28, 49, 1908.
43. **Weinberg, W.,** Über den Nachweis der Vererbung beim Menschen, *Jahresh. Ver. Vaterl., Naturkd., Wurttemb.,* 64, 368, 1908.
44. **Gill, P., Jeffreys, A. J., and Werrett, D. J.,** Forensic application of DNA "fingerprints", *Nature,* 318, 577, 1985.
45. **Comey, C. T. and Budowle, B.,** Validation studies of the analysis of the HLA-DQα locus using the polymerase chain reaction, *J. Forensic Sci.,* 36, 1633, 1991.
46. **Hochmeister, M. M., Budowle, B., Jung, J., Borer, U. V., Comey, C. T., and Dirnhofer, R.,** PCR-based typing of DNA extracted from cigarette butts, *Int. J. Legal Med.,* 104, 229, 1991.
47. **Sajantila, A., Ström, M., Budowle, B., Karhunen, P., and Peltonen, L.,** The polymerase chain reaction and post-mortem forensic identity testing: application of amplified D1S80 and HLA-DQα loci to the identification of fire victims, *Forensic Sci. Int.,* 51, 23, 1991.
48. **Helminen, P., Johnsson, V., Ehnholm, C., and Peltonen, L.,** Proving paternity of children with deceased fathers, *Hum. Genet.,* 87, 657, 1991.
49. **Kasai, K., Nakamura, Y., and White, R.,** Amplification of a variable number of tandem repeats (VNTR) locus (pMCT118) by the polymerase chain reaction (PCR) and its application to forensic science, *J. Forensic Sci.,* 35, 1196, 1990.
50. **Boerwinkle, W., Fourest, E., and Chan, L.,** Rapid typing of tandemly repeated hypervariable loci by the polymerase chain reaction: application to the apolipoprotein B 3′ hypervariable region, *Proc. Natl. Acad. Sci. U.S.A.,* 86, 212, 1989.
51. **Wu, S., Senio, S., and Bell, G. I.,** Human collagen, type II, alpha 1 (Col2A1), gene: VNTR polymorphism detected by gene amplification, *Nucleic Acids Res.,* 18, 3102, 1990.
52. **Kimpton, C., Walton, A., and Gill, P.,** A further tetranucleotide repeat polymorphism in the vWF gene, *Hum. Mol. Genet.,* 1, 287, 1992.
53. **Edwards, A., Civitello, A., Hammond, H. A., and Caskey, C. T.,** DNA typing and genetic mapping with trimeric and tetrameric tandem repeats, *Am. J. Hum. Genet.,* 49, 746, 1991.
54. **Sajantila, A. et al.,** unpublished data.

Chapter 30

PCR IN FORENSIC SCIENCE

Vincenzo L. Pascali, Marina Dobosz, and Ernesto d'Aloja

TABLE OF CONTENTS

I. PCR as a Forensic Technique 289

II. Amplifiable Markers Used in Forensics 290
    A. Unique Sequences 290
    B. Repetitive DNA 290
        1. AMP-FLPs 290
        2. Microsatellites 291
        3. Minisatellite Variant Repeats 292
    C. Mitochondrial Polymorphisms 295

III. PCR Protocols and Ancillary Techniques 296
    A. Steps of Amplification 296
    B. Detection of PCR Products 296
        1. Dot Blot Hybridization with ASO Probes 296
        2. Sequencing of Asymmetric PCR Products 296
        3. Hot PCR and PAGE 297
        4. Native Horizontal PAGE and Silver Staining 298
        5. Automated Laser Fluorescent Procedures 298

IV. Applications 298
    A. Parenthood Analysis 298
    B. Identification of Stains 301
    C. Other Applications 301
        1. Diagnosis of Species 301
        2. Controls of Animal Breeding 302
        3. RAPDs 302
        4. Miscellaneous 302

Acknowledgments 303

References 303

I. PCR AS A FORENSIC TECHNIQUE

Analysis of genetic diversity is today largely based on the use of the polymerase chain reaction (PCR). Already, a considerable number of amplifiable (PCR-based) polymorphic sequences are known, in human[1] and other mammalian[2] genomes, and the number is steadily increasing. Applications include the analysis of biological samples for forensic purposes.

Forensic science resorts to tests of biological identity (DNA profiles) in a broad range of situations. DNA profiles help identify perpetrators of crimes, victims of disasters and war casualties, parents of individuals, poachers of protected animal species, as well as detect commercial frauds. PCR suits these applications for the following reasons:

- Highly polymorphic PCR markers are available, conferring on forensic profiles a remarkable power of identification.
- Analysis of minute amounts of biological samples is possible; this is of crucial importance for the development of DNA profiles in typical forensic instances (for example, analysis of stains left at the scene of a crime).
- Severely degraded samples (such as aged samples) can be profiled with high rates of success by amplification of low-molecular-weight markers.
- Objective typing results are derived from procedures employing PCR markers; this is a major advantage over other profiling methods (for instance, Southern blotting of restriction fragments) and results in fewer controversies in courtrooms.
- PCR is a real-time procedure, used daily to help forensic investigations while they are being deployed.

This chapter is a short review of markers, procedures, and applications of PCR as a forensic technique.

II. AMPLIFIABLE MARKERS USED IN FORENSICS

A list of amplifiable polymorphisms used for human identification, given in Table 1, mentions those loci that have well-established applications in casework, through experience and collaborative efforts at validation.[3,4]

A. UNIQUE SEQUENCES

Although a few unique sequences are good candidates for forensic profiling,[5-7] a single-copy gene, DQA1 (also known as DQα[7,8]) was the first example of amplifiable polymorphism applied to forensics. The class II haplotype of human major histocompatibility complex (MHC) is composed of three different genes (DR, DQ, DP), each encoding an alpha and a beta subunit glycoprotein. The DQα gene expresses four major alleles (A1, A2, A3, A4) and some subtypes (A1*0101, A1*0102, A1*0103; A1*0401, A1*0501, A1*0601) due to sequence differences arising from mutation, recombination, or gene conversion. A simplified version of this polymorphism (comprising the four alleles and the A1 subtypes) is currently used by hundreds of laboratories in the field of criminal investigations. This success is due in part to a commercial version of a reverse dot blot hybridization procedure, which has proved to be a reliable and sensitive tool for the detection of sequence variations at this locus.

B. REPETITIVE DNA

While in principle several categories of DNA polymorphisms could be used for purposes of identification, most forensic markers come from the category of highly repetitive DNA. The acronym VNTRs (variable number of tandem repeats) was coined by Nakamura et al.[9] for noncoding sequences having polymorphism that arises from tandem iteration of a simple nucleotide motif (core sequence). Originally applied to restriction fragments polymorphisms (RFLPs), this designation extends to amplifiable repetitive polymorphisms as well.

1. AMP-FLPs

Amplifiable restriction fragments (AMP-FLPs) are restriction polymorphisms that can be subjected to PCR analysis. They are derived from restriction maps and generally have long repeat units in multiple copies and high-molecular-weight allelic fragments.

TABLE 1.
Amplifiable Polymorphisms Used for Human Identification

	Marker	Locus	Size (bp)	Heterozygosity (Caucasians)	Detection
Nuclear DNA					
Unique sequences	DQA1	Chromosome 6	239–242	0.75	Dot blot hybridization
Repetitive DNA					
	APOB 3'HVR	Chromosome 2	563–915	0.73	AGE
AMP-FLPs					
	MCT118	D1S80	401–801	0.80	PAGE
	COL2A1	12q14.3	600–700	0.81	PAGE
	YNZ22	D17S5	100–1100	0.90	PAGE
STRs	HUMTH01	11p15.5-p15	183–207	0.75	PAGE
	ACTBP2	Chromosome 6	237–312	0.93	PAGE
	VWF	12p12-12pter	135–167	0.75	PAGE
	HUMFXIIIA1	6p25-p24	182–236	0.78	PAGE
	HUMFES/FPS	15q25-qter	214–238	0.75	PAGE
Minisatellite variant repeats	MS32 MVR	D1S8	293–2004	0.991	Asymmetric PCR + S.B.A.
Mitochondrial DNA	D-loops	—	245–288	—	Asymmetric PCR + sequencing

Note: A list of loci currently employed in forensics is provided.

Boerwinkle et al.[10] first amplified a hypervariable region downstream of the apolipoprotein gene (APOB 3'HVR, APOB), in the short arm of chromosome 2. Soon afterward, Kasai et al.[11] described another VNTR (D1S80, MCT118) with equivalent polymorphic content and fragment sizes. Following the adoption of these and other AMP-FLPs[12-14] as markers in identity tests,[15,16] it soon became apparent that they were of limited applicability. In fact, their size is such that the relevant polymorphism is difficult to reproduce in classical forensic situations, such as the analysis of age-degraded samples. Although this procedure is still used in protocols of parenthood analysis (Figure 1), it is rarely used to generate profiles of criminal evidence.

2. Microsatellites

An elementary form of repetitive DNA occurs in mammalian genomes, determined by di-, tri-, and tetranucleotide repeats arranged in very short arrays (short tandem repeats, STRs; also called microsatellites). Microsatellites emerged from screenings of genomic libraries with small human DNA fragments that had motifs drawn from genome databanks.[17] These polymorphisms display several valuable properties: they are short (100 to 200 bp on average) but highly polymorphic (0.75 average polymorphic information content, PIC); distributed extensively throughout human chromosomes (a 745-kb human DNA sequence contains a microsatellite every 6 kb[18,19]); and ideally suited for PCR and automated analysis (for example, amplification of several loci at once and machine-driven analysis of fragments[20,21]).

A large number of STRs are already available and their numbers are growing quickly, since hundreds of novel loci are being discovered.[22-26] Their use is having a profound impact on genome mapping[27] and is also changing forensic DNA profiling strategies. A recent collaborative study based on STR strategies has clearly shown that tetranucleotide repeats are effective tools to profile bloodstains.[4] There are several reasons why these perform better than smaller-motif repeats. For example, dinucleotide-based differences are difficult to detect, and

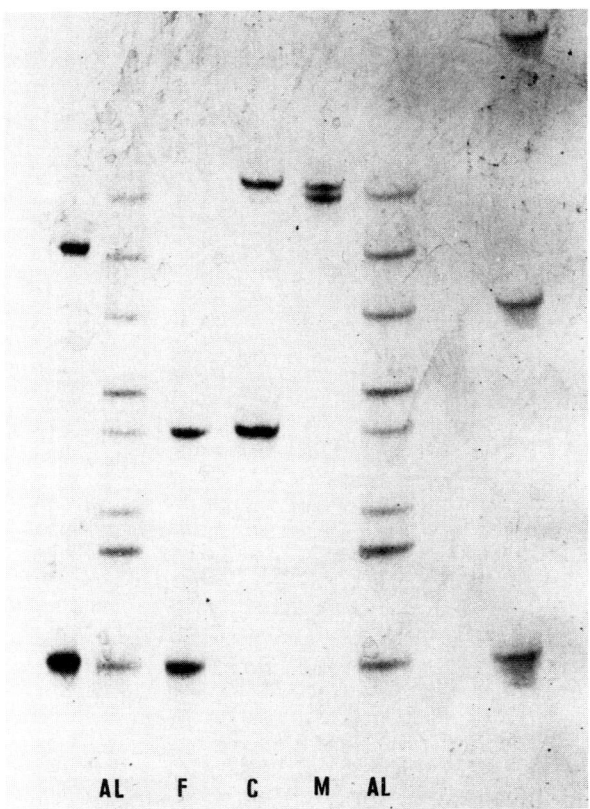

Figure 1. D1S80 is an AMP-FLP used in forensics. Two family segregations are here shown, in which the father is either compatible (1A) or excluded (1B) (F, father; C, child; M, mother; AL, allelic ladder). Samples (10 to 50 ng DNA) were amplified on a thermal cycler (Perkin Elmer, Model 480) with the following step cycle procedure: 94°C for 60 s, 65°C for 60 s, and 72°C for 60 s (28 c with a prolonged 10-min extension in the last step). Amplimers were as follows:

5'- GAA ACT GGC CTC CAA ACA CTG CCC GCC G- 3'
5' GTC TTG TTG GAG ATG CAC GTG CCC CTT GC- 3'

according to Budowle et al.[50] Five microliters of amplification product were mixed (1:1) with 2% Bromophenol in water and separated on a nondenaturing water-cooled PAGE (T, 7.5%, C, 2%; electrode distance: 200 mm; thickness 0.35 mm; power pack settings: 25 mA, 3 W). At electrophoretic completion of the gel, silver staining was carried out as follows: (1) fixation in 10% ethanol for 10 min; (2) immersion in nitric acid for 3 min; (3) two rinsing passages in double-distilled water; (4) immersion in a 0.2% silver nitrate solution for 30 min; (5) rinsing twice in double-distilled water; (6) immersion in a prewarmed 30% sodium carbonate/0.25% formaldehyde developing solution under continuous stirring; (7) change to a 5% acetic acid stop solution (all solutions are w/v or v/v in double distilled water). Flanking the samples, some homemade allelic ladders are loaded, to help genotype identification.

CA- and GT-rich dinucleotide strands have irregular mobility.[28] Dinucleotides give rise to shadowing bands derived from slippage during the amplification process ("stutter" bands[19]), an artifact that seldom occurs with tetranucleotide repeats. Several tetranucleotide-based polymorphisms are being adopted by forensic laboratories.[21,29-32] A tetranucleotide STR pattern is shown in Figure 2.

3. Minisatellite Variant Repeats

An innovative approach to PCR profiles was devised by Jeffreys et al.,[33] who described a hypervariable locus (D1S8, a minisatellite with a 29-bp core sequence repeated in long arrays) having a sequence heterogeneity superimposed on length differences (minisatellite

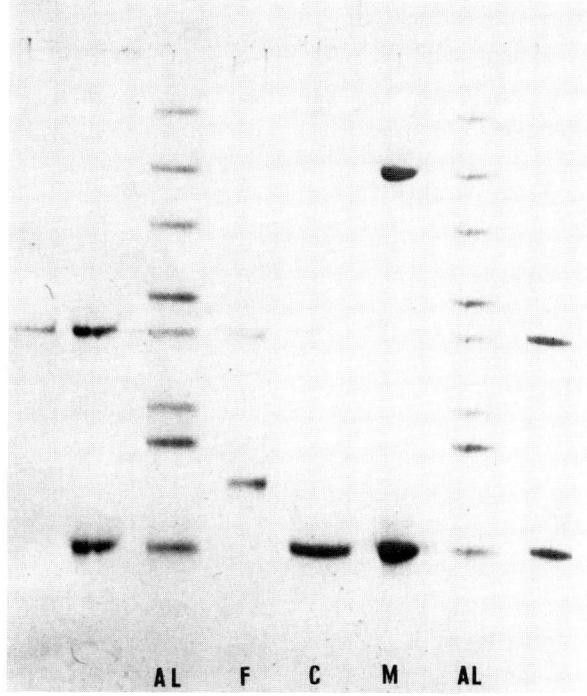

Figure 1. (continued)

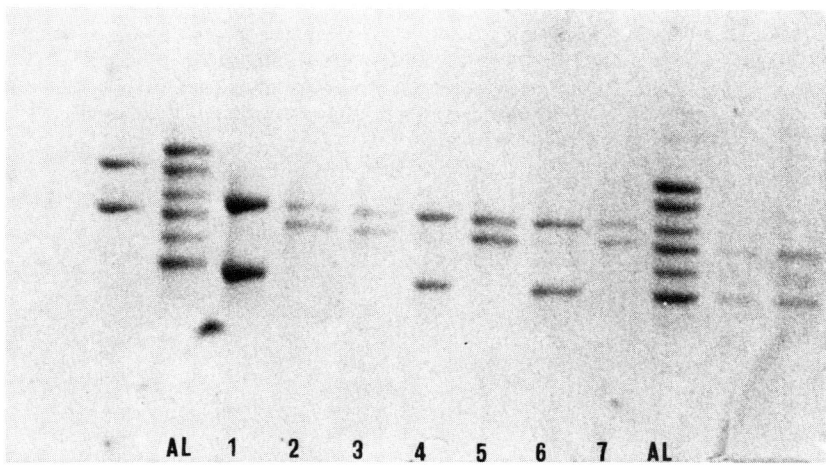

Figure 2. HUMTH01 is a short tandem repeat used for stain analysis.[21] An identification analysis is reported. DNA samples (10 ng on average) were amplified at the following condition: 94°C for 60 s, 64°C for 60 s, and 72°C for 90 s (30 c; an extension period of 5 min is required in the last cycle). Amplimers were as follows:

5' - GTG GGC TGA AAA GCT CCC GAT TAT - 3'
5'- ATT CAA AGG GTA TCT GGG CTC TGG - 3'.

A non-denaturing PAG electrophoresis (T, 8%, C, 3%) was used, followed by silver staining (as described in Figure 1). Genotyping is obtained by a side-to-side comparison and an allelic ladder. From left: 1, 4,6, victim; 2,3, sperm stain; 5,7 culprit; AL, allelic ladder.

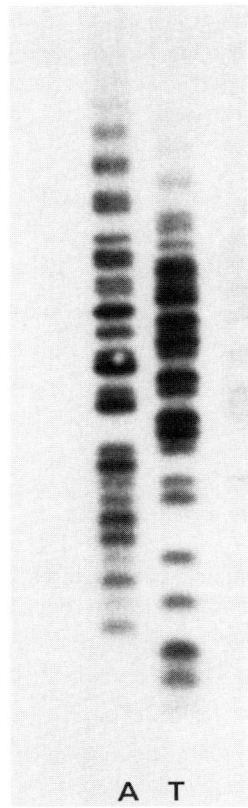

Figure 3. Minisatellite variant repeats (MVRs) within the hypervariable locus D1S8 (probe MS32, Cellmark Diagnostics). MVR essentially consists of a PCR amplification followed by an agarose separation and Southern blotting.[34] Genomic DNA samples (10 to 200 ng) were amplified at the following conditions after a denaturation step of 5 min at 94°C (hot start): 94°C for 80 s, 66°C for 60 s, and 70°C for 120 s (19 c, plus 2 c further: 60 s at 68°C, 300 s at 70°C). The following set of primers was employed:

- Common primer 32-OR
  3'- TGG TGG GAA GGG TGG TTT GAT GAG - 5'
- TAG:
  3'- AGG CCT GGT ACC TGC CGT ACT - 5'
- 32TAG-A:
  3'- CGG TCC CCA CTG AGT CTT ACA GGC CTG GTA CCT GCG TAC T- 5'
- 32TAG-T:
  3'-TGG TCC CCA CTG AGT CTT ACA GGC CTG GTA CCT GCG TAC T-5'

(final concentration: common primer 32-OR, 1 µM; TAG, 1 µM; TAG-T, 20 µM; TAG-A, 5 µM). Electrophoretic separation was on a 2% NuSieve agarose (FMC) gel 20 cm long in 1X TBE at 120 V. The gel was subsequently denatured-depurinated (acid wash for 15 min, alkaline denaturing wash for 30 min, and a final 30-min neutralizing step), then blotted to a nylon membrane. Hybridization was carried out with a radioactive labeled (random exanucleotide priming) MS32 probe, and exposure of blots was prolonged for 1 to 2 days. An individual pattern is shown with its two-track (for either 32TAG-A or 32TAG-T priming) ladder of alleles. At each position the individual is homozygous or heterozygous by the presence/absence of the relevant amplified band.

variant repeats, MVR). An A → G transition affects about half of D1S8 repeats, creating a HaeIII restriction site at 5' of each basic repeat and a highly ordered ladder of restricted fragments in partially restricted digests. However, the original procedure of identifying sequence variations by partial restriction of DNA was difficult to reproduce. Another protocol was described lately, addressing this polymorphism by PCR analysis.[34] In diploid genomes, MVR profiles are "bar codes", read as binary information (presence/absence of the mutation) in two adjacent tracks, with bands for both/either form of mutant sequence. This gives profiles an internal consistency making interpretation of patterns highly objective and unaffected by some classical uncertainties of DNA profiles, such as those brought about by electrophoretic skewedness of bands and by errors in sizing adjacent fragments (Figure 3).

MVR-PCR is not exempt from drawbacks.[35] In fact it may be difficult to score all rungs of a ladder comprising close to 40 to 50 bands. Furthermore, data on the occurrence of so complex MVR haplotypes is still missing. Finally scoring complete MVR profiles, each with dozens of bands, is a more laborious task than initially supposed (MVR requires Southern blot analysis of amplified fragments, since ethidium bromide stained bands are often faint and not clear-cut to read in agarose). Designed to fit criminal investigations, MVR has not yet found application in casework.

C. MITOCHONDRIAL POLYMORPHISMS

Mitochondrial DNA (mtDNA) is maternally inherited[36] and is present in 1,000 to 10,000 copies per mammalian cell.[37] Two highly polymorphic domains exist in the D-loop region of human mtDNA.[38] One of these domains, comprising a 228-bp region, was used by Higuchi et al.[7] for profiling DNA extracted from single hair roots. The same approach has recently been applied to other sources of mtDNA.[39] In the two D-loop domains, nucleotide substitutions occur at several positions of an invariant consensus sequence,[36] leading to considerable variability. Positioned as they are in short-size domains, the relevant base-pair transitions are closely associated. If a given haplotype is observed in x individuals of a sample of size n, then the probability of a random match is[39]

$$x/n + 2[x(n-x)/n]^{1/2}/n \tag{1}$$

Random matches at every position of the consensus sequence are arranged in a bimodal curve of probabilities, with most individuals differing by either 5 to 10 or 20 to 25 average mutations.[40]

A database of 524 individual assets, mostly of African ancestry, was reported by Vigilant et al.[41] Other data are available for Afro-Americans and Caucasians.[39,42]

MtDNA profiles have several advantages and a promising spectrum of applications:

1. Mitochondrial sequences are present in multiple copies per cell and can consequently be amplified in critical situations, where nuclear markers often fail to work (samples from old skeletal remains, teeth, hair shafts).
2. Identity tests rely on comparisons of unknown mtDNA with that of any relative available in the maternal lineage.
3. Sequence ambiguities due to heterozygote sequence patterns are absent, since mtDNA is haploid.
4. Polymorphic information is limited to two small regions, which can be sequenced in just one deoxy-chain termination reaction.

These advantages are counterbalanced by the fact that the inherent typing procedure is labor intensive. While efforts to standardize an automatic sequencing procedure for mtDNA are in progress,[40,43] the ultimate way to genotype D-loop loci seems to be the use of sequence-specific oligonucleotides in a dot-blot hybridization procedure.[44]

III. PCR PROTOCOLS AND ANCILLARY TECHNIQUES

A. STEPS OF AMPLIFICATION

Forensic protocols for amplification do not usually differ from those designed for other applications. However, since in this field degraded DNA is often analyzed, special care is taken to evaluate its rate of degradation and choose amplimers that have amplified products within the size range of the template. PCR-related procedures, suitable for criminal sample analysis or expressly designed for these purposes, include the hot-start procedure (a prolonged denaturation step administered before *TAQ* polymerase addition),[45] and heat-soaked PCR, (a more prolonged step of denaturation before addition of the entire PCR cocktail).[46] Both are thought to enhance amplification of age-degraded samples.

A procedure intended to improve the readability of PCR band patterns and increase the rate of success of PCR genotyping is the primer extension preamplification (PEP),[47,48] essentially a two-round amplification with nested priming. In the first step, standard PCR step-cycles are performed with amplimers encompassing a relatively large genomic portion. In the second round, one of the primers is replaced by an internal primer (heminesting), and further amplification is carried out with fewer cycles. PEP has been used to type haploid chromosomes of single sperm cells, allowing study of recombination at microsatellite loci. Its main advantages are the reduction of non-specific background and of PCR slipped bands, and the increased yield of specific products. Heminested primers are available for a number of potential forensic markers.[48]

B. DETECTION OF PCR PRODUCTS

1. Dot Blot Hybridization with ASO Probes

Unique sequence polymorphisms are conveniently typed by amplification of the entire locus and subsequent hybridization to several sequence-specific oligonucleotides (SSO), each containing an allele-specific motif (ASO).[8] Multiple dot blot formats with non-radioactive detection are used for typing the HLA DQA1 locus. Nine oligonucleotides, bearing allele specificities tailed with a polydT sequence, are bound to a nylon membrane by UV linking (254 nm for 15 s). Amplification of the whole region is carried out using two flanking amplimers, one of which is biotinylated. Products of amplification are first hybridized to the nylon membrane, which is then processed with an avidin-horseradish peroxidase conjugate. Presence of hybridization is revealed as a blue precipitate following the action of the enzyme on a colorless dye. A commercial version of this dot blot in reverse format (Amplitype, Cetus Perkin Elmer) is now a largely adopted procedure allowing typing with only a few nanograms of template DNA (as much as is extracted from a shed hair root).

Mitochondrial DNA is another good candidate for SSO probing. Some interesting results have been recently described by Stoneking,[44] who used a variety of short oligonucleotide sequences specifically designed to hybridize single sequence patterns of mtDNA.

2. Sequencing of Asymmetric PCR Products

Sequencing of single-strand PCR products is in principle the ideal, error-proof strategy to address sequence polymorphisms for purposes of identification. Nucleotide sequences can easily be compared in identity tests and no eye-matching of adjacent bands is required. This technique cannot be applied to identification of diploid genomes, because of possible heterozygosities occurring in the relevant sequence patterns. However, mitochondrial DNA is an ideal target for sequence analysis, because it is haploid. A two-step amplification procedure is applied to hypervariable D-loop domains of human mtDNA. In the first round of amplification a large, double-stranded portion of the domain is obtained. The second step relies on two nested primers, one of which is added to the PCR cocktail in limiting concentration (0.02 to 0.01 μM). This "asymmetric" PCR generates short, single-stranded fragments, which are

PCR in Forensic Science

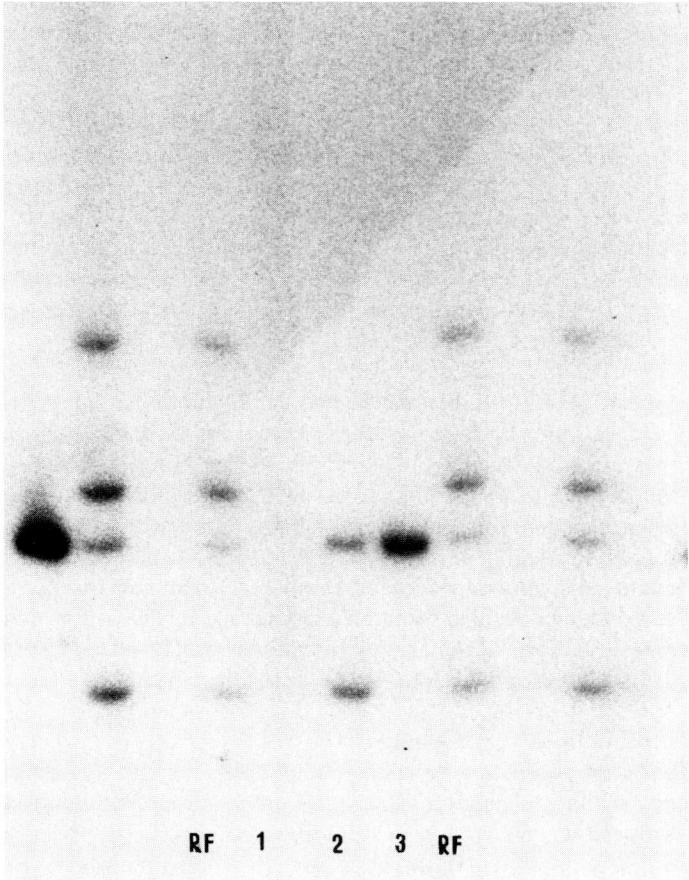

Figure 4. APOB 3'HVR[10] was the first AMP-FLP used in forensics. A procedure for amplification with radionuclide incorporation (hot PCR) is shown, which makes it easier to detect very faint bands. DNA samples were amplified using a standard mixture of "cold" nucleotides (dNTPs) plus an α32-dCTP radionuclide (0.2 µCi for each amplified sample). Amplimers were:

5'-ATG GAA ACG GAG AAA TTA TG- 3'
5'-CCT TCT CAC TTG GCA AAT AC-3'

Amplification conditions were as follows: 94°C for 60 s, 58°C for 60 s, and 59°C for 300 s (30 c with a final extension of 600 s). Separation of amplified fragments was obtained on a 25-cm-long polyacrylamide gel (T, 5%, C, 2%). At electrophoresis completion, the gel was overlaid with a X-ray film and stored at room temperature for 4 to 24 h. Specific radioactive bands correspond to amplified products from some forensic samples found at the scene of a crime. From left: RF, reference fragments; 1, a degraded sample; 2, victim; 3, bloodstain found on a suspect's cloth. Hot PCR is useful whenever age-degraded samples are to be typed. Alternatively, an ethidium bromide staining with UV transillumination allows for a clearcut detection of bands on agarose gels.

sequenced in one deoxy-chain termination reaction. Progress in the automation of sequence analysis promises to speed up this procedure.[40,43]

Minisatellite variant repeat analysis supplies an example of asymmetric PCR from nuclear DNA,[34] in which detection of some specific base transitions is sought instead of the integral sequencing of the molecule.

3. Hot PCR and PAGE

Polyacrylamide gel electrophoresis (PAGE) is used to identify size differences in VNTRs. Separation on vertical denaturing gels — 6% T, 2% C polyacrylamide (T = acrylamyde + bisacrylamide/100; C = bisacrylamide/T), 7 M urea — is the standard procedure for microsatellites typing, since difference of as little as 2 bp can be detected between allelic species. Visualization of band patterns is achieved by radioactive labeling of PCR products with subsequent autoradiography. Several labeling procedures are in use. A widespread approach is the use of a single 5′ end-labeled primer in a standard PCR reaction. Labeling (10 µCi of [γ^{32}P]dNTP, >5000 Ci/mmol) is obtained by T4 polynucleotide kinase (the reaction usually takes 30 min at 37°C). Unincorporated nucleotides are then removed by double precipitation in ethanol, and the primer is resuspended in water. Another simple way to obtain labeled PCR products is to mix aliquots of one dNTP α32 (1 to 2 µCi) into a standard reaction cocktail (Figure 4). With both procedures, PCR products are visualized as bands on dried gels after a short time exposure. However, the latter procedure generates stuttering and multiple bands, which are occasionally difficult to read and interpret.

4. Native Horizontal PAGE and Silver Staining

Since handling sequence-format gels and radioactive material may be impractical as a daily routine, simpler approaches have been tried to separate PCR products. Allen et al.[49] introduced a discontinuous horizontal polyacrylamide gel electrophoresis procedure followed by silver staining of gels. This procedure may speed up AMP-FLP genotyping considerably. Initially used for separation of large double-stranded DNA fragments,[50] this approach can be applied to tetranucleotide microsatellite analysis as well. Several forensic laboratories use this simple technique, which offers considerable advantages: simplicity, sensitivity, no radioactive hazard, brevity (one day), and the original gels remain available for permanent records. A description of the entire procedure is summarized in Figure 1.

5. Automated Laser-Fluorescent Procedures

Procedures based on polyacrylamide gel electrophoresis have been greatly improved by fluorescent labeling and laser detection of the PCR products (by an automated sequencer). This application resulted as a by-product of the process of automation of sequencing procedures. Automated sequencers improve the efficiency of typing procedures in several ways: bands are automatically revealed (no UV transillumination, nor silver staining or autoradiography of gels); signals are processed in real time, and typing can be carried out while the experiment is taking place; fluorescence gives quantitative data, making it easy to distinguish between genuine and spurious bands; multiple amplifications of PCR loci are possible if different fluorescent dyes are used; band comparisons are done on a more objective basis; and quantitation, size calculation, and storage of results are simplified by computer facilities.

There are two types of fluorescent-automated hardware based on as many fluorescent dyes: (1) a four-dye based machine (Applied Biosystem, Inc.),[51] enabling different fluorescent products to be distinguished by different wavelengths; this machine makes it possible to coamplify different microsatellites (multiplex PCR), even when their sizes overlap, and to use internal size markers in the same lane; (2) a one-wavelength fluorescence fixed-laser machine[52] (A.L.F., Pharmacia, Uppsala); this apparatus is simple to use (just one fluorescent dye is required; see Figure 5) and very accurate in sizing and comparing comigrating bands; it also makes it possible to recycle gels for several electrophoresis experiments.

A general advantage of laser fluorescent technology derives from its sensitivity: pico- to femtograms of PCR products are detectable by this procedure. This is important with respect to some forensic applications.

IV. APPLICATIONS

A. PARENTHOOD ANALYSIS

This part of forensic work underwent radical changes since the introduction of DNA markers,[53] which have polymorphic content unprecedented by that of any protein marker

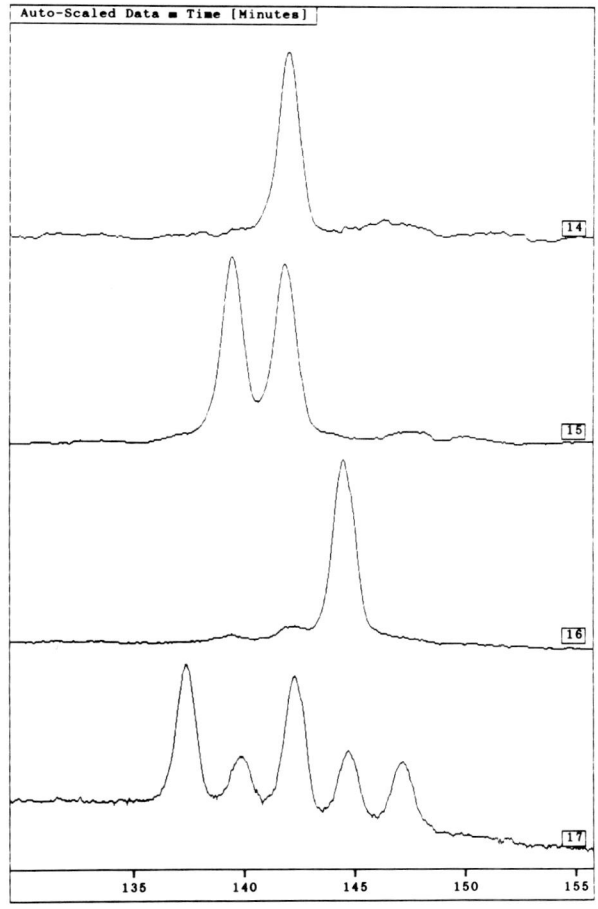

Figure 5. Automated laser-fluorescent analysis of the MIT-MH26[22], a short tandem tetranucleotide repeat polymorphism. Amplification conditions are as follows: 94°C for 60 s, 64°C for 60 s, and 72°C for 90 s (30 c). Primers are the following:

5' - AAT ACC CCA AGG GGT GGT AA - 3'
5' - CAT TGA TGA ACA GTT CAA GCA- 3'.

A fluorescent group (fluorescein amidite, FluorePrime™, Pharmacia) was incorporated in the second amplimer by a modified synthesis cycle carried out in an automated oligonucleotide synthesizer (Model 391 PCR-mate, Applied Biosystem). A sequence-format gel (250 × 400 × 0.5 mm, 7 M urea, 6% T polyacrylamide) was used to separate 0.5 µl aliquots of amplified product. Specific fluorescent bands were detected by a fixed laser source placed at the bottom of the gel. The entire electrophoretic process was carried out on an automated laser-fluorescent analyzer (A.L.F., Pharmacia). Peaks corresponding to band patterns of unknown samples are typed according to an internal ladder of fluorescent fragments (lane 17).

polymorphism (except for the major histocompatibility complex). Introduction of PCR has played an important part in this process of innovation, making molecular biology procedures accessible to a larger number of forensic laboratories.

A consequence of the transition to hyperpolymorphic markers was that some traditional guidelines had to be revised. Two possible outcomes derive from a protocol of parenthood investigation. Parenthood is excluded when inconsistencies are found in segregation between one parent and the child. With traditional protein markers, one or two incompatibilities were accepted as a solid proof to exclude a parenthood relationship. When using hypervariable DNA loci, multiple exclusions must be expected. Consequently, sporadic incompatibilities

TABLE 2.
A Tentative Protocol of Forensic Investigation Based on Solely Amplifiable Markers

Marker	P_{excl}	P_{comb}	Most common genotype frequency (×100 individuals)	Haplotype probability
DQA1	0.5565			
APOB 3'HVR	0.6085	0.8263	17	
D1S80	0.5887	0.9285	14	23.8×10^{-3}
COL2A1 3'HVR	0.6107	0.9722	19	4.5×10^{-4}
pYNZ22	0.7318	0.9925	19	8.6×10^{-5}
HUMTH01	0.5820	0.9969	15	1.3×10^{-5}
MIT-MH26	0.5400	0.9986	15	1.9×10^{-6}
VWF	0.6060	0.9994	12	2.3×10^{-7}
ACTBP2	0.8606	0.99992	2	4.6×10^{-9}
HUMFXIIIA1	0.5176	0.99996	24	1.1×10^{-9}
HUMFESFPS	0.4320	0.99998	22	2.4×10^{-10}

Note: For each marker, the theoretical chance of excluding false fathers, P(excl), and the commonest genotype are indicated (for other details on each marker, see Table 1).

arising from a highly discriminatory protocol should be regarded with suspicion (for occurrence of mutations and risks of procedural mistakes).

Compatibility throughout a protocol of analysis is the opposite case. When this occurs, positive evidence is given by weighting the relevant odds of compatibility by chance against compatibility by parenthood, according to a Bayesian probability approach based on the frequencies of each segregating allele. Calculations are done by either a likelihood ratio:

$$X/Y \qquad (2)$$

or by the Essen Moeller's formula:[54]

$$1/1 + (Y/X) \qquad (3)$$

where X is the *a priori* probability of transmission of a given character (1 for homozygosity, 0.5 for heterozygosity) and Y is the gene frequency of a given paternal character.

With highly informative codominant markers, probability estimates are greatly simplified by the fact that both parents are usually heterozygous with no shared alleles, so that there are no uncertainties on the maternal and paternal segregating characters.

The discrimination powers of PCR-based protocols are very high, ensuring that the majority of cases are solved without resorting to other sources of variability (Table 2).

The fact that a high number of hypervariable markers are available at relatively low cost makes it easy to perform investigations on father/child only, paternity analysis on relatives in the absence of both parents, analyses of consanguinity, and several other tasks that would have been difficult to accomplish earlier.

Although several categories of polymorphisms (unique sequences, AMP-FLPs, STRs) are applicable to family relationship analysis, microsatellites are now making their way into this field. Microsatellites are numerous, simple to use, and highly polymorphic, and they present a favorable balance between the number of alleles and polymorphic content. The occurrence of mutations at these systems is sufficiently low as to be practically negligible.[55]

B. IDENTIFICATION OF STAINS

Biological samples left at the scene of a crime are often in scanty amounts, and the DNA extracted from these may be degraded to a variable extent. PCR has a remedy to both problems and gives consistent results in a significant number of difficult cases. However, enzymatic amplification of forensic samples is prone to pitfalls, rendering the most classical forensic application of PCR a difficult art to practice. Degradation of stains may be a serious obstacle to the reproduction of coherent profiles, due to DNA fragmentation and to presence of molecular inhibitors.[56,57] Furthermore, deeply degraded template DNA may result in spurious amplification products. PCR sensitivity, while making possible successful amplification from single sperm cells under controlled conditions,[48,58] may give rise to artifacts of contamination when "street DNA" is targeted. To prevent errors, tight precautions have to be observed. However, if correct measures are adopted,[59] PCR is an excellent tool for the identification of an individual. While mtDNA is being increasingly used in some instances, nuclear DNA covers most current applications in the field.

An important step affecting the outcome of criminal PCR protocols is the procedure of template DNA extraction. Numerous versions of extraction procedures have been applied to a surprising number of different substrates, including blood stains, sperm and vaginal swabs,[53] hair,[7] skeletal remains,[60,61] teeth,[39] cigarette butts,[62] urine,[63] saliva-contaminated objects,[64] freeze-dried tissues,[65] paraffin-embedded tissues,[66] and ancient tissue specimens.[67] The classical lysis procedure of proteinase K/dithiothreitol/sodium dodecylsulfate cocktail in TE buffer with subsequent phenol-chloroform extraction[53] is the basis of many. A variant of this procedure is a two-step lysis used for vaginal swabs containing sperms. This relies on the property of sperm cell membrane to be refractory to a lysis solution not containing thiol groups. Another procedure is based on the use of anion exchange resin (Chelex®, Dow Chemical Co.), consisting of a styrene lattice with acid exchange group. This protocol was devised as a remedy to the analysis of samples that do not amplify due to the presence of contaminating by-products (hematin, iron ions) co-extracted by classical extraction procedures.[56,57] Samples are simply boiled in the presence of the ion exchange Chelex® slurry, then a few microliters are transferred to a fresh tube containing an amplification mixture. In some cases, samples are first Chelex treated, then dialyzed and concentrated (Centricon 100, Amicon), before amplification.

In most instances, criminal investigations consist in a diagnosis of genotype of DNA extracted from biological stains (in Table 1, a list of markers used for reporting criminal evidence is provided). In a number of forensic cases (rape cases, rapid screenings of bodies in mass disasters, and others), a diagnosis of sex is also useful. This is achieved in several ways, including amplification of multicopy or single-copy X and Y chromosome sequences.[68-70]

It is noteworthy that a movement exists among forensic laboratories (laboratories of the European DNA Profile Group; North American laboratories in the FBI-steered TWGDAM network) aimed at standardizing protocols and selecting common markers for purposes of criminal policy. Validation of PCR and quality controls has been the subject of a recent report of the National Research Council of the U.S. Academy of Sciences, centered on DNA analysis in forensics.[71]

C. OTHER APPLICATIONS

PCR analysis is useful in a broader range of forensic instances besides the classical situations previously covered in this survey, and some emerging applications should be at least mentioned.

1. Diagnosis of Species

A potentially important source of forensic evidence is the recently developed technique of forensically informative nucleotide sequences (FINS).[72] This technique is a key of access to a species-specific pattern of inheritance, located on mitochondrial DNA. Highly conserved sequences exist in cytochrome *b* (Cyt-*b*) of vertebrate mitochondria. Cyt-*b* sequences have the following features: (1) they are highly variable between animal species (even if closely related) and moderately subject to intraspecific mutations; (2) they are long enough to code for significant differences and short enough to be sequenced in one round of deoxy-chain termination. Cyt-*b* FINS form a haploid genome pattern with absent (or a very low rate of) endoplasmy. A database of such species-specific sequences is already available.

Cyt-*b* FINS could have interesting forensic applications. They may help to uncover commercial frauds (meat from cheap sources sold as a more expensive kind), poachers' trades (meat or other products derived from illegally shot animals), and illegal commerce in endangered species. FINS typing involves a two-step amplification of the Cyt-*b* region (double-strand amplification, then asymmetric PCR) followed by sequencing of short single strands of DNA (see Section III.B.2).

2. Controls of Animal Breeding

Controls of parenthood relationships in kindred animals have forensic relevance for the control of illegal trade in endangered species. Young individuals of protected species (a list was produced by an Intergovernmental Convention in Washington, 1973) are smuggled from their birthplace and disguised through false veterinary certifications. In such cases, a parenthood investigation is the only means to uncover the illegal origin of individuals. Some human PCR polymorphisms are conserved along the evolutionary scale and may be used to verify mother-child segregations. One such approach has been used to repress illegal trades of young chimpanzees[73] (Figure 6). In these situations, PCR protocols have some obvious advantages over other molecular biology procedures, including the fact that determinations are obtainable from plucked animal hairs (a blood withdrawal is avoided, which would involve serious risks for wild animals and their caretakers).

3. RAPDs

Randomly amplified polymorphic DNA sequences (RAPDs)[74] consist of variable DNA sequences amplified by ten-base-long oligonucleotide primers. These markers work with several plant and animal genomes and have promising prospects for use in forensic applications. Potential applications include the distinction between different plants/fruits of commercial value, assessment of the origin of commercial stocks of fish, and analysis of sources of water pollution. A forensic application of RAPDs has recently been reported, in what has turned out to be the first use of plant genome in a criminal case. Some seed pods found at the scene of a rape and murder case (stuck to the defendant's truck) were sent for analysis of genetic heterogeneity. These samples proved to be of a plant variety (Palo Verde tree) shown to be highly polymorphic for several randomly amplified polymorphic sites, each amplifiable by specific priming at random with short amplimers. As a result of this analysis, the defendant was associated with the crime[75] by a high degree of probability.

4. Miscellaneous

Recently, PCR helped in a few interesting cases of identification of pathogenic agent strains, the presence of which gave evidence of the consummation of a crime. A specific HIV population was identified in the peripheral blood of a woman who had been raped and had subsequently contracted AIDS.[76] Comparison with the specific strain isolated from peripheral blood of a suspect supplied strong evidence of his involvement in this rape. An analogous case was previously noticed, in which identification of HIV strain helped to prove civil damage liability.[77] In both cases, PCR served as a means to amplify viral genomes with following solid-phase sequencing of single-stranded DNA.

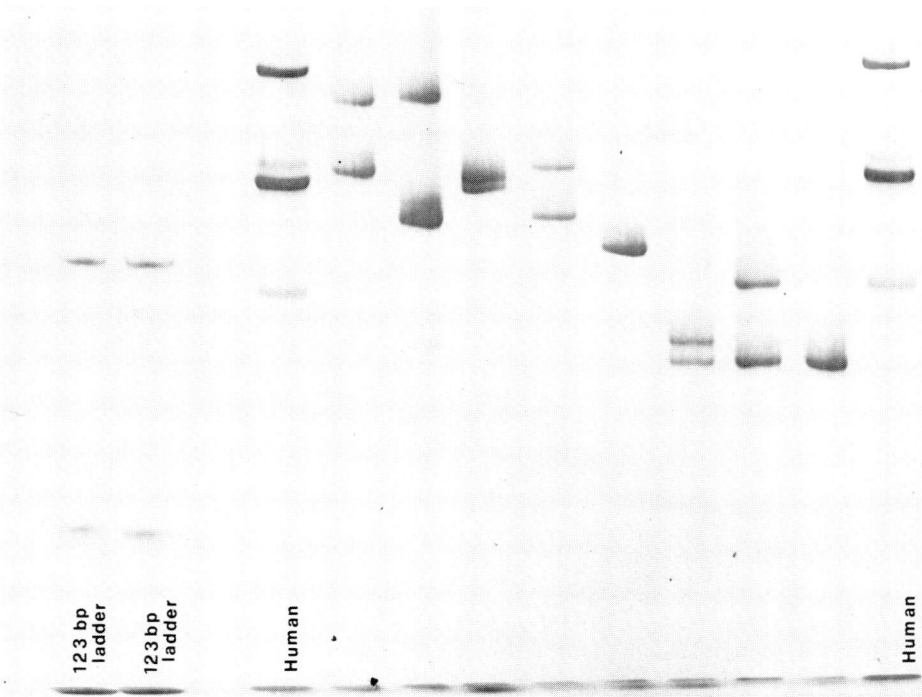

Figure 6. COL2A1 polymorphism[12] is a human VNTR whose polymorphism is conserved in chimpanzees (Pan *troglodytes*). Some chimpanzee[1] genotypes are shown in comparison with human products (at either extremity of the gel). Amplification conditions were as follows: 94°C for 60 s, 60°C for 60 s, and 70°C for 90 s for 28 c (last extension step is prolonged for 10 min); human amplimers were used according to Wu et al.:[13]

5' - CCA GGT TAA GGT TGA CAG CT - 3'
5' - GTC ATG AAC TAG CTC TGG TG - 3'.

ACKNOWLEDGMENTS

We thank Prof. A. Fiori (Department of Forensic Medicine), Prof. G. Neri, and Dr. M. Genuardi (Department of Genetics, Catholic University) for reading the manuscript and for criticism. We acknowledge the following persons from our laboratory for supplying us unpublished data and pictures from their own research work: M. Pescarmona, A. Moscetti, G. Destro-Bisol, I. Boschi, and L. Grimaldi. This work was financed in part by a M.U.R.S.T. fund (quote 40%) and by a grant from the Italian Ministry of Agriculture (Department of Mountain and Forest administration).

REFERENCES

1. **Weissenbach, J., Gyapay, G., Dib, C., Vignal, A., Morisette, J., Milasseau, P., Vaysseix, G., and Lathrop, M.,** A second-generation linkage map of the human genome, *Nature*, 359, 794, 1992.
2. **O'Brien, S. J., Womack, J. E., Lyons, L. A., Moore, K. J., Jenkins, N. A., and Copeland, N. G.,** Anchored reference loci for comparative genome mapping in mammals, *Nat. Genet.*, 3, 103, 1993.
3. **Gill, P., Woodroffe, S., Bar, W., Carracedo, A., Eriksen, B., Jones, S., Kloosterman, A. D., Ludes, B., Mevag, B., Pascali, V. L., Schmitter, H., Schneider, P. M., and Thomson, J. A.,** A report of an international collaborative experiment to demonstrate the uniformity obtainable using DNA profiling techniques, *Forensic Sci. Int.*, 53, 29, 1992.

4. **Gill, P.,** Report of the European DNA profiling group (EDNAP) towards standardization of short tandem repeat (STR) loci, in preparation.
5. **Ugozzoli, L. and Wallace, R. B.,** Application of an allele-specific polymerase chain reaction to the direct determination of ABO blood group genotypes, *Genomics,* 12, 670, 1992.
6. **Braun, A., Bichlmaier, R., and Cleve, H.,** Molecular analysis of the gene for the human vitamin-D-binding protein (group-specific component): allelic differences of the common genetic GC types, *Hum. Genet.,* 89, 401, 1992.
7. **Higuchi, R. K., von Beroldingen, C. H., Sensabaugh, G. F., and Erlich, H. A.,** DNA typing from single hairs, *Nature,* 332, 543, 1988.
8. **Saiki, R. K., Bugawan, T. L., Horn, G. T., Mullis, K. B., and Erlich, H. A.,** Analysis of enzymatically amplified beta-globin and HLA-DQα DNA with allele-specific oligonucleotide probes, *Nature,* 324, 163, 1986.
9. **Nakamura, Y., Leppert, M., O'Connel, P., Wolff, R., Holm, T., Culver, M., Martin, C., Fujimoto, E., Hoff, M., Kumlin, E., and White, R. L.,** Variable number of tandem repeat (VNTR) markers for human gene mapping, *Science,* 235, 1616, 1987.
10. **Boerwinkle, E., Xiong, W., Fourest, E., and Chan, L.,** Rapid typing of tandemly repeated hypervariable loci by the polymerase chain reaction: application to the apolipoprotein B 3′ hypervariable region, *Genetics,* 86, 212, 1989.
11. **Kasai, K., Nakamura, Y., and White, R. L.,** Amplification of a variable number of tandem repeats (VNTR) locus (pMCT118) by the polymerase chain reaction (PCR) and its application to forensic science, *J. Forensic Sci.,* 5, 1196, 1990.
12. **Stoker, N. G., Cheah, K. S. E., Griffin, J. R., Pope, F. M., and Solomon, E.,** A highly polymorphic region 3′ to the human type II collagen gene, *Nucleic Acids Res.,* 13, 4613, 1985.
13. **Wu, S., Seino, S., and Bell, G. I.,** Human collagen, type II, alpha 1, (COL2A1) gene: VNTR polymorphism detected by gene amplification, *Nucleic Acids Res.,* 18, 3102, 1990.
14. **Horn, G. T., Richards, B., and Klinger, K. W.,** Amplification of a highly polymorphic VNTR segment by the polymerase chain reaction, *Nucleic Acids Res.,* 17, 2140, 1989.
15. **Vuorio, A. F., Sajantila, A., Hamalainen, T., Syvanen, A. C., Ehnolm, C., and Peltonen, L.,** Amplification of the hypervariable region close to the apoliprotein B gene: application to forensic problems, *Biochem. Biophys. Res Commun.,* 170, 269, 1990.
16. **Sajantila, A., Budowle, B., Strom, M., Johnsson, V., Lukka, M., Peltonen, L., and Ehnholm, C.,** PCR amplification of alleles at the D1S80 locus: comparison of a Finnish and a North American Caucasian population sample, and forensic casework evaluation, *Am. J. Hum. Genet.,* 50, 816, 1992.
17. **Weber, J. L. and May, P. E.,** Abundant class of human DNA polymorphism which can be typed using the polymerase chain reaction, *Am. J. Hum. Genet.,* 44, 388, 1989.
18. **Beckman, J. S. and Weber, J. L.,** Survey of human and rat microsatellites, *Genomics,* 12, 627, 1992.
19. **Hearne, C. M., Ghosh, S., and Todd, J. A.,** Microsatellites for linkage analysis of genetic traits, *T.I.G.,* 8, 288, 1992.
20. **Edwards, A., Civitello, A., Hammond, H. A., and Caskey, C. T.,** DNA typing and genetic mapping with trimeric and tetrameric tandem repeats, *Am. J. Hum. Genet.,* 49, 746, 1991.
21. **Edwards, A., Hammond, H. A., Jin, L., Caskey, C. T., and Chakraborty, R.,** Genetic variation at five trimeric and tetrameric tandem repeat loci in four human population groups, *Genomics,* 12, 241, 1992.
22. **Hudson, T. J., Engelstein, M., Lee, M. K., Ho, E. C., Rubenfield, M. J., Adams, C. P., Housman, D. E., and Dracopoli, N. C.,** Isolation and chromosomal assignment of 100 highly informative human simple sequence repeat polymorphisms, *Genomics,* 13, 622, 1992.
23. **Engelstein, M., Hudson, T. H., Lane, J. M., Lee, M. K., Leverone, B., Landes, G. M., Peltonen, L., Weber, J. L., and Dracopoli, N. C.,** A PCR-based linkage map of human chromosome 1, *Genomics,* 15, 251, 1993.
24. **Petrukhin, K. E., Speer, M. C., Cayanis, E., de Fatima Bonaldo, M., Tantravahi, U., Soares, M. B., Fischer, S. G., Warburton, D., Gilliam, T. C., and Ott, J.,** A microsatellite genetic linkage map of human chromosome 13, *Genomics,* 15, 76, 1993.
25. **Straub, R. E., Speer, M. C., Luo, Y., Rojas, K., Overhauser, J., Ott, J., and Gilliam, T. C.,** A microsatellite linkage map of human chromosome 18, *Genomics,* 15, 48, 1993.
26. **Porter, J. C., Ram, K. T., and Puck, J. M.,** Twelve polymorphic microsatellites on human chromosome 22, *Genomics,* 15, 57, 1993.
27. **Goodfellow, P. N.,** Variation is now the theme, *Nature,* 359, 777, 1992.
28. **Hughes, A. E.,** Optimization of microsatellite analysis for genetic mapping, *Genomics,* 15, 433, 1993.
29. **Polymeropoulos, M. H., Rath, D. S., Xiao, H., and Merril, C. R.,** Tetranucleotide repeat polymorphism at the human coagulation factor XIII A subunit gene (F13A1), *Nucleic Acids Res.,* 19, 4036, 1991.
30. **Polymeropoulos, M. H., Rath, D. S., Xiao, H., and Merril, C. R.,** Tetranucleotide repeat polymorphism at the human beta-actin related pseudogene H-beta-Ac psi-2 (ACTBP2), *Nucleic Acids Res.,* 20, 1432, 1991.

31. **Kimpton, C. P., Walton, A., and Gill, P.,** A further tetranucleotide repeat polymorphism in the vWF gene, *Hum. Mol. Genet.,* 1, 287, 1992.
32. **Polymeropoulos, M. H., Rath, H., Xiao, H., and Merril, C. R.,** Tetranucleotide repeat polymorphism at the c-fes/fps proto-oncogene (FES), *Nucleic Acids Res.* 19, 4018, 1991.
33. **Jeffreys, A. J., Neumann, R., and Wilson, V.,** Repeat unit sequence variation in minisatellites: a novel source of DNA polymorphism for studying variation and mutation by single molecule analysis, *Cell,* 60, 473, 1990.
34. **Jeffreys, A. J., MacLeod, A., Tamaki, K., Neil, D. L., and Monckton, D. G.,** Minisatellite repeat coding as a digital approach to DNA typing, *Nature,* 354, 204, 1991.
35. **Farr, C. J. and Goodfellow, P. N.,** New variations on the theme, *Nature,* 354, 184, 1991.
36. **Anderson, S., Bankier, A. T., Barrel, B. G., de Bruijn, M. H. L., Coulson, A. R., Drouin, J., Eperon, I. C., Nierlich, D. P., Roe, B. A., Sanger, F., Schreier, P. H., Smith, A. J. H., Staden, R., and Young, I. G.,** Sequence and organisation of the human mitochondrial genome, *Nature,* 290, 457, 1981.
37. **Bogenhagen, D. and Clayton, D. A.,** The number of mitochondrial deoxyribonucleic acid genomes in mouse L and in human HeLa cells, *J. Biol. Chem.,* 249, 7791, 1974.
38. **Aquadro, C. F. and Greenberg, B. D.,** Human mitochondrial DNA variation and evolution: analysis of nucleotide sequences from seven individuals, *Genetics,* 103, 287, 1983.
39. **Ginther, C., Issel-Tarver, L., and King, M. C.,** Identifying individuals by sequencing mitochondrial DNA from teeth, *Nat. Genet.,* 2, 135, 1992.
40. **Sullivan, K. M., Hopgood, R., Lang, B., and Gill, P.,** Automated amplification and sequencing of human mitochondrial DNA, *Electrophoresis,* 12, 17, 1991.
41. **Vigilant, L., Pennington, R., Harpending, H., Kocher, T. D., and Wilson, A. C.,** Mitochondrial DNA sequences in single hairs from a southern African population, *Proc. Natl. Acad. Sci. U.S.A.,* 86, 9350, 1989.
42. **Greenberg, B. D., Newbold, J. E., and Sugino, A.,** Intraspecific nucleotide sequence variability surrounding the origin of replication in human mitochondrial DNA, *Gene,* 21, 33, 1983.
43. **Hopgood, R., Sullivan, K. M., and Gill, P.,** Strategies for automated sequencing of human mitochondrial DNA directly from PCR products, *BioTechniques,* 13, 82, 1992.
44. **Stoneking, M.,** Population genetics of human mitochondrial DNA, communication in 2nd Int. Symp. Forensic Aspects of DNA Analysis, FBI Academy, Quantico, VA, March 29 to April 2, 1993.
45. **Mullis, K. B.,** William Allan Award Address, 41st Annu. Meet. American Society of Human Genetics, Cincinnati, 1990.
46. **Ruano, G., Pagliaro, E. M., Schwartz, T. R., Lamy, K., Messina, D., Gaensslen, R. E., and Lee, H. C.,** Heat-soaked PCR: an efficient method for DNA amplification with application to forensic analysis, *BioTechniques,* 13, 266, 1992.
47. **Zhang, L., Cui, X., Schmitt, K., Hubert, R., Navidi, W., and Arnheim, N.,** Whole genome amplification from a single cell: implications for genetic analysis, *Proc. Natl. Acad. Sci. U.S.A.,* 89, 5847, 1992.
48. **Hubert, R., Weber, J. L., Schmitt, K., Zhang, L., and Arnheim, N.,** A new source of polymorphic DNA markers for sperm typing: analysis of microsatellite repeats in single cells, *Am. J. Hum. Genet.,* 51, 985, 1992.
49. **Allen, R. C., Graves, G., and Budowle, B.,** Polymerase chain reaction amplification products separated on rehydratable polyacrylamide gels and stained with silver, *BioTechniques,* 7, 736, 1989.
50. **Budowle, B., Chakraborty, R., Giusti, A. M., Eisenberg, A. J., and Allen, R. C.,** Analysis of the VNTR locus D1S80 by the PCR followed by high-resolution PAGE, *Am. J. Hum. Genet.,* 48, 137, 1991.
51. **Smith, L. M., Sanders, J. Z., Kaiser, R. J., Hughes, P., Dodd, C., Connell, C. R., Heiner, C., Kent, S. B. H., and Hood, L. E.,** Fluorescent detection in automated DNA sequence analysis, *Nature,* 321, 674, 1986.
52. **Ansorge W., Sproat, B. S., Stegemann, J., and Schwager, C.,** A non-radioactive automated method for DNA sequence determination, *J. Biochem. Biophys. Meth.,* 13, 315,1986.
53. **Gill, P., Jeffreys, A. J., and Werrett, D. J.,** Forensic application of DNA 'fingerprints', *Nature,* 318, 577, 1985.
54. **Essen Moeller, E.,** Die Beweiskraft der Ahnlichkeit im Vaterschaftsnachweis, teoretische Grundlagen, *Mitt. Anthropol. Ges. Wien,* 68, 9, 1938.
55. **Kwiatkowski, D. J., Henske, E. P., Weimer, K., Ozelius, L., Gusella, J. F., and Haines, J.,** Construction of a GT polymorphism map of human 9q, *Genomics,* 12, 229, 1992.
56. **Walsh, P. S., Metzger, D. A., and Higuchi, R.,** Chelex® 100 as a medium for simple extraction of DNA for PCR-based typing from forensic material, *BioTechniques,* 10, 506, 1991.
57. **Reynolds, R., Sensabaugh, G., and Blake, E.,** Analysis of genetic markers in forensic DNA samples using the polymerase chain reaction, *Anal. Chem.,* 63, 2, 1991.
58. **Li, H., Gyllensten, U. B., Cui, X., Saiki, R. K., Erlich, H. A., and Arnheim, N.,** Amplification and analysis of DNA sequences in single human sperm and diploid cells, *Nature,* 335, 414, 1988.
59. **Kwok, S. and Higuchi, R.,** Avoiding false positives with PCR, *Nature,* 339, 237,1989.

60. **Hagelberg, E., Gray, I. C., and Jeffreys, A. J.**, Identification of the skeletal remains of a murder victim by DNA analysis, *Nature,* 352, 427, 1991.
61. **Jeffreys, A. J., Allen, M. J., Hagelberg, E., and Sonnberg, A.**, Identification of the skeletal remains of Joseph Mengele by DNA analysis, *Forensic Sci. Int.,* 56, 65, 1992.
62. **Hochmeister, M. N., Budowle, B., Jung, J., Borer, U. V., Comey, C. T., and Dirnhofer, R.**, PCR-based typing of DNA extracted from cigarette butts, *Int. J. Legal Med.,* 104, 229, 1991.
63. **Brinkmann, B., Rand, S., and Bajanowsski, T.**, Forensic identification of urine samples, *Int. J. Legal Med.,* 105, 59, 1992
64. **Walsh, D. J.**, Isolation of DNA from saliva and forensic samples containing saliva, *J. Forensic Sci.,* 37, 387, 1992
65. **Huckenbeck, W. and Bonte, W.**, DNA fingerprinting of freeze-dried tissues, *Int. J. Legal Med.,* 105, 39, 1992.
66. **Gill, P., Kimpton, C. P., and Sullivan, K.**, A rapid polymerase chain reaction method for identifying fixed specimens, *Electrophoresis,* 13, 173, 1992.
67. **Paabo, S.**, Ancient DNA: extraction, characterization, molecular cloning, and enzymatic amplification, *Proc. Natl. Acad. Sci. U.S.A.,* 86, 1939, 1989.
68. **Ebensperger, C., Studer, R., and Epplen, J. T.**, Specific amplification of the ZFY gene to screen sex in man, *Hum. Genet.,* 82, 289, 1989.
69. **Witt, M. and Ericksson, R. P.**, A rapid method for the detection of Y-chromosomal DNA from dried blood specimens by the polymerase chain reaction, *Hum. Genet.,* 82, 271, 1989.
70. **Nakahori, Y., Hamano, K., Iwaya, M., and Nakagome, Y.**, Sex identification by polymerase chain reaction using X-Y homologous primers, *Am. J. Med. Genet.,* 39, 472, 1991.
71. Report of the National Research Council, Committee on DNA Technology in Forensic Science, National Academy Press, Washington DC, 1992.
72. **Bartlett, S. E. and Davidson, W. S.**, FINS (forensically informative nucleotide sequencing): a procedure for identifying the animal origin of biological specimens, *BioTechniques,* 12, 408, 1992.
73. **Pascali, V. L., Destro Bisol, G., d'Aloja, E., Dobosz, M., Paonessa, G., and Mereu, U.**, Chimpanzee DNA profiles on trial, *Nature,* in press.
74. **Hedrick, P.**, Shooting the RAPDs, *Nature,* 355, 679, 1992.
75. **Yoon, C. K.**, Botanical witness for the prosecution, *Science,* 260, 894, 1993.
76. **Albert, J., Wahlberg, J., and Uhlán, M.**, Forensic evidence by DNA sequencing, *Nature,* 361, 595, 1993.
77. **Ou, C. Y., Ciesielski, C. A., Myers, G., Bandea, C. I., Luo, C., Korber, B. T., Mullins, V. I., Schochetman, G., Berkelman, R. L., Economou, A. N., et al.**, Molecular epidemiology of HIV transmission in a dental practice, *Science,* 256, 1165, 1992; comment in *Science,* 256, 1155, 1992.

Chapter 31

APPLICATIONS OF POLYMERASE CHAIN REACTION METHODOLOGY IN CLINICAL DIAGNOSTICS

Meena H. Mahbubani and Asim K. Bej

TABLE OF CONTENTS

I. Introduction 308

II. PCR in Clinical Microbiology 308

III. Sample Processing 309
    A. Gastric Biopsy 309
    B. Paraffin-Embedded Gastric Biopsy 310
    C. Clinical Swabs 310
    D. Saliva 311
    E. Stool Samples 311
    F. Blood Samples 312
    G. Sample Processing for the Detection of RNA Viruses 312
        1. Stool Filtrate, Urine, Throat Swab Medium, Cerebrospinal Fluid, or Virus Culture Supernatant Samples 312
        2. Cryopreserved Tissue 313
        3. Formalin-Fixed Tissue 313
        4. Proteinaceous Body Fluid 313
        5. Leukocytes 313
    H. General Methods for Lysis of Microbial Cells 313
        1. Freeze-Thaw Lysis 313
        2. Freeze-Boil Lysis 313
        3. Boiling with Chelex™ 100 314

IV. PCR Amplification 314
    A. Target DNA 314
    B. Primers for DNA Amplification 314
    C. Thermostable DNA Polymerase 315
    D. Deoxyribonucleotides 315
    E. Magnesium Ion 315
    F. PCR Reaction Buffer 315
    G. Cosolvents 316
    H. Mineral Oil/Wax Overlay 316
    I. PCR Reaction Mix 316
    J. PCR Reaction Parameters 316
    K. Multiplex DNA Amplification 316
    L. Controls 317

V. Identification and Analysis of the Amplified DNA 317
    A. Probe Selection 317
    B. Detection of Amplified DNA by Immobilized Capture Probe DNA-DNA Hybridization 317
        1. Labeling Amplified DNA 317

                2. Synthesis of a Homopolymer Tail at the 3' End of the Probe 317
                3. Immobilization of the Probe onto Nylon Membrane 317
                4. DNA-DNA Hybridization Using Biotin-Labeled Amplified DNA 318
                5. Color Development 318
            C. Detection of the Amplified DNA Using Microtiter Plate Hybridization:
            Immobilization of the Hybridization Probes to Microtiter Wells 318
                1. Passive Transfer 318
                2. Chemical-Mediated Active Transfer 318
                    a. Capture Probe Synthesis 319
                    b. Activation of Wells 319
                    c. Activation of Probe 319
                    d. Probe Coupling 319
                3. Hybridization 319
                4. Color Development 319
            D. Labeling PCR-Amplified DNA with Biotin or with Digoxigenin (DIG) 320
            E. Immunogenic Detection of DIG-Labeled PCR Products 320

    VI. PCR Contamination Control 320

    VII. Enhancing the Specificity of the PCR Reaction 321

    VIII. PCR in the Diagnosis of Human Infectious Diseases 321
            A. Bacteria and Protozoa 321
            B. Viruses 322

    IX. PCR in the Diagnosis of Human Genetic Diseases 322
            A. Prenatal Diagnosis 322
            B. Preimplantation Diagnosis 322
            C. Detection of Diseases from Rare Genetic Messages 323
            D. Cancer Therapy 323

    X. Discussion 323

Acknowledgments 324

References 324

I. INTRODUCTION

Outstanding advancements have been made in the applications of the polymerase chain reaction (PCR) *in vitro* DNA amplification methodology, in the area of clinical microbiology and diagnosis of human diseases. This chapter discusses the basic criteria and sample preparation involved in the use of the PCR methodology, for rapid detection of microbial pathogens causing serious diseases in humans, and summarizes some of the recent achievements in the diagnosis of human diseases.

II. PCR IN CLINICAL MICROBIOLOGY

The conventional microbiological culture method for the detection of specific microbial infectious agents requires the use of diverse growth media and conditions, followed by identification of the pathogens by their multiple physiological characteristics through a series of biochemical tests, which may also require culturing of the organisms in specific media. The culture-based methods of identification of microbial pathogens may require several days to weeks, depending upon the nature of the infectious agent. Moreover, a number of such microbial pathogens require culturing in animals, which may take a considerable amount of time and effort. An alternate approach for detecting these microorganisms is by microscopic examination of the suspected samples using poly- or monoclonal antibody, labeled with fluorescent dye. In many instances, it has been found that such immunofluoresent-based microscopic identification methods, show cross-reactivity with other related microbial pathogens, and therefore the interpretation of positive results may be difficult and require expert judgment.

Although, the emergence of genetic-based molecular diagnostic methods for the detection of these microbial pathogens shows great promise for rapid detection, some of these methods lack the sensitivity required for such detection. Moreover, most of these methods require specific genetic information for specific microbial pathogens which is not always available. Until late 1988, most of the molecular-based clinical diagnostic methods were limited to Southern blot analysis of DNA and Northern blot analysis of RNA. Both methods are limited by the large amount of DNA or RNA required, and at least one to several thousand cells must be present in the sample to show a detectable amount of signal.

The development of the polymerase chain reaction (PCR) method[1] and its application in the diagnosis of micobial pathogens has shown its potential to change the practice of diagnosis of microbial pathogens in the clinical laboratories across the world. Following sample preparation, PCR DNA amplification itself is a fast and automated method, simply requiring addition of a standard reaction mixture in a designated reaction tube. The nature of the amplified DNAs can then be analyzed by the DNA-DNA hybridization method using the appropriate oligonucleotide probe. PCR DNA amplification coupled with the gene probe method has already proven its high specificity and sensitivity in detecting microbial pathogens in clinical samples.[2-6] This chapter discusses the protocols for PCR gene-probe-based detection of various microbial pathogens from clinical samples. In general, successful detection of microbial pathogens from clinical samples requires three steps: (1) sample processing, (2) PCR DNA amplification, and (3) identification and analysis of the amplified DNA. This chapter discusses these three major steps in the area of clinical dignostics and briefly addresses the applications of PCR methodology in the diagnosis of human genetic and infectious diseases with high specificity and sensitivity.

III. SAMPLE PROCESSING

A. GASTRIC BIOPSY

Processing of the gastric biopsy for PCR detection of various gastrointestinal related microbial pathogens, specially *Helicobacter pylori* has been described by Clayton et al.[7] The specimen from gastric biopsy is dissected with a sterilized scalpel, and 50 µl of sterile distilled water containing 0.01 mM EDTA are added to it. The sample is mixed completely by vortexing for 5 min. The sample is then boiled for 10 min, cooled to room temperature, and centrifuged in a table-top microcentrifuge (10,000 g) for 5 min to pellet the insoluble tissue material. The supernatant is transferred to a 0.5-ml microcentrifuge tube using a micropipette, for PCR amplification.

An alternative approach for processing gastric biopsy specimens for PCR detection of *H. pylori* has been described by Hammar et al.[8] Biopsy tissue samples, with the accompanying tissue media, are centrifuged in a microcentrifuge for 5 min at 10,000 g to collect the insoluble tissue material. After discarding the supernatant, the tissue material is resuspended in 70 µl extraction buffer (20 mM Tris-Cl, pH 8.0, 0.5% [w/v] Tween 20) and mixed by vortexing for 5 min. The sample is then treated with proteinase K (from a stock solution of 20 mg/ml), to a final concentration of 0.5 mg/ml and incubated at 37°C for 1 h. After incubation, the proteinase K is inactivated by heating the sample to 85°C for 10 min. The sample can be used at this stage for PCR amplification without any further purification stage. Alternatively, proteinase K-treated samples can be further purified once by adding equal volume (70 µl) of phenol:chloroform (1:1 v/v), mixing by vortexing, and centrifuging at 10,000 g for 5 min. The top aqueous layer is carefully withdrawn using a micropipette, without disturbing the bottom phenol:chloroform layer, and transferred to a 0.5-ml microcentrifuge tube for PCR amplification.

The processing of the frozen gastric biopsy specimens and gastric aspirate samples for recovery of genomic DNA from the microbial cells, for PCR-amplification detection, has been described by Valentine et al.[9] For freshly frozen gastric biopsy samples, 100 µl of tissue extraction buffer I (50 mM Tris-Cl, pH 7.5, 1 mM EDTA, 0.45% Tween 20 [v/v], 0.45% Nonidet P40 [v/v], and 100 µg proteinase K per ml) are added and incubated at 55°C for 2 h. The sample is then purified once with phenol:chloroform, and the top aqueous layer is transferred to a 0.5-ml microcentrifuge tube and directly used for PCR DNA amplification. For gastric aspirate specimens, 250 µl of the sample are treated with 20 µl of 1 N NaOH solution to neutralize acidity. They are then treated with extraction buffer II (50 mM Tris-Cl, pH 7.5, 1 mM EDTA, 1% Laureth-12 [w/v], and 100 µg of proteinase K per ml) and incubated at 55°C for 2 h. The samples are then purified once with phenol:chloroform, and the aqueous layer is directly used for PCR DNA amplification. In both frozen gastric biopsy and gastric aspirate samples, following phenol:chloroform purification, the released DNA can be precipitated by adding 2.5 vol of ice-cold ethanol, chilled by transferring to a –70°C freezer for 10 min, followed by centrifugation in a microcentrifuge for 10 min at 10,000 g. The DNA pellet is dried in vacuum and resuspended in sterile distilled water for PCR amplification.

B. PARAFFIN-EMBEDDED GASTRIC BIOPSY

Generally, biopsy specimens are first immersed into 10% (v/v) formalin saline solution (10 ml formalin solution in 90 ml 0.9% NaCl solution) to fix the tissue and the microbial cells associated with it, followed by embedding the tissue in paraffin wax for routine histological analyses.[7] The embedded biopsy specimen is subjected to histological sectioning of an approximate size of 4 × 10 µm, on a microtome using a 1 M HCl-treated microtome blade. The sections are then transferred to a 1.5-ml microcentrifuge tube, and 70 µl of sterile distilled water containing 0.01 mM EDTA are added to it. The sample is boiled for 15 min, cooled on ice for 10 min, and centrifuged in a table-top microcentrifuge for 5 min at 10,000 g. The supernatant is then tranferred to a 0.5-ml microcentrifuge tube and stored at –20°C for PCR amplification. Gastric biopsy samples are routinely used for identification of *H. pylori* infection and can be used for possible infection with other gastrointestinal microbial pathogens.

C. CLINICAL SWABS

Clinical swabs are generally performed for identification of microbial pathogens in the genital areas such as cervical, vulvar, or penile areas, for tissue culture cells, or in other situations where cells are easily collected by soft abrasion.[10,11] The process involves abrading the surface with a saline pre-wet cotton swab or a cytobrush. The specimen is then transferred to a 10- to 15-ml conical tube containing 2 ml of phosphate buffered saline plus a 2x concentration of Fungi-Bact. If the sample is to be processed within 24 h, it can be kept at room temperature. Refrigeration at 4°C is recommended for longer storage times. The cells are collected by centrifugation at 2000 to 3000 g and the supernatant is discarded by an aspirator without losing the cell pellet. If any blood is present, the cell pellet is resuspended in 1 ml of Tris-EDTA buffer, pH 8.0, transferred to a microcentrifuge tube, and centrifuged at 10,000 g for 10 s. The supernatant is removed, and the pellet is resuspended in 4 vol of (ca. 30 to 300 µl) of K buffer (50 mM KCl, 20 mM Tris-Cl, pH 8.3, 2.5 mM $MgCl_2$). The sample is then incubated at 55°C for 1 h followed by heating to 95°C for 10 min., frozen once, and 5 to 10 µl of it are used for PCR DNA amplification.

D. SALIVA

A standard method for treating saliva for detection of microbial pathogens by PCR amplification has been described by Hammar et al.[8] Typically 0.5 ml of saliva are mixed with 1 ml (2 vol) of sputolysin (Sigma Chemicals, St. Louis) in a 2-ml microcentrifuge tube and shaken gently for 20 to 30 min at room temperature. The sample is then centrifuged in a microcentrifuge at 10,000 g for 5 min, and the supernatant is discarded, using a micropipette, without disturbing the pellet. The pellet is resuspended in 70 µl of extraction buffer, containing 0.5 mg/ml proteinase K, and incubated at 37°C for 1 h. Following incubation, the proteinase K activity is destroyed by heating the sample at 85°C for 10 min and purified once with an equal volume (70 µl) of phenol:chloroform, as described earlier. The top aqueous layer is used for PCR amplification. Saliva samples are used for the detection of microbial pathogens that cause respiratory illnesses.

E. STOOL SAMPLES

Stool samples are processed for the detection of various enteric microbial pathogens such are *Vibrio cholerae*, *Clostridium difficile*, *Salmonella* spp., *Shigella* spp., *Escherichia coli*, *Giardia* spp., *Entamoeba histolytica*, and various enteric viruses. The sample processing for the enteric viruses is discussed later in this chapter. The processing of human stool samples for PCR amplification detection of microbial pathogens has been described by Pollard et al.[12] Typically 200 mg of stool sample are suspended in 1 ml of PBS, pH 7.4, and mixed well by vortexing for 5 min. The relatively large insoluble particulates in the sample are centrifuged for 10 min at 3000 g, and the supernatant is immediately transferred into a 1.5-ml microcentrifuge tube using a micropipette and without disturbing the pellet. The sample is then centrifuged at 10,000 g for 15 min to collect the bacterial cells. The collected bacterial cells are lysed by treatment with 70 µl of lysis buffer (10 mM Tris-Cl, pH 8.0, 1 mM EDTA, 0.5% SDS), in which 100 µg of lysozyme are added freshly, and incubated at 37°C for 30 min. Then 300 µl of a 50 mM NaCl solution containing 1% SDS (w/v) and 800 µg proteinase K are added to the sample, and incubation is continued at 37°C for another 1 h. Following this, the sample can be processed by two different methods for recovery of DNA for PCR DNA amplification. In one method, the released genomic DNA can be precipitated by the addition of absolute alcohol, followed by centrifugation in a microcentrifuge at 10,000 g for 10 min to pellet the DNA. Following centrifugation, the alcohol is carefully discarded by decanting or by using a micropipette without disturbing the pellet. The DNA pellet is then dried in a vacuum, resuspended in sterile distilled water, and subjected to PCR amplification. Alternatively, the sample can be incubated at 85°C for 10 min to inactivate any residual activity of proteinase K, purified once with an equal volume of phenol:chloroform (1:1 v/v), and aliquots of the aqueous supernatant can be used directly for PCR amplification.

Another simple method for processing human stool samples for detection of gastrointestinal-tract-related microbial pathogens has been described by Shirai et al.[13] The samples are kept at ambient temperature during transportation. The stool samples are then heated to 94°C for 5 min and stored at −20°C until use. The samples are then resuspended in TE buffer, pH 8.5, and 10 µl of sample are used directly for PCR amplification of the target microorganisms.

An alternative procedure for processing human stool samples for PCR amplification of target microbial pathogens has been described by Olive.[14] This method requires use of an Extractor column™* and associated reagents, which are provided in the kit available from Molecular Biosystems Syngene, Inc. In this procedure, stool samples are suspended in TE

---

* Registered trademark of Molecular Biosystems Syngene, Inc., 10030 Barnes Canyon Rd, San Diego, CA 92121-2789.

buffer, pH 7.5 to make a 10% suspension. From this suspension, 100 µl of sample are transferred to a separate tube and mixed with 1 ml of specimen dilution buffer. The sample is then mixed with 2 ml of lysing buffer, 100 µl of 2 mg/ml proteinase K is added to it, and incubated at 60°C for 30 min. This suspension is then centrifuged at 750 g for 5 min, and the clear supernatant is transferred to an Extractor column™. The sample is allowed to enter into the column by gravity flow, and the column is washed with 15 ml of wash reagent 1, followed by another washing with 5 ml of wash reagent 2 (supplied in the kit). The nucleic acid is eluted from the column with 2 ml of elution reagent (supplied in the kit). An aliquot (generally 20 µl) of the eluted nucleic acid can be used for PCR amplification, or, it can be precipitated first with 2.5 vol of ice-cold ethanol for 10 min in a –70°C freezer, centrifuged at 10,000 g for 15 min, the pellet dried with a hair dryer or in a vacuum, resuspended in sterile distilled water, and used for PCR DNA amplification.

Recently an efficient method for processing stool samples has been developed for the detection of *Salmonella* spp. and *Shigella* spp. and can be used for the detection of other related microbial pathogens.[15-17] In this approach, 1-g stool samples were resuspended in Tris-EDTA buffer, pH 8.0 and vortexed to mix. The samples were then centrifuged at 10,000 g for 5 min, and the pellets were resuspended in sterile distilled water. The samples were then treated with Chelex™* 100 to a final concentration of 5% (w/v) at 60°C for 10 min, followed by boiling for 20 min. The samples were then cooled to room temperature and Ammonium acetate (NH_4OAc) was added to a final concentration of 2.5 M. After cetrifugation at 10,000 g for 5 min, the supernatants were further purified one time with phenol:chloroform (1:1 v/v) and concentrated with a Centricon™** 100 microconcentrator. Approximately 5 to 50 µl of the concentrated samples can be used for PCR amplification detection of the target microbial pathogen(s) with appropriate primers and reaction conditions.

F. BLOOD SAMPLES

Blood samples are processed for the detection of various microbial pathogens, including viruses. Considerable effort has been devoted by many investigators to develop a method for rapid isolation and purification of target DNA from blood samples. The protocols for extraction and purification of DNA from blood samples are described in Chapter 17.

G. SAMPLE PROCESSING FOR THE DETECTION OF RNA VIRUSES

A number of RNA viruses, including some of the enteric viruses, are required to be detected by reverse transcriptase-PCR amplification method. Various samples for the detection of RNA viruses are described by Muir et al.[18]

For extraction of the viral RNA including enteroviruses, the specimens are pre-treated to release viral RNA from virions or cells. Viral RNA can then be extracted by standard RNA extraction procedure using alkaline lysis or by repeated freezing and thawing of the samples (see below) with RNAsin™*** to prevent destruction and damage of RNA.[19] The pre-treatment of various samples for the detection of RNA viruses is described below.

1. Stool Filtrate, Urine, Throat Swab Medium, Cerebrospinal Fluid, or Virus Culture Supernatant Samples

Before pre-treatment, a stool sample is filtered and a stool-filtrate is obtained. The stool samples are resuspended in ~10 vol of Eagle maintenance medium and clarified by centrifugation followed by successive filtration through syringe filters of decreasing pore sizes of 5, 2, 0.45, and 0.2 µm.**** At this point, 90 µl of the stool filtrate, urine, throat swab, CSF, or

---

*   Registered trademark of BioRad Laboratories, 3300 Regatta Blvd., Richmond, CA 94804.
**   Registered trademark of Amicon Division, W.R. Grace Co.-Coon., 72 Cherry Hill Dr., Beverly, MA 01915.
***   Registered Trademark of Promega Corp., 2800 Woods Hollow Rd., Madison, WI 53711-5399.
**** Gelman Sciences, 600 South Wagner Rd., Ann Arbor, MI 48106-1448.

virus culture supernatant samples are mixed with 10 µl of 10x boiling buffer (100 mM Tris pH 7.5, 100 mM vanadyl ribonucleoside complexes (VRC),* 1% sodium dodecyl sulfate (SDS), heated in a boiling water bath for 2 min, and then chilled on ice.[18]

2. Cryopreserved Tissue

For cryopreserved tissue, up to 100 mg of tissue was ground to a fine powder in liquid nitrogen and then homogenized in 100 µl of ice-cold extraction buffer (100 mM Tris-Cl, pH 8.0, 500 mM LiCl, 10 mM EDTA, 5 mM dithiothreitol, 10 mM VRC). The homogenate was then clarified by brief centrifugation and chilled on ice.[18]

3. Formalin-Fixed Tissue

For pretreatment of formalin-fixed tissue, 20 to 30-µm-thick sections were deparaffinized by extracting twice with 10 ml of octane and then twice with 2 ml of absolute ethanol. To approximately 10 mg of tissue were added 100 µl of proteinase K digestion buffer (50 mM Tris-Cl, pH 8.5, 1 mM EDTA, 0.5% Tween 20) and 10 mM VRC, with nuclease-free proteinase K added immediately before use to a final concentration of 200 µg/ml), and the mixture was incubated for 3 h at 55°C and then for 8 min at 99°C to inactivate proteinase K. The sample was then clarified and chilled on ice.[18]

4. Proteinaceous Body Fluid

For proteinaceous body fluids such as serum, plasma, and pericardial effusion, 100 µl of sample was mixed with 10 µl of 2x proteinase K digestion buffer containing 400 µg of proteinase K per milliliter and incubated at 55°C for 1 h and then at 99°C for 8 min. The sample was then clarified and chilled on ice.[18]

5. Leukocytes

For pretreatment of leukocytes, approximately 10^6 washed peripheral blood lymphocytes or cells from 1 ml of whole heparinized blood were pelleted by centrifugation, suspended in 200 µl of cold hypotonic lysis buffer (10 mM Tris-Cl, pH 7.5, 140 mM NaCl, 5 mM KCl, 1% Nonidet P-40, 10 mM VRC), and incubated on ice for 1 min. The cell nuclei were used for RNA extraction.[18]

H. GENERAL METHODS FOR LYSIS OF MICROBIAL CELLS

For the detection of microbial pathogens in the clinical samples, in many instances, it is necessary to further process the sample to ensure lysis of the target microbial pathogens, to release their genomic DNAs for PCR amplification. Some of the simple lysis methods that can be used conveniently for PCR DNA amplification are described below.

1. Freeze-Thaw Lysis

Microbial cells can be lysed by repeated freezing and thawing (5 to 20 c) of the sample. Rapid freezing is performed in a dry-ice methanol/ethanol bath, and thawing is performed in a 50°C water bath.[3,4] Following freeze-thaw cycles, the sample is heated to 80°C for 10 min and cooled to room temperature; PCR reagents are added, and PCR amplification is performed. Most Gram negative microorganisms are sensitive to repeated freeze-thaw lysis, and this method of lysis can be applied for their detection by PCR amplification.[20]

2. Freeze-Boil Lysis

Microbial cells can be frozen as described above, followed by boiling in a water bath or in a thermal cycler for a total of 3 to 5 times. Then PCR reagent mixture is added to the sample,

---

* GIBCO BRL, 8400 Helgerman Ct., Gaithersburg, MD 20877.

and PCR amplification is performed.[21-23] Most of the Gram negative and some Gram posititive cells such as *Staphylococcus* spp. can be lysed by this approach.[20]

3. Boiling With Chelex™ 100

Samples can be incubated at 60°C for 10 min followed by boiling for 10 min in a water bath or in a thermal cycler in the presence of 50 to 100 µl of Chelex™ 100 (BioRad, CA). Chelex™ 100 is known to stabilize the genomic DNA in boiling water by maintaining the ionic strength of the sample. Following boiling, the samples are centrifuged in a table-top microcentrifuge at 10,000 g for 3 min, and 5 to 10 µl of the supernatant can be used for PCR amplification. This approach is more effective than the freeze-boil method for releasing nucleic acids by lysis of the cells, for most of the Gram negative, Gram positive, and even some cysts such as *Giardia* spp.[20,24]

IV. PCR AMPLIFICATION

A. TARGET DNA

For amplification of DNA by the PCR method, there must be at least one intact copy of the target DNA in the sample. A greater number of target copies enhances the probability of successful DNA amplification. Any nicks in the target DNA may block PCR amplification. The target DNA sequence can be from <100 bp to a few kilobases. The total amount of DNA typically used for PCR is 0.05 to 1.0 µg; this permits detection of approximately a single copy of the target DNA, depending upon the copy number of the target gene present in a single cell. A sample need not be highly purified and may be prepared, in general, by lysing the cells, by boiling in a hypotonic solution, by freeze-thaw or by freeze-boil cycling, as described earlier. However, some impurities, such as formalin, blood, humic acids, chelating agents, detergents, and heavy metals, must be eliminated or diluted for successful amplification of the target DNA.[5,25-27]

B. PRIMERS FOR PCR AMPLIFICATION

Selection of primers for successful PCR amplification has been described in Chapter 2. In general, the primer-target melting temperature (T_m value) should be calculated by the total GC and AT content of the primer segment using the formula $T_m = 2(A+T)+4(G+C)$[28] or by using a computer-aided primer selection program such as *Oligo*.[29] The preferable T_m value for each primer is 60 to 70°C.[25,30] The primer lengths should be between 18 and 30 nucleotides. The location of the primers should not be close to or within a strong hairpin structure. It is important that the 5 to 6 nucleotides at the 3′ ends of the primers exhibit precise base pairing with the target DNA. The terminal match at the 3′ end is critical, and exact base pair match is generally required at the 3′ end for effective amplification. A mismatched T at the 3′ end, however, still allows amplification if the PCR is run in non-stringent conditions. Therefore, to avoid primer-mismatched amplifications, it is preferable to select the length and the location of the primer along the target DNA, such that the annealing temperature for each primer remains high (60 to 70°C). Also, stretches of GC or AT within the primers should be avoided. Where possible, primers should have a G + C content of around 50% and a random base distribution. The concentration of each primer in a PCR-amplification reaction is recommended to be between 0.1 and 0.5µM. Higher primer concentration may promote non-specific amplified product formation and may increase the formation of primer-dimer. For optimum results, after synthesis and standard deprotection methods, primers should be purified by HPLC or gel purification. The lyophilized primers should be stored at −20°C when they are and not in use. The lyophilized primers are resuspended in Tris-EDTA buffer, pH 8.0, and generally diluted to required concentrations, aliquoted, and stored at −20°C. The shelf life of oligonucleotide primers is at least six months when stored in liquid and 12 to 24 months when stored in lyophilized condition. The primers can also be stored for several months in 20% acetonitrile solution after HPLC purification, which prevents microbial growth.[5]

C. THERMOSTABLE DNA POLYMERASE

The thermostable DNA polymerase plays the key role in a successful PCR reaction. The most used thermostable DNA polymerse in PCR DNA amplification is *Taq* or Ampli*Taq* DNA polymerase from *Thermus aquaticus*. Recently, a number of thermostable DNA polymerases with many new features such as proofreading, reverse transcriptase activity, high thermostability, etc., have been described. Descriptions and uses of various thermostable DNA polymerases have been described in Chapter 25.

D. DEOXYRIBONUCLEOTIDES

Free deoxyribonucleotide triphosphates (dNTPs) are required for DNA synthesis. The concentrations of each of the dNTPs for PCR reaction should be 200 µM, to give optimal specificity and fidelity of the reaction. The four dNTPs (dATP, dGTP, dCTP, and dTTP) should be used at equivalent concentrations to minimize misincorporation errors, which may interfere during DNA-DNA hybridization. A final concentration of greater than 50 mM total dNTP in the PCR reaction inhibits *Taq* DNA polymerase activity.[31] A 20 µM concentration of each of the dNTPs, in a 100-µl reaction is sufficient to synthesize 2.6 µg of a 400-bp DNA sequence.[31] A stock mixture of 5 or 2.5 mM concentration of each of the dNTPs is prepared in sterile distilled water and stored in the freezer (–20°C) when not used. Also, aliquoting of this stock solution is recommended, because repeated freezing and thawing of the stock dNTP solution will decrease the shelf life of the reagent.

E. MAGNESIUM ION

One of the most important ingredients in a PCR reaction is the divalent magnesium ion. The concentration of the magnesium ion in the PCR reaction affects primer annealing, DNA melting temperatures, and enzyme activity. A range of 0.5 to 4 mM magnesium ion can be used for optimum PCR reaction.[32] The $MgCl_2$ concentration can be increased up to 10 mM when the Stoffel Fragment is used (see Chapter 25). The presence of EDTA and other chelating agents in the PCR reaction mixture may interfere with the magnesium ion concentration. The optimal magnesium ion concentration for each primer set should be determined by running replicate PCRs with varied magnesium concentrations. A concentration of 2.5 mM of $MgCl_2$ in a PCR reaction, for detection of most of the microbial pathogens, is found to be optimum for obtaining robust PCR amplification. More information about the use of $MgCl_2$ in the PCR-DNA amplification process and its role in the function of the thermostable DNA polymerases is described in Chapter 25.

F. PCR REACTION BUFFER

The recommended buffer for optimum PCR reaction is 10 to 50 mM Tris-Cl, pH 8.3 to 8.9. In many cases, changing the buffering capacity of the PCR reaction increases the specificity and yield of the amplified DNA products. For example, increasing the concentration of Tris-Cl up to 50 mM, pH 8.9, and KCl concentration to 50 mM facilitates specific primer annealing, thus increasing the specificity of the reaction. A potassium chloride concentration greater than 50 mM inhibits *Taq* DNA polymerase activity.[33] Gelatin or bovine serum albumin (100 µg/ml) and non-ionic detergent such as Tween 20 can be used to stabilize the *Taq* DNA polymerase activity. It is important to note that if gelatin is chosen as an ingredient for the PCR reaction, it has to be properly sterilized. There are instances where the presence of gelatin produced PCR false-positive results for microbial pathogens such as *E. coli*.[64] The optimum concentration and effect of KCl in the PCR reaction and its effect on the function of the thermostable DNA polymerases, in general, is described in Chapter 25.

G. COSOLVENTS

If the template DNA contains strong secondary structures (determined by a high $-\Delta G$ value), or a primer has to be chosen within or near a hairpin structure, use of a cosolvent such as 1 to 10% dimethylsulfoxide (DMSO) is recommended.[34,35] Also, use of 5 to 20% glycerol to increase the yield of amplified DNA and 15 to 20% glycerol to amplify larger DNA fragments of about >2.5 kbp is found to be useful.[34,35]

H. MINERAL OIL/WAX OVERLAY

An overlay of 50 to 80µl light mineral oil on top of the reaction mix is often used to prevent evaporation of the liquid. Such an overlay maintains heat stability and limits evaporation so that salt concentrations in the PCR reaction are maintained during amplification. It has been found that an overlay of mineral oil or wax can increase the yield of the amplified product approximately five times.[5] Recently, the use of light mineral oil has been replaced with paraffin wax to avoid further purification of the amplified DNA from the oil by using chloroform.[36,37] However, in some thermocyclers, the reaction volume is low and the total PCR-amplification time is so short that no such overlay is required.

I. PCR REACTION MIX

For PCR amplification of DNA a reaction mix containing the following reagents is required: (a) a reaction buffer containing 50 mM Tris-Cl, pH 8.9, 50 mM KCl, and 2.5 mM MgCl$_2$;[3,4] (b) 200 µM of each of the deoxynucleotide triphosphates (dATP, dTTP, dCTP, and dGTP); (c) 0.5 to 1 µM of each of the primers; and (d) 1 to 2.5 U of thermostable DNA polymerase. The volume of the reaction is adjusted with sterile distilled water to 100 µl and 50 to 80 µl of light mineral oil is overlaid on the reaction mixture to prevent evaporation during amplification.

J. PCR REACTION PARAMETERS

Initially, the samples are heated to 94 to 95°C for 3 to 5 min for denaturation of the target DNA. This step is critical because during the first few cycles the primers must synthesize enough targets, which can be used during successive cycles to yield enough amplified products at the end of the amplification. Also, the time and temperature of the initial denaturation step depends on the complexity of the target DNA, i.e., total GC content, presence of hairpin structures, or stretches of GC nucleotides. Following denaturation of the template DNA, typically 25 to 30 cycles of PCR amplification are performed. Each cycle has a denaturation step of usually 94 to 95°C for 1 min, primer-annealing step for 1 min (the temperature of which depends on the T_m values of the primers), and primer-extension step, usually 72°C for 0.5 to 5 min (depending upon the size of the template to be amplified). In many cases, where the T_m values for the primers are within the range of primer-extension temperature (68 to 72°C), the primer-extension step can be omitted (two-step PCR amplification). At the end of the amplification cycles, the reaction is incubated at 72°C for 3 to 5min to ensure that the amplified DNAs are completely synthesized. If too many amplification cycles are performed (45 to 50 c), additional amplified DNA bands along with the actual target amplified DNA may result when analyzed by agarose or polyacrylamide gel electrophoresis method.[5,25-27] Generalized parameters for PCR amplification have been illustrated in Chapter 1.

K. MULTIPLEX DNA AMPLIFICATION

In case of multiplex PCR amplification for the detection of more than one microbial pathogen in the same sample, the T_m values for all primer sets should be within the range of ±5°C. Also, the sizes of various amplified DNA products should be close to one another for efficient amplification (within the range of 200 bp).[3-5] In some instances, the extension time can be increased to 2 to 3 min for a better yield of the amplified DNA products. Also if required, each dNTP concentration can be increased to 300 μM. Although, equimolar quantities of each of the primer sets is used for multiplex PCR amplification, in case of unequal yield of one or more of the amplified DNAs, the concentration of the primer set giving a lower yield of amplified DNA product can be increased.[5,25,26]

L. CONTROLS

It is essential that appropriate control samples are kept with each set of PCR samples. To determine possibe contamination of the PCR reagent with the target DNA, a sample should be amplified without any target DNA or sample to be tested in it. Also, a serial dilution, starting from at least 10^4 cells, of the target microorganism from a pure culture should be mixed with the biological sample to be tested, and PCR amplification should be performed. This will make it easier to determine the sensitivity of detection when a target microbial pathogen is present in the sample. Additionally, this control experiment will determine the presence of inhibitory agents, if any, in the sample. These types of controls will make it easier to avoid confusion about any false positive or false negative results.

V. IDENTIFICATION AND ANALYSIS OF THE AMPLIFIED DNA

A. PROBE SELECTION

The oligonucleotide probe for the detection of the PCR-amplified DNA is generally designed by selecting short stretches of nucleotide sequence located internal to the amplified DNA.[5,25,26] In general, the oligonucleotide probes are between 18 and 50 nucleotides long, are located within the amplified DNA, and do not overlap with the primer sequences. More information about the principal criteria for selecting and generating probes for the detection of the PCR amplified DNA has been described in Chapters 2, 6, and 7.

B. DETECTION OF AMPLIFIED DNA BY IMMOBILIZED CAPTURE PROBE DNA-DNA HYBRIDIZATION

1. Labeling Amplified DNA

The PCR DNA amplification is performed with one of the two primers labeled with a hapten (usually biotin) molecule. As a result, one strand of each of the amplified DNA molecules is labeled with biotin. The amplified DNA strand that is complementary to the probe is labeled with the biotin molecule. The desired primer is labeled with a biotin molecule at its 5′ end during synthesis. More information about labeling of the probe using the PCR-amplification method has been described in Chapters 6 and 7.

2. Synthesis of a Homopolymer Tail at the 3′ End of the Probe

A homopolymer dT-tail approximately 100 to 150 nucleotides long is synthesized at the 3′ end of each probe molecule by using 100 mM dTTP, 25 U of terminal deoxyribonucleotidyl transferase (TdT) enzyme, and 1 x TdT buffer (100 mM potassium cacodylate, 25 mM Tris-Cl, pH 7.6, 1 mM cobalt chloride, and 0.2 mM dithiothreitol). The reaction mixture is incubated at 37°C for 1 h. The reaction is terminated by adding 10 mM EDTA and purified by using a Centricon™ 10 microconcentrator (Amicon, MA). The length of the homopolymer dT-tail is determined by running a polyacrylamide gel stained with ethidium bromide.[4,38]

3. Immobilization of the Probe onto Nylon Membrane

The nylon membrane is spotted with 3 pmol of homopolymer dT-tail-containing oligonucleotide probe by using a dot blot apparatus (BioRad, CA). The probes are attached to the nylon membrane by UV irradiation at 254-nm wavelength for 5 to 10 s.[4,38] Then the membrane is washed in hybridization buffer (50 mM sodium phosphate, pH 7.0, 0.9 M NaCl, 5 mM Na$_2$EDTA, 0.5% SDS [w/v]) for 30 min at 55°C to remove unbound probes.

The membrane is air dried by being kept in between two 3M Whatman filter papers; it can be stored at 4°C or at room temperature in a sealed plastic hybridization bag for months until needed.

4. DNA-DNA Hybridization Using Biotin-Labeled Amplified DNA

The nylon membrane strip containing immobilized capture probe is treated with hybridization buffer (see above) at 55°C for 30 min in a plastic hybridization bag with gentle shaking. An aliquot of the amplified DNA (usually 20 to 25 µl) is denatured using an equal volume of denaturing solution (400 mM NaOH, 10 mM Na$_2$EDTA). The hybridization buffer is discarded, new hybridization buffer is added to the bag, and hybridization is performed with the denatured amplified DNA for 3 h with gentle shaking at a specific hybridization temperature for each probe. Following hybridization, the buffer is discarded and the nylon membrane is washed twice in washing buffer (20 mM sodium phosphate, pH 7.0, 0.36 M NaCl, 2 mM EDTA, 0.1% SDS [w/v]) at room temperature, followed by washing once in the same washing buffer at a temperature 5°C below the hybridization temperature, and then quickly rinsed in PBS, pH 7.2, at room temperature.

5. Color Development

The color development is performed by using streptavidin-horseradish peroxidase (HRP-SA) conjugate. First, the hybridized nylon membrane is treated with 3% bovine serum albumin Type V (BSA) (Sigma) prepared in washing buffer at hybridization temperature for 30 min with gentle shaking. Following incubation, the BSA solution is discarded and the membrane is washed in PBS (pH 7.2) twice (15 min each) at room temperature. The membrane is treated with HRP-SA diluted 1:1000 times in PBS (pH 7.2) for 15 min at room temperature with gentle shaking. The HRP-SA solution is discarded and the membrane is washed with PBS, pH 7.2, three times (10 min each) to eliminate unreacted HRP-SA.

The color is developed by treating the membrane with TMBlue®* solution at room temperature in a plastic hybridization bag or in a small plastic or glass tray, and the purple blue color develops within 15 to 20 min. Incubation for a longer period of time (more than an hour) may result in development of background color.

C. DETECTION OF THE AMPLIFIED DNA USING MICROTITER PLATE HYBRIDIZATION: IMMOBILIZATION OF THE HYBRIDIZATION PROBES TO MICROTITER WELLS

1. Passive Transfer

The passive transfer of the oligonucleotide probe to the polystyrene microtiter plate is described by Keller et al.[39] The oliginucleotide probe is diluted in sterile distilled water to a concentration of 100 µg/ml. For the preparation of four eight-well strips, 108 µl of DNA and 1.65 µl of binding buffer (25 mM K$_2$HPO$_4$, pH 7.2, 200 mM MgCl$_2$) are mixed. Three hundred nanograms (approximately 50 µl) of the diluted DNA are added to each well and incubated at room temperature for 2 h with gentle shaking. The rest of the microtiter well surfaces are blocked with 400 µl of blocking buffer per well and incubated at room temperature for 1 h. The blocking buffer is discarded and each well is washed three times with 400 µl of sterile distilled water. The wells are air dried and used immediately for hybridization or stored at room temperature in a heat-sealed plastic bag.

2. Chemical-Mediated Active Transfer

The transfer of the oligonucleotide probes to activated polystyrene microtiter plate wells has been described by Running and Urdea.[40]

---

* Registered trademark of TSI Washington Laboratories, 5516 Nicholson Ln., Kensington, MD 20895.

a. Capture Probe Synthesis

An alkylamine linker is incorporated at the 5' or at the 3' end of the oligonucleotide probes during their synthesis. Following synthesis, standard deprotection and gel purification is performed and the oligonucleotides must be purified by reverse phase column chromatography.

b. Activation of Wells

A 20-ml solution of 200 µg/ml poly(Lys-HBr, Phe; Sigma) is prepared in sterile distilled water, and 200 µl of this solution is added to each well of a microtiter plate (Immulon 2 or Microlite 1 Removawells).* The plate is incubated at room temperature for 30 min followed by washing four times with PBS (pH 7.2).

c. Activation of Probe

A 10 ml of PD10(G-25) gel** is equilibrated with 30 ml of PBS (pH 7.2). One milligram of ethylene glycol bis-succinimidylsuccinate (EGS); Pierce Chemical, Rockford, IL) dissolved in 14 µl of dimethylformamide is added to 6.7 nmol of alkylamine probe (e.g., 1.3 OD_{260} unit of 21-mer probe), mixed by vortexing, and incubated at room temperature for 15 min with gentle shaking. The reaction is added to the column and eluted with PBS, pH 7.2, in a total of ten microcentrifuge tubes, with each fraction containing 0.5-ml elution. From each tube 2 to 3 µl of sample are spotted onto a thin layered chromatography (TLC) plate, dried, and examined with a handheld UV lamp for dark spots indicative of DNA samples. Usually, fractions 6 and 7 contain the eluted DNA. The samples containing DNA can be combined.

d. Probe Coupling

After combining the probe DNAs, the volume is brought up to 5ml with PBS (pH 7.2). In each activated microtiter well 50 µl of the probe is added, incubated at room temperature for 30 min, and washed four times with PBS (pH 7.2) at room temperature. If the plates are not used for hybridization immediately, they can be wrapped with plastic wrap and stored at 4°C in the dark for several months.

3. Hybridization

The microtiter strip coated with capture probe is hybridized with 25 µl amplified DNA, which is denatured with 2.5 µl of denaturing solution containing 2.5 M NaOH with 10 mM EDTA followed by neutralization with 15 µl of 2 M HEPES (N-2-hydroxyethylpiperazine-N'-2-ethanesulfonic acid, pH 6.5). One hundred microliters of hybridization buffer are added to the denatured probe and used for hybridization. The microtiter plate wells are treated with 150 to 200 µl of hybridization solution[39] at 42°C for 30 min with gentle shaking. Following treatment, the hybridization buffer is discarded and new hybridization buffer is added to each well along with the denatured probe; hybridization is performed at the suggested hybridization temperature for a specific probe for 4 h. Following hybridization, the microtiter wells are washed five times with washing solution.[39]

4. Color Development

Following hybridization, each well is blocked with 200 µl of 3% BSA (w/v) prepared in washing solution. The wells are treated with 100 µl of 1 µg/ml of peroxidase conjugated streptavidin prepared in 3% BSA-washing solution for 10 min at room temperature with gentle shaking. Wells are washed five times with 200 µl of washing solution at room temperature. Color development is performed by adding 100 µl of tetramethylbenzidine color reagent for

---

* Dynatech, 14340 Sullyfield Cir., Chantilly, VA 22021.
** Pharmacia LKB Technology, 800 Centennial Ave., Piscataway, NJ 08854.

30 min. Then 100 µl of 0.1 N H$_2$SO$_4$ are added to each well, and the yellow color is measured at 450 nm in a microtiter plate reader.

D. LABELING PCR AMPLIFIED DNA WITH BIOTIN OR WITH DIGOXIGENIN (DIG)

The labeling of the PCR-amplified DNA with biotin or DIG is described in Chapters 6 and 7. Briefly, the PCR-amplfied DNA can be labeled with biotin-11-dUTP (Sigma) or with digoxigenin-11-dUTP (DIG, Boehringer Mannheim Biochemicals [BMB], Indianapolis) by adding a ratio of 3:1 dTTP to biotin-11-dUTP or dCTP to DIG in the PCR reaction mixture with other 3-deoxyribonucleotide triphosphates. As a result, during PCR amplification, the amplified DNAs will be labeled with biotin or DIG due to the random incorporation of biotin-11-dUTP or DIG, respectively. Following amplification, the amplified DNA samples are purified from unincorporated biotin or DIG by using a Centricon™ 10-microconcentrator (Amicon, MA) using TE, pH 7.5, buffer. The biotin- or DIG-labeled amplified DNA is used for immobilized capture probe DNA-DNA hybridization on solid support such as nylon membrane or on microtiter plate, as described before. The color development for biotin-labeled amplified DNA is performed as described earlier, and for DIG-labeled amplified DNA an immunogenic detection is generally performed.

E. IMMUNOGENIC DETECTION OF DIG-LABELED PCR PRODUCTS

Following hybridization, the membrane or the microtiter plate surface is blocked with 3% BSA in buffer 1[41] (100 mM Tris-Cl, pH 7.5, 150 mM NaCl, and 3% BSA [w/v]) for 30 min at 50°C with gentle shaking. The hybridization solid surface is washed with buffer 1 (3% BSA [w/v] in 100 mM Tris-Cl, pH 7.5, 150 mM NaCl) for 15 min at room temperature with gentle shaking. The membrane or the microtiter plate is then incubated with a 1:5000 dilution of alkaline phosphatase-conjugated anti-digoxigenin Fab fragment (BMB, Indianapolis) in buffer 1 for 30 min at room temperature with gentle shaking. The solid support is then washed in buffer 1 for 30 min at room temperaure and equilibriated with buffer 2 (100 mM Tris-Cl, pH 9.5, 100 mM NaCl, 50 mM MgCl$_2$)[41] for 2 min. The color development is performed by treating the solid support in 40 µl of AP substrates, i.e., nitroblue tetrazolium (75 mg/ml in dimethyl formamide) and 5-bromo-4-chloro-3-indolyl phosphate (50 mg/ml in dimethyl formamide) per 10 ml of buffer 2. Color development is terminated by washing the hybridization solid surface in TE buffer, pH 8.0, for 5 min in room temperature and air dried.

VI. PCR CONTAMINATION CONTROL

Because of the amplification power of a single copy of the target DNA, it is critical to avoid even traces of contamination of the DNA containing the target sequence. Contamination of the PCR amplification reaction with products of a previous PCR reaction (PCR carryover), cross-contamination between samples, contamination with exogenous nucleic acids from the laboratory environment, and even contamination from the skin of laboratory personnel, can create a false-positive result. The general preventive measures, precautions, and reduction of such PCR-carryover contamination approaches are described in Chapter 27. Briefly, pre-aliquoting reagents, using pipettes exclusively for specific steps of the reaction, using positive-displacement pipettes, or using tips with barriers to prevent contamination of the pipette barrel, and physical separation of the amplification reaction preparation from the area of amplified product analysis can minimize the possibility of such contamination.[42,43] Addition of multiple negative control reactions without any target DNA added is essential as quality control, with every set of PCR reactions to reveal and monitor possible contamination.

Several methods have been developed to prevent amplification of contaminants.[44-47] The reaction mixture can be exposed to ultraviolet irradiation (300- or preferably 254-nm wavelength) for 5 to 20 min prior to addition of actual target DNA and thermostable DNA polymerase. This reduces contaminating DNA that could be amplified by a factor of 10^5 to 10^6.[48] Contaminating DNA can be photochemically modified using psoralen or isopsoralen.[45,46] The reaction mixture is incubated prior to addition of the template DNA with a psoralen (e.g., 8-methoxypsoralen) to a final concentration of 460 µM in the dark for 30 min to 12 h; then the reaction mixture is exposed to UV light (365-nm wavelength) for 1 h.

Contaminating DNA also can be eliminated by treating reaction mixtures with 0.5 to 1.0 U of DNAse I or 10 to 20 U of a restriction enzyme that recognizes short (e.g., 4-bp) sequences. The template DNA and the thermostable DNA polymerase are added after appropriate inactivation of the DNAse I or restriction endonuclease. Another approach can be used to make contaminating PCR products susceptible to degradation. This method involves substituting dUTP for dTTP in the PCR reaction mix (either as free nucleotides or within the primers) and treating subsequent PCR amplifications with uracil glycosylase.[47] One to three units of uracil glycosylase are added to the reaction, and the sample is incubated at 96°C for 10 min to inactivate the uracil glycosylase, to cleave the dU-containing contaminating PCR products and to denature the template DNA for PCR amplification. When using this method it is important that the primer annealing temperature be above 55°C because uracil glycosylase is active below 55°C; this avoids the risk of degradation of the newly synthesized dU-containing PCR products by residual uracil glycosylase. At the end of the PCR amplification the samples should be held at 70 to 72°C, rather than 4°C, until they are removed from the thermocycler; then they should be transferred immediately to –20°C or an equal volume of chloroform should be added to prevent degradation of dU-containing PCR products by residual uracil glycosylase.

VII. ENHANCING THE SPECIFICITY OF THE PCR REACTION

In a PCR reaction mix, at least one critical reactant, for example, thermostable DNA polymerase, can be omitted until after the reaction is heated to 94 to 95°C; this approach is called "hot start".[30] This hot-start method is based on the fact that non-specific priming and subsequent production of unwanted amplified DNA bands is generally produced due to the retention of considerable enzymatic activity at temperatures below the optimum for DNA synthesis. Therefore, in the initial heating step of the PCR reaction, primers that anneal non-specifically to a partially single-stranded template region can be extended and stabilized before the reaction reaches 72°C for extension of specifically annealed primers. If the DNA polymerase is activated only after the reaction has reached high (>70°C) temperatures, non-target amplification can be minimized. This can be done by manual addition of an essential reagent, e.g., thermostable DNA polymerase, to the reaction tube at elevated temperatures. This approach improves specificity and minimizes the formation of primer dimers. Recently, use of acetamide 5% (w/v) in the PCR reaction has been shown to increase the specificity of the primers to their targets. As a result, amplification of the specific target DNA can be achieved without much interference with the non-target DNA in the sample preventing generation of spurious amplified DNAs.[49]

VIII. PCR IN THE DIAGNOSIS OF HUMAN INFECTIOUS DISEASES

A. BACTERIA AND PROTOZOA

The application of PCR methodology has been most pronounced in the detection of bacterial and protozoan pathogens. A large number of bacterial pathogens causing human diseases have been detected in various biological matrices that have high sensitivity and specificity, and are much faster than the conventional microbiological culture assays. A multiple listing of a wide variety of important bacterial and protozoans pathogens detected by PCR, such as *Bordetella pertussis, Borrelia burgdorferi, Aeromonas hydrophila*, various pathogenic species of *Chlamydia, Clostridium difficile*, various pathogenic *Escherichia coli, Helicobacter pylori, Legionella pneumophila, Legionella* spp., *Mycobacterium leprae, M. tuberculosis, Mycoplasma genitalium, Rickettasia typhi, Treponema pallidum, Shigella* spp., *Vibrio cholerae, Salmonella typhimurium, Salmonella* spp., *Corynebacterium jeikeium, Giardia duodenalis, Cryptosporidium* spp., *Entamoeba histolytica*, etc., can be found in this book or most of the clinical microbiology and/or medical journals. These microbial pathogens have been reported for their highly specific and sensitive PCR detection of 1 to 10 cells in the sample.

B. VIRUSES

A number of important human viruses have been detected by the PCR method with high sensitivity, specificity, and with more speed than the conventional approaches. The ability of the PCR method to detect some of the viruses that are difficult to culture such as human immunodeficiency viruses (HIV), has made this methodology popular and useful in clinical diagnostics. The fact that HIV-1 can be detected by PCR six months before seroconversion has been confirmed by Westen blot analysis.[50,51] PCR can also look for the presence of HIV in at risk cases, for example, the children of seropositive mothers or those who have intimate contact with a seropostive person.[51] A number of viruses other than HIV such as Leukemia viruses, cytomegalovirus (CMV), hepatitis B and C viruses, herpes simplex virus (HSV), Epstein Bar Virus (EBV), enteroviruses, coronaviruses, dengue viruses, human papillomaviruses, rotavirus, rubella and various other viruses, have also been detected by the PCR amplification method. The multiple citations of detection of these viruses can be obtained from this book or most of the clinical microbiology, virology, or medical journals. Most of these viruses have been detected by the PCR method with the sensitivity of 10 to 100 plaque-forming units (PFU) and with a high specificity, using rapid non-radioactive detection assays.

IX. PCR IN THE DIAGNOSIS OF HUMAN GENETIC DISEASES

A. PRENATAL DIAGNOSIS

For diagnosis of genetic diseases in the fetus, DNA techniques are being used routinely at 8 to 10 weeks into pregnancy.[52] In general, two strategies are used for DNA-based prenatal diagnosis. The first one is based on mutations that alter the restriction endonuclease cleavage sites on the double-stranded DNA, thereby altering the banding patterns in an agarose gel, which is detected by Southern blot DNA-DNA hybridization method with a specific probe.[53,54] The second approach is a more indirect analysis of following the inherited patterns of restriction fragment length polymorphisms through pedigrees.[55] Both approaches are tedious and time consuming. The use of PCR amplification of DNA with primers designed from regions flanking the restriction sites, followed by treatment of the amplified DNA with the appropriate restriction endonuclease, can provide direct analysis of the mutation on the restriction endunuclease sites, simply by a gel electrophoresis analysis.[56] This approach is specific and rapid, taking only a day to complete compared to a week using conventional approaches. This is particularly important in prenatal diagnosis, where applying PCR methodology enables couples to receive test results on the same day, rather than in a week when conventional DNA technology by DNA-DNA hybridization is performed.

B. PREIMPLANTATION DIAGNOSIS

Detection of sex or inherited genetic diseases in the fetus before artificial implantaion of the embryo is an alternative and viable option to the trauma of termination of pregnancy after eight weeks, which is ethically unacceptable to many couples. One to two cells are removed from the pre-embryo at the 8- to 16-cell stage, before implantation is performed. These can be used for PCR amplification of DNA from a specific gene or genes for the detection of genetic diseases or the determination of sex (by analyzing the amplified DNA using a Y-chromosome-specific DNA probe).[57,58] Preimplantaion diagnosis now appears to be a viable option for couples known to be at risk for inherited genetic diseases.

C. DETECTION OF DISEASES FROM RARE GENETIC MESSAGES

The sensitivity of PCR methodology has been utilized to detect sequences that are expressed at very low levels in tissues, which can be obtained more easily from patients, than in samples containing an abundance of the mRNA in question (from tissues that are difficult to obtain). For example, dystrophin mRNA can be amplified from lymphoblastoid cell lines for further analysis, even though the sequences cannot be detected there by conventional DNA-DNA filter hybridization methods. Amplification of such rare sequences therefore has potential to eliminate the painful muscle biopsies used in Duchenne muscular dystrophy patients and the invasion of other deep tissues in other diseases.[59]

D. CANCER THERAPY

PCR has been applied in cancer therapy for the detection of residual tumor cells and the detection and characterization of oncogenic mutations. In patients with follicular lymphomas, residual cells with the translocation t(14;18)(q32;21) have been detected with PCR, when conventional cytogenetic techniques indicated eradication of the residual cells.[60] Oncogenic mutations occurring in the *c-Ki-ras* gene, before the development of malignancy, have been detected and characterized by PCR.[61,62] This early detection of oncogenic mutations may lead to the prevention of cancer, by making it possible to treat with specific anticancer cell antibodies.[63]

X. DISCUSSION

Although PCR amplification of DNA is automated, and the selection of targets and their primers are routine for the detection of various microbial pathogens, including viruses and a number of human genetic diseases, the success of such detection depends largely on sample processing and analyses of the amplified DNAs. Therefore, an easy but effective sample-processing method for the recovery of the target nucleic acids is an essential component of this method. An inefficient sample processing method will greatly affect the sensitivity or even the success of the amplification of the target DNA. Another important component associated with the PCR-based detection of microbial pathogens from clinical samples is a rapid and sensitive non-radioactive detection of the amplified DNAs. Although DNA-DNA hybridization followed by colorimetric detection methods has the reputation of producing high background signals with lower sensitivity than the radioactive detection, the immobilized capture probe DNA-DNA hybridization coupled with HRP-SA-TMBlue color detection method (which has specificity, sensitivity, and no background signal) is ideal for the detection of amplified DNA at this time. However, a faster and possibly automated method for analysis of the amplified DNA needs to be developed. In spite of its tremendous sensitivity, contamination of the PCR samples for diagnosis of human diseases can be a serious problem. Therefore, recommended precautions should be taken to avoid any such false-positive results, which can be costly and may produce confusion in interpreting the actual result. Appropriate control samples in each PCR set and analyses of the amplified DNA by hybridization method are essential to avoid such problems. A PCR-based, sensitive detection of multiple microbial pathogens can be an ideal method for monitoring and for diagnosis of the specific cause of an infection in the gastrointestinal tract, in a very short period of time. The power of PCR-based diagnosis of human diseases has shown great potential to be used as a routine method in the clinical laboratories across the world.

ACKNOWLEDGMENTS

Numerous research articles have been published in a variety of international peer-reviewed journals on the applications of PCR methodology in clinical microbiology, human genetic diseases, and diagnostics. The space constraints of this chapter prevent a comprehensive survey of various microorganisms, including viruses, and genetic diseases with references. However, we acknowledge all the studies in this area and apologize for our inability to specifically reference them.

REFERENCES

1. **Saiki, R. K., Gelfand, D. H, Stoffel, S., Scharf, S. J, Higuchi, R., Horn, G. T., Mullis, K. B., and Erlich, H. A.,** Primer-directed enzymatic amplification of DNA with a thermostable DNA polymerase, *Science,* 239, 487, 1988.
2. **Bej, A. K., Steffan, R. J., DiCesare, J. L., Haff, L., and Atlas, R. M.,** Detection of coliform bacteria in water by polymerase chain reaction and gene probes, *Appl. Environ. Micrbiol.,* 56, 307,1990.
3. **Bej, A. K., McCarty, S. C., and Atlas, R. M.,** Detection of coliform bacteria and *Escherichia coli* by multiplex polymerase chain reaction: comparison with defined substrate and plating methods for water quality monitoring, *Appl. Environ. Microbiol.,* 57, 2429, 1991.
4. **Bej, A. K., DiCecare, J. L., Haff, L., and Atlas, R. M.,** Detection of *Escherichia coli* and *Shigella* spp. in water by using the polymerase chain reaction and gene probes for *uid. Appl. Environ. Microbiol.,* 57, 1013, 1991.
5. **Atlas, R. M. and Bej, A. K.,** Polymerase Chain Reaction, in *Methods for General and Molecular Bacteriology,* Gerhardt, P., Ed., American Society for Microbiology, ASM Press, Washington, DC, 1993, 418.
6. **Bej, A. K. and Mahbubani, M. H.,** Detection of microbial pathogens in the gastrointestinal tract by PCR and gene probe methods, in *PCR Based Clinical Diagnostics. I. Laboratory Setup and Microbial Detection,* Greenburg, S. and Ehrlich, G., Eds., Blackwell Publishing, New York, 1993, 561–591.
7. **Clayton, C. L., Kleanthous, H., Coates, P. J., Morgan, D. D., and Tabaqchali, S.,** Sensitive detection of *Helicobacter pylori* by using polymerase chain reaction, *J. Clin. Microbiol.,* 30, 192, 1992.
8. **Hammar, M., Tysziewicz, T., Wadstrom, T., and O'Toole, P. W.,** Rapid detection of *Helicobacter pylori* in gastric biopsy material by polymerase chain reaction, *J. Clin. Microbiol.,* 30, 54, 1992.
9. **Valentine, J. L., Arthur, R. R, Mobley, H. L. T., and Dick, J. D.,** Detection of *Helicobacter pylori* by using the polymerase chain reaction, *J. Clin. Microbiol.,* 29, 689, 1991.
10. **Kawasaki, E. S.,** Sample preparation from blood, cells, and other fluids, in *PCR Protocols: A Guide to Methods and Applications,* Innis, M., Gelfand, D. H., Sninsky, J. J., and White, T. J., Eds., Academic Press, San Diego, CA, 1990, 146.
11. **Claas, H. C. J., Wagenvoort, J. H. T., Niesters, H. G.** M., Tio, T. T., van Rijsoort-vos, J. H., and Quint, W. G. V., Diagnostic value of the polymerase chain reaction for *Chlamydia* detection as determined in a followup study, *J. Clin. Microbiol.,* 29, 42, 1991.
12. **Pollard, D. R., Johnson, W. M., Lior, H., Tyler, S. D., and Rozee, K. R.,** Rapid and specific detection of verotoxin genes in *Escherichia coli* by the polymerase chain reaction, *J. Clin. Microbiol.,* 28, 540, 1990.
13. **Shirai, H., Nishibuchi, M., Ramamurthy, T., Bhattacharya, S. K., Pal, S. C., and Takeda, Y.,** Polymerase chain reaction for detection of the cholera enterotoxin operon of *Vibrio cholerae. J. Clin. Microbiol.,* 29, 2517, 1991.
14. **Olive, D. M.,** Detection of enterotoxigenic *Escherichia coli* after polymerase chain reaction amplification with a thermostable DNA polymerase, *J. Clin. Microbiol.,* 27, 261, 1989.
15. **Jones, D. D., Law, R., and Bej, A. K.,** Detection of *Salmonella* spp. in contaminated oysters using polymerase chain reaction and gene probes, *J. Food Sci.,* submitted.
16. **Mahbubani, M. H., Jones, D. D., and Bej, A. K.,** Species specific detection of *Shigella* spp. by polymerase chain reaction, in preparation.
17. **Bej, A. K., Mahbubani, M. H., Boyce, M. J., and Atlas, R. M.,** Detection fo *Salmonella* in shellfish by polymerase chain reaction, *Appl. Environ. Microbiol.,* in press.
18. **Muir, P., Nicholson, F., Jhetam, M., Neogi, S., and Banatvala, J. E.,** Rapid diagnosis of enterovirus infection by magnetic bead extraction and polymerase chain reaction detection of enterovirus RNA in clinical specimens, *J. Clin. Microbiol.,* 31, 31, 1993.

19. **Rotbart, H. A.**, PCR amplification of enteroviruses, in *PCR Protocols: A guide to Methods and Applications*, Innis, M., Gelfand, D. H., Sninsky, J. J., and White, T. J., Eds., Academic Press, San Diego, CA, 1990, 372.
20. **Cook, K. L., Mahbubani, M. H., Bej, A. K., and Gauthier, J. J.**, Rapid methods for preparation of template DNA suitable for polymerase chain reaction amplification, submitted.
21. **Mahbubani, M. H., Bej, A. K., Perlin, M., Schaeffer, F. W., III, Jakubowski, W., and Atlas, R. M.**, Detection of *Giardia* cysts by using the polymerase chain reaction and distinguishing live from dead cysts, *Appl. Environ. Microbiol.*, 57, 3456, 1991.
22. **Mahbubani, M. H., Bej, A. K., Perlin, M. H., Schaeffer, F. W., III, Jakubowski, W., and Atlas, R. M.**, Differentiation of *Giardia duodenalis* from other *Giardia* spp. by using polymerase chain reaction and gene probe methods, *J. Clin. Microbiol.*, 30, 74, 1992.
23. **Mahbubani, M. H., Schaefer, F. W., and Bej, A. K.**, Detection of *Giardia* lamblia in environmental waters by polymerase chain reaction and immunomagnetic separations, in preparation.
24. **de Lamballerie, X., Zandotti, C., Vignoli, C., Bollet, C., and de Micco, P.**, A one-step microbial DNA extraction method using "chelex 100", suitable for gene amplification, *Res. Microbiol.*, 143, 785, 1992.
25. **Bej, A. K., Mahbubani, M. H., and Atlas, R. M.**, Amplification of nucleic acids by polymerase chain reaction (PCR) and other methods and their applications, *Crit. Rev. Biochem. Mol. Biol.*, 26(3/4), 301, 1991.
26. **Bej, A. K. and Mahbubani, M. H.**, Applications of the polymerase chain reaction in environmental microbiology, *PCR Methods Appl.*, 1, 151, 1992.
27. **Atlas, R. M. and Bej, A. K.**, Detecting bacterial pathogens in environmental water samples by using PCR and gene probes, in *PCR Protocols: A Guide to Methods and Applications*, Innis, M., Gelfand, D. H., Sninsky, J. J., and White, T. J., Eds., Academic Press, San Diego, CA, 399, 1990.
28. **Girgis, S. I., Alevizaki, M., Denny, P., Ferrier, G. J. M., and Legon, S.**, Generation of DNA probes for peptides with highly degenerate codons using mixed primer PCR, *Nucleic Acids Res.*, 16, 10371, 1988.
29. **Rychilk, W. and Rhoads, R. E.**, A computer program for choosing optimal oligonucleotides for filter hybridization, sequencing and *in vitro* amplification of DNA, *Nucleic Acids Res.*, 17, 8543, 1977.
30. **Erlich, H. A., Gelfand, D., and Sninsky, J. J.**, Recent advances in the polymerase chain reaction, *Science*, 252, 1643, 1991.
31. **Innis, M., Gelfand, D. H., Sninsky, J. J., and White, T. J.**, Eds., *PCR Protocols: A Guide to Methods and Applications*, Academic Press, San Diego, CA, 1990.
32. **Oste, C.**, Optimization of magnesium concentration in the PCR reaction, *Amplifications Forum PCR Users*, 1, 10, 1989.
33. **Isaacs, S. T., Tessman, J. W., Metchette, K. C., Hearst, J. E., and Cimino, G. D.**, Post-PCR sterilization: development and application to an HIV-1 diagnostic assay, *Nucleic Acids Res.*, 19, 109, 1991.
34. **Smith, K. T., Long, C. M., Bowman, B., and Manos, M. M.**, Using cosolvents to enhance PCR amplification, *Amplifications Forum PCR Users*, 5, 16, 1990.
35. **Winship, P. R.**, An improved method for directly sequencing PCR amplified material using dimethyl sulfoxide, *Nucleic Acids Res.*, 17, 1266, 1989.
36. **Don, R. H., Cox, P. T., Wainwright, B. J., Baker, K., and Mattick, J. S.**, "Touch-down" PCR to circumvent spurious priming during gene amplification, *Nucleic Acids Res.*, 19, 4008, 1991.
37. **Sparkman, D. R.**, Paraffin wax as a vapor barrier for the PCR, *PCR Methods Appl.*, 2, 180, 1992.
38. **Saiki, R. K., Walsh, P. S., Leverson, C. H., and Erlich, H. A.**, Genetis analysis of amplified DNA with immobilized sequence-specific oligonucleotide probes, *Proc. Natl. Acad. Sci. U.S.A.*, 86, 6230, 1989.
39. **Keller, G. H., Huang, D. P., Shih, W. K., and Manak, M.**, Detection of hepatitis B virus DNA in serum by polymerase chain reaction amplification and microtiter sandwich hybridization, *J. Clin. Microbiol.*, 28, 1411, 1991.
40. **Running, J. A. and Urdea, M. S.**, A procedure for productive coupling of synthetic oligonucleotides to polystyrene microtiter wells for hybridization capture, *BioTechniques*, 8, 276, 1990.
41. **Jackson, M. P.**, Detection of Shiga toxin-producing *Shigella dysenteriae* Type 1 and *Escherichia coli* by using polymerase chain reaction with incorporation of digoxigenin-11-dUTP. *J. Clin. Microbiol.*, 29, 1910, 1991.
42. **Kwok, S. and Higuchi, R.**, Avoiding false positives with PCR, *Nature*, 339, 237, 1989.
43. **Orrego, C.**, Organizing a laboratory for PCR work, in, *PCR Protocols: A guide to Methods and Applications*, Innis, M. A., Gelfand, D. H., Sninsky, J. J., and White, T. J., Eds., Academic Press, San Diego, CA, 1990, 447.
44. **Furrer, B., Candrian, U., Wieland, P., and Luthy, J.**, Improving PCR efficiency, *Nature*, 3246, 324, 1990.
45. **Dilworth, D. D. and McCarrey, J. R.**, Single-step elimination of contaminating DNA prior to reverse-transcriptase PCR, *PCR Methods Appl.*, 1, 279, 1992.
46. **Jinno, Y., Yoshiiura, K., and Niikawa, N.**, Use of psoralen as extinguisher of contaminated DNA in PCR, *Nucleic Acids Res.*, 18, 6739, 1990.
47. **Longo, M. C., Berninger, M. S., and Hartley, J. L.**, Use of uracil DNA glycosylase to control carry-over contamination in polymerase chain reaction, *Gene*, 93, 125, 1990.

48. **Sarkar, G. and Sommer, S. S.,** More light on PCR contamination, *Nature,* 347, 340, 1990.
49. **Reysenbach, A. L., Giver, L. J., Wickham, G. S., and Pace, N. R.,** Differential amplification of rRNA genes by polymerase chain reaction, *Appl. Environ. Microbiol.,* 58, 3417, 1992.
50. **Schochetman, G., Ou, C. Y., and Jones, W. K.,** Polymerase chain reaction, *J. Infect. Dis.,* 158, 1154, 1988.
51. **Clewley, J. P.,** The polymerase chain reaction, a review of the practical limitations for human immunodefficiency virus diagnosis, *J. Virol. Methods,* 25, 179, 1989.
52. **Old, J. M., Weatherall, D. J., and Ward, R. H. T,** First trimester diagnosis of the haemoglobin disorders, *Ann. N.Y. Acad. Sci.,* 445, 349, 1985.
53. **Thein, S. L., Lynch, J. R., Old, J. M., and Weatherall, D. J.,** Direct detection of Hb E with *Mnl*I, *J. Med. Genet.,* 24, 110, 1987.
54. **Thein, S. L., Lynch, J. R., Weatherall, D. J., and Wallace, R. B.,** Direct detection of Hb E with a synthetic oligonucleotide-application to prenatal diagnosis, *Lancet,* i, 93, 1986.
55. **Brock, D. J. H.,** Prenatal diagnosis of cystic fibrosis, *Arch. Dis. Child.,* 63, 701, 1988.
56. **Kogan, S. C., Doherty, M., and Gitschier, J.,** An improved method for prenatal diagnosis of genetic diseases by analysis of amplified DNA sequences, *New Engl. J. Med.,* 317, 985, 1987.
57. **Handyside, A. H., Penketh, R. J. A., and Winston, R. M. L.,** Biopsy of human preimplantation embryos and sexing by DNA amplification, *Lancet,* i, 347, 1989.
58. **Lynch, J. R. and Brown, J. M.,** The polymerase chain reaction: current and future applications, *J. Med. Genet.,* 27, 2, 1990.
59. **Gibbs, R. A.,** DNA amplification by the polymerase chain reaction, *Anal. Chem.,* 62, 1202, 1990.
60. **Lee, M. S., Chang, K. S., Cabanillas, F., Freireich, E. J., Trujillo, J. M., and Stass, S. A.,** Detection of minimal residual cells carrying the t(14;18) by DNA sequence amplification, *Science,* 237, 175, 1987.
61. **McMahon, G., Davis, E., and Wogan, G. N.,** Characterization of c-Ki-ras oncogene alleles by direct sequencing of enzymatically amplified DNA from carcinogen-induced tumors, *Proc. Natl. Acad. Sci. U.S.A.,* 84, 4974, 1987.
62. **Bos, J. L., Fearon, E. R., and Hamilton, S. R.,** Prevalence of ras gene mutations in human colorecto cancers, *Nature,* 327, 293, 1987.
63. **Reichmann, L., Clark, M., Waldmann, H., and Winter, G.,** Reshaping human antibody for therapy, *Nature,* 332, 323, 1988.
64. **Bej, A.,** unpublished.

Chapter 32

APPLICATIONS OF THE POLYMERASE CHAIN REACTION (PCR) *IN VITRO* DNA-AMPLIFICATION METHOD IN ENVIRONMENTAL MICROBIOLOGY

Asim K. Bej and Meena H. Mahbubani

TABLE OF CONTENTS

I. Introduction 327

II. Purification of Nucleic Acids from the Environmental Samples for PCR Amplification 328
    A. From Soil and Sediments 328
    B. From Aquatic Environment 329

III. Detection of Genetically Engineered Microorganisms in the Environment 330

IV. Detection of Indigenous Microorganisms in the Environment 331

V. Detection of Indicator Microorganisms in Water 332

VI. Detection of Waterborne Microbial Pathogens 333

VII. Multiplex PCR Amplification for Environmental Monitoring of Microorganisms 333

VIII. Detection of Viable but Nonculturable Microorganisms in the Environment 334

IX. Detection of Viruses in Water 336

X. Discussion 336

References 337

I. INTRODUCTION

Since its development in 1983, the polymerase chain reaction (PCR) has gained widespread acceptance in many areas of molecular biology.[1-3] Prior to the introduction of automated PCR, nucleic acid analysis was limited to Southern blotting for DNA and Northern blotting for RNA. PCR allows the *in vitro* amplification of a specific RNA or DNA nucleotide sequence, in relatively large abundance, so that subsequent analyses are not confounded by the presence of heterologous mixtures of DNA, or a lack of specific DNA segments. Besides its applications in the areas of molecular biology, diagnostics, systematics, and forensic science, the PCR method has been used to explore various aspects of environmental microbiology. Application of PCR methodology in environmental microbiology requires extensive sample preparation to recover contaminant-free target nucleic acids. The high specificity, sensitivity, and consistency of the PCR detection of target microorganisms in the environmental samples have provided us with a useful tool in molecular microbial ecology. This chapter will discuss the methods and importance of sample preparation and the use of the PCR method for detection of various microorganisms in the environmental samples.

II. PURIFICATION OF NUCLEIC ACIDS FROM THE ENVIRONMENTAL SAMPLES FOR PCR AMPLIFICATION

A. FROM SOIL AND SEDIMENTS

Extraction and extensive purification of nucleic acids from the environmental microorganisms is necessary for successful PCR amplification. Among the general environmental contaminants, the presence of humic materials, clay, and organics potentially inhibits the PCR reaction. It has been found that the addition of as little as 0.001 µg of montomorillite humic material in the PCR reaction inhibits the amplification process.[63] Much effort has been devoted to developing convenient methods to remove humic materials from the environmental samples to achieve successful PCR amplification of the target nucleic acids.

For extraction of DNA from microorganisms present in the environmental soil or sediment, either direct lysis or isolation of bacterial cells followed by lysis has been described.[4] The purification of the released DNA can be performed by applying a combination of the various standard purification methods. Among some of the combinations are phenol-chloroform extraction followed by ammonium acetate-ethanol precipitation, repeated polyvinylpolypyrrolidone (PVPP) treatment or dialysis, hydroxyapatite or affinity chromatography, and multiple CsCl-EtBr density gradient centrifugation, all of which produce positive PCR amplification from the environmental samples. Milligram quantities of DNA were recovered by the direct lysis method, which was a yield >1-order-of-magnitude higher than the cell extraction method. Also, the purified DNA recovered by the direct lysis method was of high quality and could be used for PCR DNA amplification.[4,5] Although significantly higher amounts of nucleic acids can be recovered by following the direct lysis method, the presence of eukaryotic DNA in the sample is a possibility. A modification of the direct lysis method was described by using lysozyme followed by freeze-thaw disruption of the cells. The released DNAs were then purified by standard phenol-chloroform extraction and chromatography. Using this approach, it was possible to detect <10 cells of *Escherichia coli* per 1 g of seeded or unseeded soil sample using 16S rRNA as a target.[6-8]

In another study, bacterial cells were differentially separated from soil colloids on the basis of their buoyant densities.[9] In this method, a modified sucrose gradient centrifugation protocol has been developed to separate most of the soil colloids from the bacterial cells in the sample for PCR amplification of the target DNA. However, since this approach retained minute quantities of inhibitory colloidal soil particles in the bacterial cell suspension, it was necessary to run an additional 25 c of PCR DNA amplification, with an aliquot of the amplification product of the first 25 c ("double PCR"), to detect 1 to 10 target microorganisms per gram of soil sample.[9]

In a separate study, DNA was extracted and purified from soil by the following sequential steps: cold lysozyme-and SDS-assisted lysis with either freezing-thawing or bead beating, cold phenol extraction of the resulting soil suspension, CsCl and potassium acetate precipitation, and finally, spermine-HCl or glassmilk purification.[10] The resulting DNA was pure enough for PCR amplification of up to 20 ng of soil-derived DNA from *Pseudomonas fluorescens* (RP4::*pat*) per 50 µl of reaction mix. Recently, an effective method for removal of inhibitory humic substances in crude DNA extract from soil samples was described by using Sephadex G-200™ spun column* saturated with Tris-EDTA, pH 8.0, buffer.[6,8] This

---

* Registered trademark of Pharmacia LKB Biotechnology, Inc., 800 Centennial Ave., Piscataway, NJ 08854.

extraction method resulted in the detection of <70 cells of *Escherichia coli* by PCR DNA-amplification method from a 1/100 fraction of the purified sample using 16S rRNA as a target. Also, a simple and effective method of purification of PCR-amplifiable DNA from soil matrices has been described by separating crude DNA extracts by direct lysis method in a low-melting-point agarose gel (1.2% w/v) mixed with polyvinylpyrrolidone (PVP) (2% w/v).[11] The water-soluble PVP forms hydrogen bonds with phenolic compounds in the DNA extracts and prevents co-migration during agarose gel electrophoresis. Differentially migrated DNA bands were purified from the agarose gel, and PCR amplification yielded ~0.5 µg of amplified DNA from seeded *E. coli* cells using 16S rRNA as a target.

In another approach, total DNA was purified from low quantities (100 mg) of soil samples seeded with bacterial species such as *Agrobacterium tumifaciens* and *Frankia* spp. In this extraction procedure, 100 mg of the soil sample was resuspended in 0.5 ml of TENP (50 mM Tris-Cl, pH 8.0, 100 mM NaCl, 1% [w/v] PVPP) buffer and sonicated with a titanium microtip for 5 min at 15 W power. The sample was centrifuged, and the pellet was heated at 900 W five times and suspended in 0.1 ml of TENP buffer. The sample was then "freeze-boiled" (liquid nitrogen and boiling water) three times for lysis and washed four succesive times with TENP buffer. The final purification of the supernatant was performed in an Elutip d™ column* for PCR amplification detection of the target microorganisms with a sensitivity of 10^7 to 10^3 cells of *A. tumifaciens* and 0.2×10^5 cells of indigenous *Frankia* spp. per gram of soil sample.[12]

Another more tedious and long procedure for extraction of PCR-amplifiable DNA from soil, sediment and sand samples has been described for the detection of native bacterial populations using 16S rRNA genes and the mercury-resistance gene (*mer*).[13,14] This procedure required lysis of the cells by treatment with sodium dodecyl sulfate (SDS) followed by concentration of the nucleic acids with polyethylene glycol and NaCl, and precipitation overnight with ethanol. After CsCl density gradient centrifugation purification, the DNA was further purified by phenol:chloroform (v/v) treatment and precipitated overnight with sodium acetate and ethanol. The DNA was recoverd by centrifugation and used for PCR DNA amplification. The entire procedure of extraction and purification of DNA takes three days to complete.

For the isolation and purification of total RNA from microorganisms in the environmental samples, 10 g of soil sample seeded with *Pseudomonas aeruginosa* PU21 was treated with guanidium thiocyanate (4 M) mixed with sodium citrate (25 mM), sarcosyl (5% w/v), and 2-mercaptoethanol (0.1 M) to achieve lysis of cells, fixation of total cellular RNA, and hydrolysis of DNA. The treated sample was then purified with phenol-chloroform-isoamyl alcohol (24:24:1 v/v) followed by precipitation of the RNA with isopropanol.[15] This approach yielded 17 µg of total RNA and 0.16 µg of mRNA from 1 g of soil containing 8×10^8 *P. aeruginosa* PU21 cells. This extraction method can be completed within a few hours and has great potential for the study of gene expression in various microorganisms in the environment, when coupled with reverse transcription of the target mRNA and PCR DNA amplification with specific primers sets.

B. FROM AQUATIC ENVIRONMENT

The procedure for isolation and purification of nucleic acids for PCR amplification from microorganisms present in the aquatic environment is found to be less involved than from the more complex soil and sediment matrices. Sommersville et al.,[16] described a simple method for isolating nucleic acids from aquatic samples, which can be used for the PCR DNA amplification method. In this approach, 300 to >1000 ml of water sample were concentrated

---

* Registered trademark of Schleicher & Schuell, Kenne, NH 03431.
** Registered trademark of Millipore Corporation, 80 Ashby Rd., Bedford, MA 01730.

on a single cylindrical filter membrane (Sterivex-GS™** Filter unit type SVGS01015) to harvest cells, following which alkaline lysis and proteolysis were performed within the filter housing. Crude high-molecular-weight nucleic acids were extruded from the filter unit, and further purification was carried out by conventional methods such as CsCl density gradient centrifugation or treatment with ammonium acetate (NH_4OAc) followed by ethanol precipitation. Using this approach, it was possible to extract sufficiently pure chromosomal DNA, plasmid DNA, or various species of RNAs (5S, 16S, and 23S rRNAs). This rapid and convenient approach, filtration to concentrate relatively large volumes of environmental waters for recovery of dissolved and particulate DNA, can be used for PCR-based analysis of community diversity and microbial activity in the aquatic ecosystems.

In various other studies, CsCl-EtBr density gradient centrifugation[17] for all environmental DNAs, multiple PVPP treatment for rRNA studies,[18] and repeated phenol-chloroform extractions for total planktonic DNA[19,20] or cyanobacterial DNA[21] were found to be essential to yield sufficiently pure DNA for various molecular biological analyses and perhaps PCR analysis. In another approach to the use of PCR methodology for the detection of microbial cells (both indicator and pathogens) in drinking water or relatively clear waters, cells are concentrated on a membrane filter, lysed by repeated freeze-thaw method followed by PCR DNA amplification without removal of the filter from the reaction tube.[22] Alternatively, filtered cells can be lysed directly on the filter by lysozyme, and the released DNA can be purified by phenol-choroform extraction followed by ethanol precipitation to yield adequately purified DNA for PCR amplification and analysis.[23]

Although these methods can remove the majority of the environmental contaminants and have been shown to be useful for various molecular biological studies, a standard protocol for removing all possible inhibitors, which can be universally applied for all types of environmental samples and can be utilized for PCR amplification, is yet to be described.

III. DETECTION OF GENETICALLY ENGINEERED MICROORGANISMS (GEMS) IN THE ENVIRONMENT

The potential for the use of genetically engineered microorganisms (GEMs) in industry, agriculture, and medicine has been limited so far because of a lack of sensitive methods for the monitoring and detection of GEMs after their release in the environment. Conventional cultural methods and other genetic-based approaches such as colony/Southern hybridization which depend on the ability to recover and culture the organism from an environmental sample, lack sensitivity due to the limited efficiency of recovering bacteria from natural environments. Extraction of microbial DNA from the environmental sample, either directly or after recovery of microbial cells, followed by gene probe hybridization, though more sensitive than cultural methods, still lacks the level of sensitivity required to determine the ultimate fate of GEMs because of the limited relative numbers of target genetic sequences that may be present in the sample.[5,24] Several-million-fold amplification of the engineered genes by PCR methodology, from the released microorganisms can potentially increase the sensitivity of detection of released GEMs in the environment.

PCR methodology was first applied to monitor genetically engineered microorganisms by amplification of a portion of the 1.3-kb repeat sequence from *Pseudomonas cepacia* AC1100, a herbicide (2,4,5-T) degrading bacterium, following release in the soil microcosms, to a sensitivity of 100 GEMs in 100 g of sediment against a background of 10^{11} diverse nontarget microorganisms. This sensitivity was at least 10^3-fold higher than non-amplified conventional dot-blot hybridization detection. A 0.3-kb unique DNA sequence from *Pennisetum purpureum* (napier grass) has been cloned into pRC10, a derivative of 2,4-dichlorophenoxyacetic degrading plasmid, and tranferred into *E. coli*.[25] This genetically altered microbe was released into filter-sterilized lake and sewage-water samples to a concentration of 10^4 cells per milliliter and detected by PCR at a sensitivity several times higher than the conventional plating technique, even after 10 to 14 days of incubation, using the unique cloned DNA sequence as a target. From these studies it can be concluded that the PCR method can be used for monitoring released GEMs in an environment consisting of a complex habitat of diverse microorganisms, where it may be tedious and time consuming to discriminate the GEMs from the indigenous microorganisms.

In another study, a single copy of the transposon *Tn5* was transferred into the genomic DNA of *Rhizobium leguminosarum* released in the soil. These genetically altered microorganisms were detected by double PCR amplification using transposon *Tn5* as target to a sensitivity of 1 to 10 colony-forming units (CFU) per gram of soil.[9] Although it is adequate to use *Tn5*, which contains an antibiotic resistance gene as a model target for PCR detection, it may not be an appropriate marker for releasing GEMs in the environment because of its ability to be transferred into indigenous microorganisms, making them antibiotic resistant.

IV. DETECTION OF INDIGENOUS MICROORGANISMS IN THE ENVIRONMENT

A number of microorganisms present in the environment are involved in the biodegradation of various pollutants and toxic wastes. It is possible to monitor such degrading microorganisms in the polluted and toxic waste sites by application of the PCR method using conserved regions of the genes that are involved in such function. In a study, using the nucleotide sequence information of a chlorocatechol dioxygenase degrading gene (*tfdC*) from *Alcaligenes eutrophus* JMP134 (pJP4), it was possible to design oligonucleotide primers for the detection of various chloroaromatic-degrading bacteria by PCR amplification.[26] In a very short period, PCR amplification using such oligonucleotide primers gives information on the variations, similarities, and functional aspects of various pollutant-degrading genes present in closely or distantly related microorganisms in the environment. It allows detection of the specific microorganism carrying the degrading gene, from a complex mixed population in the environment. When such a polluted site is identified, it is important to investigate the possibility of the presence of various degrading microbes at that site.

In addition, by using the mRNA-reverse transcriptase-cDNA-PCR approach, it is possible to determine the degrading activities of these microorganisms in the contaminated sites. In another study, specific detection of the same species of the herbicide (2,4-dichlorophenoxyacetic acid)-degrading bacterium, *A. eutrophus*, was achieved by PCR amplification of a region of *tfd*B gene from pJP4 and its derivative plasmid pRO103.[27] In this study, by direct PCR-amplified DNA analysis, it was possible to detect approximately 3000 CFU or 15.6 pg of plasmid DNA. The sensitivity of such detection was onefold higher when DNA-DNA hybridization was performed with an oligonucleotide probe internal to the amplified DNA. Native bacterial populations were detected from various samples of soil, sediment, and sand by PCR amplification of a conserved region of the 16S rDNA segment and the mercury resistance (*mer*) gene.[13] Since the rDNA target is present in the chromosome and the *mer* gene is plasmid-borne, amplification of both targets simultaneously has the potential to serve as a model system to study the microbial interactions and gene transfer in the natural environment.

The ribulose biphosphate (*rbcL*) gene was used as a target to amplify planktonic DNA and analyze the microbial community in the aquatic environment.[17] Also, using the same target, dissolved DNA associated with the phytoplankton in the aquatic environment was determined by PCR amplfication of the extracellular DNA fraction.[17] By use of this conserved gene as a target for PCR amplification, it is possible to analyze important ecological functions in the aquatic environments.

Using such oligonucleotide primers and PCR amplification, it is possible to detect the specific microorganisms carrying the degrading gene from a complex mixed microbial population in the environment. Formation of biofilms on various surfaces by microorganisms in the environment can be beneficial or detrimental. For example, microbial aggregation or attachment is required for various water treatments, while on the other hand, extensive corrosion and biodeterioration can be caused due to the formation of such microbial biofilms. Characterization and ecology of microbial populations in biofilms have been hindered due to the available determinative techniques, which require culture of microorganisms in selective media. These methods eliminate many of the important microbes from the biofilms since they survive only in a mixed culture and live on the cometabolism. Use of PCR amplification of specific targets makes it possible to identify a group of microbes in such a biofilm, which may have been missed by the conventional technique. To determine the feasibility of such use, a sulfidogenic biofilm has been established in an anaerobic fixed-bed bioreactor, and PCR amplification was performed for detection of the population architecture, of all the Gram-negative sulfate-reducing bacteria, using a region of the 16S RNA conserved in the resident sulfate reducing bacteria.[28]

V. DETECTION OF INDICATOR MICROORGANISMS IN WATER

Coliform bacteria are monitored in water supplies to test their bacteriological safety. The presence of coliform bacteria in water indicates potential human fecal contamination and the possibility of the presence of enteric pathogens. The conventional method for the detection of coliform bacteria is culturing on media such as McConkey, m-Endo, eosin methylene blue, or brilliant-green-lactose-bile media.These media are selective for Gram-negative bacteria and differentially detect lactose-utilizing bacteria. At an incubation temperature of 37°C total coliform bacteria are enumerated, while at 44.5°C fecal coliforms, mainly *E. coli*, are enumerated. *Escherichia coli* is primarily associated with human feces and is therefore a useful indicator of human fecal contamination.

There are several problems with the culture method for monitoring *E. coli* in environmental and potable waters. The conventional confirmative tests for the detection of *E. coli*, all of which require culturing of the organism, are time consuming and do not detect viable but nonculturable bacteria, which may occur due to the chlorine injury during the process of water purification and treatment. Also, the cells may die between the time of collection and the test. The Colilert test is a a colorimetric method for the detection of *E. coli*, based on the detection of the β-D-glucuronidase enzyme produced by the *uidA* gene. This test requires the culturing of bacteria; moreover, it fails to detect β-D-glucuronidase-negative *E. coli*.

A PCR-gene-probe-based method has been developed by Bej et al.,[23] for the detection of coliform bacteria. *Escherichia coli* and other coliform bacteria, including *Shigella* spp., were detected by PCR amplification of a segment of the *lacZ* gene, while *E. coli*, *Salmonella* and *Shigella* spp. were detected by amplification of a portion of the *lamB* gene. *Escherichia coli* and *Shigella* spp. were also detected by amplification of four different regions of the *uidA* gene, which codes for the β-glucuronidase enzyme and part of the *uidR* gene, which is the regulatory region of the *uidA* gene.[29] This method can detect the *uidA*-negative *E. coli*, which do not show a positive signal with the conventional tests because of the lack of the β-glucuronidase enzyme. Besides, it is not as time consuming, and it has higher specificity and sensitivity. The sensitivity of the method is 1 to 10 fg of genomic DNA and one to five viable *E. coli* cells.

Amplification of the *uid* gene of *E. coli* for the detection of *E. coli* and *Shigella* spp. has also been reported by other workers.[30] A field evaluation of the PCR amplification detection of enteric pathogens and indicator microorganisms has been reported using *uidA* and *lacZ* as targets.[31] Current data suggest that this PCR-gene-probe-based method has the required specificity and sensitivity for monitoring coliforms as indicator organisms in water. Although the target for specific detection of *E. coli* has not yet been reported, target genes for the specific detection of *Salmonella* spp.[32,34] and for species-specific detection of *Shigella*[35] have been identified. PCR-gene-probe-based specific detection of *Salmonella* spp. was developed using the flagellin gene as a target, with a sensitivity of 1 CFU, from environmental waters, after 50 c of amplification.[34]

VI. DETECTION OF WATERBORNE MICROBIAL PATHOGENS

It is important to detect various waterborne microbial pathogens with high sensitivity and specificity. *Legionella* spp. causes Legionnaires' disease in humans via aerosol. *Legionella pneumophila* has been detected by PCR amplification of a fragment of DNA of unknown function.[36] The sensitivity of detection was equivalent to 35 CFU detected by viable plating. Mahbubani et al.[37] have developed a method for the detection of *Legionella* in environmental water sources, based upon PCR and gene probes. In another study, a 0.104-kbp DNA sequence, which codes for a region of 5S rRNA, was amplified to detect all species of *Legionella*, including all 15 serogroups of *L. pneumophila*. Specific detection of *L. pneumophila* (all serogroups) was achieved by amplification of a portion of the coding region of the macrophage-infectivity potentiator (*mip*) gene. *Pseudomonas* spp., which exhibit antigenic cross-reactivity in serological detection methods, did not produce positive signals in the PCR-gene-probe-based method using Southern blot analyses.

In another study, 27 cooling tower water samples were subjected to detection of *Legionella* spp. or *L. pneumophila* by PCR amplification method[38] using 5S rDNA and *mip* gene, respectively. Of the 27 samples, 25 were positive for *Legionella* spp. and 14 were positive for *L. pneumophila*. Detection of *L. pneumophila* and *L. dumoffii* in water by PCR-amplification method has been compared with the conventional culture method, and the sensitivity of detection by PCR method was found to be no greater than by the culture method.[39] In contrast to this study, Nowicki et al.[40] showed that the sensitivity of detection of *L. pneumophila* in environmental water samples was one order of magnitude higher than the conventional culture method (10^2 cells per liter by PCR vs. 10^3 cells per liter by conventional culture methods). The use of PCR DNA amplification of a portion of the *mip* gene coupled with a number of restriction endonuclease analyses differentiated serogroups 1, 5, 6, 8, and 11 of *L. pneumophila* among all the 14 serogroups, which can be useful for epidemiological study during an outbreak.[41]

Giardia lamblia causes defined waterborne diarrhea in the U.S. and in many other parts of the world. Environmental samples are tested for *G. lamblia* by concentrating 100 gal of water followed by microscopic examination using a fluorescent dye. *Giardia lamblia* was differentiated from *G. muris* by PCR amplification of different segments of the giardin gene of *G. lamblia*.[42] Single-cyst detection has also been achieved by PCR amplification.[43] The specificity and sensitivity of detection of *Giardia* by PCR shows great promise for the monitoring of this pathogen in water rapidly and reliably. Also, *G. lamblia* cysts were isolated from concentrated river water samples by using immunomagnetic beads and PCR amplification detection was performed with a sensitivity of $\leq 10^1$ cysts per liter of water.[44]

VII. MULTIPLEX PCR AMPLIFICATION FOR ENVIRONMENTAL MONITORING OF MICROORGANISMS

Environmental samples and drinking water samples may contain more than one type of microbial pathogen in addition to the indicator microorganism. Multiplex PCR is a useful tool for the detection of more than one target in a single PCR reaction, and it can be useful for monitoring multiple microbial pathogens in a single environmental or water sample. This method was first described by Chamberlain et al.[45] to detect human genes.

A modification of this approach of simultaneous PCR amplification of multiple genes from different bacteria has been applied to environmental samples.[46] Both the genus *Legionella* and the species *L. pneumophila* were detected in one sample by multiplex amplification of the *mip* gene, which is specific for *L. pneumophila*, and the 5S rRNA gene, specific for the genus *Legionella*.[46] The two sets of primers were added sequentially at different concentrations. Multiplex PCR amplification was applied to *lacZ* and *uidA* targets in a field study of water-quality monitoring.[31] In this study, it was posssible to detect, in one sample, total coliform bacteria by amplification of the *lacZ* gene, the indicator microorganism *E. coli*, and a pathogen *Shigella* spp. by amplification of the *uidA* gene.[31] Also, in this study the *lacZ* PCR detection gave results statistically equivalent to those of the conventional plate count and defined substrate methods accepted by the U.S. Environmental Protection Agency for water-quality monitoring. The *uidA* PCR method was more sensitive than the 4-methylumbelliferyl-β-D-glucuronide-based defined substrate test for the specific detection of *E. coli*. In another study multiplex amplification of five different targets in a single PCR reaction has been achieved for the detection of non-pneumophila *Legionella* spp., *L. pneumophila*, total coliforms, *E. coli* and *Shigella* spp., and total eubacterial species.[47]

When several genetically engineered microorganisms are released together for the degradation of complex hazardous wastes and pollutants, they may be monitored together, possibly both qualitatively and quantitatively, in a single PCR reaction by amplifying a unique segment of the DNA of each of the GEMs, and together by amplifying a common segment of all the GEMs that is not present in other eubacterial species.

VIII. DETECTION OF VIABLE BUT NONCULTURABLE MICROORGANISMS IN THE ENVIRONMENT

Some microorganisms, including human pathogens, exist in the environment in a viable but nonculturable, i.e., dormant stage[48-50] and can become potentially infectious when suitable conditions prevail.[48] The routine microbiological methods do not allow these nonculturable cells to grow (on agar media) and will not distinguish them from the dead cells (by microscopic technique). Live cells are considered to be those capable of cell division, metabolism (respiration), or gene transcription (mRNA production).[49] Microbial cells that are in a viable but nonculturable state in the environment, can be detected by targeting mRNA rather than DNA, first for cDNA synthesis followed by PCR amplification. The potential problem of this approach would be that most of the prokaryotic mRNAs have half-lives of only a few minutes. It has been shown that the mRNA of the *mip* gene of *L. pneumophila* can be stabilized simply by growing the cells for 10 to 15 min in the presence of chloramphenicol before harvesting.[51] The PCR amplification of the *mip* mRNA could be a potential means of detecting of metabolically active *L. pneumophila* cells. Use of chloramphenicol for increasing the stability of bacterial mRNA is yet to be tested in other microorganisms.

Another perplexing issue that may create additional problems in such an approach is the efficiency of gene expression of these dormant microbial pathogens. It is possible that the transcription or regulatory systems of the target genes in these microbial pathogens are inhibited by various environmental factors and inhibitors when they are present in the natural environment. Therefore, in this situation the quantity of the target mRNA level may be so low that it remains undetected even by the most sophisticated methods like PCR. However, targeting DNA for PCR amplification may be sufficient for the detection of culturable and nonculturable microbial pathogens.[52,53] Both viable culturable and viable nonculturable cells of *L. pneumophila*, formed during exposure to hypochlorite, showed positive PCR amplification, whereas nonviable cells did not. A field verification of this approach for the detection of metabolically active (viable vs. dead) *L. pneumophila* from contaminated environmental samples has not yet been done.[52]

Recently, the efficiency of PCR detection of *L. pneumophila* serogroup 1 in sterile creek and drinking water was compared with conventional viable plating (CFU) and direct fluorescent antibody (DFA) method for 535 days at 15°C, in a standing culture.[54] The cells (10^4 cells per milliliter) in the culture entered into viable but nonculturable state after 51 days of incubation, which was detectable only by DFA method but not by the CFU and PCR methods. It was predicted that the chromosomal DNA undergoes damage during longer incubation of the cells in the culture, giving negative PCR results.[54] The size of the PCR-amplified DNAs (0.65 vs. 0.168 kbp) of the *mip* gene in *L. pneumophila*, after treatment with sodium hypochlorite to generate nonviable cells, showed that the larger amplicon may serve as a better indicator to determine the viability of *L. pneumophila* in water than smaller amplicons.[55]

The marine waterborne microbial pathogen *Vibrio vulnificus*, which can cause fatal infections in humans when ingested with contaminated raw oyster, has been found to enter into a viable but nonculturable state during the colder months and resuscitate from the nonculturable state when a suitable environment prevails.[56] PCR amplification of the hemolysin gene has been used to detect 72 pg of DNA from culturable and 31 ng from nonculturable cells.[53] Although the decreased sensitivity of detection of nonculturable cells by the PCR method is not well understood at this time, several possible explanations have been described.[53] Among these possibilities, the important criteria that may be of concern in applying the PCR methodology for the detection of viable but nonculturable microorganisms are (1) less DNA content per cell, (2) difficulty in breaking open because of the change in the cell wall, which may occur due to the carbon or nitrogen starvation or changes in environmental conditions, and (3) modification of the target gene due to genetic rearrangement.

Another study has shown that mRNA-PCR alone is not sufficient to distinguish live from dead *Giardia* cysts, since cysts killed by heat treatment or monochloramination also give positive mRNA PCR amplification.[43] In this organism, using the giardin mRNA as a target for PCR amplification, it is necessary to include an mRNA induction step in the procedure, to determine the viability of the cysts. Since in the viable but nonculturable stage there may be changes in the structure as well as in gene expression in many microorganisms, it may be necessary to develop a modified version of the PCR method for the detection of such microbial pathogens in the environment.

An important issue in environmental microbial molecular genetics is how various genes are regulated and expressed under various environmental conditions. One of the known facts is that some of the environmental microbial pathogens such as *L. pneumophila* and *V. vulnificus* alter their gene expression and remain in a dormant stage as nonculturable organisms in the environment. It has also been predicted that several biodegradative microorganisms may not express their degrading genes in this state. As a result, one may not be sure whether the released GEMs or indigenous microorganisms are degrading the pollutants at a contaminated site. By using a specific mRNA as target for PCR amplification and developing a quantitative assay for such a method, it is possible to detect the level of mRNA production with high sensitivity in the environmental samples. A promising method for extraction of specific mRNA from soil seeded with naphthalene-degrading and mercury-resistant bacterial cells has been described.[15] This method can be completed within a few hours, and approximately 17 µg of total RNA per gram (wet weight) of soil, containing 8.0×10^8 bacterial cells, can be purified with a DNA-RNA hybridization detection sensitivity of 160 ng of specific target mRNA. Although this method has potential for studying *in situ* gene expression, the humic acid compounds may precipitate with samples containing high-cation exchange capacity, e.g., some sediments, which will greatly reduce the total RNA recovery efficiency and sensitivity of detection. Application of PCR for detecting specific mRNA, extracted from various environmental samples by this method, has yet to be evaluated.

IX. DETECTION OF VIRUSES IN WATER

Outbreaks of enteric viral diseases due to the drinking of contaminated water has increased significantly during the past decade.[57,58] Conventional monitoring for the presence of enteric viruses in the environment and in drinking water requires animal cell culture and is technically difficult and time consuming. Since the enteric viruses are relatively resistant to the process used for treatment and disinfection of wastewater, a rapid and efficient method is necessary for periodic, perhaps routine monitoring of the drinking water and finished water sources, for possible viral contamination. Detection of viruses, especially enteroviruses, from the environmental samples requires sample cleanup for successful PCR DNA amplification. In one approach, 97% of the enteroviruses were recovered routinely from the environmental waters by concentrating the samples by the bioflocculation method followed by precipitaion with polyethylene glycol and sodium chloride.[59] The mRNA-reverse transcriptase-cDNA-PCR-gene probe (RT-PCR-gene probe) detection of the enteric viruses required further purification of the concentrated viral particles by using Sephadex G-200 spin column™* to remove interfering factors.[58,59] This RT-PCR approach, following sample processing, has been shown to detect <1 plaque-forming units (PFU) of most of the known enteroviruses.[58-60]

More recently, a related procedure for purification of entroviruses from groundwater samples and samples containing humic acids has been described.[61] In this approach, the samples were treated with Sephadex 100 or 200 spin columns™* in combination with Chelex 100™** following conventional filter elution-adsorption of the viral particles to remove interfering factors for RT-PCR detection. This approach effectively removed the contaminants, including the humic materials, from the water sample, for detection of enteroviruses by RT-PCR with the sensitivity of 0.1 PFU.[61] Presence of human immunodeficiency virus type 1 (HIV-1) in wastewater, sludge, final effluent, soil, and pond water, released by infected individuals and hospital wastes, and their possible role in the spread of the disease is a major concern. Total nucleic acids, directly from environmental samples or from viral concentrates, from various sources, was extracted by conventional alkaline lysis method or by aluminum flocculation method, respectively.[62] By using the RT-PCR DNA-amplification method, the presence of HIV-1 was detected in several wastewater samples with the viral particle amounts equivalent to 0.04 and 0.4 pg of P24 antigen.[62] However, the study did not show any evidence of the infectivity of these viruses in the wastewaters or their possible role in spreading disease to the human population.

X. DISCUSSION

Rapid progress of the applications of PCR methodology in the area of environmental microbiology has shown great promise in solving various difficult problems that had remained unsolved for years due to limitations of the conventional methods. One of the important problems in environmental microbiology is the detection and monitoring of released GEMs in the environment with high sensitivity. The application of the PCR methodology coupled with several improved sample-processing approaches have been shown to detect 1 to 100 GEMs per gram of soil or sediment, which is a level of sensitivity several orders of magnitude higher than that of the conventional DNA-DNA hybridization method. The application of PCR technology for monitoring pathogens and indicator microorganisms in water and in the environmental samples has shown great potential to be used as an alternative to the conventional methods, with greater specificity and sensitivity. However, a considerable amount of effort may be needed to investigate the presence of inhibitors and contaminants, and their

---

* Registered trademark of Pharmacia LKB Biotechnology, Inc., 800 Centennial Ave., Piscataway, NJ 08854.
** Registered trademark of BioRad Laboratories, 1414 Harbour Way S., Richmond, CA 94804.

effective removal, to establish consistent reliability of the PCR method, in a variety of environmental samples. The quantitation of microbial populations by the PCR method needs to be investigated to study microbial succession, competetion, and community structure in an ecosystem, including the microorganisms living in extreme environments.

Another potential application of PCR methodology in the area of environmental microbiology is in distinguishing the live and dead cells in a given environmental sample. Although a few investigations have shown that the PCR method can be used for such purposes, more research is warranted at this time, before application of the technique to actual environmental samples.

Alteration of gene expression in many microbial pathogens and other pollutant degrading microbes due to various environmental stresses is a growing concern. Application of the RT-PCR approach by targeting specific mRNA in the environment has the potential to provide the information on *in situ* microbial activities. Also, the detection of viruses in the environmental samples, and its application to drinking water safety by RT-PCR, have shown great promise. Moreover, introduction of thermostable DNA polymerase (*Tth* from *Thermus thermophilus*), which has both RT and polymerizing activities, may simplify the RT-PCR approach. PCR shows promise in cloning genes from environmentally important microorganisms, including those organisms that have not been cultured yet. Overall, technological improvements and subsequent new development of the PCR method will solve many unanswered questions in the area of microbial ecology, microbial community structure, environmental health, and environmental analyses of molecular microbiology.

REFERENCES

1. **Mullis, K. B. and Faloona, F. A.**, Specific synthesis of DNA *in vitro* via a polymerase-catalyzed chain reaction, *Methods Enzymol.*, 155, 335–351, 1987.
2. **Mullis, K. B.**, The unusual origin of the polymerase chain reaction, *Sci. Am.*, 262(4), 56–65, 1990.
3. **Saiki, R. K., Gelfand, D. H., Stoffel, S., Scharf, S. J., Higuchi, R., Horn, G. T., Mullis, K. B., and Erlich, H. A.**, Primer-directed enzymatic amplification of DNA with a thermostable DNA polymerase, *Science*, 239, 487–494, 1988.
4. **Steffan, R. J., Goksoyr, J., Bej, A. K., and Atlas, R. M.**, Recovery of DNA from soils and sediments, *Appl. Environ. Microbiol.*, 54, 2908–2915, 1988.
5. **Steffan, R. J. and Atlas, R. M.**, DNA amplification to enhance the detection of genetically engineered microorganisms in environmental samples, *Appl. Environ. Microbiol.*, 54, 2185–2191, 1988.
6. **Tsai, Y. and Olson, B.**, Rapid method of separation of bacterial DNA from humic substances in sediments for polymerase chain reaction, *Appl. Environ. Microbiol.*, 58, 2292–2295, 1992.
7. **Tsai, Y. and Olson, B. H.**, Rapid method for direct extraction of DNA from soil and sediments, *Appl. Environ. Microbiol.*, 57, 1070–1074, 1991.
8. **Tsai, Y. and Olson, B. H.**, Detection of low numbers of bacterial cells in soils and sediments by polymerase chain reaction, *Appl. Environ. Microbiol.*, 58, 754–757, 1992.
9. **Pillai, S. D., Josephson, K. L., Bailey, R. L., Gerba, C. P., and Pepper, I. L.**, Rapid method for processing soil samples for polymerase chain reaction amplification of specific gene sequences, *Appl. Environ. Microbiol.*, 57, 2283–2286, 1991.
10. **Smalla, K., Cresswell, N., Mendonca, L. C., Wolters, A., and van Elsas, J. D.**, Rapid DNA extraction protocol from soil for polymerase chain reaction-mediated amplification, *J. Appl. Bacteriol.*, 74, 78-85, 1993.
11. **Young, C. C., Burghoff, R. L., Keim, L. G., Minak-Bernero, V., Lute, J., R., and Hinton, S. M.**, Polyvinylpyrrolidone-agarose gel electrophoresis purification of polymerase chain reaction-amplifiable DNA from soils, *Appl. Environ. Microbiol.*, 59, 1972–1974, 1993.
12. **Picard, C., Ponsonnet, C., Paget, E., Nesme, X., and Simonet, P.**, Detection and enumeration of bacteria in soil by direct DNA extraction and polymerase chain reaction, *Appl. Environ. Microbiol.*, 58, 2717–2722, 1992.
13. **Bruce, K. D., Hiorns, W. D., Hobman, J. L., Osborn, A. M., Strike, P., and Ritchie, D. A.**, Amplification of DNA from native populations of soil bacteria by using polymerase chain reaction, *Appl. Environ. Microbiol.*, 58, 3413–3416, 1992.
14. **Selenska, S. and Klingmüller, W.**, DNA recovery and direct detection of *Tn5* sequences from soil, *Lett. Appl. Microbiol.*, 13, 21–24, 1991.

15. Tsai, Y., Park, M. J., and Olson, B. H., Rapid method for direct extraction of mRNA from seeded soils, *Appl. Environ. Microbiol.*, 57, 765–768, 1991.
16. Sommerville, C. C., Knight, I. T., Straub, W. L., and Colwell, R. R., Simple, rapid method for direct isolation of nucleic acids from aquatic environments, *Appl. Environ. Microbiol.*, 55, 548–554, 1989.
17. Paul, J. H., Cazares, L., and Thurmomd, J., Amplification of the *rbcL* gene from dissolved and particulate DNA from acquatic environments, *Appl. Environ. Microbiol.*, 56, 1963–1966, 1990.
18. Weller, R. and Ward, D., Selective recovery of 16S rRNA sequences from natural microbial communities in the form of cDNA, *Appl. Environ. Microbiol.*, 55, 1818–1822, 1989.
19. Fuhrman, J. A., Comeau, D. E., Hagstrom, A., Cham, A. M., Extraction from natural planktonic microorganisms of DNA suitable for molecular biological studies, *Appl. Environ. Microbiol.*, 54, 1426–1429, 1988.
20. Lee, S. and Fuhrman, J. A., DNA hybridization to compare species compositions of natural bacterioplankton assemblages, *Appl. Environ. Microbiol.*, 56, 739–746, 1990.
21. Zehr, J. P. and McReynold, L. A., Use of degenerate oligonucleotides for the amplification of the *nifH* gene from the marine cyanobacterium *Trichodesmium thiebautii*, *Appl. Environ. Microbiol.*, 55, 2522–2526, 1989.
22. Bej, A. K., Mahbubani, M. H., DiCesare, J. L., and Atlas, R. M., PCR-Gene probe detection of microorganisms using filter-concentrated samples, *Appl. Environ. Microbiol.*, 57, 3529–3534, 1991.
23. Bej, A. K., Steffan, R. J., DiCesare, J. L., Haff, L., and Atlas, R. M., Detection of coliform bacteria in water by polymerase chain reaction and gene probes, *Appl. Environ. Microbiol.*, 56, 307–314, 1990.
24. Steffan, R. J. and Atlas, R. M., Polymerase chain reaction: applications in environmental microbiology, *Ann. Rev. Microbiol.*, 45, 137–161, 1991.
25. Chaudhry, G. R., Toranzos, G. A., and Bhatti, A. R., Novel method for monitoring genetically engineered microorganisms in the environment, *Appl. Environ. Microbiol.*, 55, 1301–1304, 1989.
26. Greer, C. W., Beaumier, D., Bergeron, H., and Lau, P. C. K., Polymerase chain reaction isolation of a chlorocatechol dioxygenase gene from a dichlorobenzoic acid degrading Alcaligenes denitrificans, in 91st Gen. Meet. American Society for Microbiology, Abstract Q-99, 1991, 292.
27. Neilson, J. W., Josephson, K. L., Pillai, S. D., and Pepper, I. L., Polymerase chain reaction and gene probe detection of the 2,4-dichlorophenoxyacetic acid degradation plasmid, pJP4, *Appl. Environ. Microbiol.*, 58, 1271–1275, 1992.
28. Amann, R. I., Stromley, J., Devereux, R., Keryl, R., and Stahl, D. A., Molecular and microscopic identification of sulfate-reducing bacteria in multispecies biofilms, *Appl. Environ. Microbiol.*, 58, 614–623, 1991.
29. Bej, A. K., DiCesare, J. L., Haff, L., and Atlas, R. M., Detection of *Escherichia coli* and *Shigella* spp. in water by using polymerase chain reaction (PCR) and gene probes for *uid*, *Appl. Environ. Microbiol.*, 57, 1013–1017, 1991.
30. Cleuziat, P. and Baudouy-Robert, J., Specific detection of *Escherichia coli* and *Shigella* species using fragments of genes coding for beta-glucuronidase, *FEMS Microbiol. Lett.*, 72, 315–322, 1991.
31. Bej, A. K., McCarty, S. C., and Atlas, R. M., Detection of coliform bacteria and *Escherichia coli* by multiplex polymerase chain reaction: comparison with defined substrate and plating methods for water quality monitoring, *Appl. Environ. Microbiol.*, 57, 2429–2432, 1991.
32. Bej, A. K., Mahbubani, M. H., Boyce, M. J., and Atlas, R. M., Detection of *Salmonella* in shellfish by using polymerase chain reaction, in press.
33. Jones, D. D., Law, R., and Bej, A. K., Detection of *Salmonella* spp. in contaminated oysters using polymerase chain reaction (PCR) and gene probes, 58, 523–532.
34. Way, J. S., Josephson, K. L., Pillai, S. D., Abbaszadegan, M., Gerba, C. P., and Pepper, I. L., Specific detection of *Salmonella* spp. by multiplex polymerase chain reaction, *Appl. Environ. Microbiol.*, 59, 1473–1479, 1993.
35. Mahbubani, M. H., Jones, D. D., and Bej, A. K., Species-specific detection of *Shigella* spp. by polymerase chain reaction, in preparation.
36. Starnbach, M. N., Falkow, S., and Tompkins, L. S., Species specific detection of *Legionella pneumophila* in water by DNA amplification and hybridization, *J. Clin. Microbiol.*, 2, 1257–1261, 1990.
37. Mahbubani, M. H., Bej, A. K., Miller, R., Haff, L., DiCesare, J., and Atlas, R. M., Detection of *Legionella* with polymerase chain reaction and gene probe methods, *Mol. Cell. Probes*, 4, 175–187, 1990.
38. Koide, M., Saito, A., Kusano, N., and Higa, F., Detection of *Legionella* spp. in cooling tower water by polymerase chain reaction method, *Appl. Environ. Microbiol.*, 59, 1943–1946, 1993.
39. Loutit, J. S. and Tompkins, L. S., Evaluation of a DNA amplification procedure for detection of *Legionella pneumophila* and *Legionella dumoffii* in water, in *Legionella: Current Status and Emerging Perspective*, Barbaree, J. M., Breiman, R. F., and Dufor, A. P., Eds., ASM Press, Washington, DC, 1993, 176–178.
40. Nowicki, M., Bornstein, N., Jaulhac, B., Piemont, Y., Monteil, H., and Fleurette, J., Rapid detection of Legionellae in clinical and environmental samples by polymerase chain reaction, in *Legionella: Current Status and Emerging Perspective*, Barbaree, J. M., Breiman, R. F., and Dufor, A. P., Eds., ASM Press, Washington, DC, 1993, 178–181.

41. **Bej, A. K., Mahbubani, M. H., and Atlas, R. M.,** Detection and molecular serogrouping of *Legionella pneumophila* by polymerase chain reaction amplification and restriction enzyme analysis, in *Legionella: Current Status and Emerging Perspective,* Barbaree, J. M., Breiman, R. F., and Dufor, A. P., Eds., ASM Press, Washington, DC, 1993, 173–174.
42. **Mahbubani, M. H., Bej, A. K., Perlin, M. H., Schaefer, F. W., Jakubowski, W., and Atlas, R. M.,** The differentiation of *Giardia duodenalis* from other *Giardia* spp. based on the polymerase chain reaction and gene probes, *J. Clin. Microbiol.,* 30, 74–78, 1992.
43. **Mahbubani, M. H., Bej, A. K., Perlin, M. H., Schaefer, F. W., Jakubowski, W., and Atlas, R. M.,** Detection of *Giardia* using the polymerase chain reaction and distinguishing live from dead cysts, *Appl. Environ. Microbiol.,* 57, 3456–3461, 1991.
44. **Mahbubani, M. H., Schaefer, F. W., and Bej, A. K.,** Detection of *Giardia lamblia* in environmental waters by polymerase chain reaction and immunomagnetic separation, in preparation.
45. **Chamberlain, J. S., Gibbs, R. A., Ranier, J. E., Nguyen, P. N., and Radolf, J. H.,** Deletion screening of the Duchenne muscular dystrophy locus via multiplex DNA amplification, *Nucleic Acids Res.,* 16, 11141–11156, 1988.
46. **Bej, A. K., Mahbubani, M. H., Miller, R., DiCesare, J. L., Haff, L., and Atlas, R. M.,** Multiplex PCR amplification and immobilized capture probes for detection of bacterial pathogens and indicators in water, *Mol. Cell. Probes,* 4, 353–365, 1990.
47. **Bej, A. K. and Atals, R. M.,** Bacterial detection using PCR and colorimetric gene probe methods, in 91st Gen. Meet. American Society for Microbiology, Abstract Q-144, 1991, 300.
48. **Colwell, R. R., Brayton, P. R., Grimes, D. J., Roszak, D. B., Huq, S. A., and Palmer, L. M.,** Viable but nonculturable *Vibrio cholerae* and released pathogens in the environment: implication for release genetically engineered microorganisms, *Bio/Technology,* 3, 817–820, 1985.
49. **Hussong, D., Colwell, R. R., O'Brien, M. O., Weiss, E., Pearson, A. D., Eiener, R. M., and Burge, W. D.,** Viable *Legionella pneumophila* not detectable by culture on agar media, *Bio/Technology,* 5, 947–950, 1987.
50. **Roszak, D. B. and Colwell, R. R.,** Survival strategies of bacteria in the natural environment, Microbiol. Rev., 51, 365–379, 1987.
51. **Mahbubani, M., Bej, A., DiCesare, J., Miller, R., Haff, L., and Atals, R. M.,** Detection of bacterial mRNA using PCR, *BioTechniques,* 10, 48–49, 1991.
52. **Bej, A. K., Mahbubani, M. H., and Atlas, R. M.,** Detection of viable *Legionella pneumophila* in water by polymerase chain reaction and gene probe methods, *Appl. Environ. Microbiol.,* 57, 597–600, 1991.
53. **Brauns, L. A., Hudson, M. C., and Oliver J. D.,** Use of the polymerase chain reaction in detection of culturable and nonculturable *Vibrio vulnificans* cells, *Appl. Environ. Microbiol.,* 57, 2651–2655, 1991.
54. **Paszko-Kolva, C., Yamamoto, H., Shahamat, M., and Colwell, R. R.,** Polymerase chain reaction, gene probe, and direct fluorescent antibody staining of *Legionella pneumophila* serogroup 1 in drinking water and environmental samples, in *Legionella: Current Status and Emerging Perspective,* Barbaree, J. M., Breiman, R. F., and Dufor, A. P., Eds., ASM Press, Washington, DC, 1993, 181–183.
55. **McCarty, S. C. and Atlas, R. M.,** Effect of amplicon size on polymerase chain reaction detection of bacteria exposed to chlorine, submitted.
56. **Linder, K. and Oliver, J. D.,** Membrane fatty acids and virulence changes in the viable but nonculturable state of *Vibrio vulnificans, Appl. Environ. Microbiol.,* 55, 2837–2842, 1989.
57. **Craun, G. F.,** Surface water supplies and health, *J. Am. Water Works Assoc.,* 80, 40–52, 1988.
58. **DeLeon, R., Shieh, C., Baric, R. S., and Sobsey, M. D.,** Detection of entroviruses and hepatitis A virus in environmental samples by gene probes and polymerase chain reaction, in Proc. 1990 Water Quality Technology Conf., American Water Works Association, Denver, 1990, 833–853.
59. **Schwab, K. J., DeLeon, R., Baric, R. S., and Sobsey, M. D.,** Detection of rotaviruses, enteroviruses, and hepatitis A virus by reverse transcriptase-polymerase chain reaction, in Proc. 1991 Water Quality Technology Conf., American Water Works Association, Denver, in press.
60. **Chapman, N. M., Tracy, S., Gauntt, C. J., and Fortmueller, U.,** Molecular detection and identification of enteroviruses using enzymatic amplification and nucleic acid hybridization, *J. Clin. Microbiol.,* 28, 843–850, 1990.
61. **Abbaszadegan, M., Huber, M. S., Gerba, C. P., and Pepper, I. L.,** Detection of enteroviruses in groundwater with the polymerase chain reaction, *Appl. Environ. Microbiol.,* 59, 1318–1324, 1993.
62. **Ansari, S. A., Farrah, S. R., and Chaudhry, G. R.,** Presence of human immunodeficiency virus nucleic acids in wastewater and their detection by polymerase chain reaction, *Appl. Environ. Microbiol.,* 58, 3984–3990, 1992.

Chapter 33

DETECTION OF FOODBORNE MICROBIAL PATHOGENS USING POLYMERASE CHAIN REACTION METHODS

Daniel D. Jones and Asim K. Bej

TABLE OF CONTENTS

I. Introduction 341

II. Advantages of PCR 342

III. Concerns Using PCR for Identifying Microbial Pathogens in Food 343

IV. Multiplex PCR Applications 358

V. Selected Foods and Protocols 358
    A. Soft Cheeses 359
    B. Milk 359
    C. Meats 360

References 361

I. INTRODUCTION

More than 250 foodborne diseases are now recognized, and most require specific laboratory diagnosis. Reported illnesses due to contaminated food are usually sporadic, isolated cases, but an increase in the number of people eating at restaurants, with relatively large volumes of centralized food preparation, has resulted in more outbreaks.[1,2] Higher population densities contribute to microbial contamination of water resources and of inhabiting shellfish that are often destined for human consumption. Also, increasingly larger segments of the population are more susceptible to opportunistic pathogens, e.g., the elderly and those who are immunocompromised. Thus, rapid and accurate methods of detection of foodborne microbial pathogens are needed for protecting the health of the public.

Traditionally, the strategy for identifying most microbial pathogens in food involves enrichment culture, cultivation on selective agar media, and ultimately a series of biochemical tests on presumptive colonies to identify the organisms.[3,4] Such standard microbiological techniques are slow and laborious, often requiring several days to weeks to perform. Enrichment protocols also may fail to detect strains of virulent bacteria present in foods at low levels. Such strains, especially those pathogenic for humans, do not compete effectively in physiologically demanding enrichments and can be overgrown by nonpathogenic members of the same genus or species.[5,6] For some important pathogens, such as strains of enterotoxigenic *Escherichia coli*, no satisfactory culturing technique exists to allow direct discrimination between nonpathogenic and pathogenic strains.[3,7] Also, selective conditions of conventional culturing may lead to the loss of plasmids with toxin genes.[8,9]

The enforcement of laws by public health agencies and the implicit obligation of producers to provide consumer safety have stimulated the search for faster, more specific, and less costly methods to detect and identify microbial pathogens in food that often cause outbreak of diseases in humans. During the last decade use of immunological assays and nucleic acid probes have been added to the repertoire of diagnostic methods to detect microbial pathogens in food.[3,10-12] However, both have deficiencies. Immunological assays are influenced by variability of gene expression and antibody response. Cross-reactivities of antisera exist.[13,14] Antibodies to cell surface antigens allow only genus-level specificity, making it necessary to perform confirmative biochemical tests on presumptive isolates. Furthermore, at low concentrations of specific toxins other chemical factors may conceal the presence of toxins in immunological tests, resulting in false-negative reactions.[15-18] There is also evidence that specific enterotoxin production by pathogens, such as *Staphylococcus aureus*, may differ when growth is in natural substrate compared to growth in laboratory media.[19] The relatively limited availability of specific serological reagents also restricts their widespread use.

Since nucleic acid probes directly associate with genetic material, hybridization-based tests are generally more specific and reliable than immunoassays. However, gene probe techniques for foodborne pathogens often require 10^5 to 10^6 copies of the target sequence to provide a dependable, positive signal.[11,20-25] Therefore, the organisms must be incubated for several hours to obtain adequate copies of the target for hybridization protocols. Such incubations add considerably to the total time needed for diagnosis. Also, as noted by Wernars et al.,[26] in their study of enterotoxigenic *E. coli*, both immunological and DNA colony hybridization procedures require the identification of random isolates. Such techniques are laborious, and the probability is low of identifying a pathogenic *E. coli* among a majority of nonpathogenic isolates.

An increasing number of reports indicate that application of PCR to reliably and quickly detect pathogens in food is a promising new diagnostic tool to monitor food safety (Tables 1 to 3). Organisms detected by PCR that are responsible for foodborne disease outbreaks, PCR specific primers, probes of amplified target DNA utilized (Table 1), amplification protocols (Table 2), and foods and sensitivity levels (Table 3) reported are summarized in this review. Note that in several of the studies included, PCR was performed on only pure cultures of pathogens that were isolated from food (compare Tables 1 and 3). Fewer reports describe a PCR system for direct detection of microbial pathogens present in food samples (see Table 3). The majority involve an enrichment culture from which the DNA is extracted and on which the PCR assay performed. Also, only a few studies used oligonucleotide probes to detect PCR-amplified DNA specific for microbial pathogens isolated from food on gels (Table 1). Wang et al.[27] have reported a convenient and simple dot blot procedure with an internal probe hybridization on nitrocellulose filters instead of gels. They spotted 1 ml from a PCR sample with the dye loading directly onto the nitrocellulose, resulting in sensitivity equivalent to that obtained in gels. Often adequate amplification has been achieved to allow direct observation on gels of bands coinciding with amplified DNA of predicted size. To confirm the identity of amplified target, some investigators have coupled PCR DNA amplification with restriction fragment length polymorphism (RFLP) analyses.[28-38]

II. ADVANTAGES OF PCR

The PCR method offers several advantages for rapid, reliable detection of microbial pathogens in food and promises to be a valuable addition and complement to the toolbox of food microbiologists. Several investigators have noted certain advantages offered by PCR for diagnosis, but Candrian et al.[34] have summarized most of the important ones: (1) a short time requirement improves public health security and minimizes personnel costs; (2) the method is able to identify microorganisms that are difficult to culture; (3) the culture and enrichment of pathogens are not necessary for quality control; (4) PCR reagents are more readily available and easier to store than to those required for serological procedures; (5) animal models are not needed; (6) the choice of primers determines specificity, which contrasts with the frequent cross-reactivities of antisera utilized in immunoassays; (7) elaborate diagnostic equipment and media are not required, thereby increasing the flexibility as to locations where PCR may be performed and eliminating many sample transport problems; and (8) automation of PCR in the future should result in excellent cost efficiency.

As investigators continue to optimize conditions for the PCR and increase the production of the number of copies of DNA, use of highly sensitive radioactively labeled probes can be replaced by less sensitive nonradioactively labeled probes. Such probes can detect as little as 50 to 100 ng of DNA.[3] Indeed, DNA labels are often not necessary because most molecules generated by PCR are of uniform size and adequate copy numbers make gel electrophoresis sufficiently sensitive for the diagnostic step.[24,39-49] Using nested primers (two sets of primers with one set directing amplicon production internal to the other set), Wilson et al.[50] improved sensitivity of detection of purified target DNA of *S. aureus* from 100 pg to about 1 fg.

Deneer and Boychuk[41] reported that two rounds of 35 PCR cycles each could increase the sensitivity from 542 bacteria used as starting material to about 54 cells. However, they did not apply the procedure to DNA extracted from bacteria in food samples. Wang et al.[27] developed a protocol that had a shorter denaturing time, a shorter annealing time, a rapid transition, and an increase in the number of cycles compared to other reports (Table 2). The rapid transitions (heating at 48°C/min and cooling with tap water at 2.65 l/min) for each step of a PCR cycle proved an effective protocol to avoid inactivation of *Taq* DNA polymerase. Furthermore, they utilized the rapid freeze-boiling approach to denature and release nucleic acids from microbial cells; originally described by Mahbubani et al.[51,52] By heating the *Listeria* cells in 1% Triton X-100 for 5 min at 100°C and immediately cooling them in ice water, a shorter time (3 min) at 95°C in the PCR program was adequate and minimized *Taq* enzyme denaturation.[27] This procedure was quite effective for detecting *L. monocytogenes* in a variety of foods (Table 3).

III. CONCERNS USING PCR FOR IDENTIFYING MICROBIAL PATHOGENS IN FOOD

Notwithstanding the numerous advantages of PCR methodology, problems have been encountered when it is applied to food, slowing its widespread adoption: (1) *Taq* DNA polymerase in the PCR assay may be inhibited by compounds that can occur in samples, especially those such as foods that represent complex compositions and diverse matrices.[27,53-55] Consequently, extraction and purification of the DNA from the food or culture broth prior to PCR have been necessary to remove inhibitory compounds. However, as noted below, some PCR protocols have made DNA extraction unnecessary for selected samples. (2) Another complication is that the number of contaminating pathogenic microorganisms in food samples is often relatively low, particularly compared to pathogens in clinical samples, and enrichment cultures may still be required for adequate sensitivity (see Table 3). (3) An additional point of concern relates to the potentially high sensitivity of PCR and its inability to discriminate between live pathogens and those that have already been killed before processing.[34,53] The presence of PCR-amplifiable target DNA from dead pathogens may lead to a false-positive PCR product. Although false-positives can mislead one as to health hazards of foods, there are situations where detection of target DNA in even dead cells would be valuable. For example, staphylococcal enterotoxins A to E are capable of surviving extreme conditions that may kill the staphylococcal cells.[50] Thus, PCR allows recognition of the potential for enterotoxins that may have survived heat processing by identifying the inherent capability to produce the toxins in dead cells. Some bacteria such as *Vibrio cholerae*,[56,57] *Salmonella* spp.[58,59] *Campylobacter jejuni*,[60] *E. coli*,[57] and *Vibrio vulnificus*[56] can enter a nonculturable state, although still be viable. For example, *V. vulnificus* becomes dormant during the cold seasons and remaining viable but not culturable. Such organisms may pose a health hazard if not detected.[56,61-63] Since

TABLE 1.
Microbial Pathogens Isolated from Food and PCR Primer Sequences, Target DNA, Sizes of Amplified DNA Sequences and Oligonucleotide Probes

Organism	Primer sequence	Target	Size of amplified DNA (bp)	Probe	Ref.
Bacteria					
Campylobacter coli, C. jejuni, C. Lari	C442: 5'-GGAGGATGACACTTTTCGGAGC-3' C490: 5'-ATTACTGAGATGACTAGCACCCC-3'	16S rNA	426	C631: 5'-GGAAGAA TTCTGACGGTACCT-3'	77
Escherichia coli (enteroinvasive) and Shigella boydii, S. dysenteriae, S. flexneri, and S. sonnei	KL1: 5'-TAATACTCCTGAACGGCG-3' KL8: 5'-TTAGGTGTCGGCTTTTCTG-3'	Most of ial gene associated with 220-kbp invasive	748	210-bp AccI-PstI fragment	7
Escherichia coli (enteroinvasive)					44
Escherichia coli (enteroinvasive)					29
Escherichia coli (enteroinvasive)	Left 5'-CCTCTCTATATGCACACGGAGCTCCCCAG-3' Right: 5'-CTATATGTTGACTGCCCGGGACTTCGACC-3'	LT Gene (heat-labile enterotoxin)B subunit	195	5'-ATACGGGAATCGA TGGCAGGC-3'	26
Escherichia coli (enterotoxigenic)	1: 5'-TCGCCACACGCTGACGCTGACCA-3' 2: 5'-TTACATGACCTCGGTTTAGTTCACAGA-3'	malB Operon	595	331 bp from malB-PCR product (strain JM 101)	33
Escherichia coli (enterotoxigenic)	1: 5'-TTACGGCGTTACTATCCTCTCTA-3' 2: 5'-GGTCTCGGTCAGATATGTGATTC-3'	LT1 Gene, (LTlh and LTIp alleles) (heat-labile enterotoxin) type I	275	5'-CATTTCAGGTCGA AGTCCCG-3'	37, 33
Escherichia coli (enterotoxigenic)	1: 5'TTTTTTCTGTATTTCTT TTICIICTTTIIITCAG-3'[a] 2: 5'-GCAGGATTACAACAIAI[a] TTCACAGC-3'	STI Gene, (STh and STIp alleles) (heat-stable enterotoxin)	175	STIh; alleles estA2 and estA3/4: 5'-GCTACTAT TCATGCTTTCAGGA-3' STIp; allele estA1: 5'-CATTAGAGACTA AAAGTGTGAT-3'	33

Organism	Primer sequences	Target	Product size	Ref.
Escherichia coli (Shiga-like toxin; I and II producing) (Multiduplex PCR)	SLTI-F: 5'-ACACTGGATGATCTCAGTGG-3' SLTI-R: 5'-CTGAATCCCCCTCCATTATG-3' and SLTII-F: 5'-CCATGACAACGGACAGCAGTT-3' SLTII-R: 5'-CCTGTCAACTGAGCACTTTG-3'	Shiga-like toxin genes (SLT I and SLT II)	614 for SLTI and 779 for SLTII	42
Lactobacillus brevis (Also, L. brevis, L. pastorianus, L. casei, and Pediococcus damnosus recognized by primers)	Lb5S-1: 5'-TGTGGTGGCGATAGCCTGAA-3' Lb5S-2: 5'-GCGTGGCAACGTCCTATCCT-3'	5S rRNA of lactic acid bacteria	117	49
Listeria monocytogenes (Maritime)	Pair of 24-mer oligonucleotides (sequences not reported)	Listeriolysin O gene (hlyA)	606	76
Bacteria (universal target) (Multiplex with U1/L11 and LM1/LM2)	U1: 5'-CAGC(A or C)GCCGCGGTA AT(A or T)C-3' U2: 5'-CCGTCAATTC(A or C)T TT(A or G)AGTTT-3'	16S rRNA	408	40
Listeria spp.	U1: 5'-CAGC(A or C)GCCGCG GTAAT(A or T)C-3' LI1: 5'-CTCCATAAAGGTGACCCT-3'	16S rRNA	938	
Listeria monocytogenes	LM1: 5'-CCTAAGACGCCAATCGAA-3' LM2: 5'-AAGCGCTTGCAACTGCTC-3'	Listeriolysin O gene (hlyA)	702	
Listeria monocytogenes	Set A (preferred): 5'-CCGGGAGCTGCTAAAGCGGT-3' 5'-GCCAAACCACCGAAAAGACC-3' Set B: 5'-GAAGCACCTTTTGACGAAGC-3' 5'-GCTGGTGTCTACAGGTGTTTC-3'	Dth18 gene	326 122	53
Listeria monocytogenes	LM 14 and LM 16; sequences not reported	Downstream of hlyA gene	Not reported	46
Listeria monocytogenes	Lis-1: 5'-GCATCTGCATTCAATAAAGA-3' Lis-2: 5'-TGTCACTGCATCTCCGTGGT-3'	Listeriolysin O gene (hlyA)	174	41

Probes:
- ^{32}P-labeled 24-mer (sequence not reported)
- A': 5'-AAGATACAG TTGGCGGATGG-3'
- B': 5'-GACGGAATGG GCGCGCTTGT-3'

TABLE 1 (continued).
Microbial Pathogens Isolated from Food and PCR Primer Sequences, Target DNA, Sizes of Amplified DNA Sequences and Oligonucleotide Probes

Organism	Primer sequence	Target	Size of amplified DNA (bp)	Probe	Ref.
Listeria monocytogenes (All primer pairs specific and directed yields of equal amounts of amplicons)	LL1 (Forward orientation): 5′-GACATTCAAGTTGTGAA-3′ LL4 (reverse orientation): 5′-CGCCACACTTGAGATAT-3′	Listeriolysin O gene (hlyA)	560		47
	LL1 (Forward orientation): 5′-GACATTCAAGTTGTGAA-3′ LL6 (Reverse orientation): 5′-CTGTAAGCCATTTCGTC-3′		299		
	LL3 (Forward orientation): 5′-ATTGCGAAATTTGGTAC-3′ LL4 (Reverse orientation): 5′-CGCCACACTTGAGATAT-3′		240		
	LL5 (Forward orientation): 5′-AACCTATCCAGGTGCTC-3′ LL4 (Reverse orientation): 5′-CGCCACACTTGAGATAT-3′		520		
	LL5 (Forward orientation): 5′-AACCTATCCAGGTGCTC-3′ LL6 (Reverse orientation): 5′-CTGTAAGCCATTTCGTC-3′		267		
Listeria monocytogenes	Sequences not reported	mpl or prt A (Metalprotease gene)	342		34
Listeria monocytogenes (Serotypes 4a and 4c not recognized by iap primers) (Multiplex PCR)	1: 5′-CGGAGGGTTCCGCAAAAGATG-3′ 2: 5′-CCTCCAGAGTGATCGATGTT-3′	hlyA (α-Hemolysin) gene	234	5′-CCATCTGTATA AGCTTTTGAAG-3′ (Furrer et al.[72] only)	72
	1: 5′-ACAAGCTGCACCTGTTGCAG-3′ 2: 5′-TGACAGCGTGTGTAGTAGCA-3′	iap (β-Hemolysin) gene	131	5′-GGCGCAGGTGTAG TTGCTTG-3′ (Furrer et al.[72] only)	73

Organism	Primer sequences	Target	Size (bp)	Notes	Ref.
Listeria monocytogenes	Sequences not reported	Listeriolysin O gene (hlyA)	924		24
Listeria monocytogenes (Some serovar 4c not recognized by HylA primers; Five of ten L innocua and one L ivanovii recognized only by ImaA primers) (Multiplex PCR)	hlyA-a: 5'-ATTGCGAAATTTGGTACAGC-3' hlyA-b: 5'-ACTTGAGATATATGCAGGAG-3' ImaA-a: 5'-AACAAGGTCTAACTGTAAAC-3' ImaA-b: 5'-ACTATAGTCAGCTACAATTG-3'	Listeriolysin O gene (hlyA) Listeria antigen (ImaA) gebe	234 257		38
Listeria monocytogenes	Set 1 (More yield or "sensitive" than Set 2): L-1: 5'-CACGTGCGACAATGGATAG-3' L-2: 5'-AGAATAGTTTTATGGGATTAG-3' Set 2: L1: 5'-CACGTGCTACAATGGATAG-3' RL-2: 5'-ATAGTTTTATGGGATTAGC-3'	16s rRNA	L-1/L-2: 70 L-1/RL-2: 67	RL-3: 5'GTCGCAAGC CGCGAGGT-3'	27
Listeria monocytogenes	Primer 234: 5'-CATCGACGGCAACCTCGGAGA-3' Primer 319: 5'-ATCAATTACCGTTCTCCACCATTC-3'	Listeriolysin O gene (hylA)	417		36
Listeria monocytogenes	Primer A: 5'-CTGTTGGAGCTCTTCTTG GTGAAGCAATCG-3' Primer B: 5'-AGCAACCTCGGTACCA TATACTAACTC-3'	prfA, A regulator gene	1060	1060-bp Segment corresponding to entire prfA gene used as colony probe	86
Listeria monocytogenes	PCTG0: 5'-GAATGTAAACTTCG GCGCAATCAG-3' PCRDO: 5'-GCCGTCGATGATTT GAACTTCATC-3'	Listeriolysin O gene (hlyA)	388	MBA: 5'GAATGTAAAC TTCGGCGC-3' MBB: 5'-CGATGATTTG AACTTCATC-3'	31
Salmonella typhimurium	A86, B83, and C84; sequences not reported	Not reported	Not reported	P47, Between primers A86 and B83, sequence not reported	74
Staphylococcus aureus (Primer pairs used individually)	SEA-1: 5'-TTGGAAACGGTTAAAACGAA-3' SEA-2: 5'-GAACCTTCCCATCAAAAACA-3' SEB-1: 5'-TCGCATCAAACTGACAAACG-3' SEB-2: 5'-GCAGGTACTCTATAAGTGCC-3' SEC-1: 5'-GCACATAAAAGCTAGGAATTT-3' SEC-2: 5'-AAATCGGATTAACATTATCC-3'	Enterotoxin A (sea) Enterotoxin B (seb) Enterotoxin C (sec-1)	120 478 257		43

TABLE 1 (continued).
Microbial Pathogens Isolated from Food and PCR Primer Sequences, Target DNA, Sizes of Amplified DNA Sequences and Oligonucleotide Probes

Organism	Primer sequence	Target	Size of amplied DNA (bp)	Probe	Ref.
Staphylococcus aureus (Primer sets 1 and 2 for *sec-1* were nested; primers for *nuc* and *seb* used individually)	SED-1: 5′-CTAGTTTGGTAATATCTCCT-3′ SED-2: 5′-TAATGCTATATCTTATAGGG-3′	Enterotoxin D (*sed*)	317		
	SEE-1: 5′-TAGATAAAGTTAAAACAAGC-3′ SEE-2: 5′-TAACTTACCGTGGACCCTTC-3′	Enterotoxin E (*see*)	170		50
	Set 1: Primer 1: 5′-ATGAATAAGAGTCGATTTA TTTCAT-3′ Primer 2: 5′-TTATCCATTCTTTTGTTGTA AGGTGG-3′ Set 2, internal to Set 1 (nested): Primer 3: 5′-ACACCCAACGTATTAGCAG AGAGCC-3′ Primer 4: 5′-CCTGGTGCAGGCATCATAT CATACC-3′	Enterotoxin C1 (*sec-1* or *entC1*)	Primers 1 and 2: 801 Primers 3 and 4: 631	Probe 3: 5′-AAATTTGACCAATC TAAATATTTAATGAT GTACAATGAC-3′ (Also see Notermans[87])	
	Primer 1: 5′-AGTATATAGTGCAACTTCAA CTAAA-3′ Primer 2: 5′-ATCAGCGTTGTCTTCGCTCC AAATA-3′	Thermonuclease (*nuc*)	450		
	Primer 1: 5′-GAGAGTCAACCAGATCCTAAACCAG-3′ Primer 2: 5′-ATACCAAAAGCTATTCTCATTTTCT-3′	Enterotoxin B (*seb* or *entB*)	593		
Staphlococcus aureus (Primer pairs used individually)	SEA-A1: 5′-AAAGTCCCGATCAATTTATGGCTA-3′ SEA-A2: 5′-GTAATTAACCGAAGGTTCTGTATA-3′	Enterotoxin A (*sea* or *entA*)	210		48
	SED-D1: 5′-GCAGATAAAAATCCAATAATAGGA-3′ SED-D2: 5′-ATCTAAAGAAACTTCTTTTGTAC-3′	Enterotoxin D (*sed* or *entD*)	333		

Organism	Primer sequences	Size (bp)	Probe	Ref.
	SEE-E1: 5'-TTACAAAGAAATGCTTTAAGC-3' SEE-E2: 5'-TAAACCAAATTTTCCGTG-3'	456	Enterotoxin E (see or entE)	
Vibrio cholerae O1	CT-1: 5'-TCAAACTATATTGTCTGGTC-3' CT-2: 5'-CGCAAGTATTACTCATCGA-3'	380	Cholerae toxin A subunit (ctxA)	78
Vibrio cholerae O1	1: 5'-CTCAGACGGGATTTGTTAGGCACG-3' 2: 5'-TCTATCTCTGRAGCCCTATTACG-3' (Primers 1 and 2 were best, but four other primers and four other combinations were tried)	302	Cholerae toxin (ctx)	80 5'-ACTATATTGT CTGGTCATTCTACT-3'
Vibrio cholerae O1	CTX2: 5'-CGGGCAGATTCTAGACCTCCTG-3' CTX3: 5'-CGATGATCTTGGAGCATTCCCAC-3'	564	Cholerae toxin A subunit (ctxA)	35
Vibrio cholerae O1	P1: 5'-TGAAATAAAGCAGTCAGGTG-3' P2: 5'-GTGATTCTGCACACAAATCAG-3' P3: 5'-GGTATTCTGCACACAAATCAG-3' (P3 corrects two mismatches in P2 and is used in later experiments)	777	Cholerae toxin (ctxAB gene)	79 ctxA-Specific 5'-GCAAGAGGAAC TCAGACGGG-3'
Vibrio vulnificus	VVp1: 5'-CCGGCGGTACAGGTTGGCGC-3' VVp2: 5'-CGCCACCCACTTTCGGGCC-3'	519	Cytotoxin-hemolysin gene	28
Vibrio vulnificus	Vv oligo 1: 5'-CGCCGGCTCACTGGGGCAGTGGCTG-3' Vv oligo 3: 5'-GCGGGTGGTTCGGTTAACGGCTGG-3'	340	Cytotoxin-hemolysin gene	32 25-bp probe: 5'-CGGCTCTGCGGGC TCGTCAACCAAC-3'

Yeast

Organism	Primer sequences	Size (bp)	Probe	Ref.
Saccharomyces cerevisiae	P$_1$: 5'-AAACTTTCAACAACGGATCTCTTGG-3' P$_4$: 5'-GAGCTCTTGCCGCTTCACTCGCCG-3'	0.5	Multigeneric rDNA (between 18S and 58S)	39
Saccharomyces cerevisiae 49 (Multiplex PCR with Lactobacillus brevis (Lb5S-1 and Lb5S-2)	Sc5S-3: 5'-GAGACATTGTGAGACCCTCC-3' Sc5S-4: 5'-CGATGCCGCCACGTGCAAAG-3'	100	5S rDNA	
Saccharomyces cerevisiae (Multiplex PCR with Z. bailii and Z. rouxii)	Primer 1: 5'-GGAGCACCTAATAACATTCT-3' Primer 2: 5'-ACAGTTTATATCGGTTGCATT-3'	310	Strain-specific plasmid	45

TABLE 1 (continued).
Microbial Pathogens Isolated from Food and PCR Primer Sequences, Target DNA, Sizes of Amplified DNA Sequences and Oligonucleotide Probes

Organism	Primer sequence	Target	Size of amplied DNA (bp)	Probe	Ref.
primers					
Zygosaccharomyces bailii	Primer 3: 5'-CATTGTTGCTGAGAATGCTT-3' Primer 4: 5'-TCTCTCTCAATACGTCGAT-3'	Strain-specific plasmid	603		
Zygosaccharomyces rouxii	Primer 5: 5'-AGTTCTGTACTTCCACAGAACT-3' Primer 6: 5'-TTCTCACTCTGTCATATGCT-3'	Strain-specific plasmid	872		
		Viruses			
Polio	Upstream: 5'-AGCACTTCTGTTTCCC-3' Downstream: 5'-ACGGACACCCAAAGTA-3'	5'-noncoding region	394		30
Hepatitis A	Upstream: 5'-ACAGGTATACAAAGTCAG-3' Downstream: 5'-CTCCAGAATCATCTCC-3'	VP3 Region	207		
Norwalk	Upstream: 5'-ATAAAAGTTGGCATGAACA-3' Downstream: 5'-CTTGTTGGTTTGAGGCCATAT-3'	Polymerase region	470		

[a] I = inosine.

TABLE 2.
PCR Parameters to Amplify Unique DNA Sequences for Detection of Food-Related Microbial Pathogens

Organism	Primers	PCR Parameters				Ref.
		Denaturation	Primer annealing	Primer extension	Total cycles	
Bacteria						
Campylobacter coli, C. jejuni, C. lari	C442 and C490	1 min at 94°C	1 min at 52°C	1 min at 74°C	40	77
Escherichia coli (enteroinvasive) and *Shigella boydii, S. dysenteriae, S. flexneri*, and *S. sonnei*	KL1 and KL8	1 min at 94°C	1.5 to 2.0 min at 55 to 65°C	1.5 to 2.0 min at 72°C	30	7
Escherichia coli (enteroinvasive)		1 min at 94°C	2.0 min at 62°C	2.0 min at 72°C	30 to 45	44
Escherichia coli (enteroinvasive)		First for 1 min at 94°C, then 0.5 min	0.5 min at 60°C	0.5 min at 72°C	50	29
Escherichia coli (enterotoxigenic)	Left and right	1 min at 94°C	3 min at 54°C	3 min at 72°C	40	26
Escherichia coli (enterotoxigenic)	*mal*B: 1 and 2	30 s at 94°C	1 min at 65°C; annealing and extension cycles combined		40 or 50	33
Escherichia coli (enterotoxigenic)	LT1: 1 and 2	1 min at 94°C	1 min at 55°C	1 min at 72°C; final for 5 min	35	37
Escherichia coli (enterotoxigenic)	LT1: 1 and 2 and ST1: 1 and 2	30 s at 94°C	1 min at 50°C	1 min at 72°C	40 or 50	33
Escherichia coli Shiga-like toxin (SLT I) *Escherichia coli*, SLT II (Multiplex with SLT I)	SLT1-F and SLT-R SLTII-F and SLTII-R	1 min at 94°C	1 min at 60°C	2 min at 72°C; final for 5 min	35	42
Lactobacillus brevis (Also, *L. brevis, L. pastorianus, L. casei,* and *Pediococcus damnosus* recognized by primers)	Lb5S-1 and Lb5S-2 (Multiplex with Sc5S-3 and Sc5S-4)	1 min at 94°C	2 min at 50°C	1.5 min at 72°C	35	49

TABLE 2 (continued).
PCR Parameters to Amplify Unique DNA Sequences for Detection of Food-Related Microbial Pathogens

Organism	Primers	PCR Parameters			Total cycles	Ref.
		Denaturation	Primer annealing	Primer extension		
Listeria monocytogenes (Maritime)	24-mer pair	2 min at 94°C	2 min at 51°C	2.5 min at 74°C	30	76
Bacteria (universal target) (Multiplex with LI1/U1 and LM1/LM2)	U1 and U2	4 min at 95°C for first, then for 1 min	2 s at 50°C	1 min at 72°C	30	40
Listeria	LI1 and U1					
Listeria monocytogenes	LM1 and LM2					
Listeria monocytogenes	Set A and Set B	3 min at 94° for first, then for 1 min	2 min at 54°C	3 min at 72°C	30	53
Listeria monocytogenes	LM14 and LM16	30 s at 94°C	30 s at 55°C	1 min at 72° C; final for 10 min	30	46
Listeria monocytogenes	Lis-1 and Lis-2	4 min at 95°C for first, then 45 s	45s at 60°C	1 min at 72°C; final for 5 min	Two rounds of 35	41
Listeria monocytogenes	LL1 and LL4 LL1 and LL6 LL3 and LL4 LL5 and LL4 LL5 and LL6	1 min at 94°C	1 min at 55°C	2 min at 72°C; final for 5 min	30	47
Listeria monocytogenes	hlyA and iap (α– and β-hemolysin) target sets	0.5 min at 95°C	1 min at 55°C; except 1 min at 50°C for strains from cooked sausage	1 min at 72°C; final for 5 min	40	72, 73
Listeria monocytogenes	Unspecified pair	2 min at 92°C	1 min at 56°C	1 min at 72–74°C	25	24
Listeria monocytogenes (Selected serovar 4c strains not recognized by hlyA primers; selected L. innocula and L. ivanovii recognized by ImaA primers	hlyA: hlyA-a and hlyA-b ImaA: ImaA-a and ImaA-b	5 min at 94°C first, then 2 min	2 min at 55°C	1 min at 72°C	30	38

Organism	Primers	Denaturation	Annealing	Extension	Cycles	Ref.
Listeria monocytogenes	L-1 and L-2; L-1 and RL-2	3 min at 95°C for first, then 20 s	20 s at 48°C	40 s at 73°C; final for 3 min followed by 5 s at 25°C	40	27
Listeria monocytogenes	Primers 234 and 319	7 min at 94°C for first, then 1 min	30 s at 62°C	30 s at 72°C; final for 1 min	30	36
Listeria monocytogenes	Primers A and B	15 s at 95°C	30 s at 60°C	90 s at 72°C	30	86
Listeria monocytogenes	PCRGO and PCRDO	5 min at 94°C for first, then 1 min	1 min at 65°C	2 min at 70°C	30	31
Salmonella typhimurium	A86, B83, and C84	1 min at 94°C	1 min at 65°C	2.5 min at 72°C	35	74
Staphylococcus aureus (Primer pairs used individually)	Five pairs used individually: SEA-1 and SEA-2 SEB-1 and SEB-2 SEC-1 and SEC-2 SED-1 and SED-2 SEE-1 and SEE-2	2 min at 94°C	2 min at 55°C	1 min at 72°C	Not reported	43
Staphylococcus aureus	A: Nested primer pairs used simultaneously: entC1 primers 1 and 2 and primers 3 and 4 and B: entB primers 1 and 2 C: nuc primers 1 and 2	30 s at 94°C	30 s at 50°C	30 s at 72°C	50	50
Staphylococcus aureus (Nested primers for Set A: primer pairs B and C used individually)						
Staphylococcus aureus (Primer pairs used individually)	A1 and A2 D1 and D2 E1 and E2	1.5 min at 94°C	2 min at 67°C 2 min at 63°C 2 min at 62°C	2 min at 72°C 2.5 min at 72°C 3 min at 72°C	35	48
Vibrio cholerae O1	1 and 2	1 min at 94°C	1.5 min at 60°C	1.5 min at 72°C	30–40	80
Vibrio cholerae O1	CTX2 and CTX3	5 min at 95°C for first; then 1 min	1 min at 60°C; 50° and 55°C in some experiments	1 min at 72°C final for 10 min	25	35
Vibrio cholerae O1	P1 and P2 or P3	1 min at 94°C	1 min at 55°C	1 min at 72°C	25–35	79
Vibrio vulnificus	VVp1 and VVp2	10 min at 94°C before first then 1.75 min/c	2 min at 67–69°C	2 min at 72°C	30	28
Vibrio vulnificus (culturable cells)	Vv oligo 1 and 3	1 min at 94°C	1 min at 65°C; annealing and extension cycles combined; last cycle to 10 min		40	32

TABLE 2 (continued).
PCR Parameters to Amplify Unique DNA Sequences for Detection of Food-Related Microbial Pathogens

Organism	Primers	PCR Parameters			Total cycles	Ref.
		Denaturation	Primer annealing	Primer extension		
Vibrio vulnificus (nonculturable cells)	Vv oligo 1 and 3	30 s at 94°C	30 s at 65°C; annealing and extension cycles combined; last cycle to 10 min		50	
Yeast						
Saccharomyces cerevisiae	P_1 and P_4	1.5 min at 94°C	2 min at 42°C	2.5 min at 72°C; final for 7 min	25	39
Saccharomyces cerevisiae (Multiplex with *Lactobacillus brevis*)	Sc5S-3 and Sc5S-4 (Multiplex with Lb5S-1 and Lb5S-2)	1 min at 94°C	2 min at 50°C	1.5 min at 72°C	35	49
Saccharomyces cerevisiae (Multiplex with *Z. bailii* and *Z. rouxii*)	Primers 1 and 2	2 min at 92°C	3 min at 55°C	2 min at 72°C	30	45
Zygosaccharomyces bailii *Zygosaccharomyces rouxii*	Primers 3 and 4 Primers 5 and 6					
Polio	Upstream and downstream	4 min at 94°C for first, then 1 min	1 min 30 s at 49°C	1 min at 72°C; final for 5 min	40	30
Hepatitis A	Upstream and downstream	3 min at 94°C for first, then 1 min	1 min 20 s at 49°C	40 s at 72°C; final for 15 min	40	
Norwalk	Upstream and downstream	4 min at 94°C for first, then 1 min	1 min 30 s at 55°C	1 min at 72°C; final for 15 min	40	

TABLE 3.
Foods, Enrichment Cultures, and Sensitivity of Detection for Specific Microbial Pathogens

Organism	Culture or food(s)	Enrichment culture before DNA extraction	Sensitivity	Ref.
Campylobacter coli, C. jejuni, Jejuni, C. lari	Inoculated and naturally contaminated chicken skin	18 h, resulting in 500 CFU/ml	12.5 CFU in presence of 10^6 to 10^8 CFU/g	77
Escherichia coli (enteroinvasive), *Shigella boydii, S. dysenteriae, S. flexneri,* and *S. sonnei*	Inoculated lettuce	None required with 10^4 inoculate; but also cultured for 4 and 24 h	10^4 cells per gram (at time 0 after inoculation)	7
Escherichia coli (enteroinvasive)	Inoculated raw, whole milk	Seeded milk incubated up to 24 h, but target DNA extracted directly from milk	No incubation of seeded milk: 10^4 cells per milliliter 4 h incubation of seeded milk: 10^2 cells per milliliter	44
Escherichia coli (enterotoxigenic)	Bean sprouts, lettuce, raw oyster meat, terrific broth, cooked shrimp, Havarti cheese, crab meat, coconut milk, soft tofu, raw egg and milk powder, mushroom, and macadamia nut meat	None; direct extraction of plasmid DNA from foods	10^3 CFU/ml; foods seeded with 10^5 CFU/ml; suggested with larger sampling volume added to column sensitivity could be improved manyfold	29
Escherichia coli (enterotoxigenic)	Tested minced meat from local butcher shops; also inoculated meat samples	4 h, plus 20 h	2×10^4 CFH/ml after incubation with initial inoculation of 3 CFU	26
Escherichia coli (enterotoxigenic strains LTI and STI)	Swiss-Italian soft cheese, homemade mayonnaise	Strains isolated, cultured, and 10^5 cells used in PCR	10 cells among 10^5 background population	33
Escherichia coli (Shiga-like toxin; I and II producing)	Inoculated ground beef	6 h	After incubation, 1 CFU/g of SLT strains	42
Lactobacillus brevis (also *Saccharomyces cerevisiae*)	Inoculated pasteurized beer	None; direct extraction	30 cells per 250 ml beer	49
Listeria monocytogenes (Maritime)	Inoculated whole homogenized milk	None; direct extraction	10^5 CFU/ml	76

TABLE 3 (continued).
Foods, Enrichment Cultures, and Sensitivity of Detection for Specific Microbial Pathogens

Organism	Culture or food(s)	Enrichment culture before DNA extraction	Sensitivity	Ref.
Listeria spp. and *Listeria monocytogenes*	Isolates from wheat, cream, potatoes, cucumber, lettuce, mushrooms, and radish	Crude cell lysates made from isolates from food and maintained on media	Not reported	40
Listeria monocytogenes	Inoculated soft cheeses	None; direct extraction	10^3 to 10^8 CFU/0.5 g, depending on type of cheese	53
Listeria monocytogenes	Soft cheese, mayonnaise salads, ham, smoked tenderloin, sausages	40–48 h	2×10^5 CFU/ml for soft cheese; (the only food for which limit was reported)	46
Listeria monocytogenes	Inoculated pasteurized 2% milk and ground beef	Overnight	0.1 CFU/ml or g of milk and beef, respectively	47
Listeria monocytogenes	Cooked sausage and inoculated milk	24 h for sausage	Sausage: 10 bacteria with α-hemolysin primers Milk: 10 bacteria/10 ml by direct detection	72
Listeria monocytogenes	330 Naturally contaminated foods (dairy products, raw meat, cooked meat, poultry, vegetables, seafood)	Procedure A: ~48 h Procedure B: ~24 h	10 CFU/10 g (except for raw meat about 10 times less sensitive)	73
Listeria monocytogenes	Inoculated chicken, pork, turkey, and beef franks; chicken breast, drumsticks, and nuggets; soft ripened cheese; fermented sausage; yogurt; cooked crawfish	None, direct extraction	2 to 20 CFU of pure culture; 4 to 40 CFU of inoculated (10^8 CFU), diluted food; exception was soft cheese (Camembert) which interfered with PCR	27
Listeria monocytogenes	Inoculated chicken skin and pate, yogurt, skim milk, mature cheese, soft cream cheese	18 h	10–100 CFU/g food inoculated with 10^4 CFU/g	36
Listeria monocytogenes	180 Natural and spiked foods, including milk, meat, ice cream, sausage, chicken	48 h	10 CFU/25 g food before enrichment; 10^2 CFU/ml of enrichment broth	31
Salmonella typhimurium	Inoculated milk and beef steak	Direct on inoculated milk; overnight enrichment for beef	2.5 CFU/ml milk; 2×10^4 cells/ml beef homogenate	74

Organism	Sample	Conditions	Detection limit	Ref.
Staphylococcus aureus	Inoculated dried skim milk	24 h after inoculation	10^5 CFU/ml	50
Staphylococcus aureus	Inoculated raw beef, pork, chicken, fish, milk	None on spiked food homogenates	10^0 to 10^1 cells/g food	48
Vibrio cholerae O1	Inoculated pasteurized blue crabmeat, shucked raw oysters, fresh shrimp, lettuce, strawberries	6 to 8 h Alkaline peptone water (APW) homogenates of shellfish, APW washes of fresh fruits and vegetables	10 CFU/g of oyster; 1 CFU/10 g of lettuce	79
Vibrio vulnificus	Inoculated oyster homogenates	24 h	10^2 CFU/g oyster meat	28
Vibrio vulnificus	Culturable and nonculturable cells	Overnight; cells from nonculturable microcosm filtered prior to incubation	72 pg DNA: culturable cells; 31 pg DNA: nonculturable	32

Viruses

Organism	Sample	Conditions	Detection limit	Ref.
Polio	Oysters allowed to accumulate added viruses	Virus added to processed oysters; also, viruses allowed to bioaccumulate for 15–16 h in oysters	10 PFU: seedes whole oyster; 38 PFU: seeded extract; 164 PFU: bioaccumulated oyster	30
Hepatitis A			Not detected in whole oysters but only in spiked oyster extracts	
Norwalk			50 to 500 Virus particles	

PCR methodology depends only on the presence of target DNA and not on culturable cells, the nonculturable cells are more likely to be detected by PCR than by conventional methods.[32,64]

One common solution to the aforementioned concerns is to incorporate a brief growth step (three to five cell doublings) followed by dilution before PCR is conducted. By diluting endogenous inhibitors of *Taq* DNA polymerase, as noted by Lampel et al.,[7] the effect of false-positive results may be reduced since DNA not biologically duplicated would be diluted out. This method would be useful, however, only if the level of dead cells is relatively constant.

Increasing the primer annealing temperature to about 68°C has been demonstrated to improve specificity of PCR.[20,21,65] Candrian et al.[33] eliminated an additional PCR product when the annealing temperature for primers of *mal*B in *E. coli* was raised from 55 to 65°C (Table 2). Similarly, Fields et al.[35] eliminated nonspecific PCR products by increasing the annealing temperature from 50 to 60°C.

IV. MULTIPLEX PCR APPLICATIONS

The simultaneous use of more than one pair of carefully designed primers (multiplex) allows identification of multiple organisms in one PCR assay.[52,64,66-69] In the food industry, where contamination of surfaces, equipment, and food with nonpathogens is common, the increased specificity and ability to rapidly distinguish between nonpathogenic and pathogenic species by proper design of multiple primers should be particularly valuable.[70,71] Border et al.,[40] using three pairs of primers were able to simultaneously detect the presence of bacteria and to distinguish between the pathogen *L. monocytogenes* and other *Listeria* spp. (Tables 1 to 3).

Gannon et al.[42] used two pairs of oligonucleotide primers to distinguish between Shiga-like toxin (SLT) producing *E. coli* strains SLTI and SLTII (Tables 1 to 3). Similarly, Furrer et al.[72] and Niederhauser et al.[73] could specifically identify *L. monocytogenes* (rare serotypes 4a and 4c excepted since they are not recognized by *iap* primers) in a multiplex assay and use simple size determination of the amplification fragment doublet to ascertain specificity of the fragment (Tables 1 to 3). It was noted, however, that fresh cultures needed to be used since storage for a week or more at 4°C did not yield adequate PCR products. Similarly, Johnson et al.[38] found that only *L. monocytogenes* (some strains of serovar 4c excepted) was recognized by both of two pairs of primers (for genes *hlyA* and *ImaA*), and some strains of *L. innocua* and *L. ivanovii* by only one pair (i.e., for *ImaA*) (Table 1).

Pearson and McKee[45] designed three sets of primers that were specific for three strains of yeast that are known for their spoilage of food products (Tables 1 and 2). In a multiplex PCR, the primer sets for the yeast produced three species-specific amplicons of distinct size that were easily distinguishable in an agarose gel (Table 1).

Tsuchiya et al.[49] used a mixture of two sets of primers to detect and identify two important organisms responsible for beer spoilage, *Lactobacillus brevis* and *Saccharomyces cerevisiae* (Tables 1 to 3). DNAs from other lactic acid bacteria, such as innocuous *S. casei*, also produced PCR product; hence, primers more specific for organisms typical of beer spoilage are needed.[49]

Also, simultaneous detection of four microbial pathogens from oyster meat by amplifying four targets in a single PCR reaction has been accomplished (see Section V.C).[71]

V. SELECTED FOODS AND PROTOCOLS

Attention has been directed toward increased efficiency in cell lysis and better DNA-purification protocols to adequately remove factors from food samples that inhibit PCR. Most reports have evaluated PCR assays with only a limited number of food types, but a few of the most recent studies have tested a spectrum of foods (Table 3). The protocols generally include an enrichment step of 24 to 48 h (Table 3),[28,31,73] but there are exceptions.[27,74] For examples, see Section V.C. Also, using a plasmid target instead of a genomic target for PCR, Andersen and Omiecinski[29] successfully isolated target plasmid DNA directly from a variety of food homogenates without enrichment (Table 3). DNA affinity columns (Magic Minipreps, Promega, Madison, WI) were used and extractions with organic solvents, such as chloroform and phenol, and ethanol precipitations were eliminated.

A. SOFT CHEESES

Protocols that have not included an enrichment culture for detecting the PCR target in soft cheese have been unsuccessful.[27,53] Wernars et al.[53] encountered difficulties in adequately eliminating inhibiting substances from soft cheeses, and the detection limit (see Table 3) of *L. monocytogenes* depended on the brand of cheese used. DNA was precipitated from inoculated cheese samples that had been heated (40°C preferred) and centrifuged. The ethanol precipitate was further purified by dialysis, by phenol extraction, or by ion exchange chromatography (Qiagen-columns, Qiagen, Inc., Chatsworth, CA), with the latter providing the best results. The phenol extraction usually removed inhibitors that were co-isolated with DNA, but resulted in a marked reduction in recovered DNA. Wang et al.[27] used several washing steps, followed by heating to lyse cells, and successfully detected *L. monocytogenes* in several foods, such as chicken, but not in soft cheese, confirming that this food contains significant inhibitors of PCR.

Those protocols that have included an enrichment step and dilution step have achieved positive PCR with soft cheese. For example, Rossen et al.[46] used a simplified procedure that eliminated centrifugation and precipitation steps, but included a 40-h enrichment. To extract DNA from *L. monocytogenes*, 5 µl of enrichment broth were mixed with 50 µl 0.05 M NaOH, 0.25% SDS, and heated to 90°C for 15 min; a 5-µl aliquot was used directly in the PCR assay (see Table 3 for sensitivity). Cell lysis in a dry-heater block was found to be more reproducible than microwave treatment. Also, Triton X-100 was less efficient than SDS for lysis, and lysozyme did not improve lysis efficiency.[46] Tween-20 was found to offset inhibitory effects of ionic detergents (0.01% SDS), but in the absence of SDS, was inhibitory itself.[20,46,75]

Fitter et al.[36] achieved a high level of sensitivity (Table 3) for inoculated soft cheese by including a limited (18-h) enrichment step. In brief, the DNA preparations included a low-speed pelleting of food, a subsequent higher speed pelleting of cells of *L. monocytogenes*, and then a washing of the cells twice in sterile saline, resuspension in sterile water, and lysing by microwaves. The procedure was also effective for detecting *L. monocytogenes* from raw chicken skin, which has a high fat content.[36]

B. MILK

Keasler and Hill[44] evaluated two approaches for milk sample preparations. One protocol was designed to extract bacterial DNA directly from the seeded milk, and the second to eliminate interfering food matrix prior to extracting bacterial DNA. The second approach was most successful. The greatest sensitivity (Table 3) was obtained by (1) incubating supernatant from centrifugation (175 g) at 37°C for 30 min to liquefy fats, (2) treating the pellet from a second centrifugation (5000 g) with proteinase K (0.1 mg/ml) for 60 min, (3) boiling for 8 to 10 min, (4) precipitating DNA with 0.1 vol of 3-M sodium acetate and 2 vol of ethanol, and (5) washing the pellet (7850 g) with ethanol. Several alternative protocols, including treatment with proteinase K and detergent, yielded less amplified target after PCR.[44]

Pahuski et al.[74] have filed a patent on a protocol for milk (and other foods, including beef) that detects *S. typhimurium* at a relatively high level of sensitivity (Table 3) without any special steps taken to lyse the bacteria or extract the DNA. The procedure utilizes a "clearing solution" consisting of 0.25 M EDTA, 0.5% Triton X-100, and 0.01% microparticulate carrier (such as surfactant-free polystyrene beads). Following centrifugation, the cream layer removed, supernatant aspirated, and the pellet washed and resuspended in water. PCR was performed directly on an aliquot of the resuspended cells.

Furrer et al.[72] washed cells of *L. monocytogenes* from inoculated milk five times in PCR reaction buffer before lysing them with lysozyme and proteinase K and boiling to achieve a sensitivity of 10 cells in 10 ml. Tsen and Chen[48] achieved high sensitivities of 10^0 to 10^1 *S. aureus* cells per gram of inoculated food. Homogenates of food were spiked with *S. aureus*, then incubated in SDS, followed by proteinase K, before the mixture was centrifuged and pelleted. The pellet was then treated with lysostaphin and the DNA extracted with phenol-chloroform and subsequently washed with ethanol.

Bessesen et al.[76] successfully detected *L. monocytogenes* after pelleting 10^5 cells per milliliter of milk and washing the cells in saline and water. The microorganisms were lysed by microwaving, which proved more effective than sonication (60 s) or treatment with 1% SDS and lysozyme (1 mg/ml) (Table 3).

Some investigators utilizing PCR to specifically detect bacteria isolated from milk choose to incorporate an enrichment step (Table 3). Thomas et al.[47] lysed cells isolated on plating media with lysozyme and mutanolysin. Proteinase K and sodium lauryl sarcosinate were subsequently added, and DNA precipitated with sodium acetate and ethanol. Wilson et al.[50] used heat to lyse cells pelleted from solubilized milk and purified the DNA with a binder (Isogene, ILS, Ltd., London) prior to performing PCR.

C. MEATS

Many of the studies evaluating PCR detection of organisms in milk and cheeses also included meat (Table 3). Pahuski et al.[74] performed DNA extractions directly on milk, but for meat, a broth was inoculated and then processed to yield a distinct amplicon. After multiple washings of pelleted cells and heat lysis, Wang et al.[27] successfully detected *L. monocytogenes* in meat products without including an enrichment or dilution step.

Others have included an enrichment step of varying duration (Table 3).[26,28,42,77-79] Giesendorf et al.[77] determined that samples inoculated with 500 CFU/ml required a minimum culturing time of 14 to 18 h before a positive PCR signal was detected. Koch et al.[79] noted that increasing the number of PCR amplification cycles often leads to the formation of nonspecific amplification products due to mispriming or primer-dimer formation; thus, it is preferable to allow bacterial growth to amplify target copy number as opposed to the use of an increase in amplification cycles. They obtained a positive amplification of *V. cholerae ctx* only after 6 h of enrichment, and observed an even stronger amplification signal after 8 h. Koch et al.[79] also found that amplification always occurred when performed directly on enrichment cultures prepared from washes of seeded fruits and vegetables, but not on those from seeded shellfish homogenates — unless the proportion of homogenate were reduced from 10 to 1% (w/v). DNA extraction on the enrichment cultures was unnecessary, and aliquots of the culture broth that had been boiled 5 min were added directly to the PCR reaction mixture.[79] This parallels the approaches previously reported by Kobayashi et al.[78] and Shirai et al.[80] Likewise, Wernars et al.[26] found pretreatment of enrichment cultures of *E. coli* inoculated with meat unnecessary prior to performing PCR. Recently, the "hot start" approach[65] was found to be useful for eliminating nonspecific amplified products due to mispriming. This approach has proven effective for detection of *V. vulnificus*, *Salmonella* spp., and other microbial pathogens from oyster and chicken meat.[28,81,82]

Gannon et al.[42] and Giesendorf et al.[77] successfully detected *E. coli* and *Campylobacter* spp. by subjecting culture broths to standard nucleic acid extraction procedures involving organic solvent extractions and DNA precipitation. Hill et al.[28] were able to specifically detect by PCR methodology the human pathogen *V. vulnificus* in seeded oysters. DNA purified from cells lysed by guanidine isothiocyanate and extracted with chloroform gave the best amplification.

Cold temperatures induce vibrios to enter a viable but nonculturable state.[56,63,83] Brauns et al.[32] determined that PCR could also detect target DNA extracted from microcosms of such cells that are not recoverable by standard microbiological techniques. However, the PCR conditions found to be optimal for DNA extracted from culturable cells did not yield observable amplicon unless an additional 40 PCR cycles were employed. But by decreasing the duration of the denaturing, annealing, and extension steps each to 30 s and increasing the number of cycles to 50 (Table 2), a clearly observable DNA band was produced from 31 ng of DNA extracted from nonculturable cells.[32] As noted above, Koch et al.[79] did not require DNA extractions to detect *V. cholera ctx* directly in diluted shellfish homogenates.

Jones et al.[70] used a gene which encodes a small DNA-binding protein as a target for PCR amplification to detect with high specificity total *Salmonella* as a genus at a sensitivity of <10 cells per gram of oyster meat. Bej et al.[71] used a combination of four sets of primers to detect simultaneously in oyster meat *Salmonella* spp., pathogenic *Salmonella*, *V. vulnificus*, and *V. cholerae* in a single multiplex PCR. In 1 g of contaminated oyster meat fewer than 10 cells of each of the four microbial pathogens could be detected in the presence of 10^6 cells of the other three pathogens. Rapid and efficient methods for purification of target nucleic acids from the complex matrices of oyster meat were developed in these two studies. The purification protocols included treatment of samples with Chelex 100®, then ammonium acetate, followed by purification with chloroform extraction.[81] Purifying the samples further with a microconcentrator improved the level of detection. Furthermore, chicken contaminated with *Salmonella* spp. required additional purification with a protein-precipitating reagent, ProCipitate®, to achieve amplification of the target DNA.[82]

Viruses such as hepatitis A and Norwalk or Norwalk-like viruses are accumulated by shellfish and are responsible for nonbacterial diseases which can cause disease outbreaks and erode public confidence in consuming shellfish.[84,85] PRC promises to be the first economical and practical approach available to detect and monitor enteric viruses in shellfish. Using cetyltrimethylammonium bromide precipitation, Atmar et al.[30] removed inhibitors of reverse transcription-polymerase chain reaction, and concentrated and detected poliovirus (as a model virus) and Norwalk virus in oysters by PCR. However, using the same purification procedures, hepatitis A virus was not concentrated and detected in whole oysters — only in spiked oyster extracts.[30] The sensitivity of detection for poliovirus was high, and the estimated level of detection for Norwalk virus was good (Table 3).

As methods and protocols are refined, the advantages offered by PCR as an investigatory and monitoring tool will provide marked improvements in quality control of food products.

REFERENCES

1. **Bean, N. H., Griffin, P. M., Golding, J. S., and Ivey, C. B.,** Foodborne disease outbreaks, 5-year summary, 1983-1987, *Morbidity Mortality Wkly. Rep.,* 39(SS-1), 15, 1990.
2. Annual Summary of Foodborne Disease, Unpublished Data from 1983 to 1986, Centers for Disease Control, U.S. Department of Health and Human Services, Atlanta, 1989.
3. **Vanderzant, C. and Splittstoesser, D. F.,** Eds., *Compendium of Methods for the Microbiological Examination of Foods,* American Public Health Association, Washington DC, 1992.
4. **Food and Drug Administration,** Bacteriological Analytical Manual, Association of Official Analytical Chemists, Arlington, VA, 1992.
5. **Hill, W. E., Ferreira, J. L., Payne, W. L., and Jones, V. M.,** Probability of recovering pathogenic *Escherichia coli* from foods, *Appl. Environ. Microbiol.,* 49, 1374, 1985.
6. **Mehlman, I. J. and Romero, A.,** Enteropathogenic *Escherichia coli*: methods for recovery from foods, *Food Technol.,* 36(3), 73, 1982.
7. **Lampel, K., Jagow, J., Trucksess, M., and Hill, W.,** Polymerase chain reaction for detection of invasive *Shigella flexneri* in food, *Appl. Environ. Microbiol.,* 56(6), 1536, 1990.

8. **Chosa, H., Makin, S., Sasakawa, C., Okada, N., Yamada, M., Komatsu, K., Suk, J. S., and Yoshikawa, M.,** Loss of virulence in *Shigella* strains preserved in culture collections due to molecular alteration of the invasion plasmid, *Microb. Pathogen.*, 6, 337, 1989.
9. **Evans, D. G. and Evans, D. J., Jr.,** New surface associated heat-labile colonization factor antigen (CGA/II) produced by enterotoxigenic *Escherichia coli* of serogroups O6 and O8, *Infect. Immun.*, 21, 638, 1978.
10. **Hill, W. E.,** Detection of bacteria in foods using DNA hybridization, in *DNA Probes for Infectious Diseases*, Tenover, F. C., Ed., CRC Press, Boca Raton, FL, 1989, 43.
11. **Hill, W. and Keasler, S.,** Identification of foodborne pathogens by nucleic acid hybridization, *Int. J. Food Microbiol.*, 12(1), 67, 1991.
12. **Gavalchin, J., Landy, K., and Batt, C. A.,** Rapid methods for the detection of *Listeria*, in *Molecular Approaches to Improving Food Quality and Safety*, Bhatnagar, D. and Cleveland, T. E., Eds., Van Nostrand Reinhold, New York, 1992, 189.
13. **Ito, H., Terai, A., Takeda, Y., and Nishibuchi, M.,** Cloning and nucleotide sequencing of vero toxin 2 variant genes from *Eshcerichia coli* O91:H21 isolated from a patient with hemolytic uremic syndrome, *Microb. Pathogen.*, 8, 47, 1990.
14. **Willshaw, G. A., Smith, H. R., Scotland, S. M., Field, A. M., and Rowe, B.,** Heterogeneity of *Escherichia coli* phages encoding vero cytotoxins: comparison of cloned sequences determining VT1 and VT2 and development of specific gene probes, *J. Gen. Microbiol.*, 133, 1309, 1987.
15. **Peterkin, P. I. and Sharpe, A. N.,** Rapid enumeration of *Staphylococcus aureus* in foods by direct demonstration of enterotoxigenic colonies on membrane filters by enzyme immunoassay, *Appl. Environ. Microbiol.*, 47, 1047, 1984.
16. **Rose, S. A., Bankes, P., and Stringer, M. F.,** Detection of staphylococcal enterotoxins in dairy products by the reversed passive latex agglutination (SET-RPLA) kit, *Int. J. Food Microbiol.*, 8, 65, 1989.
17. **Wieneke, A. A.,** The detection of enterotoxin and toxic shock syndrome toxin-1 production by strains of *Staphylococcus aureus* with commercial RPLA kits, *Int. J. Food Microbiol.*, 7, 25, 1988.
18. **Windemann, H., Luthy, J., and Maurer, M.,** ELISA with enzyme amplification for sensitive detection of staphylococcal enterotoxin in foods, *Int. J. Food Microbiol.*, 8, 25, 1989.
19. **Gomez-Lucía, E., Goyache, J., Orden, J. A., Blanco, J. L., Ruiz-Santa-Quiteria, J. A., Domínguez, L., and Suárez, G.,** Production of enterotoxin A by supposedly nonenterotoxigenic *Staphylococcus aureus* strains, *Appl. Environ. Microbiol.*, 55, 1447, 1989.
20. **Bej, A. K., Mahbubani, M. H., and Atlas, R. M.,** Amplification of nucleic acids by polymerase chain reaction (PCR) and other methods and their applications, *Crit. Rev. Biochem. Mol. Biol.*, 26(3/4), 301, 1991.
21. **Bej, A. K. and Mahbubani, M. H.,** Applications of the polymerase chain reaction in environmental microbiology, *PCR Methods Appl.*, 1, 151, 1992.
22. **Hill, W. E., Payne, W. L., Zon, G., and Moseley, S. L.,** Synthetic oligodeoxyribonucleotide probes for detecting heat-stable enterotoxin-producing *Escherichia coli* by DNA colony hybridization, *Appl. Environ. Microbiol.*, 50, 1187, 1985.
23. **Jagow, J. A. and Hill, W. E.,** Enumeration of virulent *Yersinia enterocolitica* colonies by DNA colony hybridization using alkaline treatment and paper filters, *Mol. Cell. Probes*, 2, 189, 1988.
24. **McKee, R. A., Gooding, C. M., Garrett, S. D., Powell, H. A., Lund, B. M., and Knox, M.,** DNA probes and the detection of food-borne pathogens using the polymerase chain reaction, *Biochem. Soc. Trans.*, 19(3), 698, 1991.
25. **Steffan, R. J. and Atlas, R. M.,** Polymerase chain reaction: applications in environmental microbiology, *Ann. Rev. Microbiol.*, 45, 137, 1991.
26. **Wernars, K., Delfgou, E., Soentoro, P., and Notermans, S.,** Successful approach for detection of low numbers of enterotoxigenic *Escherichia coli* in minced meat by using the polymerase chain reaction, *Appl. Environ. Microbiol.*, 57(7), 1914, 1991.
27. **Wang, R., Cao, W., and Johnson, M.,** 16S rRNA-based probes and polymerase chain reaction method to detect *Listeria monocytogenes* cells added to foods, *Appl. Environ. Microbiol.*, 58(9), 2827, 1992.
28. **Hill, W. E., Keasler, S. P., Trucksess, M. W., Feng, P., Kaysner, C. A., and Lampel, K. A.,** Polymerase chain reaction identification of *Vibrio vulnificus* in artificially contaminated oysters, *Appl. Environ. Microbiol.*, 57(3), 707, 1991.
29. **Andersen, M. R. and Omiecinski, C. J.,** Direct extraction of bacterial plasmids from food for polymerase chain reaction amplification, *Appl. Environ. Microbiol.*, 58(12), 4080, 1992.
30. **Atmar, R. L., Metcalf, T. G., Neill, F. H., and Estes, M. K.,** Detection of enteric viruses in oysters by using the polymerase chain reaction, *Appl. Environ. Microbiol.*, 59(2), 631, 1993.
31. **Bohnert, M., Dilasser, F., Dalet, C., Mengaud, J., and Cossart, P.,** Use of specific oligonucleotides for direct enumeration of *Listeria monocytogenes* in food samples by colony hybridization and rapid detection by PCR, *Res. Microbiol.*, 143(3), 271, 1992.
32. **Brauns, L., Hudson, M., and Oliver, J.,** Use of the polymerase chain reaction in detection of culturable and nonculturable *Vibrio vulnificus* cells, *Appl. Environ. Microbiol.*, 57(9), 2651, 1991.

33. **Candrian, U., Furrer, B., Hofelein, C., Meyer, R., Jermini, M., and Luthy, J.,** Detection of *Escherichia-coli* and identification of enterotogigenic strains by primer-directed enzymatic amplification of specific DNA sequences, *Int. J. Food Microbiol.*, 12(4), 339, 1991.
34. **Candrian, U., Niederhauser, C., Hoefelein, C., Buehler, H., Mueller, U., and Luethy, J.,** Polymerase chain reaction in bacteriological food analysis *Listeria monocytogenes* as an example, *Mitt. Gebiete Lebensmitteluntersuchung Hygiene*, 82(6), 539, 1991.
35. **Fields, P., Popovic, T., Wachsmuth, K., and Olsvik, O.,** Use of polymerase chain reaction for detection of toxigenic *Vibrio cholerae* 01 strains from the Latin American cholera epidemic, *J. Clin. Microbiol.*, 30(8), 2118, 1992.
36. **Fitter, S., Heuzenroeder, M., and Thomas, C.,** A combined PCR and selective enrichment method for rapid detection of *Listeria monocytogenes*, *J. Appl. Bacteriol.*, 73(1), 53, 1992.
37. **Furrer, B., Candrian, U., and Lüthy,** Detection and identification of *E. coli* producing heat-labile enterotoxin type I by enzymatic amplification of a specific DNA fragment, *Lett. Appl. Microbiol.*, 10, 31, 1990.
38. **Johnson, W., Tyler, S., Ewan, E., Ashton, F., Wang, G., and Rozee, K.,** Detection of genes coding for Listeriolysin and *Listeria monocytogenes* antigen A IMAA in *Listeria* spp. by the polymerase chain reaction, *Microb. Pathogen.*, 12(1), 79, 1992.
39. **Bertin, B. and Van Hoegaerden, M.,** Colorimetric detection of yeasts by in vitro amplification of DNA, incorporation of biotin-labeled deoxynucleotide, and hybridization, *Methods Mol. Cell. Biol.*, 2, 112, 1991.
40. **Border, P., Howard, J., Plastow, G., and Siggens, K.,** Detection of *Listeria* spp. and *Listeria monocytogenes* using polymerase chain reaction, *Lett. Appl. Microbiol.*, 11(3), 158, 1990.
41. **Deneer, H. and Boychuk, I.,** Species-specific detection of *Listeria monocytogenes* by DNA amplification, *Appl. Environ. Microbiol.*, 57(2), 606, 1991.
42. **Gannon, V. P. J., King, R. K., Kim, J. Y., and Golsteyn Thomas, E. J.,** Rapid and sensitive method for detection of shiga-like toxin-producing *Escherichia coli* in ground beef using the polymerase chain reaction, *Appl. Environ. Microbiol.*, 58(12), 3809, 1992.
43. **Johnson, W. M., Tyler, S. D., Ewan, E. P., Ashton, F. E., Pollard, D. R., and Rozee, K. R.,** Detection of genes for enterotoxins, exfoliative toxins, and toxic shock syndrome toxin 1 in *Staphylococcus aureus* by the polymerase chain reaction, *J. Clin. Microbiol.*, 29(3), 462, 1991.
44. **Keasler, S. and Hill, W.,** Polymerase chain reaction identification of enteroinvasive *Escherichia coli* seeded into raw milk, *J. Food Protect.*, 55(5), 382, 1992.
45. **Pearson, B. and McKee, R.,** Rapid identification of *Saccharomyces cerevisiae Zygosaccharomyces bailii* and *Zygosaccharomyces rouxii*, *Int. J. Food Microbiol.*, 16(1), 63, 1992.
46. **Rossen, L., Holmstrom, K., Olsen, J., and Rasmussen, O.,** A rapid polymerase chain reaction PCR-based assay for the identification of *Listeria monocytogenes* in food samples, *Int. J. Food Microbiol.*, 14(2), 145, 1991.
47. **Thomas, E. J. G., King, R. K., Burchak, J., and Gannon, V. P.,** Sensitive and specific detection of *Listeria monocytogenes* in milk and ground beef with the polymerase chain reaction, *Appl. Environ. Microbiol.*, 57(9), 2576, 1991.
48. **Tsen, H. and Chen, T.,** Use of the polymerase chain reaction for specific detection of type A D and E enterotoxigenic *Staphylococcus aureus* in foods, *Appl. Microbiol. Biotechnol.*, 37(5), 685, 1992.
49. **Tsuchiya, Y., Kaneda, H., Kano, Y., and Koshino, S.,** Detection of beer spoilage organisms by polymerase chain reaction technology, *J. Am. Soc. Brewery Chem.*, 50(2), 64, 1992.
50. **Wilson, I. G., Cooper, J. E., and Gilmour, A.,** Detection of enterotoxigenic *Staphylococcus aureus* in dried skimmed milk: use of the polymerase chain reaction for amplification and detection of staphylococcal enterotoxin genes entB and entC1 and the thermonuclease gene nuc, *Appl. Environ. Microbiol.*, 57(6), 1793, 1991.
51. **Mahbubani, M. H., Bej, A. K., Perlin, M., Schaefer III, F. W., Jakubowski, W., and Atlas, R. M.,** Detection of *Giardia* cysts by using the polymerase chain reaction and distinguishing live from dead cysts, *Appl. Environ. Microbiol.*, 57(12), 3456, 1991.
52. **Mahbubani, M. H., Bej, A. K., Perlin, M. H., Schaefer III, F. W., Jakubowski, W., and Atlas, R. M.,** Differentiation of *Giardia duodenalis* from other *Giardia* spp. by using polymerase chain reaction and gene probes, *J. Clin. Microbiol.*, 30(1), 74, 1992.
53. **Wernars, K., Heuvelman, C., Chakraborty, T., and Notermans, S.,** Use of the polymerase chain reaction for direct detection of *Listeria monocytogenes* in soft cheese, *J. Appl. Bacteriol.*, 70(2), 121, 1991.
54. **Mercier, B., Gaucher, C., Feugeas, O., and Mazurier, C.,** Direct PCR from whole blood, without DNA extraction, *Nucleic Acids Res.*, 18, 5908, 1990.
55. **Olive, D. M.,** Detection of enterotoxinogenic *Escherichia coli* after polymerase chain reaction amplification with a thermostable DNA polymerase, *J. Clin. Microbiol.*, 27, 261, 1989.
56. **Colwell, R. R., Brayton, P. R., Grimes, D. J., Roszak, D. B., Huq, S. A., and Palmer, L. M.,** Viable but non-culturable *Vibrio cholerae* and related pathogens in the environment: implications for release of genetically engineered microorganisms, *BioTechnology*, 3, 817, 1985.

57. **Xu, H. S., Roberts, N., Singleton, F. L., Atwell, R. W., Grimes, D. J., and Colwell, R. R.,** Survival and viability of nonculturable *Escherichia coli* and *Vibrio cholerae* in estuarine and marine environments, *Microb. Ecol.*, 8, 313, 1982.
58. **Knight, I. T., Shults, S., Kaspar, C. W., and Colwell, R. R.,** Direct detection of *Salmonella* spp. in estuaries by using a DNA probe, *Appl. Environ. Microbiol.*, 56, 1059, 1990.
59. **Roszak, D. B., Grimes, D. J., and Colwell, R. R.,** Viable but nonrecoverable stage of *Salmonella enteritidis* in aquatic systems, *Can. J. Microbiol.*, 30, 334, 1984.
60. **Rollins, D. M. and Colwell, R. R.,** Viable but nonculturable stage of *Campylobacter jejuni* and its role in survival in natural aquatic environments, *Appl. Environ. Microbiol.*, 52, 531, 1986.
61. **Johnston, J. M., Becker, S. F., and McFarland, L. M.,** Gastroenteritis in patients with stool isolates of *Vibrio vulnificus*, *Am. J. Med.*, 80, 336, 1986.
62. **Klontz, K. C., Lieb, S., Schreiber, M., Janowski, H., Baldy, L. M., and Gunn, R. A.,** Syndromes of *Vibrio vulnificus* infections: clinical and epidemiologic features in Florida cases, *Ann. Intern. Med.*, 109, 318, 1988.
63. **Oliver, J. D. and Wanucha, D.,** Survival of *Vibrio vulnificus* at reduced temperatures and elevated nutrient, *J. Food Safety*, 10, 79, 1989.
64. **Bej, A. K., Mahbubani, M. H., and Atlas, R. M.,** Detection of viable *Legionella pneumophila* in water by polymerase chain reaction and gene probe methods, *Appl. Environ. Microbiol.*, 57(2), 597, 1991.
65. **Erlich, H. A., Gelfand, D., and Sninsky, J. J.,** Recent advances in the polymerase chain reaction, *Science*, 252, 1643, 1991.
66. **Bej, A. K., Mahbubani, M. H., Miller, R., DiCesare, J. L., Haff, L., and Atlas, R. M.,** Multiplex PCR amplification and immobilized capture probes for detection of bacterial pathogens and indicators in water, *Mol. Cell. Probes*, 4, 353, 1990.
67. **Bej, A. K., McCarty, S. C., and Atlas, R. M.,** Detection of coliform bacteria and *Escherichia coli* by multiplex polymerase chain reaction: comparison with defined substrate and plating methods for water quality monitoring, *Appl. Environ. Microbiol.*, 57(8), 2429, 1991.
68. **Bej, A. K., Mahbubani, M. H., Dicesare, J. L., and Atlas, R. M.,** Polymerase chain reaction-gene probe detection of microorganisms by using filter-concentrated samples, *Appl. Environ. Microbiol.*, 57(12), 3529, 1991.
69. **Chamberlain, J. S., Gibbs, R. A., Ranier, J. E., Nguyen, P. N., and Caskey, C. T.,** Deletion screening of the duchenne muscular distrophy locus via multiplex DNA amplification, *Nucleic Acids Res.*, 16, 11141, 1988.
70. **Jones, D. D., Law, R., and Bej, A. K.,** Detection of *Salmonella* spp. in contaminated oysters using polymerase chain reaction (PCR) and gene probes, submitted.
71. **Bej, A. K., Law, R., and Jones, D. D.,** Species specific detection of *Vibrio vulnificus*, *V. cholerae*, *Salmonella* spp. and pathogenic *Salmonella* in oysters using polymerase chain reaction (PCR) method, submitted.
72. **Furrer, B., Candrian, U., Hoefelein, C., and Luethy, J.,** Detection and identification of *Listeria monocytogenes* in cooked sausage products and in milk by in-vitro amplification of hemolysis gene fragments, *J. Appl. Bacteriol.*, 70(5), 372, 1991.
73. **Niederhauser, C., Candrian, U., Hofelein, C., Jermini, M., Buhler, H., and Luthy, J.,** Use of polymerase chain reaction for detection of *Listeria monocytogenes* in food, *Appl. Environ. Microbiol.*, 58(5), 1564, 1992.
74. **Pahuski, E. E., Dimond, R. L., Priest, J. H., Martin, L. S., Stebnitz, K. K., and Mendoza, L. G.** (inventor), Method and kit for the separation, concentration and analysis of cells, WO 92/00317 A1, PCT International, Japan, January 9, 1992.
75. **Gelfand, D. H.,** *Taq* DNA polymerase, in *PCR Technology: Principles and Applications for DNA Amplification*, Erlich, H. A., Ed., Stockton Press, New York, 1989, 17.
76. **Bessesen, M. T., Luo, Q., Rotbart, H. A., Blaser, M. J., and Ellison, R. T., III,** Detection of *Listeria monocytogenes* by using the polymerase chain reaction, *Appl. Environ. Microbiol.*, 56(9), 2930, 1990.
77. **Giesendorf, B. A. J., Quint, W. G. V. H., M. H. C., Stegeman, H., Huf, F. A., and Niesters H. G. M.,** Rapid and sensitive detection of *Campylobacter* spp. in chicken products by using the polymerase chain reaction, *Appl. Environ. Microbiol.*, 58(12), 3804, 1992.
78. **Kobayashi, K., Seto, K., Akasaka, S., and Makino, M.,** Detection of toxigenic *Vibrio cholerae* O1 using polymerase chain reaction for amplifying the cholera enterotoxin gene, *J. Jpn Assoc. Infect. Dis.*, 64(10), 1323, 1990.
79. **Koch, W. H., Payne, W. L., Wentz, B. A., and Cebula, T. A.,** Rapid polymerase chain reaction method for detection of *Vibrio cholerae* in foods, *Appl. Environ. Microbiol.*, 59(2), 556, 1993.
80. **Shirai, H., Nishibuchi, M., Ramamurthy, T., Bhattacharya, S. K., Pal, S. C., and Takeda, Y.,** Polymerase chain reaction for detection of the cholera enterotoxin operon of *Vibrio cholerae*, *J. Clin. Microbiol.*, 29(11), 2517, 1991.
81. **Jones, D. D., Law, R., and Bej, A. K.,** An efficient method for detection of microbial pathogens in oysters and chicken meats, in preparation.
82. **Bej, A. K. and Jones, D. D.,** Detection of *Salmonella* spp. in contaminated chicken meat using polymerase chain reaction (PCR) and gene probes, in preparation.

83. **Roszak, D. B. and Colwell, R. R.,** Survival Stategies of bacteria in the natural environment, *Microbiol. Rev.*, 51, 365, 1987.
84. **Morse, D. L., Guzewich, J. J., Hanrahan, J. P., Stricof, R., Shayegani, M., Deibel, R., Grabau, J. C., Nowak, N. A., Herrmann, J. E., Cukor, G., and Blacklow, N. R.,** Widespread outbreaks of clam- and oyster-associated gastroenteritis. Role of Norwalk virus, *N. Engl. J. Med.*, 314, 678, 1986.
85. **Institute of Medicine,** Microbiological and parasitic exposure and health effects, in *Seafood Safety*, Ahmed, F. E., Ed., National Academy Press, Washington, DC, 1991, 30.
86. **Wernars, K., Heuvelman, K., Notermans, S., Domann, E., Leimeister-Wachter, M., and Chakraborty, T.,** Suitability of the *prfA* gene, which encodes a regulator of virulence genes in *Listeria monocytogenes*, in the identification of pathogenic *Listeria* spp., *Appl. Environ. Microbiol.*, 58(2), 765, 1992.
87. **Notermans, S., Heuvelman, K. J., and Wernars, K.,** Synthetic enterotoxin B DNA probes for detection of enterotoxigenic *Staphylococcus aureus* strains, *Appl. Environ. Microbiol.*, 54(2), 531, 1988.

INDEX

3SR, see Self-sustained sequence replication
3' Alu PCR, 133–140

Aerosol-resistant tips (ARTs), 268
Agarose separation, 89
Alfalfa, 201–214
  genetics, 202
Alignments of sequences, 54
Allele-specific amplification (ASA), 111
Allele-specific oligonucleotides, 101, 263, 296
Allele-specific PCR (ASPCR), 111
Alu PCR, 2, 133–140
AMP-FLPs, see Amplified fragment length polymorphisms
Amplification refractory mutation system (ARMS), 111
Amplified fragment length polymorphisms (AMP-FLPs), 261, 278, 282, 290
Anchored PCR, 144
Ancient DNA, 187
Animal breeding, 302
Anti-bacteria antibody, 160
Anti-histone antibody, 160
AP-PCR, see Arbitrarily primed PCR
Applications of different thermostable polymerases, 219–237
Applications of PCR
  in clinical diagnosis, 307–326
  in environmental microbiology, 327–339
  in forensic science, 259–306
Aquatic samples, processing for PCR, 329
Arbitrarily primed PCR (AP-PCR), 122
ARTs, see Aerosol-resistant tips
ASO, see Allele-specific oligonucleotides
Asymmetric PCR, 93, 296
Automated DNA sequencing, 85–100

Bacillus spp., 219–237
Biallelic DNA markers, 282
Biotin labeling, 168
Biotin-labeled probes, 44
Blood samples, processing for PCR, 312
Brassica spp., 182
Bst polymerase, 219–237

Cancer therapy, 323
Carbodiimide labeling, 115
cDNA amplifications, 8
Centricon microconcentrator, 14
Cheese, processing for PCR, 359
Chelex 100, 314, 336
Classification by RAPD PCR, 179–191, 201–214
Clinical diagnosis, 307–326
Clinical microbiology, 308
Clone identification by PCR, 53–57
Cloning PCR products, 21–27, 216
Codon usage table, 54
Computer generated alignments, 54

Computer-aided primer design, 9
Contamination in PCR, 249–258, 267, 320
Controls in PCR, 254, 317
Cosmid libraries, direct screening, 53–57
Cosolvents, 316
Cultivar classification, 184
Cycle sequencing, 85–100

Database, 54, 108
Degenerate primers, 8, 54
Denaturing gradient gel electrophoresis (DGGE), 101, 185
Detection of product, 317–320
DGGE, see Denaturing gradient gel electrophoresis
Diagnosis of disease, 307–326
Digoxigenin-labeled probes, 39, 45
Disease resistance, 187
DNA sequencing, 85–100, 243, 282, 296
  automated, 85–100, 298
  manual, 94, 243, 282
  solid-phase, 282
DNA recombination during PCR, 215–218
DNA polymerases, 219–237
DNA extraction from plants, 193–200
DNAase I digestion, 253
Dot-blot analysis, 263
Dot-blot hybridization, 296
Double-stranded probes, 39

EIPCR, see enzymatic inverse PCR
Electrophoretic separation, 89
End-labeled probes, 39
Endonuclease digestion, 253
Enrichment cultures, 355
Environmental microbiology, 327–339
Enzymatic inverse PCR (EIPCR), 60
Enzymes for PCR, 219–237
Ethidium bromide substitutes, 173
Ethidium homodimer, 170
Evolution, 179–191
Exonuclease, 91–93
Exonuclease digestion, 253

False-positives in PCR, 249–258
Filter hybridization, 45–46
FINS, see Forensically informative nucleotide sequences
Flanking sequence, methods of amplification, 29–36
  cDNA, 32
  genomic, 30
Fluorescent DNA sequencing, 85–100
Fluorescent procedures, 298
FoLT PCR, see Formamide low temperature PCR
Food, 341–362
Foodborne pathogens, 341–362
Forensic analysis, 268
Forensic casework, 283–284
Forensic science, 259–306

Forensically informative nucleotide sequences (FINS), 301
Formamide low temperature PCR (FoLT PCR), 151–157
Fragment size determination, 1
Freeze-spin purification, 90

Gastric biopsy sample, 309
GEMs, see Genetically engineered microorganisms
Gene amplification, 7
Gene flow, 186
Gene manipulation, 7
GeneAmpT, 37
Geneclean, 14, 90–91
Genetic disease diagnosis, 322
Genetic diversity, 187
Genetic linkage, 186, 201–214
Genetically engineered microorganisms (GEMs), detection, 330
GeneWorks, 9
Genome mapping, 239–247
Genomic DNA extraction from plants, 193–200
Glassmilk, 14, 90–91

Hot-start PCR, 250, 261, 283
Hybridization studies, 186

Identification of human remains, 285
Identification of individuals, 277–287
Identification of species, 301
Immuno-PCR, 159–163
In situ hybridization, 46
Infectious disease diagnosis, 321
Introgression, 186
Inverse PCR (IPCR), 60
Ionizing radiation, 253
IPCR, see Inverse PCR
Isopsoralen, 254
Isozymes, 179

Juniperus spp., 181

LA-PCR, see Ligation-anchored PCR
Labeling and detection of DNA, 108
Labeling of primers, 134
Lactococcus lactis, 56
Lambda exonuclease, 91
Lambda libraries, direct screening, 53–57
LCR, see Ligase chain reaction
LDA, see Limiting dilution analysis
Leptospira spp, 127–128
Library screening, 1
Ligase chain reaction (LCR), 251
Ligation of PCR products, 24
Ligation-anchored PCR (LA-PCR), 141–145
Limiting dilution analysis (LDA), 147–150
Linkage, see Genetic linkage

Magic PCR Preps, 90–91; See also Wizard PCR Preps
Magnesium ion, 315
Manual sequencing, 94

Mapped restriction site polymorphisms (MRSPS), 121–131
Mapping by STS content, 239–247
Mapping, genome, see Genome mapping
Meat, processing for PCR, 360
Melting temperature (T_m), 6, 54
Metaphor agarose, 55
Methanobacterium thermoautotrophicum, 219–237
Microbial pathogen detection, 333, 341–362
Microorganism detection, 331–336
Microsatellite DNA, 278, 291
Milk, processing for PCR, 359
Minisatellite DNA, 278, 292
Minisatellite variant repeat PCR (MVR-PCR), 266, 292
Mitochondrial polymorphisms, 295
MRSPS, see Mapped restriction site polymorphisms
Mth polymerase, 219–237
Multiallelic satellite DNA, 278
Multiplex PCR, 282, 316, 333, 358
Mutagenesis by PCR, 59–67
    enzymatic inverse PCR (EIPCR), 60
    recombinant circle PCR (RCPCR), 60, 64
    recombinant PCR (RPCR), 60
    site-directed mutagenesis, 59–67
    tagged PCR, 65
Mutation detection, 107–120
MVR-PCR, see Minisatellite variant repeat PCR

N terminal sequence, 54
Nested PCR, 250
Non-isotopic detection, 168
Non-isotopic probe generation, 39, 43–52
Non-isotopic SSCP, 165–177; See also Single-strand conformation polymorphism
Nucleic acid sequence-based amplification (NASBA), 251
Numerical analyses, 188

OLIGO software, 9
Optimization for RADP PCR in plants, 193–200
Oxazole yellow dimer (YOYO), 173

Paternity testing, 283, 298
Pathogen detection, 333, 341–362
Pathogens in food, 341–362
PCGene software, 9
PCR
    3' Alu PCR, 133–140
    allele-specific amplification (ASA), 111
    allele-specific PCR (ASPCR), 111
    Alu PCR, 133–140
    amplification of specific alleles (PASA), 111
    amplification refractory mutation system (ARMS), 111
    amplified fragment length polymorphisms (AMP-FLPs), 261, 278, 282
    anchored PCR, 144
    arbitrarily primed PCR (AP-PCR), 122
    ARMS, see Amplification refractory mutation system

Index

ASA, see Allele-specific amplification
ASPCR, see Allele-specific PCR
asymmetric, 93, 296
buffer, 55
clinical, 151
clinical diagnosis, 307–326
cloning products, 21–27, 216
contamination, 249–258, 267, 320
controls, 254, 317
design of primers, 5–11
detection of mutations, 107–120
detection of pathogens in food, 341–362
direct screening of libraries, 53–57
distinction between similar sequences, 101–106
DNA polymerases, 219–237
DNA recombination, 215–218
DNA sequencing of products, 85–100
environmental microbiology, 327–339
enzymatic inverse PCR (EIPCR), 60
false-positives, 249–258
food safety, 341–362
forensic science, 259–306
formamide low temperature (FoLT PCR), 151–157
from blood, 151–157
general techniques, 1–4
generation of labeled probes, 37–52
hot-start, 250, 261, 283
immuno-PCR, 159–163
inverse PCR (IPCR), 60
LDA, see Limiting dilution analysis
ligation-anchored PCR (LA-PCR), 141–145
limiting dilution analysis (LDA), 147–150
minisatellite variant repeat PCR (MVR-PCR), 266, 292
multiplex, 282, 316, 333, 358
mutagenesis, 59–67
nested, 250
optimization for RADP PCR in plants, 193–200
PASA, see PCR amplification of specific alleles
polymerases, 219–237
precautions, 249–258
primers, design of, 5–11
probes, generation of, 37–52
product cloning, 21–27
product ligation, 24
purification of products, 13–19, 87–91
random amplified polymorphic DNA (RAPD), 179–191, 193–214, 269, 302
reaction buffer, 55, 315
reaction conditions, 55
recombinant circle PCR (RCPCR), 60, 64
recombinant PCR (RPCR), 60
reverse transcriptase PCR (RT-PCR), 312, 336
sequence-tagged site (STS) mapping, 239–247
sequencing of products, 85–100
single primer, 29–36
site-directed mutagenesis, 59–67
tagged PCR mutagenesis, 65
techniques, 1–4
thermostable DNA polymerases, 219–237

variable number of tandem repeats (VNTR), 133, 260, 290
PEG precipitation, 91
Perfect Match polymerase enhancer, 105
Pfu polymerase, 219–237
pGEM T, see T vector
Plant DNA extraction, 193–200
Plant evolution, 179–191
Plant taxonomy, 179–191
Plants, 179–214
Plasmid libraries, direct screening, 56
Plasmodium falciparum, 86
Polymerase chain reaction, see PCR
Polymerases, 219–237
Population genetics, 284
Precautions in PCR, 249–258
Preimplantation diagnosis, 322
Prenatal diagnosis, 322
Primer
  degenerate, 54
  design, 5–11, 314
    computer aided, 9
    from amino acid sequence, 54
    gene amplification, 7
    gene manipulation, 7
    general rules, 6
  extension, 113
  labeling, 134
  removal from reactions, 16
  single primer PCR, 29–36
  truncated, 215
Primer-dimers, 14–16, 54
Probe
  double-stranded, 39
  generation, 2, 37–52
    biotin-labeled probes, 44
    digoxigenin-labeled probes, 39, 45
    end-labeled probes, 39
    labeled probes, 37–42
    nonisotopic probes, 39, 43–52
    radioactive probes, 38
  single-stranded, 38–39, 46
Product tagging, 8
Product purification, see Purification of PCR products
Psoralen, 254
Purification of nucleic acid from foods, 358–361
Purification of nucleic acid from environmental samples, 328
Purification of PCR products, 13–19, 87–91
  agarose separation, 89
  Centricon method, 14, 89
  direct purification, 15
  electrophoretic separation, 89
  elimination of non-specific amplification products, 18
  freeze/spin method, 90
  gel purification, 15
  Geneclean, 14, 90–91
  glassmilk, 14, 90–91
  Magic PCR Preps, 90–91
  PEG precipitation, 91

phenol-chloroform method, 14
primer removal, 16
primer-dimer removal, 16
Qiagen, 14
Sephacryl, 88
Sepharose, 88
size exclusion chromatography, 88
spin columns, 88
ultrafiltration methods, 17
Wizard PCR Preps, 14
Pyrococcus furiosus, 219–237

Qiagen purification, 14
Quality assurance, 270
Quality control, 270
Quantitation of primer, 93
Quantitation of sequencing template, 93, 97

Radioactive probes, 38
Random amplified polymorphic DNA (RAPD), 179–191, 193–214, 269, 302
RCPCR, see Recombinant circle PCR
Reaction buffer, 55, 315
Reaction conditions, 54
Recombinant circle PCR (RCPCR), 60, 64
Recombinant PCR (RPCR), 60
Recombination during PCR, 215–218
Repetitive DNA, 290
Restriction fragment length polymorphism (RFLP), 112, 133, 179, 260, 290
Restriction site engineering, 8
Reverse transcriptase PCR, 312, 336
RFLP, see Restriction fragment length polymorphism
RPCR, see Recombinant PCR
RT-PCR, see Reverse transcriptase PCR

Sac polymerase, 219–237
Saliva, processing for PCR, 311
Sample processing, 309, 328
SCARs, See Sequence characterized amplified regions
Screening by PCR, 53–57
Sediments, processing for PCR, 328
Self-sustained sequence replication (3SR), 251
Sephacryl, 88
Sephadex, 336
Sepharose, 88
Sequenase, 95
Sequence characterized amplified regions (SCARs), 185
Sequence polymorphism, 255; See also SSCP
Sequence specific oligonucleotides (SSO), 296
Sequence tagged site, see STS
Sequence tagged site (STS) mapping, 1, 239–247
Sequencing, see DNA sequencing
Short tandem repeats (STRs), 291; See also Microsatellite DNA
Similarity determination by RAPD PCR, 209
Single primer PCR, 29–36

Single-strand conformation polymorphism (SSCP), 101, 116, 165–177, 256
Single-stranded probes, 38–39, 46
Site-directed mutagenesis, 8, 59–67
    amplification of plasmid, 60
    amplification of short fragments, 62
Size exclusion chromatography, 88
Soil, processing for PCR, 328
Solid-phase sequencing, 282
Somaclonal variation, 186
Somatic embryogenesis, 206
Spin columns, 88
SSCP, see Single-strand conformation polymorphism
SSO, see Sequence specific oligonucleotides
Stoffel fragment, 219–237
Stool samples, processing for PCR, 311
STRs, see short tandem repeats
STS, see Sequence-tagged site (STS) mapping
Stylosanthes spp., 183
Sulfolobus acidocaldarius, 219–237
Swabs, processing for PCR, 310

T vector, 22–25
T7 gene 6 exonuclease, 92
Tac polymerase, 219–237
Tagged PCR mutagenesis, 65
Taq polymerase, 219–237
Taxonomy, 179–191
Tetramethylammonium chloride, see TMAC
Tfl polymerase, 219–237
Thermococcus litoralis, 219–237
Thermoplasma acidophilium, 219–237
Thermostable DNA polymerases, 219–237
Thermotoga spp., 219–237
Thermus aquaticus, 219–237
Thermus flavus, 219–237
Thermus ruber, 219–237
Thermus thermophilus, 219–237
Thiazole orange dimer (TOTO), 173
Tli polymerase, 219–237
T_m, 6, 54
TMAC, 105
Tru polymerase, 219–237
Truncated primers, 215
Tsp polymerase, 219–237
Tth polymerase, 219–237
Tub polymerase, 219–237

Ultraviolet irradiation, 253
Uridine incorporation, 254

Variable number of tandem repeats (VNTR), 133, 260, 290
Vent polymerase, 219–237
Viruses, detection in water, 336

Wizard PCR preps, 14; See also Magic PCR Preps

Yeast artificial chromosomes (YACs), 239, 242